AF658400

Lecture Notes in Mechanical Engineering

For further volumes:
http://www.springer.com/series/11236

About this Series

Lecture Notes in Mechanical Engineering (LNME) publishes the latest developments in Mechanical Engineering - quickly, informally and with high quality. Original research reported in proceedings and post-proceedings represents the core of LNME. Also considered for publication are monographs, contributed volumes and lecture notes of exceptionally high quality and interest. Volumes published in LNME embrace all aspects, subfields and new challenges of mechanical engineering. Topics in the series include:

- Engineering Design
- Machinery and Machine Elements
- Mechanical Structures and Stress Analysis
- Automotive Engineering
- Engine Technology
- Aerospace Technology and Astronautics
- Nanotechnology and Microengineering
- Control, Robotics, Mechatronics
- MEMS
- Theoretical and Applied Mechanics
- Dynamical Systems, Control
- Fluid Mechanics
- Engineering Thermodynamics, Heat and Mass Transfer
- Manufacturing
- Precision Engineering, Instrumentation, Measurement
- Materials Engineering
- Tribology and Surface Technology

Sehijpal Singh Khangura
Paramjit Singh · Harwinder Singh
Gurinder Singh Brar
Editors

Proceedings of the International Conference on Research and Innovations in Mechanical Engineering

ICRIME-2013

Editors
Sehijpal Singh Khangura
Paramjit Singh
Harwinder Singh
Gurinder Singh Brar
Mechanical Engineering
Guru Nanak Dev Engineering College
Ludhiana
Punjab
India

ISSN 2195-4356 ISSN 2195-4364 (electronic)
ISBN 978-81-322-1858-6 ISBN 978-81-322-1859-3 (eBook)
DOI 10.1007/978-81-322-1859-3
Springer New Delhi Heidelberg New York Dordrecht London

Library of Congress Control Number: 2014933570

Printed on acid-free paper

Springer is part of Springer Science+Business Media (www.springer.com)

Preface

The International Conference on Research and Innovations in Mechanical Engineering (ICRIME) was held on the campus of Guru Nanak Dev Engineering College (GNDEC), Ludhiana, Punjab, India during 24–26 October 2013. Notably, this was the 60th anniversary of Guru Nanak Dev Engineering College, Ludhiana.

The 63 scientific participants, 28 of whom were students, had many fruitful discussions and exchanges that contributed to the success of the conference. There were seven plenary lectures, each focusing on a different focus-area of the conference: Prof. Saji Gopinath (IIM, Kozhikode), Prof. Subhash Chandra (NIT, Jallandhar), Dr. Inderdeep Singh (IIT, Roorkee), Prof. R. S. Jaduaon (GB Pant Agricultural University, Pant Nagar), Dr. A. Manna (PEC, Chandigarh), Dr. Anupam Aggarwal (IIT, Ropar) and Er. Lakhinder Singh (Sawraj Mazda, Ropar) talked about innovative ideas for improvement of engineering processes and equipment.

Generous support for the conference was provided by All India Council for Technical Education, New Delhi; TEQIP; Department of Science and Technology, New Delhi; and Punjab Technical University, Jalandhar. The funds were sizeable, timely, and greatly appreciated. Besides exhibiting their products, local participants from the industry also came forward to support the event in the form of finance, logistics, etc. Industry participants included Guru Nanak Industrial Corporation, Ludhiana; Technic Fabricators Pvt. Ltd., Mohali; Testing and Consultancy Cell, GNDEC, Ludhiana; EduTek Equipments (I) Pvt. Ltd., Delhi; Hi Tech Castings, Ludhiana; Qualitech Systems, Ludhiana; 3DWorks Design Solutions, Ludhiana; Gate Revolution, Chandigarh; New Swan Enterprises, Ludhiana; G. S. Radiators Ltd., Ludhiana; Ideas Design Solutions (P) Ltd., Chandigarh; and Gurminder Agriculture Engineering Industry, Ludhiana. All in all, ICRIME-2013 held at GNDEC, Ludhiana was a very successful event. The plenary lectures bridged the gap between the different fields of Mechanical Engineering making it possible for non-experts to gain insight into new areas. Also included among the speakers were several young scientists, namely post-docs and students, who brought new perspectives to their fields. Given the rapidity with which research is advancing in all of the areas covered by ICRIME, we expect that future ICRIME conferences will be as stimulating as this most recent one was, as indicated by the contributions presented in this proceedings volume.

Sehijpal Singh Khangura
Paramjit Singh Bilga
Harwinder Singh
Gurinder Singh Brar

Contents

About the Editors

Dr. Sehijpal Singh is currently Professor and Head in the Department of Mechanical Engineering at Guru Nanak Dev Engineering College, Ludhiana (Punjab) India. He holds a Ph.D. from IIT Roorkee. He has teaching experience of over 21 years. He has guided five Ph.D. and 50 M.Tech. students in the areas of non-conventional machining processes, metal cutting, etc. He has published 75 papers in several journals and conferences of national and international repute.

Dr. Paramjit Singh Bilga is presently Professor in the Department of Mechanical Engineering at Guru Nanak Dev Engineering College, Ludhiana (Punjab) India. He received his Ph.D. from the Indian Institute of Technology Delhi, India. He has teaching and research experience of more than 23 years.

Dr. Harwinder Singh holds a Ph.D. from Punjabi University, Patiala. Currently, he is Professor in the Department of Mechanical Engineering at Guru Nanak Dev Engineering College, Ludhiana, Punjab, India. He has close to 15 years of teaching and research experience. He has contributed 80 research papers in international and national journals and conferences. He is also the author of one book.

Dr. Gurinder Singh Brar is Assistant Professor in the Department of Mechanical Engineering at Guru Nanak Dev Engineering College, Ludhiana (Punjab) India. He holds a Bachelor's degree in Mechanical Engineering, a Master's degree in Production Engineering, a Master of Business Administration degree, and a Ph.D. in Mechanical Engineering. He has contributed a significant number of research papers at the international/national level and is working in the areas of mechanics, finite element analysis, and non-traditional manufacturing systems.

About the Book

This book comprises the "Proceedings of International Conference on Research and Innovations in Mechanical Engineering (ICRIME 2013)" organized by Guru Nanak Dev Engineering College, Ludhiana with support from AICTE, TEQIP, DST, and PTU, Jalandhar. This international conference served as a premier forum for communication of new advances and research results in the fields of mechanical engineering. The proceedings reflect the conference's emphasis on strong methodological approaches and focus on applications within the domain of mechanical engineering. The contents of this volume aim to highlight new theoretical and experimental findings in the fields of mechanical engineering and closely related fields, including interdisciplinary fields such as robotics and mechatronics.

Comparative Study of Laminated Composite Beam Using Narrow- and Wide-Beam Criterion

Arpit Kumar Srivastava and Anand Kumar

Abstract The primary objective of the work is to effectively conduct the stress analysis of a laminated composite beam having rectangular sections. An analytical expression is used to calculate stress at each ply considering both narrow- and wide-beam assumption. The material for composite rectangular beam is AS4/3501–6 carbon/epoxy. Three types of lay-ups, chosen for the synthesis, are described collectively for both narrow- and wide-beam criterion. Beam taken here is simply a supported type along with uniformly distributed load, which is kept same for all cases.

Keywords Laminated composite beam · Bending stiffness · Narrow beam · Wide beam · Fibrous composite

1 Introduction

Fiber composites are generally used as beam structures. The current text is focused on implications of lamination on the basic ideas of determining stiffness and stress distributions in beams. The analysis of beam is based upon the basic fundamentals as generally given in strength of materials book for isotropic materials although it involves important modification. The expressions for stress distribution and stiffness are of quite different forms for composite than isotropic beams (Swanson 1997). It should also be noted that the most common result of isotropic beam $\sigma = \frac{Mz}{I}$ does not valid for composite beams because the formula is based on uniformly varying stress distributions and the stresses in composite beams can be very non-uniform because of the changes in stiffness from layer to layer.

A. K. Srivastava (✉) · A. Kumar
Department of Mechanical Engineering, HBTI, Kanpur 208002, Uttar Pradesh, India
e-mail: arpit910@gmail.com

S. S. Khangura et al. (eds.), *Proceedings of the International Conference on Research and Innovations in Mechanical Engineering*, Lecture Notes in Mechanical Engineering, DOI: 10.1007/978-81-322-1859-3_1, © Springer India 2014

Table 1 Lay-ups considered for the study

S. No.	Case	Lay-up
1	I	$[0_4/45_4/-45_4]_s$
2	II	$[(0_2/45_2/-45_2)_s]_s$
3	III	$[0_4/45_4/90_4]_s$

In this study, the value of stresses and strains for location at the farthest distance away from center line for each of the ply groups and for three lay-ups shown in Table 1 are calculated using narrow- and wide-beam assumptions. A laminate stress analysis is the process employed to obtain a complete state of stress in all the laminae (ply) for a given laminate loads and laminae elastic properties. This can be summarizing in following steps (Agrawal and Broutman 2006):

Step 1: Calculate laminae stiffness matrix from the elastic properties.
Step 2: Transform stiffness matrix to different ply orientations.
Step 3: Calculate laminate stiffness matrices [A], [B], and [D].
Step 4: Calculate mid-plane strains and plate curvatures for the given loads using narrow- and wide-beam criterion.
Step 5: Calculate laminae strains.
Step 6: Transform laminae strains from arbitrary directions to the longitudinal directions.
Step 7: Calculate laminae stresses.

The beam theory for symmetric rectangular composite beams is as follows:

$$\begin{Bmatrix} M_x \\ M_y \\ M_{xy} \end{Bmatrix} = [D] \begin{Bmatrix} k_x \\ k_y \\ k_{xy} \end{Bmatrix} \tag{1}$$

In the above equation, the moment resultants in this equation are on unit width basis and must be multiplied by the width of the beam to get the total moment used in beam theory. The structural response of the beam is dependent on the ratio of the width to height of the beam cross section. For a beam subjected to bending, the induced lateral curvature is insignificant if the width to height ratio is large. This kind of beam is so-called wide beam. Conversely, if the width to height ratio of the cross section is small, the beam is called narrow beam. For this case, the lateral curvature is induced due to the effect of Poisson's ratio. As the result, lateral moment is zero. This statement can be expressed mathematically as

$$\text{Narrow Beam} \quad M_y = 0 \quad k_y \neq 0 \tag{2}$$

$$\text{Wide Beam} \quad M_y \neq 0 \quad k_y = 0 \tag{3}$$

So for narrow beam and Eq. (2), we can say

$$\begin{Bmatrix} k_x \\ k_y \\ k_{xy} \end{Bmatrix} = [D^{-1}] \begin{Bmatrix} M_x \\ 0 \\ 0 \end{Bmatrix} = \begin{bmatrix} D_{11}^{-1} & D_{12}^{-1} & 0 \\ D_{21}^{-1} & D_{22}^{-1} & 0 \\ 0 & 0 & D_{33}^{-1} \end{bmatrix} \begin{Bmatrix} M_x \\ 0 \\ 0 \end{Bmatrix} \tag{4}$$

Also, the overall moment–curvature relationship for the beam can be written as

$$M_{\text{beam}} = bM_x = \frac{b}{D_{11}^{-1}} k_x = \overline{EI}\, k_x \tag{5}$$

Here, b is the width of the cross section of beam. So for the narrow beam, flexural rigidity is related to the reciprocal of the inverse bending stiffness from lamination theory and is given by

$$\overline{EI} = \frac{b}{D_{11}^{-1}} \tag{6}$$

Also the curvature in transverse direction k_y is given by

$$k_y = D_{21}^{-1} M_x = \frac{D_{21}^{-1}}{D_{11}^{-1}} k_x \tag{7}$$

Again the strain in longitudinal and transverse direction is given by

$$\varepsilon_x = z k_x \tag{8}$$

$$\varepsilon_y = z k_y \tag{9}$$

Here, z is the distance of respective ply from neutral axis. Again for wide beam and Eq. (3) we can say beam flexural rigidity is related to the bending stiffness obtained from lamination theory as shown and is given by

$$\begin{Bmatrix} M_x \\ M_y \\ M_{xy} \end{Bmatrix} = \begin{bmatrix} D_{11} & D_{12} & 0 \\ D_{21} & D_{22} & 0 \\ 0 & 0 & D_{33} \end{bmatrix} \begin{Bmatrix} k_x \\ 0 \\ 0 \end{Bmatrix} \tag{10}$$

$$M_{\text{beam}} = bM_x = bD_{11} k_x = \overline{EI}\, k_x \tag{11}$$

$$\overline{EI} = bD_{11} \tag{12}$$

The curvature is in only x direction which is as given in Eq. (11).

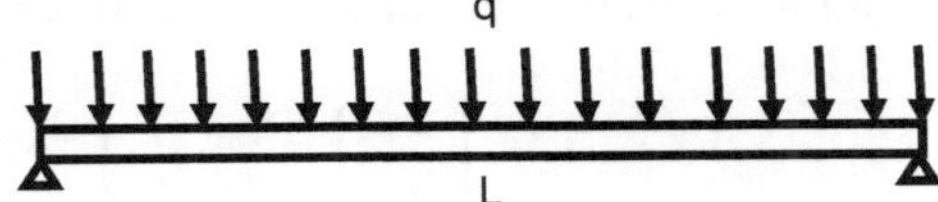

Fig. 1 Simply supported laminated beam with UDL

Table 2 Properties of AS4-3501 carbon/epoxy laminate composite (Soden 1998)

S. No.	Properties of AS04-3501 carbon/epoxy	Corresponding values
1	Longitudinal tensile strength	1,950 GPa
2	Longitudinal compressive strength	1,480 GPa
3	Transverse tensile strength	48 MPa
4	Transverse compressive strength	200 MPa
5	Shear strength	79 MPa

2 Description of Problem

The laminated composite beam to be considered is made of 24 plies of carbon/epoxy. The beam is simply supported at the ends and subject to a uniformly distributed load q as shown in Fig. 1. The desired answer is the stresses and strains at each and every lamina.

The data for the problem are summarized as: length of beam (L) = 254 mm, width of beam (b) = 50.8 mm, thickness (t) = 3.150 mm, and distributed load (q) = 3,500 N/m = 3.5 N/mm. Also, from the properties of carbon/epoxy, E_{11} = 126 GPa, E_{22} = 23.3 GPa, G_{12} = 6.55 GPa, and $\nu_{12} = 0.28$ (Swanson 2001; Soden 1998) (Table 2).

3 Result and Discussions

Tables 3, 4, 5, 6, and 7 show the value of stresses and strains for locations at the farthest distance away from center line for each of the ply groups. It is important to note from the results that the stress distribution is very non-uniform across the cross section of beam. The reason for this is the change in stiffness properties with the different angular orientations of the plies. Figure 2 shows the variation of longitudinal stresses using both narrow- and wide-beam assumption.

If stresses of wide beam in longitudinal direction are compared with the stresses of narrow beam for same lay-up, it can be easily concluded that stresses at different ply groups for wide-beam assumption are lesser than those for narrow-beam assumption. The reason for this can be easily understand by studying both the assumptions. In narrow beam for calculating curvature in x direction i.e., k_x term, D_{11}^{-1} is used while wide-beam assumption used $1/D_{11}$, where D is the bending stiffness matrix for the laminated composite beam.

Table 3 Value of stresses and strains for $[0_4/-45_4/45_4]_s$ lay-up (narrow beam)

Z (mm)	Ply group (°)	Strain (%)		Stress (MPa)		
		ε_x	ε_y	σ_x	σ_y	τ_{12}
0.528	45	0.04	−0.02	15.55	1.70	−4.24
1.057	−45	0.09	−0.04	33.33	3.64	9.09
1.585	0	0.14	−0.07	181.86	−2.85	0

Table 4 Value of stresses and strains for $[(0_2/45_2/45_2)_s]_s$ lay-up (narrow beam)

Z (mm)	Ply group (°)	Strain (%)		Stress (MPa)		
		ε_x	ε_y	σ_x	σ_y	τ_{12}
0.2	45	0.03	−0.01	6.34	0.69	−2.50
0.5	−45	0.06	−0.03	14.78	1.61	5.83
0.7	0	0.09	−0.05	109.17	−3.00	0
1.0	45	0.12	−0.07	31.68	3.46	−12.50
1.3	−45	0.15	−0.09	40.13	4.38	15.83
1.5	0	0.18	−0.11	228.26	−6.28	0

Table 5 Value of stresses and strains for $[0_4/45_4/90_4]_s$ lay-up (narrow beam)

Z (mm)	Ply group (°)	Strain (%)		Stress (MPa)		
		ε_x	ε_y	σ_x	σ_y	τ_{12}
0.528	90	0.04	−0.01	0.528	90°	0.04
1.057	45	0.10	−0.03	1.057	45°	0.10
1.585	0	0.15	−0.04	1.585	0°	0.15

Table 6 Value of stresses and strains for $[0_4/-45_4/45_4]_s$ lay-up (wide beam)

Z (mm)	Ply group (°)	Strain (%)		Stress (MPa)		
		ε_x	ε_y	σ_x	σ_y	τ_{12}
0.528	45	0.04	0	26.83	2.93	−2.72
1.057	−45	0.09	0	57.49	6.27	5.84
1.585	0	0.14	0	172.10	4.21	0

Also, if look at the obtained stress in third case i.e., for $[0_4/45_4/90_4]_s$ lay-up, the stress at 90° ply group is opposite to 0° ply group for narrow beam and in same direction for wide beam.

The reason for this lies in the assumption of both. In wide beam, curvature in transverse direction is assumed zero while for narrow beam, there is some curvature due to Poisson effect. Due to this, transverse curvature is opposite to longitudinal for narrow beam. A first look on the curves provides that stresses are

Table 7 Value of stresses and strains for $[(0_2/45_2/45_2)_s]_s$ lay-up (wide beam)

Z (mm)	Ply group (°)	Strain (%)		Stress (MPa)		
		ε_x	ε_y	σ_x	σ_y	τ_{12}
0.2	45	0.02	0	13.70	1.49	−1.39
0.5	−45	0.05	0	31.97	3.49	3.25
0.7	0	0.08	0	98.07	2.40	0
1.0	45	0.11	0	68.50	7.47	−6.96
1.3	−45	0.13	0	86.77	9.46	8.81
1.5	0	0.16	0	205.06	5.01	0

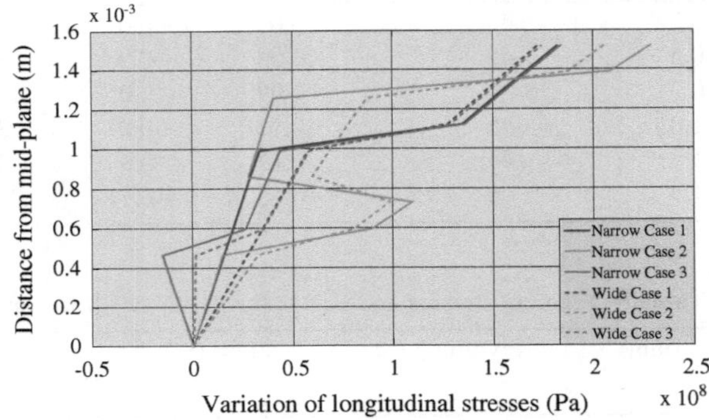

Fig. 2 Longitudinal stresses for the three lay-ups using both narrow- and wide-beam assumption

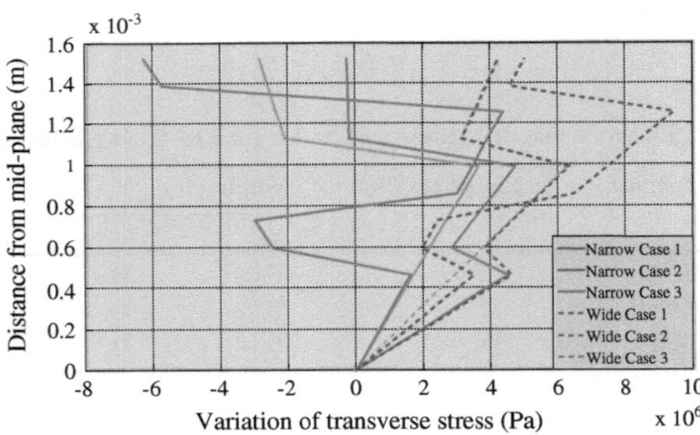

Fig. 3 Transverse stresses for the three lay-ups using both narrow- and wide-beam assumption

non-uniform across the cross section which is quite obvious and is shown in Fig. 3. These stresses are important in determining the composite material failure.

For narrow beam, the transverse stress in the 0° ply groups is opposite to the other 45° and −45° ply groups. The stresses in 0° ply groups are opposite to 90° ply groups. But for wide beam, the stresses in all the ply groups are same. Also the transverse stresses in 90° plies are greater than in other plies although they are near the mid-plane. The reason for difference lays in the assumptions of narrow and

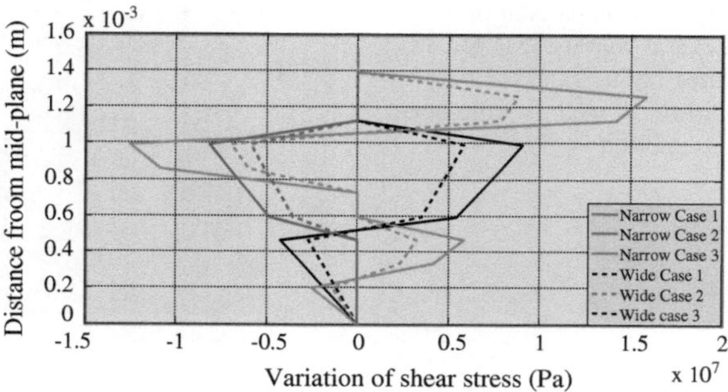

Fig. 4 Shear stresses for the three lay-ups using both narrow- and wide-beam assumption

wide beam and the stiffness properties of the various ply groups. The stresses are calculated using the following equation from classical lamination theory,

$$\begin{bmatrix} \sigma_1 \\ \sigma_2 \\ \tau_{12} \end{bmatrix} = \begin{bmatrix} Q_{11} & Q_{12} & 0 \\ Q_{12} & Q_{22} & 0 \\ 0 & 0 & Q_{66} \end{bmatrix} \begin{bmatrix} \varepsilon_1 \\ \varepsilon_2 \\ \gamma_{12} \end{bmatrix} \tag{13}$$

$$\begin{bmatrix} \varepsilon_x \\ \varepsilon_y \\ \gamma_{xy} \end{bmatrix} = z \begin{bmatrix} k_x \\ k_y \\ k_{xy} \end{bmatrix} \tag{14}$$

as k_y is equal to zero for wide beam, but for narrow it has some value and also it may be noted that the fibers are much less effective in raising the composite modulus in transverse direction than in the longitudinal direction. The direction of strains in x and y directions is opposite to each other for narrow beam, and for wide beam, strain is only in x direction. From Eq. (13), we can say

$$\sigma_2 = Q_{12} \times \varepsilon_1 + Q_{22} \times \varepsilon_2 \tag{15}$$

The sign of $Q_{12} \times \varepsilon_1$ and $Q_{22} \times \varepsilon_2$ quantities are opposite to each other. This is why the stresses in 0° plies are opposite in sign for both beam. Figure 4 shows the variation of shear stresses obtained in beam for both narrow and wide beam and for all cases. Shear stress in 0° ply group and 90° ply group is zero as curvature in shear direction is taken zero. The nonzero value for shear stress is obtained for ply groups other than 0° and 90° and for this case 45° and −45° ply groups are opposite in sign. The reason for this is the transformation of stress matrixes from arbitrary x and y axes to respective ply fibers longitudinal and transverse axes direction and is given as

$$[T_\sigma] = \begin{bmatrix} m^2 & n^2 & 2mn \\ n^2 & m^2 & -2mn \\ -mn & mn & m^2 - n^2 \end{bmatrix} \tag{16}$$

where $m = \cos\theta$ and $n = \sin\theta$ and θ is the fiber orientation of unidirectional laminate. It is quite obvious that for 0° and 90° mn term is zero while for 45° and

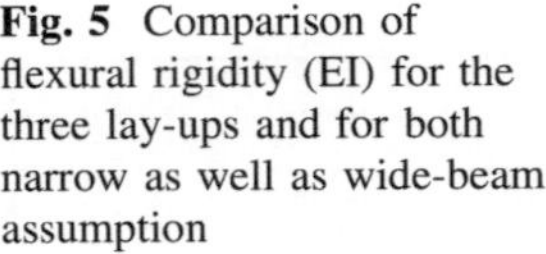

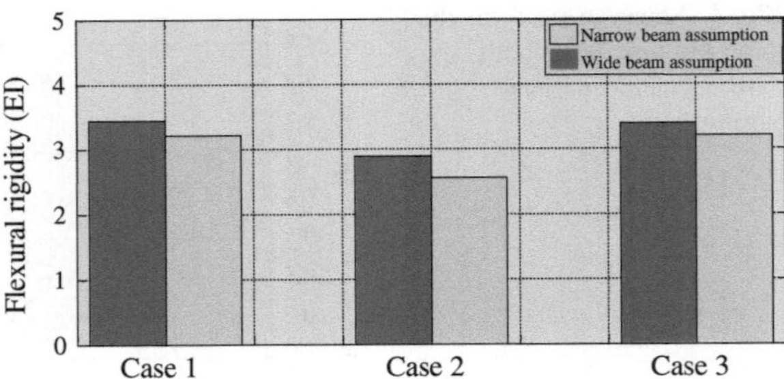

Fig. 5 Comparison of flexural rigidity (EI) for the three lay-ups and for both narrow as well as wide-beam assumption

−45° is nonzero. That is why these ply groups are stressed in shear direction of their fiber orientation and are opposite to each other. Also, the stresses are more non-uniform for Case 2 i.e., for $[(0_2/45_2/-45_2)_s]_s$ lay-up because of stacking sequence. The stresses in this lay-up are greater than all other lay-ups because bending stiffness i.e., flexural rigidity is lowest for these lay-ups. A comparative plot for flexural rigidity for all the cases is shown in Fig. 5. The flexural rigidity is minimum for Case 2 when the beam is assumed narrow and greatest for Case 1, and the wide-beam assumption is considered. The flexural rigidity for the first case and third case are approximately same. The strains in x and y directions show linear behavior. The strains are greatest for Case 2, and other two cases have approximately equal maximum strain.

4 Conclusions

The stress and strain in each ply of laminates are calculated using sectional properties and the assumptions of narrow as well as wide beams in this work. Three different cross section configurations are used to compare solutions. It can be concluded from the study and results obtained that stresses are very non-uniform across the cross section of beam because of the change in stiffness properties due to change in angular orientation. For narrow-beam assumption, flexural rigidity is lower than wide-beam assumption which is quite obvious because aspect ratio for wider beam is high. Axial stress is highest for Case 2 of narrow-beam assumption for which flexural rigidity is also lowest. The axial, transverse, and shear stresses obtained in Case 2 for both narrow- and wide-beam assumptions are more non-uniform, which is the results of their ply orientations. These types of orientations may be used for small loads. The strains that produced due to one-dimensional loading are also greatest for Case 2 of narrow-beam assumption. As the stresses produced in the longitudinal direction are higher, strains are also greater in this direction. For the wide beam, strain is produced in only one direction that is the direction of loading. All other strains are neglected.

References

Agrawal BD, Broutman LJ (2006) Analysis and performance of fiber composites, 2nd edn. Wiley, New Jersey

Soden PD (1998) Lamina properties, lay-up configurations and loading conditions for a range of fibre-reinforced composite laminates. Compos Sci Technol 58:1011–1022

Swanson SR (1997) Introduction to design and analysis with advanced composite materials. Prentice Hall, New Jersey

Swanson SR (2001) Anticlastic effects and the transition from narrow to wide behavior in orthotropic beams. Compos Struct 53:449–455

Design and Analysis of Submarine Radome

N. V. Srinivasulu, Safeeruddin Khan and S. Jaikrishna

Abstract Submarine antenna is used for communication by RF system. It is protected by radomes. Radomes are the electromagnetic (EM) windows that protect microwave subsystems from the environmental effects. Low-observable radomes are usually made of E-glass/epoxy composite due to its low dielectric constant which is necessary not to interfere EM wave transmission characteristics. Aramid fibers have lower dielectric constant and higher strength than those of E-glass fiber. The dielectric strength constant and loss tangent were measured of the E-glass epoxy and aramid epoxy materials. Increasing the performance of antenna depends upon the proper selection of material to withstand under the water applications, composite materials owing to their high strength to weight ratio, high stiffness and better corrosion resistance are potential source for under water applications. ANSYS, a Finite Element software package was used to analyze the problem. The radome design and finite element analysis validation concluded by conducting the pressure test on radome. The modal analysis is also carried out on radome to check for the natural frequency of the radome. So that resonance does not occur if the natural frequency of the radome coincides with the excitation frequency of the submarine.

Keywords Submarine radome · Finite element analysis · Composites

N. V. Srinivasulu (✉) · S. Khan
Department of Mechanical Engineering, Chaitanya Bharathi Institute of Technology, Hyderabad 500075, Andhra Pradesh, India
e-mail: vaastusrinivas@gmail.com

S. Jaikrishna
Narasimhareddy Engineering College, Hyderabad, Andhra Pradesh, India

S. S. Khangura et al. (eds.), *Proceedings of the International Conference on Research and Innovations in Mechanical Engineering*, Lecture Notes in Mechanical Engineering, DOI: 10.1007/978-81-322-1859-3_2, © Springer India 2014

1 Introduction

Radar dome, or usually called radome, is usually placed over the antenna as an antenna protector. The basic function of a radome is to form a protective cover between an antenna and the environment with minimal impact on the electrical performance of the antenna. Under ideal conditions, a radome is electrically invisible. How well a radome accomplishes this depends on matching its configuration and materials composition to a particular application and radio frequency range.

Radomes can be found protecting a wide range of outdoor terrestrial and shipboard communications systems and radar installations as well as airborne avionics system antennas. The proper selection of a radome for a given antenna can actually help to improve overall system performance by doing the following:

1. Maintaining alignment by eliminating wind loading and allowing for all-weather operations by protecting the system from rain, snow, hail, sand, salt spray, insects, animals, UV damage, and wide temperature fluctuations.
2. Providing shelter for installation and maintenance personnel.
3. Preventing visual observation of system (security).
4. Minimizing downtime and extending component and system operating life.

Radomes can be classified into ground-based, naval, and airborne radomes.

Historically, a variety of materials have been used for constructing radomes, including balsa and plywood in early structures. Modern ground-based and ship-based radomes are manufactured using composite materials such as fiberglass, quartz, and aramid fibers held together with polyester, epoxy, and other resins, such as the one shown. Foam and honeycomb cores are often added between inner and outer "skins" of the radome to function as a low-dielectric-constant spacer material providing structural strength and rigidity.

2 Radome

2.1 Radome Configuration

Several radome configurations are used to minimize RF reflections, including electrically thin, half-wave, A-sandwich, C-sandwich, and others. The best configuration for a particular application depends on the mechanical requirements and operating frequency.

A radome that is electrically thin (less than 0.1 wavelengths) as shown will generally deliver good RF performance. This is because signal reflections at the free-space/dielectric boundary are canceled out by out-of-phase reflections from the dielectric/free-space boundary on the other side of the dielectric material.

Signal losses are low, and the net transmission from an electrically thin dielectric laminate is very high. Unfortunately, electrically thin radomes provide very little thermal insulation and are not suitable for locations with wide temperature extremes and a requirement for controlled temperatures.

Another radome approach that works well is a configuration based on the half-wavelength-thick solid laminate shown in Fig. 5. It is similar to the electrically thin configuration because the reflections cancel out. The wave travels 180° through the laminate, is reflected with a phase shift of −180°, and travels another 180° on the return trip to achieve the net 180° phase shift required for cancelation. Performance of the same laminate is described in Fig. 4 at higher frequencies (through 35 GHz) where it is 0.5 wavelengths thick.

A-sandwich radome configuration consists of low dielectric foam or honey-comb core sandwiched between two thin laminates. Its operation is similar to the half-wavelength-thick solid laminate. However, it is 0.25 wavelengths thick because the reflection coefficients from the skins have the same amplitude and phase. The round trip for the reflection from the second skin is 0.5 wavelengths. The reflections, which are 180°, are out of phase.

A C-sandwich radome consists of three skin layers and two foam layers. The thickness of each foam layer, and possibly the skins, can be tuned for optimal RF performance in the bands of interest. This can lead to many potential construction combinations that can provide good RF performance and high mechanical strength. C-sandwich constructions provide better performance than A-sandwich radomes; however, the added complexity increases material and labor costs.

2.2 *Radome Wall Diffraction*

The uniform radome wall represents 90–96 % of a radome surface. Therefore, it has to be designed carefully. Any A-sandwich consists typically of three layers as demonstrated in Fig. 1:

(a) inside skin
(b) foam core
(c) outside skin (with hydrophobic coating)

Figure 1 shows that the electromagnetic (EM) wave can easily transmit throughout the material layers of foam core in sandwich construction. The transmission of EM waves through any sandwich can be calculated as the diffraction of EM energy at boundaries between areas of different dielectric properties. Based on the knowledge of the EM properties of the materials like dielectric constant and loss factor, it is possible to develop a transmission matrix T for every single boundary depending on the incidence angle, frequency, and polarization of the EM wave and to calculate the transmission loss. Figure 1 shows the one-way transmission loss for an EM wave. Over frequency at 0°, 20°, and 40° incidence angle and linear TE polarization for a typical A-sandwich designed for C-band weather

Fig. 1 Installation of radome over the antenna

radar applications. In Fig. 1, the parabolic antenna is protected by the radome. These radome only can help to improve the performance of radome underwater-depth applications and protects from the environment.

2.3 Structural Support

Although radomes are used extensively on airframes and missiles, this section focuses specifically on support structures for terrestrial and shipboard systems. Ground and shipboard radomes can range in size from very small antenna covers to massive structures.

Self-supporting radomes are usually based on an A-sandwich configuration. They are made of rigid sections that are bolted or latched together. If phase delay and insertion loss through the seam is matched to the rest of the radome, the seam becomes largely invisible to the EM wave front. Unlike other radome types mentioned in this article, A-sandwich radomes require no air blowers to maintain pressure and are not dependant on electrical power to maintain their electromagnetic or structural performance. A-sandwich radomes generally have lower overall operation and maintenance costs.

Inflatable radomes are made of electrically thin dielectric cloth. By being electrically thin, they are capable of achieving very low loss over wide bandwidths. The trade-off for high performance, however, is that they require a constant supply of air. Inflatable radomes must be supported by internally generated air pressure, which is supplied by air blowers or air compressors. In order to maintain adequate air pressure, inflatable radomes must be equipped with airlocks at all doors and a standby power supply to operate the blowers at all times and under all environmental conditions. Should the membrane suffer damage or whether power is interrupted, it is possible for the radome to deflate and collapse. Operating and maintenance costs for this type of radome usually exceed those all other radome types. Metal space frame radomes support the window portion of the radome consisting of the electrically thin, half-wave, or A-sandwich configuration, often in the shape of a geodesic dome. The window portion typically has very low loss. However, 10 signal blockages from the frame reduce system gain and reflect noise back into the system. Because the frame reflects and refracts the RF wave front, it increases side lobe levels. A method used to prevent large side lobes is the use of a quasi-random frame pattern. The quasi-random pattern is also used to minimize side lobes for the other support structure types. In contrast to metal space frame radomes, dielectric space frame radomes are supported by dielectric members who are somewhat electrically transparent. However, the wave front is phase delayed as it passes through the dielectric support, alternating between in and out of phase, depending on frequency. If the delay is 180° out of phase, with the phase of the incident signal, the energy that passes through the frame subtracts from the gain. This leads to a frequency-dependant sinusoidal ripple in the insertion loss, and the lost energy goes into the side lobes. This makes dielectric space frame radomes best suited to systems that operate at less than 1 GHz. Both types of space frame radomes usually require the use of air blowers or compressors in order to maintain and enhance the structural integrity of their thin membrane coverings during windy conditions. Failure to maintain positive pressure can result in membrane damage and failure.

2.4 *Impact of Incident Angle*

All of the plots and explanations thus far show reflections at normal incidence. Typically, an EM wave hits the radome surface at an oblique angle, or in the case of a spherical radome a continuous range of oblique angles. The transmission characteristics of the radome change with the wave incidence angle and polarization. Electric fields that are parallel to the plane of incidence have much higher transmission than fields that are perpendicular to the plane of incidence. Aerodynamic radomes used on aircraft and missiles often see high incidence angles. This can result in large amounts of axial ratio degradation for circularly polarized antennas and higher insertion loss. EM wave fronts from parabolic antennas located inside spherically shaped radomes see low incident angles at the center of the wave front. Out on the edges, however, the incident angle becomes higher. If the antenna illumination pattern is symmetric and the antenna is placed at the center of the spherical radome, the symmetric shape of the radome cancels out axial ratio degradation from the oblique incidence angles seen by the antenna. Composites are gaining wider acceptance for use on onboard warships and submarines due to number of advantages, viz., high strength-to-weight ratio, ability to be molded into complex shapes, better EMI performance, and absence of corrosion palliatives that otherwise are source for electronic and magnetic signature. Composite materials made from E-glass fibers and epoxy resins have become very popular as a radome material due to its outstanding transparency to microwaves and having good mechanical properties. The increasing popularity of the material for underwater application are posing great difficulties to the designer to select right combination of composition and shape of radome due to the complex nature of the structure and the loading conditions for the useful operation life.

Mechanical properties of composite materials are influenced by several factors like reinforcement, fiber orientation, adhesion, composition, manufacturing process, etc. Conducting the tests on standard specimens and evaluating mechanical properties is the most important aspect in the design of composite material applications. The ASTM guidelines were followed in testing and preparation of standard test specimens. The micromechanics and failure mechanism of composite material is very complex compared to the conventional isotropic materials. Depending on the reinforcement, composition content, and its percentage, appropriate theory and failure mechanism can be considered for designing the radome.

Finite element analysis of radome design is carried out using (Analysis System) ANSYS, a FEA software package. Geometrical model of radome is generated as per radome sketch. Suitable elements are selected, and optimum size of mesh is generated. Material properties, evaluated from tests, are assigned. Boundary conditions and load cases are applied to complete the preprocessing stage. The post-results obtained after FE analysis are compared with design requirements. The main objective of this project is to develop composite radome that protects the electronic equipment from high water pressure and transparent to EM waves.

In Fig. 2, we can see the shape of radome assembled with the bolts and nuts by holes.

The geometric shape of the radome is a cylindrical barrel covered with a hemispherical dome at the top. It has a circular plate at the bottom end of the cylinder having M6-size holes which acts as a flange. The radome is secured to the submarine structure with M6 bolts on its flange. Radome is made of sandwiched construction with glass-reinforced plastic (GRP) as sheet material and syntactic foam as core. E-glass woven fabric and epoxy resin are used after FE analysis compared with design requirements.

2.5 Functions of the Radome

The functions of the radome are as follows:

1. The radome protects the installation from the deteriorating effects of environment and extends the durability of antenna and other equipment.
2. The overall performance of the antenna will be increased with the use of radome.
3. FRP radome helps to have overall economy and weight reduction.
4. A radome permits the airborne antenna to function with good efficiency under high head of the water over the submarine.

Fig. 2 Submarine radome

3 Construction and Materials

Advanced composites and special products are made from reinforcements such as fiberglass, quartz, graphite, and Kevlar® along with matrices such as polyester, epoxies, and cyanate ester. We also use core materials such as honeycomb (e.g., fiberglass, aluminum, and graphite) and foams (e.g., polyisocyanate and thermoformable cores). Depending on the application, these parts are oven-cured at temperatures up to 400 °F or in autoclaves, which require high-pressure cures at high temperatures. Other materials are also available for special applications. Regardless of the application(s), we can select the right combination of reinforcement and matrix to meet requirements.

All of these products include excellent EM performance for their intended applications, providing up to 98 % transmission efficiency depending on frequency. Some typical examples include the following: Naval radomes are used in shipboard radar applications, high-data-rate communications systems, gunfire control, and high-bandwidth data link terminals.

A probe by accident into the field of thermosetting polymers has brought about a quantum growth in its basic as well as technological aspects. The synthetic thermosetting polymers with the combinational properties of the existing conventional high-strength polymers and glass fibers with a variety of filler materials have altogether offered a new field of research. The review of work presented here reveals that large effort has gone into the understanding of the mechanical, thermal, and physical properties of thermosets. A thorough literature search reveals that there are no systematic studies on mechanical properties of thermosetting composites. There is ample scope for fabrication of newer composites with different weight fractions of glass fiber and PET in polymers and their characterization for physical, mechanical, and thermal properties. A variety of filler materials have altogether offered a new field of research.

4 Hybrid Composites

A thorough literature search reveals that there are no systematic studies on mechanical properties of thermosetting composites. There is ample scope for fabrication of newer composites with different weight fractions of glass fiber and fillers in polymers and their characterization for physical, mechanical, and thermal properties.

In this thesis, a wealth of data on mechanical properties of polymer glass filler composites has been generated. These data are useful for material technologists, mechanical engineers, and defense engineers, who can make use of this database for the generation of new materials for specific application. In that respect, it has been used in GF and virgin PET fibers in the form of woven mat and epoxy as matrix. Laminates are obtained from vacuum bag molding technique. Tests were carried out to evaluate physic-mechanical and thermal properties according to ASTM standards Table 1.

5 Analysis of Submarine Radome

The following different load cases are considered for designing radome:

Case (i) Water head pressure acting on radome (due to under water)

Water head pressure acting on radome (p) = ρgh.

Density of seawater at average temperature of 25 °C (ρ) = 997.0479 kg/m^3

Radome depth in water (h) = 490 m, P = 997.0479*9.91*490 = 4.7920*10E6/Nm^2 = P = 48 bar

The maximum operating depth (popularly called the never-exceed depth) is the maximum depth at which a submarine is allowed to operate under any conditions. World War II German U-boats generally had collapse depths in the range of 200–280 m (660–920 ft). Modern nuclear attack submarines like the American Seawolf class are estimated to have a test depth of 490 m which would imply (see above) a collapse depth of 730 m (2,400 ft)

Following assumptions are made to analyze the model.

1. Water pressure acting on the periphery of the radome.

Table 1 Physical dimensions of radome

Physical dimensions	Values in mm
Diameter	1,651
Height	1,384.3
Hole diameter	50
Thickness	275

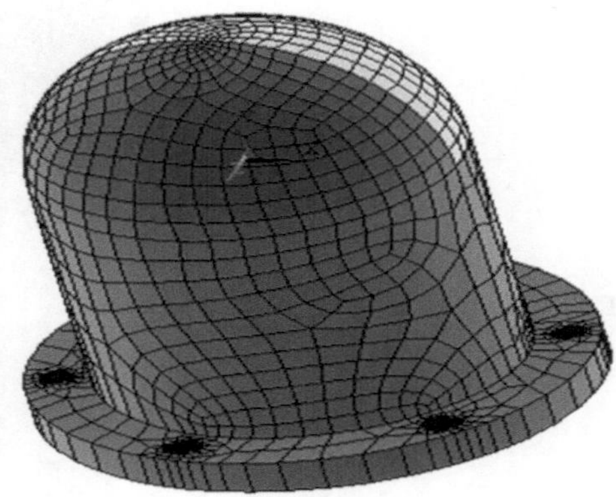

Fig. 3 Meshing of radome

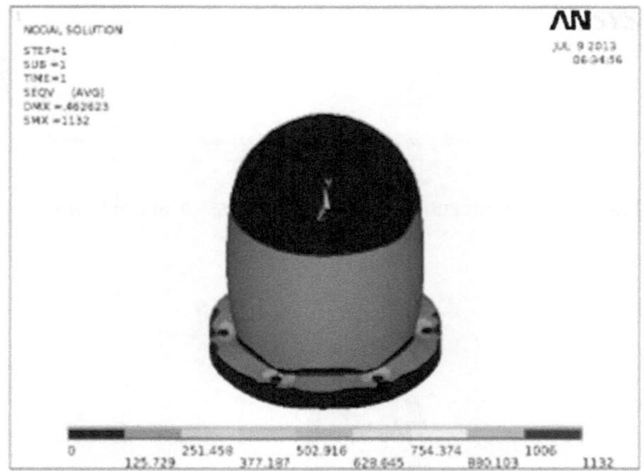

Fig. 4 Displacement vector sum

2. Material properties taken for E-glass/epoxy fiber reinforced plastic with fiber orientation of 0 and 90°.
3. Mounting flange of radome of assumed rigid body.

Static analysis:

ANSYS has been used for the finite element analysis of the radome. Linear static analysis is carried out to find out the structural response of the model. The area of radome is meshed with shape tetra and free meshed and is shown in Fig. 3.

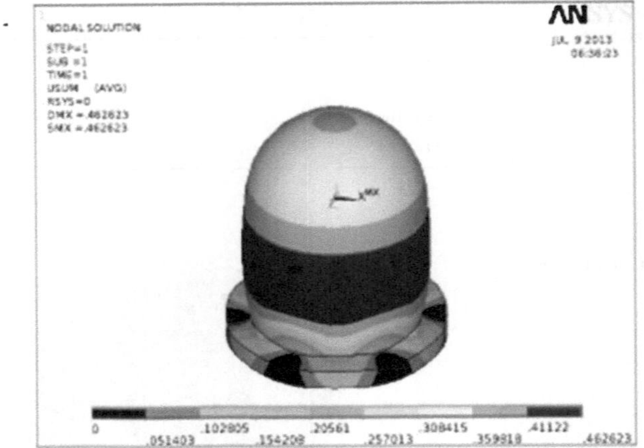

Fig. 5 Von Mises stress of aramid epoxy

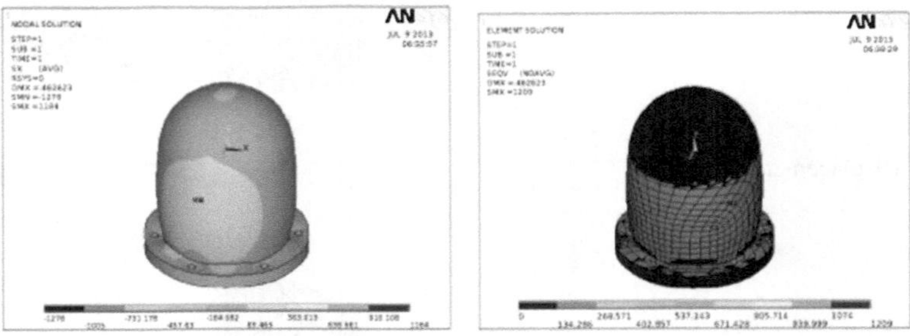

Fig. 6 *X*-component and von Mises stress of aramid epoxy

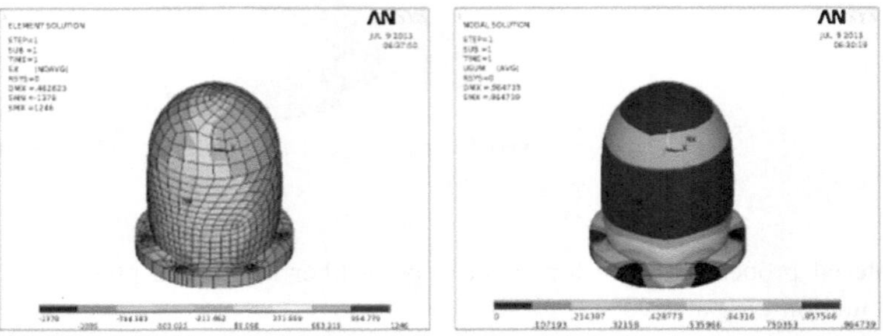

Fig. 7 Element stress in *x*-direction and displacement vector sum of E-glass epoxy

Figure 4 shows the maximum displacement vector sum is 0.462623 mm of aramid epoxy material. Maximum displacement occurs at the center of the dome

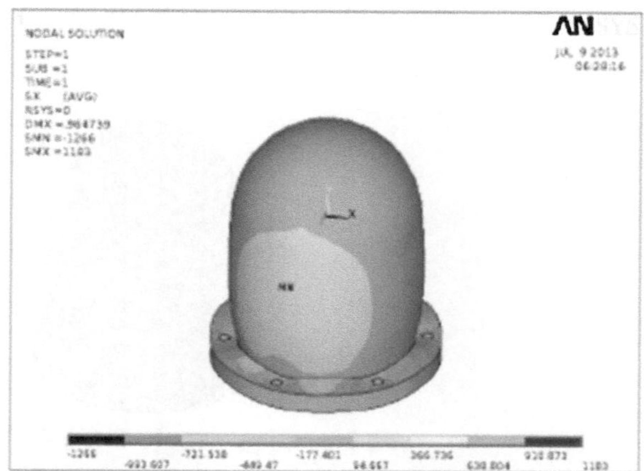

Fig. 8 Stress component in x-direction

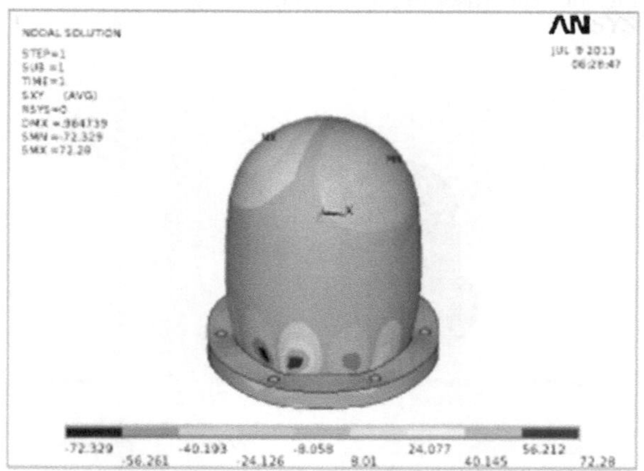

Fig. 9 Shear stress in x-direction

which is shown in figure with the indication of MX. This deflection does not have much effect on the radome.

Figure 5 shows the maximum stress occurs at the corners of the hole which is 1,132 MPa. The ultimate tensile strength of material is 1,377 MPa which is less than resulted stress. So this design is in safe limit in static mode. The high stresses are occurring only at the corner of hole. The corner hole stress can be reduced by stress concentration factor. The thickness of the material decreases which also reduces the weight.

Figure 6 shows the tensile stress in x-direction of the aramid epoxy is 1,184 MPa, but the actual strength of tensile stress in x-direction is 1,377 MPa. In case of comparing tensile stress in x-direction, also the design is in safe limit. Figure 7 shows the element stress in x-direction of aramid epoxy is 1,246 MPa. It is lower compared to ultimate tensile strength of aramid epoxy resin (Fig. 7). The elemental von Mises stress is 1,209 MPa which is below than ultimate tensile

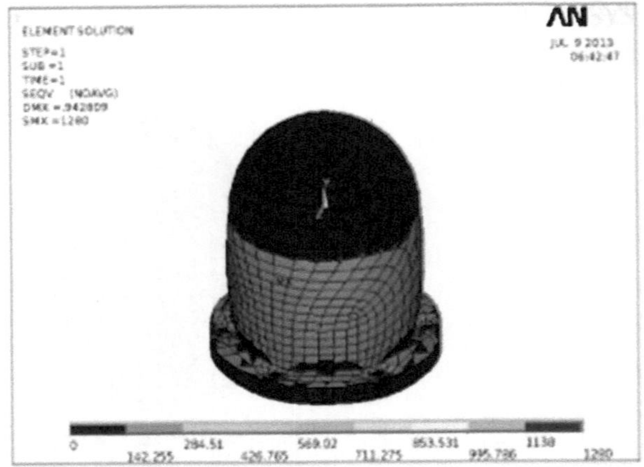

Fig. 10 Von Mises stress of E-glass epoxy

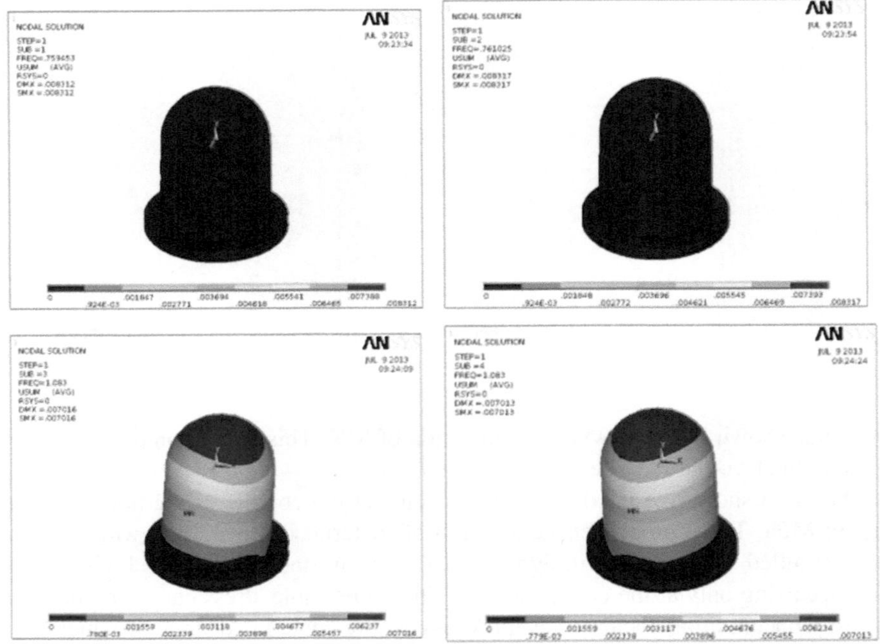

Fig. 11 Frequency and mode shape of substeps 1, 2, 3, and 4

strength of aramid epoxy. So this design is safe. The displacement vector sum of E-glass epoxy is 0.964739 mm, which is under the pressure of 48 bar. The maximum deformation occurs at the top and center of the radome. The constrained portion of hole has minimum deflection.

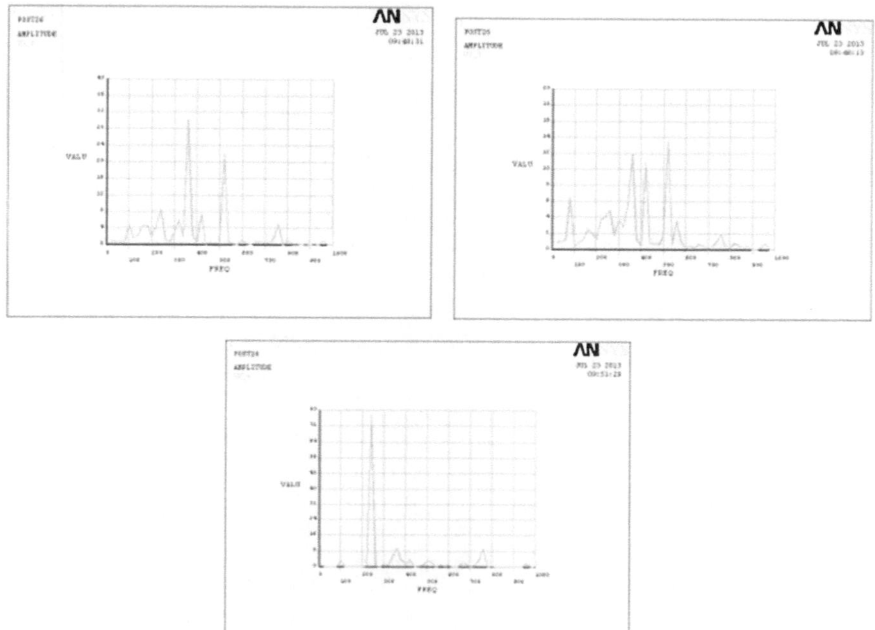

Fig. 12 Displacement (UX) graph between amplitude versus frequency for first step at node 665 for frequency 360 Hz

Figure 8 shows the resulted stress in x-direction is 1,183 MPa which lower than the ultimate tensile strength of E-glass epoxy (1,243 MPa). The maximum stress occurs at the edges of the hole.

The result of shear stress in x-direction is 72.28 MPa, but the actual shear strength capability of E-glass epoxy is 73 MPa. By comparing the shear stresses, both are closest with each other. It cannot be say as a safe limit. It may have chance to failure (refer Fig. 9). Figure 10 shows the von Mises stress of E-glass epoxy is 1,280, but the ultimate tensile strength of E-glass epoxy is 1,243. It cannot be safe design because the resulted stress reaches the ultimate tensile strength.

Modal analysis was used to determine the vibration characteristics (natural frequencies and mode shapes) of a structure or a machine component while it is being designed. It also can be a starting point for dynamic analysis, such as a transient dynamic analysis, a harmonic response analysis, or a spectrum analysis.

Figure 11 shows the frequency of aramid epoxy is 0.759453 Hz and deflection is 0.008312 for step 1; frequencies and modes of shape for aramid epoxy are 0.751025 and 0.008317 for step 2; frequency and modes of shape for aramid epoxy are 1.083 and 0.007016 for step 3; and frequency and modes of shape for aramid epoxy are 1.083 and 0.007013 for step 4. Maximum deflections occur at the top of the dome. The frequencies of each substep increase slightly. The frequencies of

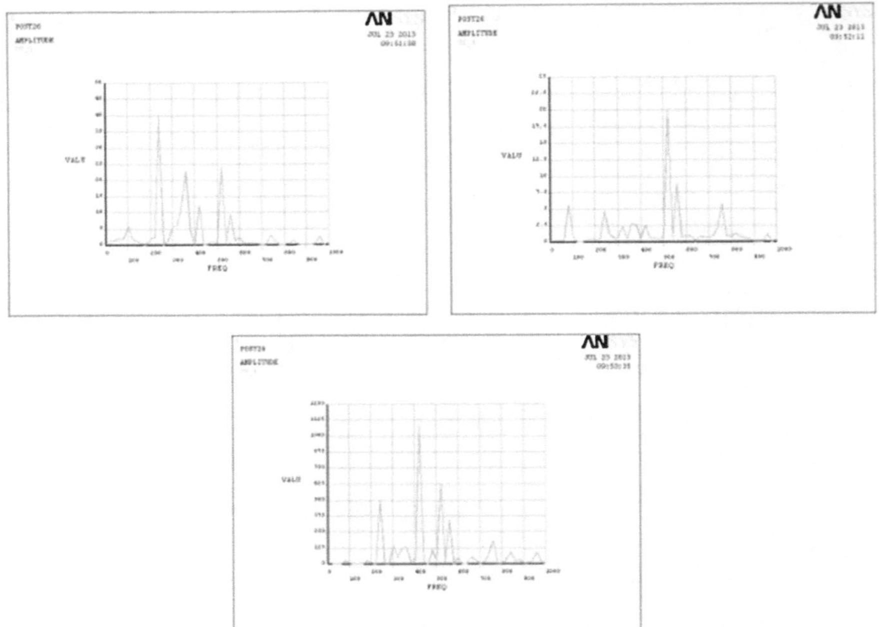

Fig. 13 Displacement in *y*-direction is for last substep at node 1,269

aramid epoxy are low and mode shapes deflections. The maximum frequency in all substep is 2.3489 which less than four times of natural frequencies.

Harmonic analysis is a branch of mathematics concerned with the representation of functions or signals as the superposition of basic waves, and the study of and generalization of the notions of Fourier series and Fourier transforms. In the past two centuries, it has become a vast subject with applications in areas as diverse as signal processing, quantum mechanics, and neuroscience. The term “harmonics” originated in physical eigenvalue problems, to mean waves whose frequencies are integer multiples of one another, as are the frequencies of the harmonics on stringed musical instruments, but the term has been generalized beyond its original meaning.

6 Results and Discussions

The von Mises stress of element is 1,280 which reaches the ultimate tensile strength of E-glass epoxy which may fail. So aramid epoxy material is under the hydrostatic pressure of 48 bar. The dielectric constant and loss tangent of the aramid/epoxy composite measured by the free-space measurement method were 3.742 and 0.018 E-glass/epoxy composite (4.686 and 0.015) aramid preferable because it has low dielectric constant and loss tangent. Maximum displacement

Table 2 Summary of results for epoxy material

Components	E-glass epoxy		Aramid epoxy	
	Ultimate tensile strength (MPa)	Resultant stress (MPa)	Ultimate tensile stress (MPa)	Resultant stress (MPa)
Stress in *x*-direction	1,243	1,183	1,377	1,184
Von Mises stress of nodal	1,260	1,173	1,392	1,132
Von Mises stress of element	1,243	1,280	1,377	1,209
Shear stress in *xy*-direction	73	72.8		

vector sum is 0.964 mm of E-glass epoxy material. Maximum displacement sum of aramid epoxy is 0.462 mm. which is less than that of E-glass epoxy material. In modal analysis, the maximum frequency value is 2.018 Hz. The natural frequency of the radome is less than four times the excitation frequency so that resonance does not occur. Displacement in *z*-direction is for substep 1 at node 1,269. Maximum amplitude is 13 mm at the frequency of 520 Hz as shown in Fig. 12. Figure 13 shows displacement in *x*-direction is for last substep. Maximum amplitude is 78 mm at the frequency of 250 Hz.

Maximum amplitude is 40 mm at the frequency of 250 Hz. Displacement in *z*-direction is for last substep at node 1,269. Maximum amplitude is 20 mm at the frequency of 520 Hz. Figure 13 shows stress in *y*-direction is for substep 1 at node 156. Maximum amplitude is 1,000 mm at the frequency of 420 Hz Table 2.

7 Conclusions

1. The high stresses are occurring only at the corner of hole. The corner hole stress can be reduced by stress concentration factor method.
2. Deformation and stress values obtained of E-glass epoxy from FE analysis are within the safe limits.
3. Conducting pressure test on radome verified the design aspects and validated the FE analysis.
4. To improve the electrical performance of the radome without compromising the mechanical properties hybrid composites to be considered in futuristic radome development.

References

Bourasseaua N, Moulina E, Delebarrea C, Bonniaub P Radome health monitoring with Lamb waves: experimental approach. aIEMN, UMR CNRS 9929, OAE Department, Universite′ de Valenciennes

Cady WM, Karelitz MB, Turner LA (1948) Radar scanners and Radomes. MIT University, New York
Choi I, Kim JG, Lee DG, Seo S (2011) Aramid/epoxy composites sandwich structures for low-observable radomes. School of Mechanical Aerospace and Systems Engineering, Korea Advanced Institute of Science and Technology
Chandrupatla TR, Belegundu AD (2011) Introduction to finite elements in engineering, 4th edn
DiCarlo A, Chung KH (2013) Radome-enclosed antenna's temperature and velocity fields. MITRE Corporation, Bedford
Gates PJ, Lynn NM (1990) Ships, submarines & the sea, vol 2. Brassey's, London
Harris B (1999) Engineering composite materials, 2nd edn
Joubert PN (2004) Some aspects of submarine design. Australian Government, Department of Defense
Kim PC, Lee DG (2009) Composite sandwich constructions for absorbing the electromagnetic waves. Department of Mechanical Engineering, Korea Advanced Institute of Science and Technology
Naito K, Kagawa Y, Kurihara K Dielectric properties and noncontact damage detection of plain-woven fabric glass fiber reinforced epoxy matrix composites using millimeter wavelength microwave. Research Center for Advanced Science and Technology, The University of Tokyo
Oh J-H, Oh K-S, Kim C-G, Hong C-S (2004) Design of radar absorbing structures using glass/epoxy composite containing carbon black in X-band frequency ranges. Division of Aerospace Engineering, Department of Mechanical Engineering, Korea Advanced Institute of Science and Technology, p 373
Satheesh Kumar Reddy P, Nagaraju C, Hari Krishna T (2012) Optimum design and analysis of filament wound composite tubes in pure and combined loading, vol 1(8). ISSN: 2278-0181
Sun CT (2000) Strength analysis of unidirectional composite laminates
Timoshenko SP, Gere JM (1963) Theory of elastic stability, 2nd edn. Mc Graw-Hill Book Company, New York

Development of Lance Clamping Device for Basic Oxygen Furnace

Asim Kumar Sahu

Abstract Two basic oxygen furnaces (converter # 4 and 5) each of 60T capacity are operative in Steel Melting Shop no. 1 of Rourkela Steel Plant. These converters had been provided with lance clamping system. The clamps were consisting of a set of linkages, which used to be actuated through pneumatic cylinders. These clamps subsequently became non-functional, and converters were being operated without lance clamping device. Unclamped lances used to dangle and vibrate, causing impingement of oxygen jet over the refractory. This resulted in poor lining life. Incorporation of conventional design of lance clamp with moving trolley and guide was not feasible due to space constraint. An innovative design of lance clamp has been developed, in which actuation of jaws is effective through motor, gear box, and lead screws. This design is compact and has been accommodated without any problem in the existing available space. Lance clamping device has been commissioned, and it is in regular operation. The design aspects of the lance clamping system have been dealt with in this paper.

Keywords Clamping device · Lead screw · Gear box · Lance

1 Introduction

Two basic oxygen furnaces each of 60T capacity (converter # 4 and 5) are operative in Steel Melting Shop of Rourkela Steel Plant (RSP). In these converters, refinement of crude molten iron is done by blowing oxygen through refractory-coated lance. In the original design, converters were provided with lance clamping system for guiding the lance during blowing. These clamps consisted of a number of linkages, which

A. K. Sahu (✉)
Research and Development Centre for Iron and Steel, Steel Authority of India Ltd., Doranda, Ranchi 834002, Jharkhand, India
e-mail: aksahu@sail-rdcis.com

S. S. Khangura et al. (eds.), *Proceedings of the International Conference on Research and Innovations in Mechanical Engineering*, Lecture Notes in Mechanical Engineering,
DOI: 10.1007/978-81-322-1859-3_3, © Springer India 2014

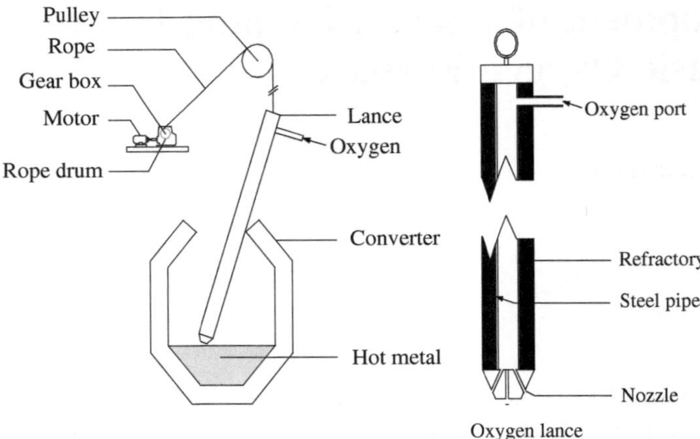

Fig. 1 Unguided lance converter system

were actuated through pneumatic cylinders. Due to capacity enhancement of the converter and space constraint, the clamps became inoperative and were removed. Converters were being operated without lance clamping device for quite long. Unclamped lances used to dangle and vibrate causing erratic blowing operation. This also caused impingement of oxygen jet over the refractory lining of the converter and thus decreasing the converter lining life. RSP requested Research and Development Centre for Iron and Steel (RDCIS) that a lance clamp should be designed and developed by in-house efforts keeping the space constraint in mind. A design of lance clamp system was developed, in which actuation of clamping mechanism was through motor, gear box, and lead screws. This design is compact and has been accommodated without any problem in the available space. The system has been installed and commissioned in converter # 5. Since then, the blowing operation has become quite smooth and converter lining life has also improved substantially.

2 Problem Formulation

Oxygen lance of converter design is hanged by steel rope and the lance is moved up and down through rope and winch system as shown in Fig. 1.

Unguided lance was subjected to vibration, and it dangled during blowing. Due to this, the following problems took place:

- Blowing was non-uniform.
- Damage of lance refractory was frequent.
- Erosion of refractory of one side of the converter becomes faster resulting in shorter converter life.
- Loss of oxygen and metal was prominent due to splashing non-uniform blowing.

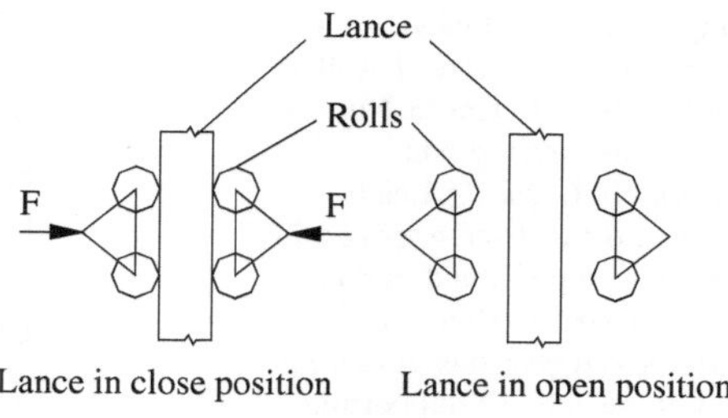

Fig. 2 Basic kinematic diagram

To minimizing the above problems, positioning the lance centrally is essential and necessity was felt to design a suitable clamping device for the lance.

3 Design Consideration

Basic purpose of the clamp is to hold the lance centrally with some amount of tightness during operation of oxygen blowing. The space for the clamp was limited. The zone is hot and full of dust. During blowing, occasionally the flame comes to that zone. The lance in the unclamped condition has a horizontal play of 1,600 at the blowing zone and 1,400 mm at the spout. This play is to be restricted to almost nil at clamp level and 300 mm maximum at the blowing level. This 300 mm play may arise due to deflection caused by various forces at the lance tip. The forces are mainly due to non-uniform blowing. The tip of the lance is having a number of nozzles. Metal and slag may stick to some holes and cause non-uniform blowing. This may result a complex kind of vibration giving rise to dynamic load to the lance. This can cause fast erosion of refractory and breakage of spout of the converter.

Following are the design bases:

- Minimum gap between clamp rollers in clamping position is 180 mm.
- Maximum gap between clamp rollers in fully open position is 650 mm.
- Vertical distance between top and bottom rolls is 300 mm.
- Clamping force on lance on lance pipe is 1,000 kg.
- Time for clamp/de-clamp is 30 s.
- Design is compact due to space constraint.
- Clamp parts are protected from heat and dust.
- Equipment by electrical motor and control system.

The basic kinematic diagram (Newell 1977) is shown in Fig. 2. The lance is hold in position using clamp during blowing. During parking condition, it is de-clamped.

During clamping condition, it is required that the lance move up or down freely. The lance will be guided by rolls. Hence, the contact force of the lance should be

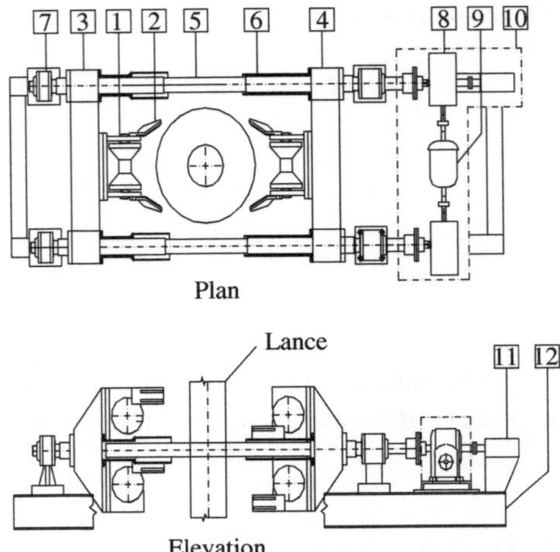

Fig. 3 Plan and elevation of the clamping device. *1* Roller assembly. *2* Finger. *3* Nut assembly (*left*). *4* Nut assembly (*right*). *5* Lead screw. *6* Lead screw cover. *7* Bearing support and bearing. *8* Gear box. *9* Motor. *10* Motor and gear box cover. *11* Limit switch. *12* Supporting frame

minimum. However, the force will come in action as soon as there will be a tendency of dangling of the lance. The lance is made of mild steel pipe and coated with refractory with weight of about 200 kg_f. The diameter of the lance is 180 mm. There are three holes at the tip of the lance 120° angle apart. Oxygen is blown through these holes inside the hot metal. Since hot metal temperature is about 1,600 °C, the lance is water-cooled. During blowing, disturbance inside the hot metal is created and churning of metal takes place. Force at the lance tip due to asymmetrical blowing was calculated to be 25–30 kg_f from impulse momentum due to oxygen flow, and it is in the horizontal direction. This gives rise to a bending moment at the clamping position and corresponding horizontal force at the rollers comes to about 1,000 kg. The bouncy force on the lance balances its weight to some extent. Dynamic force due to vibration also comes in picture and about 30 % extra of the horizontal force has been taken for design purpose. All parts of the equipment have been designed on the basis of this force. At the position of the clamp, the environment temperature is about 250 °C, and it fluctuates. Thermal expansion and contraction corresponding to this temperature have been incorporated in the design. The plan and elevation of the clamping device that was developed are shown in Fig. 3.

Roller assembly: Dumbbell-shaped rollers (Fig. 4) have been provided for holding the lance centrally and guiding them. They are antifriction bearing mounted. Fingers of special shape have been provided to bring the lance to the center while clamping. Material (Budinski 1979) for different components of this assembly is from boiler quality plates as per IS 2002–1992 and steel 40 Cr 1 Mo 28 IS:1570-1988. Provision has been made for regular lubrication of the bearings of the rollers with high heat synthetic grease.

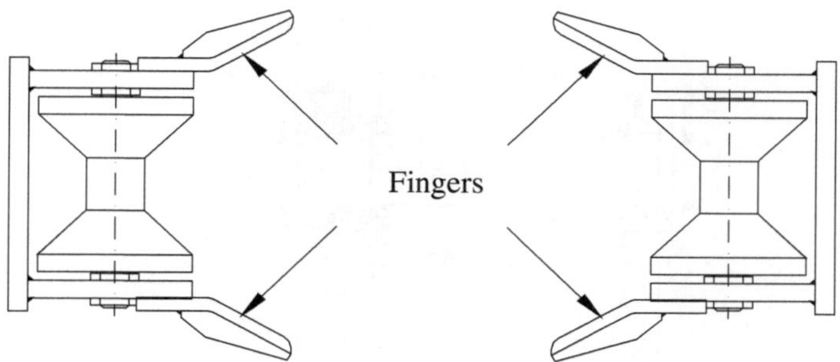

Fig. 4 Dumbbell-shaped rollers

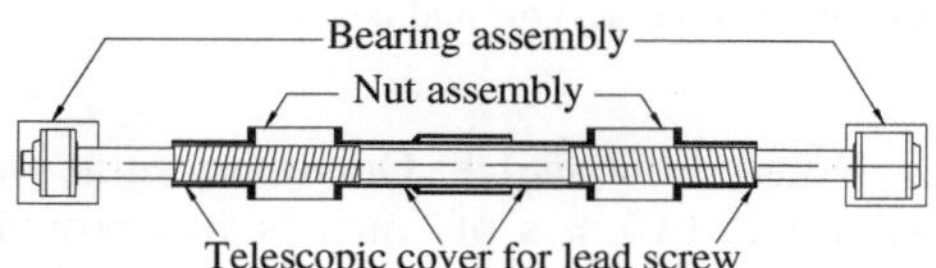

Fig. 5 Lead screw assembly

Lead screw assembly: This assembly consists of nut, lead screw (Schubert 1979) with cover and bearing subassembly. Nut is having left-hand trapezoidal thread and right-hand trapezoidal thread of 70 $\times$ 10 mm as per IS:7685-1975. The material for lead screws are forged steel 40Cr M028 and also having left- and right-hand trapezoidal threads. Nuts are made of bronze. Bearing assembly is having antifriction radial and thrust bearings. The lead screw is provided with steel pipe cover, which is telescopic in nature. These covers are meant for protecting lead screws from dust. The assembly is shown in Fig. 5.

Bearing support and bearing: There are two bearing supports, one in each side of the lead screw. The non-drive side bearing support contains one radial ball bearing. The drive side contains a combination of double row thrust bearing and radial bearing. Cover for the bearings are provided to protect from dust. Provision for lubrication with synthetic grease has been provided. General arrangement of this assembly is shown in Fig. 6.

Supporting frame: This frame is made of structural members mainly I-beams and is meant for supporting the main equipment, gearbox, motor, and limit switches.

Worm and worm wheel gear boxes: Two numbers of worm and worm wheel gear boxes with gear ratio 30:1 are provided out of which one is having extended output shaft for limit switch mounting.

Heat protection cover: This cover is meant for protection of motor and gearbox. Cold compressed air is poured inside the cover to keep it cool always.

Motor: The motor used is a squirrel cage 415 V AC, three phase, 1 HP, 1,500 rpm corrosion proof, dust proof, totally enclosed fan cooled with protection class IP:55, class F insulation.

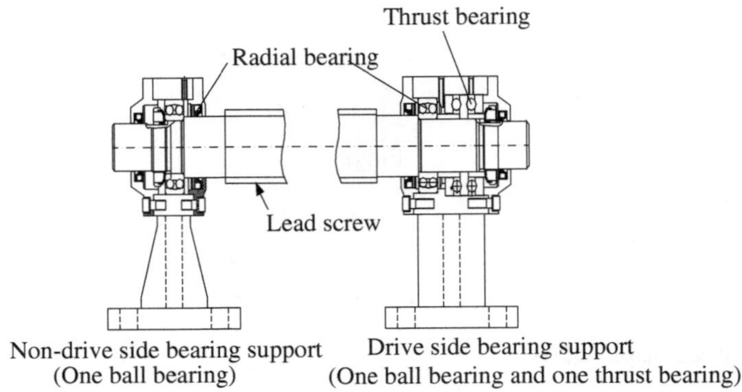

Fig. 6 Bearing support and bearing

Rotary limit switch: One rotary limit switch with reduction ratio 1:50 and with minimum two sets of contacts was provided and directly coupled with the gear system for making/breaking contacts for sensing the extreme positions of the motor for maximum and minimum opening/closing of the clamp. The limit switch is suitable for working in hot and dusty environment where ambient temperature is around 250 °C.

Snap action limit switch: One snap action limit switch has also been provided to have positive sensing of de-clamp and clamp position and to facilitate interlock with lance movement circuit.

Interlock: Arrangement for interlocking and controls are as follows:

- Lance lowering and hoisting are possible above 2.5 m level from the blowing position, while clamp is in open condition.
- After attaining 2.5 m level or below by the lance, clamping can be done from the panel by pressing "Clamp" button. In this zone, inching of the lance is possible irrespective of clamped condition.
- De-clamping of the lance is done by pressing "De-clamp" button in the panel. If de-clamping is not done fully, the lance cannot be hoisted beyond 2.5 m level. However, in case of problem such as water failure and GCP failure, de-clamping is automatically done and lance can be hoisted.

Control panel: Control panel has been installed near the converter control pulpit at shop floor level from where the lance can be clamped and de-clamped as required. Indications are available to the operator regarding clamped/unclamped position of the clamp.

Mounting of the equipment: Above the converter, there is a platform at a height of 700 mm from converter spout. The platform gets hot during blowing. Hence, for protection from heat, heat-shield made of ceramic blanket has been provided at the bottom of the platform. For further protection from heat, a water jacket has been provided on the platform. The platform was made stronger by welding stiffeners

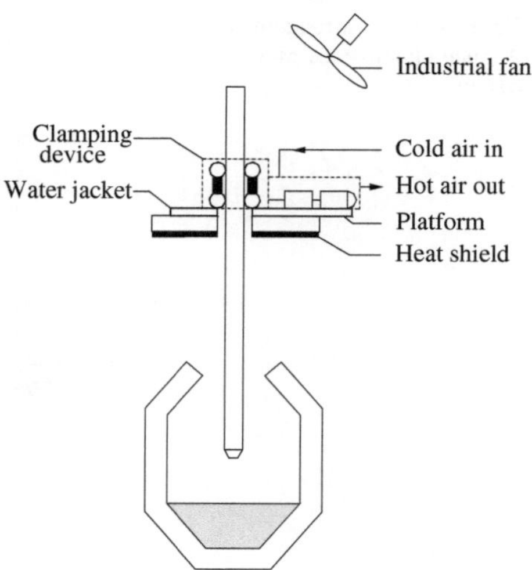

Fig. 7 Mounting arrangement

wherever possible. Two long I-beams with cross-beams were laid on the platform over water jacket, and the clamping system has been mounted over it using nuts and bolts. The arrangement is shown in Fig. 7.

4 Operation of the Equipment

Before operation, following conditions are checked physically at the site of clamping platform. The cooling water inlet and outlet valves as well as flow of water are checked. The industrial cooling fan is made on, and the compressed airline valve is opened. Greasing of the roller bearings and lead screw are done. When site conditions are alright, following operations are performed:

- Lance is operated from control panel located in the control pulpit of the converter at shop floor level. Control pulpit houses the control panel of the converter and is used by the converter operators on daily basis.
- In the control panel of the clamping system, there are incoming power switch, start/stop switches, and an emergency stop button. Lighting indications are also provided to this panel.
- Lance lowering and hoisting are possible above 2.5 m level from the blowing position, while clamp is in open condition.
- After attaining 2.5 m level or less by the lance, clamping can be done from the panel by pressing clamp button.
- When the lance is near the hot metal surface, it requires inching for adjustment of height. Inching of lance is possible irrespective of clamped condition.
- De-clamping of lance is possible by pressing de-clamp button in the panel.
- If de-clamping is not done fully, the lance cannot be hoisted beyond 2.5 m level.

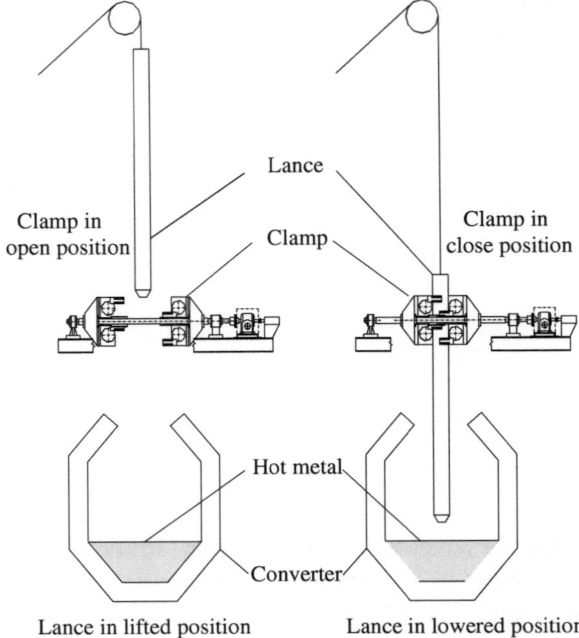

Fig. 8 General arrangement of lance clamp–converter system

- However, in case of any problem, such as water failure, de-clamping is automatically done, and lance can be hoisted. In this condition, a hooter will come in operation making operation personnel alert.
- General arrangement of lance clamp–converter system is shown in Fig. 8.

5 Conclusion

The design is cost-effective and compact. The design can be used in any converter with minor modification depending upon the size of the converter. Dangling and vibration of lance while blowing have been considerably decreased. Life of the lance has improved. Converter blowing operation has become quite smooth with steel and slag chemistry improvement. The lining life of the converter has improved substantially. Loss of oxygen and metal has reduced considerably.

Acknowledgment Author expresses his gratitude to the management of Rourkela Steel Plant and Research and Development Centre for Iron and Steel for allowing taking up this project and implementing it in Steel Melting Shop. Co-operation provided by Operation, Electrical and mechanical maintenance of Steel Melting Shop, Design department, RSP, is thankfully acknowledged without whose support execution of the project would not have been possible.

References

Budinski KG (1979) Engineering materials, properties and selection. Prentice Hall, Englewood Cliffs, NJ
Newell JA (1977) Indigenous mechanisms for designers and inventers vol-IV
Schubert PB (1979) Machinery's hand book. (ISBN 0-8311-1129-1)

Development and Performance Evaluation of Water-Resistant Corrugated Board for Packaging

Ankit Saini, Anand Kumar and S. B. Yadaw

Abstract The aim of the work is to develop lightweight and cost-effective water-resistant corrugated board for application in packaging industry. A variety of corrugated board is formed by gluing one face of corrugated board with plastic-coated paper, to analyze the performance for strength, water resistance, moisture content, bursting strength, and overall effectiveness as packaging material. In this variety of board, two types of specimen are prepared, i.e., single wall and triple wall.

Keywords Corrugated board · Packaging · Plastic-coated paper · Grammage · Bursting strength

1 Introduction

Corrugated cardboard packaging is produced out of renewable raw materials, which can be optimally integrated into cycles of valuable substances. The packaging material connects effective product protection with very low material usage and low weight as well as their outstanding mechanical processability. Therefore, their economical and ecological advantages are joint ideally. The good physical characteristics enable a wide range of applications due to the flute profiles and their combinations. Corrugated cardboard is cardboard made up of one or more layers of corrugated paper, which is glued to one layer or between several layers of

A. Saini (✉) · A. Kumar
Department of Mechanical Engineering, HBTI, Kanpur 208002, Uttar Pradesh, India
e-mail: hbti.ankit11@gmail.com

S. B. Yadaw
Defense Materials and Stores Research and Development Establishment, Kanpur 208002, Uttar Pradesh, India

S. S. Khangura et al. (eds.), *Proceedings of the International Conference on Research and Innovations in Mechanical Engineering*, Lecture Notes in Mechanical Engineering,
DOI: 10.1007/978-81-322-1859-3_4,

paper or cardboard. Corrugated cardboard is distinguished between single-layer or multi-layer cardboard. Corrugated cardboard is manufactured in a continuous automated process from pre-made paper and cardboard. A standard sheet of corrugated board is made from three components: A sheet of corrugated fluted paper sandwiched between an outside liner and an inside liner. Each side of the fluting is glued to a sheet of flat liner paper.

The basic material for the production of corrugated board is paperboard. There are two main types of paperboard:

- Kraft paperboard.
- Test paperboard.

1.1 Corrugated Wall

So far, we have looked at corrugated board made from a single fluting with two liners. This is referred to as single wall. An additional fluting and wall make a stronger board known as double wall. In double wall, if we add an additional fluting and wall, make triple wall which is used for extremely heavy goods.

Corrugated fiberboard, or combined board, has two types of flutes; both are made of a special kind of heavy paper called containerboard. Linerboard is the flat facing that adheres to the medium. The medium is wavy, fluted paper in between the liners.

1.2 Nomenclature of Corrugated Board

Combining all of the above, it is possible to specify a simple short description that defines the board grade of a box. For example, 125 K/B/125 T means a corrugated box, made from 125 gsm Kraft outer liner, *B* fluting, and 125 gsm test inner lining.

1.3 Manufacturing

Corrugated board is manufactured from a number of specially conditioned layers, called fluting medium and linerboard. This type of board consists of one sheet of fluted paper sandwiched between and glued to two facings called liners. Apart from single wall corrugated board, different structures exist as well, for instance single face and double or triple wall. Also, different wave shapes exist, and these lead to different heights and periods of the corrugated layer. The most commonly used flute type is designated *B*-flute, with a height of 2.4 mm and roughly

Table 1 Various flutes type and size

Flute type	Flute height (mm)	Flutes/30 cm
K	5.5–6.0	28–32
A	4.2–4.6	32–38
C	3.4–3.8	38–44
B	2.4–2.8	50–56
E	1.2–1.6	90–96
F	0.7–1.1	110–126
N	0.4–0.6	150–167

150 flutes/m. The production of corrugated board starts with reels of fluting and liner being fed into a machine called a corrugator (Table 1).

The fluting paper is conditioned with heat, steam, and fed between large corrugating rolls that give the paper its fluted shape. Starch is applied to the peaks of the flute on one side and the inner liner is glued to the fluting. The corrugated fluting with one liner attached to it is called single-face web and travels along the machine toward the double backer, where the single-face web meets the outer liner and forms corrugated board. The corrugated board is slit into the required widths and cut into sheets which are then stacked or palletized liner fluting.

2 Experimental Evaluation of Performance of Corrugated Boards

Procedure, experimentation, and testing of plastic-coated corrugated board for single wall and triple wall are given as follows:

2.1 Using Plastic-coated Paper

Plastic-coated paper corrugated board was made by using hand lay-up technique (Manual). Fresh piece of pre-made single-face corrugated board of standard quality was taken in order to attach plastic-coated paper to the other face. Plastic-coated paper is easily available and is of very low cost. By using plastic-coated paper, we get more strength than kraft paper (Fig. 1).

Adhesive used for the purpose of gluing plastic-coated paper to the other face of board was polyvinyl acetate (PVA). It is a good adhesive and easily available in the market within a reasonable price. For three layers, attach 3 layer of single-face corrugated board then at last end attach plastic-coated paper with the help of PVA. In this way, triple-wall corrugated board was prepared.

Fig. 1 Sample of plastic-coated paper board and single-face corrugated board

2.2 Grammage of Corrugated Fiberboard

The test of grammage was performed as per "Federation of Corrugated Box Manufactures in India (FCBM 10:93 1995)." This test is applicable to all type of corrugated fiberboards. The grammage of substance is defined as the weight of one square meter of corrugated fiberboard under standard test conditions and expressed as grams per square meter.

Test specimen was taken from a sample of corrugated fiberboard, were brought into equilibrium with standard atmospheric condition, and then weighed on a suitable accurate balance was its principle. Test result expressed as gsm or g/m^2. A sensitive balance minimum 0.01 g or more was used to make the determinations. Test specimen of 80 mm × 100 mm was used for testing. A relative humidity of 65 ± 2 % and temp of 27 °C ± 2° were taken as the standard atmospheric condition for the purpose of testing.

2.3 Calculation of Grammage

The grammage weight was calculated by the formula:

$$G = \frac{g \times 10^6}{a \times b} \tag{1}$$

where

G grammage (gsm or g/m^2)

g weight of test specimen in grams

a length of test specimen in mm

b width of test specimen in mm

2.4 Moisture Content in Corrugated Fiberboard

The test of moisture content was performed as per "Federation of corrugated box manufactures of India (FCBM 21:99 1995)." The test is applicable to all paper, paperboards, and paper products except those containing significant quantities of materials other than water that are volatile at 105 ± 2 °C.

Moisture content percent by weight was calculated by

$$\%w = \frac{w_1 - w_2}{w_1} \times 100 \quad (2)$$

where

w_1 original weight of the conditioned specimen before drying.

w_2 weight of specimen after drying.

2.5 Water Absorption (COBB Method)

The test of water absorption was performed as per "Federation of Corrugated Box Manufactures in India (FCBM 8:92 1995)." This test is applicable to all type of corrugated fiberboards.

$$\%w = \frac{w_2 - w_1}{w_1} \times 100 \quad (3)$$

2.6 Bursting Strength of Corrugated Board

The test of grammage was performed as per "Federation of Corrugated Box Manufactures in India (FCBM 1:00 1995)." Bursting strength is defined as the hydrostatic pressure required produce rupture of the material when the pressure is applied at a specific controlled increasing rate through a rubber diaphragm to a specific circular area of the material under test.

2.7 Thickness of Corrugated Fiberboard

The test of thickness was performed as per "Federation of Corrugated Box Manufactures in India (FCBM 29:04 1995)." This test is applicable for measuring thickness of all type of corrugated fiber boards.

Table 2 Grammage of plastic-coated corrugated board

S. no.	Single wall (g)	Gp_1 (g/m^2)	Triple wall (g)	Gp_2 (g/m^2)
1	3.1948	399.3500	8.4310	1053.8750
2	3.1290	391.1250	8.5687	1071.0875
3	3.1096	388.7000	8.4458	1055.7250
4	3.2539	406.7375	8.4908	1061.3500
5	3.2947	411.8375	8.5482	1068.5250
Ā	3.1964	399.5500	8.4969	1062.1125

Table 3 Moisture content plastic-coated corrugated board

S. no.	Weight (w_1)		Weight (w_2)		Weight (%)	
	Single wall	Triple wall	Single wall	Triple wall	w_{ps} (%)	w_{pt} (%)
1	3.194	8.568	3.095	8.039	3.12	6.17
2	3.129	8.431	2.991	7.996	4.40	5.15
3	3.109	8.445	2.981	7.944	4.11	5.93
Ā	3.144	8.481	3.022	7.993	3.88	5.75

The thickness of a single sheet of corrugated fiberboard is the distance between two plane parallel surfaces of micrometer by way of which the test pieces are subjected to the specified pressure. The thickness of fiberboard is measured with the test pieces under a specified pressure is its principle.

Dial gauge micrometer (200 gm/cm^2) with a plane circular anvil and a concentric plane plunger. The instrument used was accurate to permit measurement to be made to the nearest 0.01 mm. Test specimen of 80 mm × 100 mm was used for testing. A relative humidity of 65 ± 2 % and temp of 27 °C ± 2° was taken as the standard atmospheric condition for the purpose of testing.

Tables 2, 3, 4, 5, and 6 show the various observations of different test performed on the specimens.

3 Result and Discussions

Plastic-coated corrugated boards are fabricated from pre-made single-face liner corrugated board by gluing plastic-coated paper as a liner to one face of board, and testing is done according to FCBM test standard.

Figure 2 shows the graphical representation for moisture content value of the plastic-coated corrugated board. The value of three specimens for single wall changes from 3.1206 to 4.4039 %, and for triple wall, the value changes from 5.1524 to 6.1724 %.

Table 4 Water absorption in plastic-coated paper corrugated board

S. no.	Weight (w_1)		Weight (w_2)		Weight (%)	
	Single wall	Triple wall	Single wall	Triple wall	w_{ps} (%)	w_{pt} (%)
1	3.097	8.502	3.098	8.503	0.022	0.009
2	3.116	8.563	3.116	8.565	0.016	0.017
3	3.105	8.450	3.106	8.451	0.028	0.014
Ā	3.106	8.505	3.107	8.506	0.022	0.013

Table 5 Bursting strength of plastic-coated paper corrugated board

Bursting strength of plastic-coated paper C.B. (kg/cm^2)		
S. no.	Single wall	Triple wall
1	5.1	13.3
2	5.4	13.2
3	5.25	13.25

Table 6 Thickness of plastic-coated board

S. no.	Single wall	Triple wall
1	3.97	7.25
2	4.12	7.53
3	4.08	7.35
4	4.05	7.60
5	3.99	7.42
6	4.042	7.43

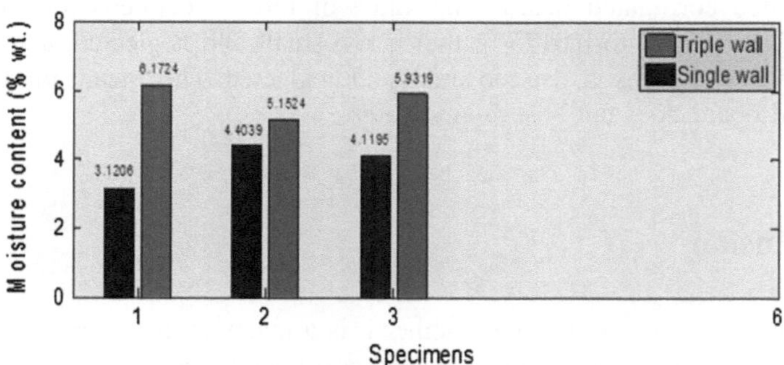

Fig. 2 Moisture content in plastic-coated corrugated board

Figure 3 shows the graphical representation for grammage value of the plastic-coated corrugated board. The value of five specimens for single wall changes from 388.7000 to 411.8375 g/m^2, and for triple, the value changes from 1053.8750 to 1071.0875 g/m^2.

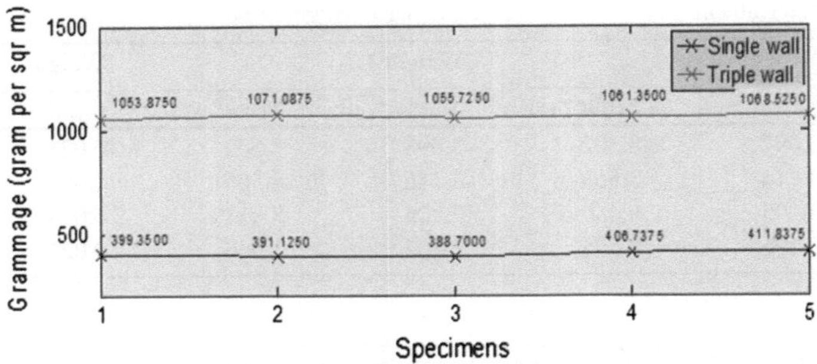

Fig. 3 Grammage value of plastic-coated corrugated board

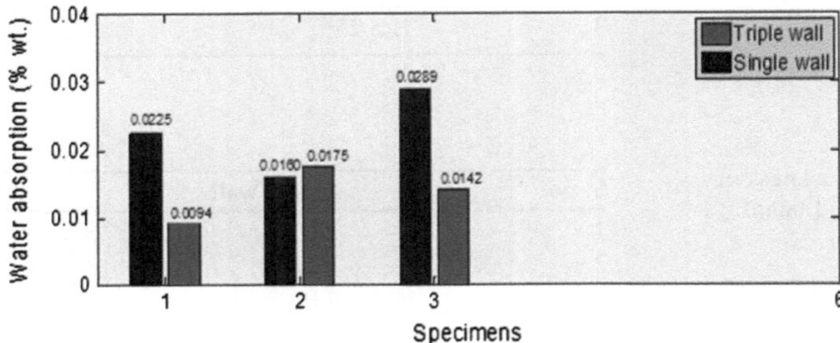

Fig. 4 Water absorption in plastic-coated corrugated board

Figure 4 shows the graphical representation for water absorption value of the plastic-coated corrugated board. The value of three specimens for single wall changes from 0.0094 to 0.0175 % that is too small and neglected, and for triple wall, the value changes is also too small and neglected. This means plastic-coated corrugated board does not absorb any water.

4 Conclusion

Every product require packaging whether it is a machine, food, fruits, beverages, or else. In the present work, plastic-coated corrugated board which will used as a water-resistant corrugated board for packaging applications is fabricated. The primary objective of the work is to develop lightweight and cost-effective hybrid composite in the form of water-resistant corrugated board for application in packaging industry. Several tests on plastic-coated corrugated boards were performed; following major conclusions are drawn:

From the grammage test results, it is clear that grammage arithmetic mean value for triple-wall plastic-coated paper corrugated board is 1,062.1125 g/m^2, and for single wall, it is 399.5500 g/m^2. The arithmetic mean value of moisture content for single wall is 3.88 %, and for triple wall, the value is 5.7522 % by weight. Water absorption value is very small and neglected means it does not absorb water. Burst test determines resistance to shock and penetration. For bursting strength, the arithmetic mean value for single wall is 5.25 kg/cm^2, and for triple wall, it is 13.25 kg/cm^2. This shows it has sufficient strength for packaging applications. The observation obtained after thickness measurements test are as follows: for single wall, the arithmetic mean value is 4.02 mm; and for triple wall, it is 7.43 mm.

From the above observations, it is clear that plastic-coated corrugated board does not absorb any water and has sufficient strength to resistant and shocks. Its overall performance considering water-resistant packaging application is good, and in packaging industry, it is a good alternative where water-related problem prevailed, and the main issue is to stop water penetration or absorption.

References

FCBM R-1:00 (1995) Federation of corrugated box manufactures India
FCBM 10:93 (1995) Federation of corrugated box manufactures of India (second print)
FCBM 29:04 (1995) Federation of corrugated box manufactures of India (second print)
FCBM 21:99 (1995) Federation of corrugated box manufactures of India (second print)
FCBM 8:92 (1995) Federation of corrugated box manufactures of India (second print)

Mathematical Modelling Approach for Optimization of a Rotary Hook Feeder

Ritika Punn, Shruti Bansal, Sakshi Jain, Meenakshi Goel and Pradeep Khanna

Abstract Modern production system requires a high degree of automation for the feeding and handling of small, individual components, which tends to reduce both the production time and human labour. Mechanized feeders are used to bring components to be fed, from a disorderly state into the required orientation in required quantity for a particular process. Present study has been carried out with regard to the analysis of performance of a Rotary Hook feeder under certain variable factors. For this, the above-mentioned feeder has been developed. A series of experiments were carried out on cylindrical parts, and effect of different input parameters on the feed rate was studied graphically. Experimentation shows that the performance of this feeder depends upon various factors such as part population, number of hooks, speed, part sizes and blind/through-holed parts. However, only the first three parameters have been studied in the present case, and a mathematical model has been formulated. The design of experiments is done using 2^3 full-factorial method. Significance of individual factors as well as their interactions is examined using Design Experts Software. Analysis of Variance (ANOVA) method is used to predict proficiency of the model. This model also provides optimized feed rate for appropriate process parameters under given constraints.

Keywords Feeder parts · Production · Hooks (spokes)

R. Punn · S. Bansal · S. Jain (✉) · M. Goel
Department of Mechanical and Automation Engineering, Indira Gandhi Delhi Technical University for Women, Delhi 110006, India
e-mail: sakshi.jain612@gmail.com

P. Khanna
Department of Manufacturing Processes and Automation Engineering, Netaji Subhash Institute of Technology, Delhi 110078, India

S. S. Khangura et al. (eds.), *Proceedings of the International Conference on Research and Innovations in Mechanical Engineering*, Lecture Notes in Mechanical Engineering, DOI: 10.1007/978-81-322-1859-3_5,

1 Introduction

Application of automation to assembly processes is important to meet the requirements of any manufacturing system. Feeders form a critical part of automated assembly lines (Stefan 2000). An assembly line is an arrangement of workers, machines and equipment in which the product being assembled passes consecutively from operation to operation until completed (Singh et al. 2009).

Feeders are designed to provide a constant feed rate to the assembly lines at a specific flow rate and orientation. Most of the time, these parts are added to other parts to become the finished product at the end of an assembly line (Mitchell and John 2010). In the automated feeders, the feed rate should be in accordance with the machine rate in order to prevent the condition of starvation or saturation of the parts.

Apart from feeding parts from a bulk supply, it is used to convert the randomness of the parts in a specified geometrical orientation in order to feed them at a predetermined rate.

The present work aims at fabricating, experimenting and analysing a rotary hook feeder under different values of the part population (PP), speed and number of hooks.

2 Principle of Working

The basic principle of working of a rotary hook feeder is the relative motion between the stationary hemispherical container and the rotary hooks.

A cylindrical hub is directly mounted on the motor shaft and contains eight threaded holes on its periphery to accommodate a maximum eight number of wire hooks with one end fitted in these holes and the other bent at 90° for parts pickup. The hemispherical container acts as a temporary reservoir for the bulk of cylindrical parts. This reservoir has a perfectly curved surface to ensure a constant gap between the hook and the surface of container throughout the traversing length of the hook, in order to avoid jamming of parts and to ensure picking up of even the bottom most part. As the hooks rotate, the parts from the container get picked up by them, taken through the circle and finally delivered to the chute. The chute is designed in such a manner that the part attains desired orientation by the time it reaches the exit of the chute (Figs. 1, 2).

The picking of parts is a random phenomenon, which depends upon many factors such as the orientation of the parts with respect to the spoke, part population, speed and part size.

Fig. 1 Top view of feeder

Fig. 2 Front view of feeder

3 Driving Mechanism

The rotary motion to the hooks of the feeder has been given by an electrical drive. The drive selected in this case is a geared DC motor because of its feature of delivering high torque at low speed, with a step less speed controller. Since the drive is transferred directly to the hooks from the motor, the losses in this case are almost negligible. The complete assembly is mounted on an elevated platform that is further firmly secured on a rigid wooden base making the overall dimensions of the feeder very compact and rendering the system with simplicity (Table 1).

4 Experimental Work

In the present work, the performance of the rotary feeder is analysed. The project primarily aimed at developing the feeder and analysing the same under different parameters. A series of experiments were carried out under different operating conditions. The investigation has been done to study the effect of three parameters:

Table 1 Specifications of feeder

S. No.	Name	Specifications
1	Part	Length—25 mm Outer diameter—12 mm Inner diameter—10 mm Material used—Nylon
2	Container	Diameter—150 mm Material used—mild steel
3	Spoke	Length—135 mm Diameter—5 mm Material used—mild steel
4	Drive	DC-geared motor

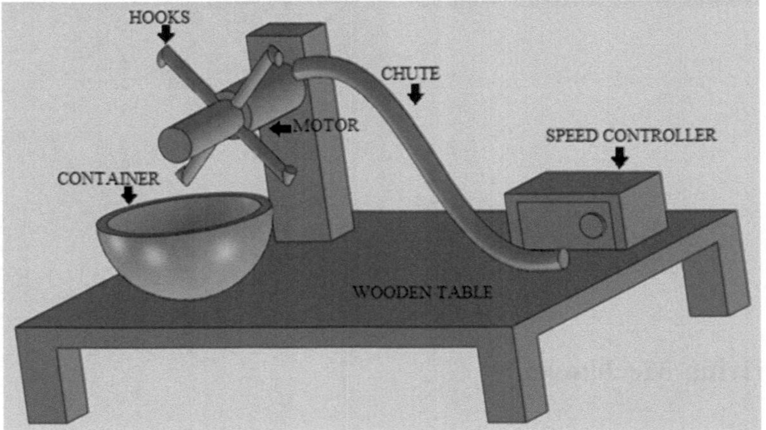

Fig. 3 Labelled side view of feeder

1. Part population (PP)
2. Speed of the motor
3. Number of hooks (Fig. 3).

5 Range of Parameters

1. Part population is varied from 10 to 100.
2. The speed of the motor varies from a minimum of 4 rpm to a maximum of 31 rpm.
3. Data is obtained using 1–7 spokes.

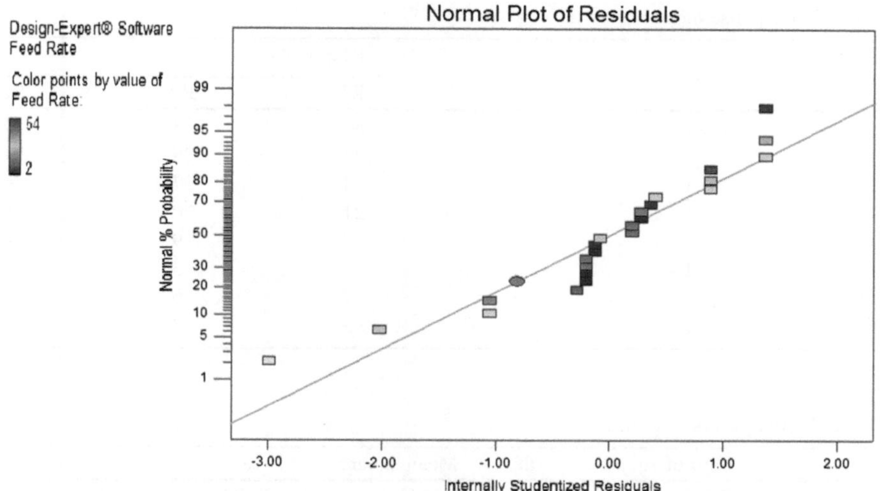

Fig. 4 Normal plot of residuals

Table 2 Process parameters

Process parameters	Low level	High level
Speed (A) rpm	4	31
Part population (B)	10	100
Spokes (C)	1	7

6 Mathematical Modelling

To study the effect of process parameters and their interaction on the response, i.e. feed rate, 2^3 full-factorial method is used for design of experiment (DOE).

The higher and lower settings of factors included in this experiment are tabulated in Table 2.

Series of experiments were performed to obtain three sets of response data for each run as shown in Table 3.

Analysis of variance using partial sum of squares-type III is carried out on the 2^3 factorial model using Design Expert® Software. The results are displayed in Table 4.

Table 5 enlists the change that response data undergoes as factors change from their low (−1) to higher (+1) level.

Since the 95 % CI (confidence interval), high and low ranges for BC and ABC span around 0. So, the coefficient could be taken as 0, indicating no effect of these

Table 3 Response data

Coded factor			Responses		
A	B	C	R1	R2	R3
1	1	1	54	53	49
1	1	−1	26	28	22
1	−1	1	44	37	46
1	−1	−1	21	20	16
−1	1	1	8	9	8
−1	1	−1	4	3	3
−1	−1	1	7	6	8
−1	−1	−1	3	2	2

Table 4 Analysis of variance

Source	Sum of squares	df	Mean square	*F*-value	*p* value Prob > *F*
Model	7,282.54	5	1,456.51	255.99	<0.0001
A-Speed	5,133.37	1	5,133.37	902.20	<0.0001
B-Parts	117.04	1	117.04	20.57	0.0003
C	1,365.04	1	1,365.04	239.91	<0.0001
AB	77.04	1	77.04	13.54	0.0017
AC	590.04	1	590.04	103.70	<0.0001
Residual	102.42	18	5.69		
Lack of fit	8.42	2	4.21	0.72	0.5036
Pure error	94.00	16	5.88		

Table 5 Effect estimate summaries

Factor	Effect estimate	Sum of squares	Percentage contribution
A–A	29.25	5,133.37	69.51
B–B	4.42	117.04	1.58
C–C	15.08	1,365.04	18.48
AB	3.58	77.04	1.04
AC	9.92	590.04	7.99
BC	0.75	3.38	0.046
ABC	0.92	5.04	0.068
Pure error		94.00	1.27

factors. If Variance Inflation Factor is orthogonal to all other factors in the model, then it is 1. Values greater than 10 are indicated that coefficients are poorly estimated due to multi-collinearity. The regression coefficients (Table 6) are exactly one-half of the usual effect estimates.

Table 6 Coefficient estimates

Factor	Coefficient estimate	95 % CI	95 % CI	VIF
		Low	High	1
Intercept	20.04	19.02	21.06	1
A–A	14.62	13.60	15.65	1
B–B	2.21	1.19	3.23	1
C–C	7.54	6.52	8.56	1
AB	1.79	0.77	2.81	1
AC	4.96	3.94	5.98	1

Table 7 Regression values

Standard deviation	2.39
Mean	20.04
C.V. %	11.90
PRESS	182.07
R-squared	0.981
Adjusted *R*-squared	0.9823
Predicted *R*-squared	0.9753
Adequate precision	40.875

The relationship between the factors and the response is given below by the coded equation, which is useful for identifying the relative significance of factors.

$$\text{Feed rate} = +20.04 + 14.62 * A + 2.21 * B + 7.54 * C + 1.79 * A * B + 4.96 * A * C$$

6.1 Significance and Diagnostics of Regression Model

R-squared tells us how our model fits the observation. It always increases when a new term is added whether it is significant or not. Hence, adjusted R^2 is calculated from Table 7. Adjusted *R*-squared is 0.9823 that is close to ordinary R^2. Also, Predicted *R*-squared and adjusted *R*-squared should be within 0.20 of each other (Singh et al. 2009).

Otherwise, there may be a problem with either the data or the model. Predicted residual error sum of squares (PRESS) is a measure of how the model fits each point in the design. It is used to calculate predicted *R*-squared. Adequate precision

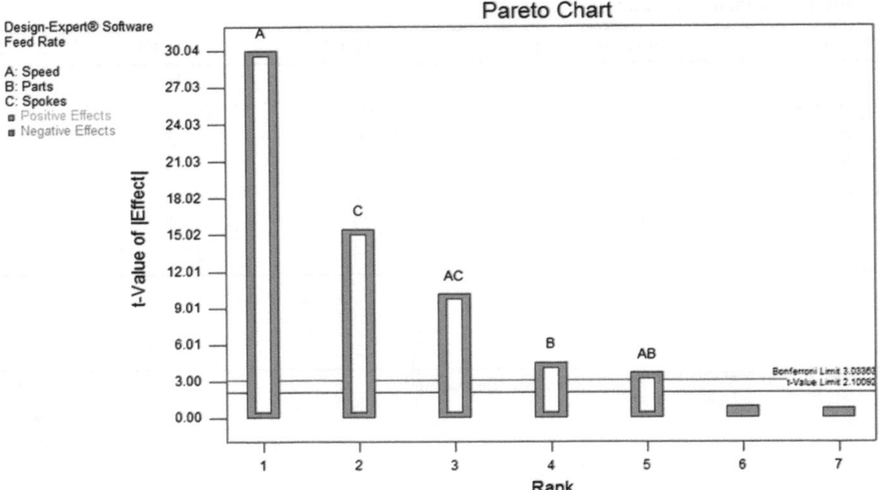

Fig. 5 Pareto chart

measures the signal-to-noise ratio. A ratio greater than 4 is desirable. Together, these statistics prevent overfitting the model.

A normal probability plot of residuals is as shown in Fig. 4. If the residuals are normally distributed, the points generally form a straight line. An adequate approximation to the true system is provided by the model as shown by the plot.

The Pareto chart (Fig. 5) displays the various factors affecting the feed rate in order of their significance, i.e. A, C, AC, B, AB, BC and ABC. From the chart, we can conclude that the factor A is most significant while BC and ABC does not affect much.

6.2 Optimization

In numerical optimization of the model, the goal is to maximize the feed rate while allowing the factors to assume any value in between their levels. To specify all the goals, an overall desirability function is used. The goal seeking begins at a random starting point and proceeds up to a maximum point. The model here is optimized considering the feed rate to be of critical importance, while the factors are considered to be of medium importance. Table 8 shows the 38 solutions in order of their feed rate.

Table 8 Optimization

No.	Speed	Parts	Spokes	Desirability	Feed rate
1	1.00	1.00	1.00	1.000	51.1667
2	1.00	−1.00	1.00	1.000	43.1667
3	0.71	−0.37	0.53	1.000	34.9665
4	0.34	−0.44	0.91	1.000	32.1438
5	0.39	0.28	0.45	1.000	30.8638
6	0.69	0.36	−0.29	1.000	28.2075
7	1.00	1.00	−1.00	1.000	26.1667
8	0.45	−0.71	0.09	1.000	25.3766
9	0.35	−0.05	−0.16	1.000	23.5122
10	0.88	0.33	−0.93	1.000	23.1338
11	0.19	0.49	−0.23	1.000	22.0600
12	−0.14	−0.67	0.64	1.000	21.1052
13	0.79	−0.19	−0.86	1.000	20.9908
14	−0.30	0.08	0.75	1.000	20.3716
15	−0.19	−0.75	0.67	1.000	20.2990
16	0.33	−0.81	−0.30	1.000	19.8806
17	0.16	0.80	−0.54	1.000	19.7920
18	0.60	−0.99	−0.65	1.000	18.7266
19	1.00	−1.00	−1.00	1.000	18.1667
20	0.00	0.51	−0.45	1.000	17.7748
21	0.14	−0.80	−0.35	1.000	17.3160
22	−0.38	0.73	0.17	1.000	16.5085
23	0.10	−0.77	−0.51	1.000	15.5139
24	−0.62	0.11	0.88	1.000	15.0373
25	−0.61	−0.48	0.86	1.000	14.4032
26	−0.64	−0.84	0.61	1.000	12.4990
27	−0.61	0.97	−0.21	1.000	11.3466
28	−0.37	0.72	−0.84	1.000	10.9265
29	−0.79	−0.25	0.45	1.000	9.97619
30	−1.00	1.00	1.00	1.000	8.41667
31	−0.68	−0.44	−0.40	1.000	8.02044
32	−0.88	0.20	0.12	1.000	7.69448
33	−1.00	−1.00	1.00	1.000	7.58330
34	−0.98	0.50	0.49	1.000	7.18735
35	−0.83	0.11	−0.03	1.000	7.18207
36	−0.91	0.36	−0.16	1.000	6.40108
37	−1.0	1.00	−1.00	1.000	3.2500
38	−1.00	−1.00	−1.00	1.000	2.41667

7 Conclusion

1. The Model *F*-value of 255.99 implies the model is significant. There is only a 0.01 % chance that a "Model *F*-value" this large could occur due to noise.
2. Values of "Prob > *F*" less than 0.0500 indicate model terms are significant. In this case, A, B, C, AB and AC are significant model terms.
3. The "predicted *R*-squared" of 0.9753 is in reasonable agreement with the "adjusted *R*-squared" of 0.9823.
4. "Adequate precision" measures the signal-to-noise ratio. A ratio greater than 4 is desirable. The ratio of 40.875 indicates an adequate signal. This model can be used to navigate the design space.

References

Mitchell JF Jr (2010) Variation in types of part feeders for manufacturing automation, 11 Nov 2010

Singh SP, Ghosh S, Khanna P, Tanwar A (2009) Mathematical performance analysis of rotary fork hopper feeder. In: Proceedings of 2009 IEEE student conference on research and development (SCORED). ISBN 978-1-4244-5186-9, UPM Serdang, Malaysia, Nov 2009

Stefan H (2000) Rationalisation of Small work piece feeding. Festo AG & Co, Germany

Tool Path Generation for Free-Form Surfaces Using B-Spline Surface

Simranpreet Singh Randhawa and Jaswinder Singh Saini

Abstract Tool path generation is an important step for the machining of the free-form surfaces. The accuracy of machining free-form surfaces greatly depends upon the tool path. A number of algorithms had been proposed by different researchers for the accurate and efficient tool path generation for the machining of the free-form surfaces. The present study is focused on implementation of one of the algorithm that generates tool paths for free-form surfaces based on the accuracy of a desired manufactured part. This algorithm includes two components. First is the forward-step function that determines the maximum distance between two cutter contact points with a given tolerance. The second component is the side-step function which determines the maximum distance between two adjacent tool paths with a given scallop height. These functions are independent of the surface type and are applicable to all continuous parametric surfaces that are twice differentiable. This algorithm reduces cutter contact (CC) points while keeping the given tolerance and scallop height in the tool paths. The algorithm is implemented using the B-spline surface. It is then used to machine a wax component which is compared with the desired surface.

Keywords Free-form surfaces · NC/CNC machines · Tool path generation · B-spline surface

S. S. Randhawa (✉)
Department of Mechanical Engineering, Guru Nanak Dev Engineering College, Ludhiana, Punjab, India
e-mail: sps171989@gmail.com

J. S. Saini
Department of Mechanical Engineering, Thapar University, Patiala, Punjab, India
e-mail: Jsaini@thapar.edu

S. S. Khangura et al. (eds.), *Proceedings of the International Conference on Research and Innovations in Mechanical Engineering*, Lecture Notes in Mechanical Engineering, DOI: 10.1007/978-81-322-1859-3_6,

1 Introduction

The surfaces that are not regular in their shape and are made by joining different patches together by defining some continuity at points of joining are called the free-form surfaces. Free-form surfaces are usually designed to meet or improve an aesthetic and/or functional requirement. Often, they are defined as surfaces containing one or more non-planar, non-quadratic surfaces generally represented by parametric and/or tessellated models free-form surface, or free-form surfacing, is used in CAD and other computer graphics software to describe the skin of a 3-D geometric element. Free-form surfaces do not have radial dimensions, unlike regular surfaces such as cylinders, cones and spheres. They are used to describe forms such as turbine blades, car bodies and boat hulls. Initially developed for the automotive and aerospace industries, free-form surfacing is now widely used in all engineering design disciplines from consumer goods products to ships.

These days CNC machines are mostly used for accurate machining of parts. The codes used on CNC machines are capable of machining only regular profiles. The free-form profiles cannot be machined using available preparatory codes on the CNC machines. Many researchers have worked on different algorithms to machine free-form surfaces on CNC machines. Some of the researchers have used parametric curves and surfaces for the development of their algorithms, given in the first part of the review.

Barnhill (1985) focused on the representation and design of surfaces in a computer graphics environment. The subject was approached from two points of view: The design of surfaces which included the interactive modification of geometric information and the representation of surfaces for which the geometric information was relatively fixed. Design takes place in 3-D space, whereas representation can be higher dimensional. Triangular patches were used to interpolate and approximate the arbitrarily located data and required the preprocessing steps of triangulation and derivative estimation. Chen and Shi (2008) presented a new method for tool path generation which was based on approximating free-form surfaces by triangular meshes. The vertices of triangular meshes were offset along surface normal. All offset vertices were then connected together to form offset triangular meshes sliced by a group of planes to obtain tool paths.

Ma and Peng (1995) presented a unified method for smoothing free-form surfaces with Bezier patches. A non-smooth surface consisting of bicubic patches was turned into a G1 surface. Using this method, solid objects bounded with bicubic Bezier patches were G1 smoothed with bicubic patches. Ge et al. (1997) presented a special class of rational Bezier curves that correspond to low-harmonic trajectory patterns. For given high-speed machinery, a significant source of the internally induced vibrational excitation was the presence of high-frequency harmonics in the trajectories that the system was forced to follow. Choi et al. (2007) presented the tool path generation method for multi-axis machining of free-form surfaces using Bezier curves and surfaces.

Konno and Chiyokura (1996) proposed a surface representation that enables the smooth interpolation of an irregular curve mesh with NURBS curves and surfaces. Designers required a means of designing complex free-form surfaces easily and intuitively. Nasri (1997) described an algorithm to generate recursive subdivision surfaces that interpolate B-spline curves. The control polygon of each curve is defined by a path of vertices of the polyhedral network describing the surface. Lartiguea et al. (2001) presented an accurate and efficient method to generate a tool path for a smooth free-form surface using planar cubic B-spline. A three-axis CNC machine tool with a ball-end mill cutter was used. The break points were generated by computing the offset surface driving plane intersection curve reflecting the curvature by a planar cubic B-spline curve. Maximum scallop height was calculated by computing the stationary points of the distance function between the scallop curve and the design surface. Narayanaswami and Pang (2003) investigated the applicability of multi-resolution analysis using B-spline wavelets to NC machining of contoured 2-D objects. A complex curve was decomposed using wavelet theory into lower-resolution curves. The low-resolution (coarse) curves were similar to rough-cuts and high-resolution (fine) curves to finish cuts in NC machining.

Huang and Oliver (1994) presented an algorithm for three-axis NC tool path generation on sculptured surfaces. Non-constant parameter tool contact curves were defined on the part by intersecting parallel planes with the part model surface. Wang (1996) presented a method to calculate intersection curves of offsets of two parametric surfaces. The method was based on the concept of normal projection. Lin and Koren (1996) presented an analytical method for planning an efficient tool path in machining free-form surfaces on 3-axis milling machines. The approach used a non-constant offset of the previous tool path which guaranteed that the cutter moving in an un-machined area of the part surface and without redundant machining. Ding et al. (2003) presented an algorithm that overcomes the disadvantage of iso-planar method while keeping its advantage of robustness and simplicity. The isophote concept was used for the partition of different regions of the surface. As compared with the conventional iso-planar tool path generation algorithm, the presented algorithm improved machining efficiency by reducing total tool path length needed to machine a free-form surface. Wei and Lin (2005) established a systematic general analytical method for CNC machining of the free-form surfaces and developed the postprocessor to obtain the NC code. Choi and Banerjee (2006) focused on generating tool path for free-form surfaces based on accuracy of desired manufactured part. The mathematical curves and surfaces used to represent a specific manufactured part were used to generate near optimal tool paths and cutter location files.

The second part of the review is based on algorithms given by researchers using some miscellaneous techniques.

Loney and Ozsoy (1987) discussed the development of an interactive computer-aided design and manufacturing system used in defining and machining of free-form surfaces. The system used a menu-driven front end with graphical feedback to guide a user through curve and free-form surface definition resulting in a mathematical model which was used to generate NC machine cutter location source file. Choi et al. (1988) presented a method for modelling and machining compound surfaces commonly found in die cavities and punches. A constructive solid geometry scheme was employed to model a compound surface that consists of planar surface elements, general quadratic surface elements and composite parametric surfaces. Suh and Lee (1990) presented a procedure to machine a pocket with a convex or concave free surface bounded by lines, circular arcs and free curves. The cutter location data were computed directly with better computational efficiency than normal, without using an iterative method. Leon and Trompette (1995) described a method to provide deformation methods involving simultaneous movements of control vertices. Deformation of free-form surfaces encounters difficulties because more than one control point must be moved to achieve satisfying results. The method used an analogy between the control polyhedron of a surface and the mechanical equilibrium of a bar network. Surface deformation was then carried out through changes of mechanical parameters. Yang and Han (1999) presented a systematic tool path generation methodology which incorporates interference detection and optimal tool selection for machining free-form surfaces on 3-axis CNC machines using ball-end cutters. Feng and Li (2002) presented a new approach for determination of efficient tool paths in the machining of sculptured surfaces using three-axis ball-end milling. The objective was to keep the scallop height constant across the machined surface so that the redundant tool paths were reduced. Yin (2004) proposed progressive fitting and multi-resolution tool path generating techniques, by which multi-level models fitting for different subsets of sampled points were obtained, and then multi-resolution rough-cut and finish-cut tool paths were generated based on the models.

Pham (1992), Dragomatz and Mann (1997), Lo (1998), Maekawa (1999) and Lasemi et al. (2010) provided a state-of-the-art review on research development in CNC machining of free-form surfaces.

As per the literature review, a large number of algorithms represent the free-form surface in the form of Bezier and B-spline surface. One such algorithm given by Choi et al. (2007) uses the Bezier surface for representing the free-form surface. This algorithm generates tool path based on the accuracy of a desired manufactured part. The purpose of the study is to implement the algorithm given by Choi et al. (2007) using the B-spline surface representation.

2 Mathematical Formulation

2.1 B-Spline Curve and Surface

The main advantage of B-Spline over Bezier is that it provides local control, which means that if we change the location of one point it cannot change the shape of whole curve, it will only effect the area of the curve corresponding to that point. The mathematical representation of the B-Spline curve (Ibrahim 2010) is given by (1).

$$p(u) = \sum_{i=0}^{n} p_i N_i^k(u),\ 0 \le u \le u_{\max} \tag{1}$$

where, $n + 1$ are the number of control points, N_i^k is the B-Spline function, k is the parameter which defines the degree of the curve.

The B-spline function is given by,

$$N_i^1 = \begin{cases} 1, & t_i \le u \le t_{i+1} \\ 0, & \text{otherwise} \end{cases} \tag{2}$$

$$N_i^k = (u - t_i)\frac{N_i^{k-1}}{t_{i+k-1} - t_i} + (t_{i+k} - u)\frac{N_{i+1}^{k-1}}{t_{i+k} - t_{i+1}} \tag{3}$$

In (2) and (3), t_i are the knot values given by

$$t_i = \begin{cases} 0, & i < k \\ i - k + 1, & k \le i \le n \\ n - k + 2, & i > n \end{cases}$$

with $0 \le i \le n + k$, and $u_{\max} = n - k + 2$.

The mathematical representation of B-Spline surface is the extension of (1), given by (4)

$$p(u, v) = \sum_{i=0}^{n} \sum_{j=0}^{m} p_{i,j} N_i^k(u) N_j^l(v), \quad 0 \le u \le u_{\max} \quad \text{and} \quad 0 \le v \le v_{\max} \tag{4}$$

where, $n + 1$ and $m + 1$ are the number of control points in u and v directions, respectively, k and l are the parameters to define the degree of the surface in the u and v directions, respectively, N_i^k and N_j^l are the B-Spline functions, $u_{\max} = n - k + 2$ and $v_{\max} = m - l + 2$.

2.2 First and Second Fundamental Form of a Surface

Given an embedded curve $p(u(t), v(t))$, the quadratic form

$$\mathrm{I} = p^t \cdot p^t = E\frac{du}{dt}\frac{du}{dt} + 2F\frac{du}{dt}\frac{dv}{dt} + G\frac{dv}{dt}\frac{dv}{dt}$$

is known as the first fundamental form with t as the independent variable along the path and,

$$E = p^u \cdot p^u$$
$$F = p^u \cdot p^v$$
$$G = p^v \cdot p^v.$$

These are the coefficients of the first fundamental form, and p^u and p^v are the partial derivatives along the u and v direction at point p on the surface.

The quadratic,

$$\mathrm{II} = L\frac{\mathrm{d}u}{\mathrm{d}t}\frac{\mathrm{d}u}{\mathrm{d}t} + 2M\frac{\mathrm{d}u}{\mathrm{d}t}\frac{\mathrm{d}v}{\mathrm{d}t} + N\frac{\mathrm{d}v}{\mathrm{d}t}\frac{\mathrm{d}v}{\mathrm{d}t}$$

is referred as second fundamental form where,

$$L = p^{\mathrm{uu}} \cdot n; \quad M = p^{\mathrm{uv}} \cdot n; \quad N = p^{\mathrm{vv}} \cdot n$$

are the coefficients of the second fundamental form.

3 Proposed Approach

The accuracy of the machined surface depends greatly on the tool path. In order to have good accuracy, the tool path generation should be accurate and proper. In tool path generation, the main job is to calculate the cutter location points by specifying forward step and side step, the basic movements of the tool to reach the next cutter location point.

3.1 Cutter Contact and Cutter Location Point

The cutter location (CL) point is a reference point on the cutter, with which the machine tool moves linearly along the tool path. The cutter contact (CC) point is the point on the cutter that touches the work piece surface. Ideally, it is desired that the CC point lies on the work piece surface and is an addressable point which is a point along the axis at which the tool can be placed so as to minimize the manufacturing error. The CC and CL points are shown in Fig. 1. The surface normal n at an arbitrary point $p(u, v)$ on the surface is expressed as:

$$n = \frac{p^u \times p^v}{|p^u \times p^v|}$$

where p^u and p^v are the partial derivatives along u and v direction at a point p on the surface.

3.2 Calculation of Forward Step

Each tool path is approximated by a series of line segments. The accuracy of tool path is controlled by deviation. Each segment amounts is a forward step and the maximum deviation is called tolerance. To calculate forward step, first and second derivatives are used. Therefore, this function is independent of the surface type and is applicable to all continuous parametric surfaces that are twice differentiable.

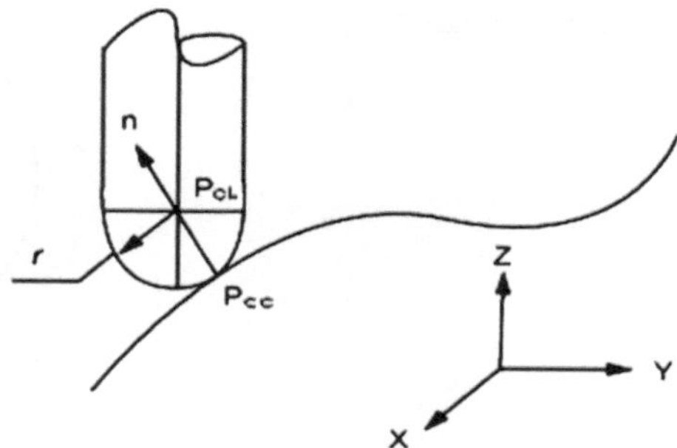

Fig. 1 CC and CL point (Choi et al. 2007)

The mathematical formula for forward step is derived from iso-parametric curve on the surface given in Choi et al. (2007).

The forward-step size, *s*, is calculated using Eq. (5)

$$s^2 = \frac{k_2}{e \times 8} \tag{5}$$

where, e is the given tolerance and k_2 be the maximum second derivative of current curve.

3.3 Calculation of Side Step

The side step, g, is a function of the scallop height h, tool-radius r, and the local radius of the curvature, R. The designed part's surface can be classified into convex, concave and flat surface. The curvatures of convex, concave and flat surfaces are positive, negative and zero, respectively. Therefore, these three different cases are considered to calculate side-step size.

For a flat surface, shown in Fig. 2, the scallop height h is given by (6)

$$h = r - \sqrt{r^2 - \left(\frac{g}{2}\right)^2} \tag{6}$$

where r the radius of the tool and g is the side step. The value of the side step is given by (7).

$$g = 2\sqrt{r^2 - (r - h)^2}. \tag{7}$$

Secondly, a convex curvature shown in Fig. 3 is considered. To find the side step for a convex surface, the difference between a designed curve and a linear tool path, δ, is found.

From Fig. 3, $\delta =$ OB – OA.

To find OA, we first calculate step size, g, using (7). As the step size is very small, we can use $p = g/2$ as the initial value. From the Fig. 3.

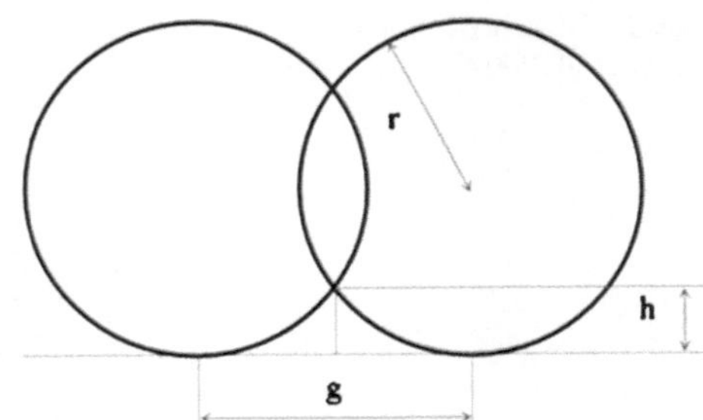

Fig. 2 Flat surface

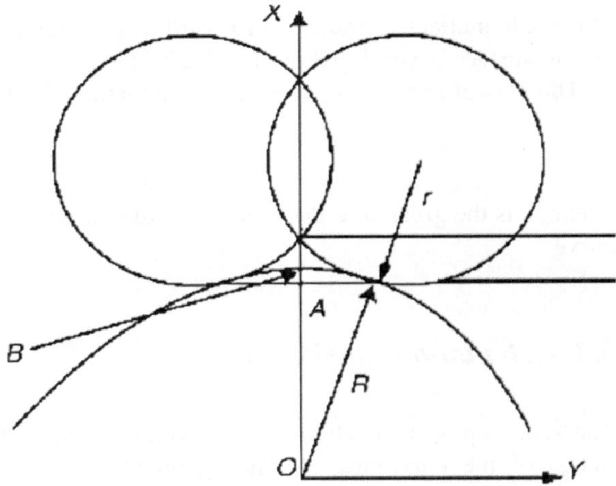

Fig. 3 Convex surface

$$\text{OA} = \sqrt{r^2 - p^2}; \quad \text{OB} = R$$

So the scallop height h and side step g is given by (8) and (9), respectively,

$$h = r - \sqrt{r^2 - p^2} + \delta \tag{8}$$

$$g = 2\sqrt{r^2 - (r + \delta - h)^2}. \tag{9}$$

In the similar manner, we can calculate the side step for a concave surface, as shown in Fig. 4.

From Fig. 4,

$$\delta = \text{OB} - \text{OA},$$

$$\text{OA} = \sqrt{r^2 - p^2},$$

$$\text{OB} = R,$$

where R is the local radius of the curvature of the surface.

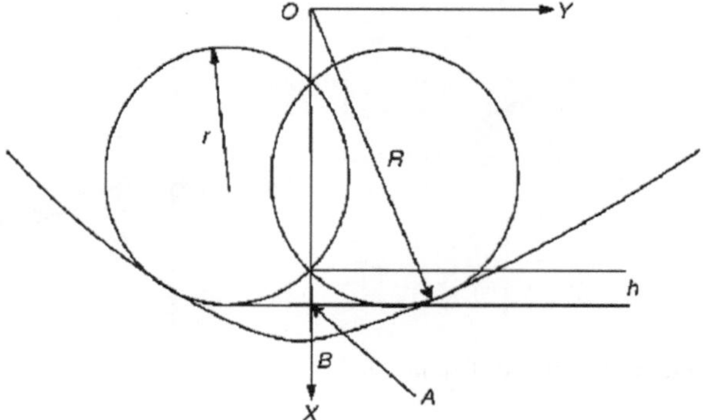

Fig. 4 Concave surface

The scallop height, *h,* and side step, *g,* for a concave curvature are given by (10) and (11), respectively,

$$h = r - \sqrt{r^2 - p^2} - \delta \tag{10}$$

$$g = 2\sqrt{r^2 - (r - \delta - h)^2} \tag{11}$$

Hence, the value of side step can be calculated for any type of the surface using the discussed equations.

4 Implementation and Results

To verify the algorithm, the cutter contact points are generated for a surface. The surface is then machined over a wax part using a 3-axis milling machine. After machining, the surface is compared with the desired one.

The proposed algorithm as discussed in the previous section for the generation of tool path is coded in MATLAB. The bicubic B-spline surface is defined by 4 × 4 control matrix, i.e.,

$$\begin{array}{cccc} (0.0\,0.0\,1.5) & (0.0\,1.0\,1.2) & (0.0\,2.0\,1.2) & (0.0\,3.0\,1.5) \\ (0.7\,0.0\,1.2) & (0.7\,1.0\,0.9) & (0.7\,2.0\,0.9) & (0.7\,3.0\,1.2) \\ (1.4\,0.0\,1.5) & (1.4\,1.0\,1.2) & (1.4\,2.0\,1.2) & (1.4\,3.0\,1.5) \\ (2.0\,0.0\,1.2) & (2.0\,1.0\,0.9) & (2.0\,2.0\,0.9) & (2.0\,3.0\,1.2) \end{array}$$

The B-spline surface generated by coding its algorithm in MATLAB is shown in Fig. 5.

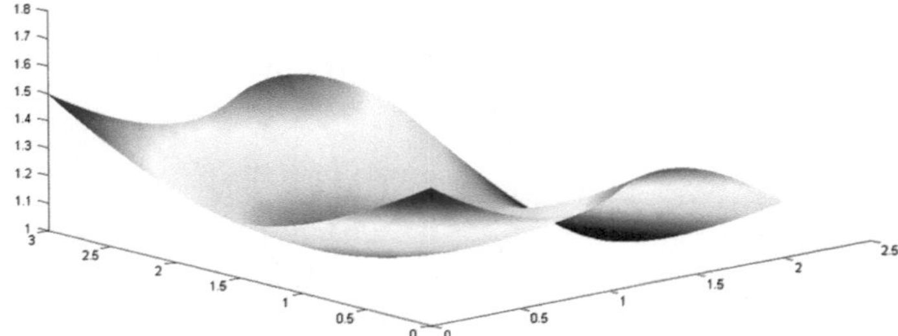

Fig. 5 Generated B-spline surface

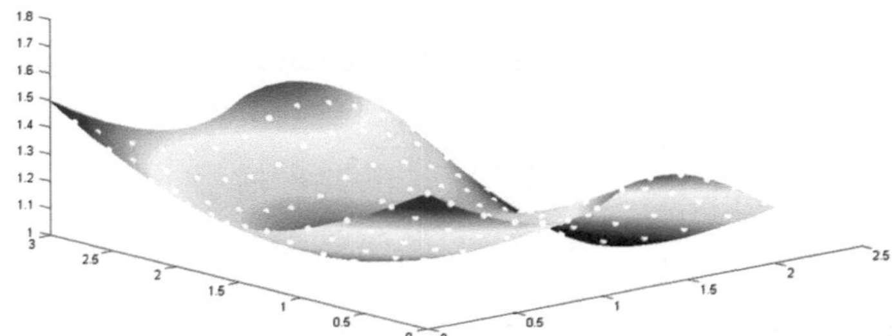

Fig. 6 Cutter contact points on B-spline surface

4.1 Generation of Cutter Contact Points

The forward and side steps are calculated as per the equations discussed in previous section. The x, y and z coordinates of all the cutter contact points are stored to generate the CNC program. The generated cutter contact points on B-spline surface are shown in Fig. 6.

4.2 Modelling of Desired Surface from Generated Points

The cutter contact points generated in MATLAB are used to model the desired surface in RHINOCEROS 4.0, a surface modelling software. Figure 7 shows the generated B-spline surfaces.

4.3 Machining

The CNC program is generated using calculated cutter contact points using word address manual part programming (Koren 2010). The program is simulated on a vertical milling centre shown in Fig. 8. The written code is transferred on to the milling centre by using a flash card.

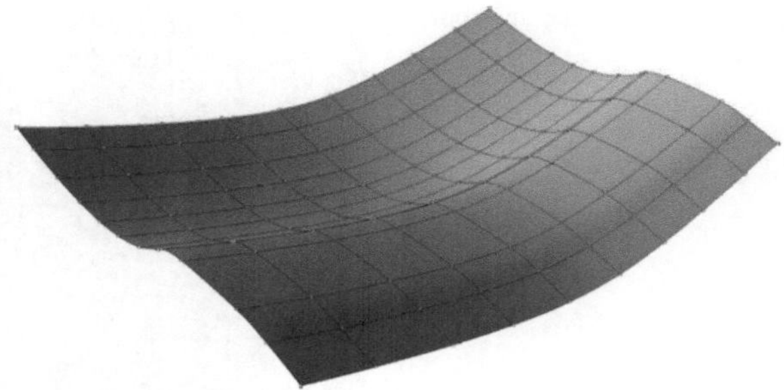

Fig. 7 B-spline surface generated using RHINOCEROS

Fig. 8 Vertical milling centre

The wax piece is used to machine the generated surface on the milling machine. The machining is done using a ball-end milling tool of 4 mm diameter with speed of 200 rpm, feed of 800 mm/min and coolant in mild flow. The work piece and the tool setup are shown in Fig. 9.

After machining the wax piece using the CNC machine, the generated B-spline surface is shown in Fig. 10.

Thereafter, the coordinate measuring machine (CMM) with point probe is used to make a point cloud of the machined B-spline surface. This point cloud is used to model the surface in RHINOCEROS 4.0. The surfaces generated from the point cloud of machined B-spline surface are shown in Fig. 11.

4.4 Comparison of Desired and Machined Surface

To check the accuracy of the machined surface, the desired and the machined surfaces are overlapped to observe the regions of the surfaces which are merging within one another. Thin overlapping will clearly show the difference between

Fig. 9 Work piece and tool setup

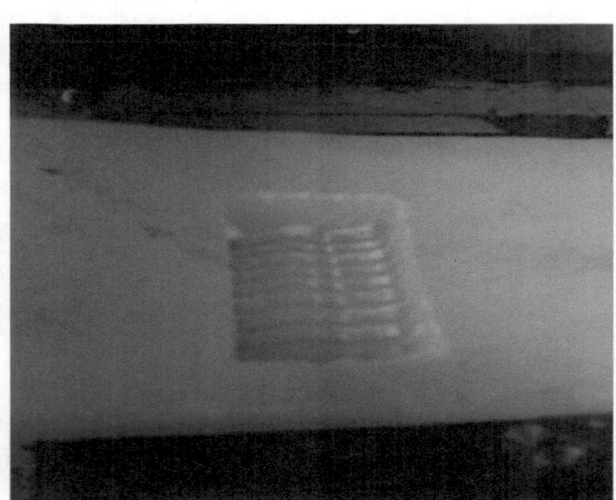

Fig. 10 Generated B-spline surface

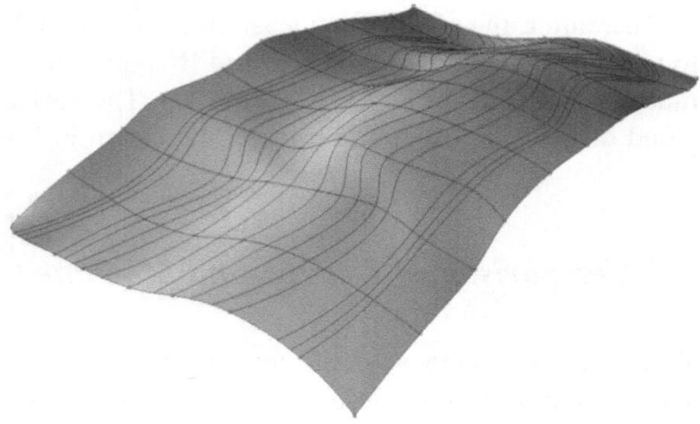

Fig. 11 Surface from point cloud for B-spline surface

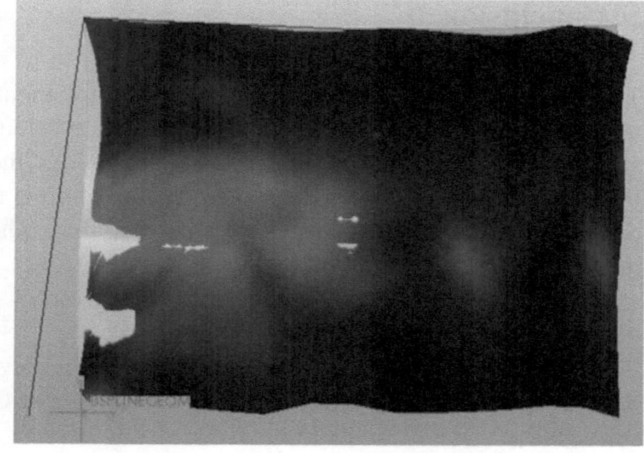

Fig. 12 Overlapping of B-spline machined and desired surface

Table 1 Scallop height for B-spline surface

Assumed scallop height (mm)	Maximum scallop height of machined surface (mm)	Average scallop height of machined surface (mm)	Number of cutter location points (mm)
0.254	1.19	0.96	130

machined surface and desired surface. The surfaces are shown in different colours to make a more effective view. In the Fig. 12, the green and blue colour shows the desired and the machined surface, respectively. It can be seen from the figure that the two surfaces overlap each other.

4.5 Comparison on Scallop Height Basis

The comparison between desired surface and machined surface is also done by measuring the scallop height of the machined surface. The scallop height of the machined surface should be within range of the assumed scallop height. The assumed minimum and maximum scallop heights are 0.254 and 1.27 mm, respectively. Table 1 shows the maximum scallop height of machined B-spline surface, number of cutter location points and average scallop height of the machined B-spline surface for assumed scallop height.

Thus, the Table 1 shows that the maximum scallop height and average scallop height of the machined surface lie within the limits of the assumed scallop height for B-spline surface. In case of B-spline surface, the number of cutter location points corresponding to 0.254 mm is 130. If the scallop height is increased to 1.27 mm, the number of cutter location points reduces to 40.

5 Conclusion

The algorithm using the B-spline surface is implemented successfully through the integration of mathematical modelling used for calculating the forward-step and side-step size into the core of the algorithm. As the algorithm is successfully implemented using B-spline surface, the algorithm is applicable to the continuous parametric surfaces that are twice differentiable. Therefore, it can be used for machining of sculptured and analytic surfaces.

Thereafter, the verification of machining errors is done by comparing the desired and the machined surface.

The present work is being further extended by developing a canned cycle that can be used directly by the part programmer on a CNC machine for free-form surfaces.

References

Barnhill RE (1985) Surfaces in computer aided geometric design: a survey with new results. Comput Aided Geom Des 2(1):1–17

Chen T, Shi Z (2008) A tool path generation strategy for three-axis ball-end milling of free-form surfaces. J Mater Process Technol 208:259–263

Choi YK, Banerjee A (2006) Tool path generation and tolerance analysis for free-form surfaces. Int J Mach Tools Manuf 47(3–4):689–696

Choi BK, Lee CS, Hwang JS, Jun CS (1988) Compound surface modeling and machining. Comput Aided Des 20(3):127–136

Choi YK, Banerjee A, Lee W (2007) Tool path generation for free form surfaces using Bezier curves/surfaces. Comput Ind Eng 52:486–501

Ding S, Mannon MA, Poo AN, Yang DCH, Han Z (2003) Adaptive iso planar tool path generation for machining of free form surfaces. Comput Aided Des 35:141–153

Dragomatz D, Mann S (1997) A classified bibliography of literature on NC milling path Generation. Comput Aided Des 29(3):239–247

Feng HY, Li H (2002) Constant scallop height tool path generation for three axis sculptured surface machining. Comput Aided Des 34:647–654

Ge QJ, Srinivasan L, Rastegar J (1997) Low-harmonic rational Bezier curves for trajectory generation of high-speed machinery. Comput Aided Geom Des 14:251–271

Huang Y, Oliver J (1994) Non-constant parameter NC tool path generation on sculptured surfaces. Int J Adv Manuf Technol 9(2):281–290

Ibrahim Z (2010) Introduction to CAD/CAM. Tata McGraw Hill, New Delhi

Konno K, Chiyokura H (1996) An approach of designing and controlling free-form surfaces by using NURBS boundary Gregory patches. Computr Aided Geom Des 13:825–849

Koren Y (2010) Numerical Control of Machine Tools. Tata McGraw Hill, New Delhi

Lartiguea C, Thiebauta F, Maekawa T (2001) CNC tool path in terms of B-spline curves. Comput Aided Des 33:307–319

Lasemi A, Deyi X, Peihua G (2010) Recent development in CNC machining of freeform surfaces: a state-of-the-art review. Comput Aided Des 42:641–654

Leon JC, Trompette P (1995) A new approach towards free-form surfaces control. Comput Aided Geom Des 12:395–416

Lin R, Koren Y (1996) Efficient tool-path planning for machining free-from surfaces. Trans ASME 118:20–28

Lo C (1998) A new approach to CNC tool path generation. Comput Aided Des 30(8):649–655

Loney GC, Ozsoy TM (1987) Machining of free form surface. Comput Aided Des 19(2):85–89

Ma L, Peng Q (1995) Smoothing of free-form surfaces with Bezier patches. Comput Aided Geom Des 12:231–249

Maekawa T (1999) An overview of offset curves and surfaces. Comput Aided Des 31:165–173

Narayanaswami R, Pang J (2003) Multiresolution analysis as an approach for tool path planning in NC machining. Comput Aided Des 35:167–178

Nasri A (1997) Curve interpolation in recursively generated B-spline surfaces over arbitrary topology. Comput Aided Geom Des 14:13–30

Pham B (1992) Offset curves and surfaces: a brief survey. Comput Aided Des 24(4):223–229

Suh YS, Lee K (1990) NC milling tool-path generation for arbitrary pockets defined by sculptured surfaces. Comput Aided Des 22(5):273–283

Wang Y (1996) Intersection of offsets of parametric surfaces. Comput Aided Geom Des 13:453–465

Wei E, Lin MC (2005) Study on general analytical method for CNC machining the free-form surfaces. J Mater Process Technol 168:408–413

Yang DCH, Han Z (1999) Interference detection and optimal tool selection in 3-axis NC machining of free-form surfaces. Comput Aided Des 31:303–315

Yin Z (2004) Rough and finish tool-path generation for NC machining of freeform surfaces based on a multiresolution method. Comput Aided Des 36:1219–1231

Response Surface-Based Modeling of Mechanical System

Shankar Sehgal and Harmesh Kumar

Abstract In this paper, response surface method has been implemented for modeling a mechanical spring-mass system. Simulated experiments based results are used for developing a design matrix. The design matrix is then used in conjunction with central composite design method so as to develop the response surface for first natural frequency using regression analysis. Analysis of variance is also performed in order to check for statistic significance of response surface model. Results show that response surface model produced during this work is statistically significant and hence can be used for further model updating related research work.

Keywords Response surface model · Regression analysis · Analysis of variance · Model updating

1 Introduction

Model updating involves the correction of the finite element model of a mechanical system so that the theoretically predicted structural dynamic response matches with its experimentally observed counterpart (Friswell and Mottershead 1995). Theoretical structural dynamic responses are obtained either through classical method (Den Hartog 1934) or through finite element method (Petyt 1998). However latter method is generally used for modal analysis of real life structures (Silva and Maia 1997; Sehgal and Kumar 2012a). Experimental

S. Sehgal (✉) · H. Kumar
University Institute of Engineering and Technology, Panjab University,
Chandigarh 160014, India
e-mail: sehgals@pu.ac.in

H. Kumar
e-mail: harmesh@pu.ac.in

S. S. Khangura et al. (eds.), *Proceedings of the International Conference on Research and Innovations in Mechanical Engineering*, Lecture Notes in Mechanical Engineering,
DOI: 10.1007/978-81-322-1859-3_7,

dynamic responses are obtained through modal testing method (Ewins 2000; Sehgal and Kumar 2012b).

However in real life, the mechanical systems are generally very complex in shape and large in size. This is why processing of finite element models of such systems consumes a lot of time thereby making the model updating a very time consuming and difficult task. In order to avoid such time constraints, one can use response surface models instead of conventionally used finite element models for complex and big mechanical systems.

Section 2 of this paper describes the theoretical background of response surface method. Application of response surface method in developing a response model for a mechanical system is presented in Sect. 3 followed by conclusion in Sect. 4 of this paper.

2 Response Surface Method

Response surface method is a collection of mathematical and statistical techniques that are useful for the modeling and analysis of problems in which a response of interest is influenced by several input variables and the objective is to optimize this response (Montgomery 2004; Cochran and Cox 1962). It is a sequential experimentation strategy for empirical model building and optimization. By conducting experiments and applying regression analysis, a model of the response to some independent input variables can be obtained. Based on the model of the response, a near optimal point can then be deduced. Response surface method is often applied in the characterization and optimization of processes (Kansal et al. 2005). In response surface method, it is possible to represent independent process parameters in quantitative form as written in (1).

$$Y = f(X_1, X_2, X_3, \ldots, X_m) \pm \varepsilon \tag{1}$$

where, Y is the response, f is the response function, ε is the experimental error, and $X_1, X_2, X_3, \ldots, X_m$ are independent parameters. By plotting the expected response of Y, a surface, known as the response surface is obtained. The form of f is unknown and may be very complicated. Thus, response surface method aims at approximating f by a suitable lower ordered polynomial in some region of the independent process variables. If the response can be well modeled by a linear function of the m independent variables, the function Y can be written as:

$$Y = C_0 + C_1X_1 + C_2X_2 + \cdots + C_mX_m \pm \varepsilon \tag{2}$$

However, if a curvature appears in the system, then a higher order polynomial such as the quadratic model as shown in (3) may be used.

$$Y = C_0 + \sum_{i=1}^{m} C_iX_i + \sum_{i=1}^{m} D_iX_i^2 \pm \varepsilon \tag{3}$$

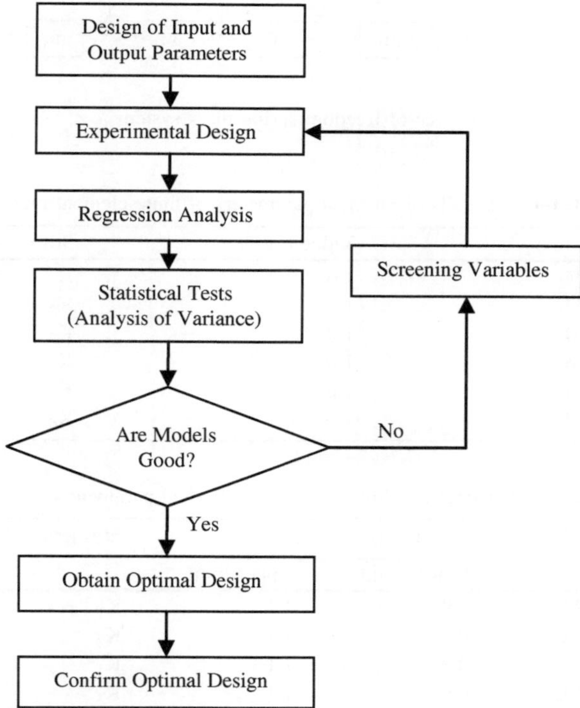

Fig. 1 Procedure of response surface method

The objective of using response surface method is not only to investigate the response over the entire factor space, but also to locate the region of interest where the response reaches its optimum or near optimal value. By studying carefully the response surface model, the combination of factors, which gives the best response, can then be established. The response surface method is a sequential process and its procedure can be summarized as shown in Fig. 1.

3 Response Model of Mechanical System

Mechanical system considered for response surface modeling was composed of six lumped masses and six springs interconnected as drawn in Fig. 2. Initial finite element model of mechanical system was developed in Matlab (2004) as per the data shown in Table 1. This model was then processed in Matlab to produce theoretical structural dynamic response that is first natural frequency as 3.84 Hz. This theoretically predicted value of first natural frequency might be an erroneous value due to some errors in input physical parameters of finite element model. Thus the quality of finite element model needed to be improved so as to minimize the error in predicted natural frequency. Therefore, the finite element model

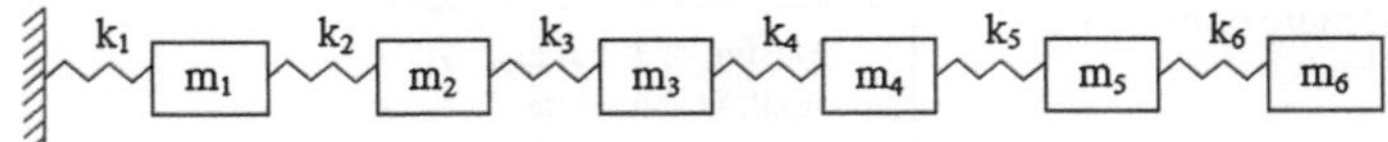

Fig. 2 Six degree-of-freedom spring-mass system

Table 1 Details of physical parameters of finite element model

Mass	Value (kg)	Stiffness	Value (N/m)
M_1	1.0	K_1	10,000
M_2	1.0	K_2	10,000
M_3	1.0	K_3	10,000
M_4	1.0	K_4	10,000
M_5	1.0	K_5	10,000
M_6	1.0	K_6	10,000

Table 2 Upper and lower limits for physical parameters

Mass	Value (kg)		Stiffness	Value (N/m)	
	Lower limit	Upper limit		Lower limit	Upper limit
M_1	0.9	1.1	K_1	8,000	12,000
M_2	0.9	1.1	K_2	8,000	12,000
M_3	0.9	1.1	K_3	8,000	12,000
M_4	0.9	1.1	K_4	8,000	12,000
M_5	0.9	1.1	K_5	8,000	12,000
M_6	0.9	1.1	K_6	8,000	12,000

Table 3 Comparison of different types of fits

Source	Sequential *p* value	Adjusted R-squared	Predicted R-squared
Linear	<0.0001	0.9869	0.9867
2FI	<0.0001	0.9943	0.9950
Quadratic	<0.0001	0.9997	0.9993
Cubic	<0.0001	1.0000	0.9977

needed updating also. In order to increase the computational efficiency of model updating process, the finite element model should be replaced by a meta-model known as response surface model of mechanical system.

In the present research work, while developing the response model of mechanical system, firstly the range of each input physical parameter was decided as per data provided in Table 2. Relationship between input variables (m_1, m_2, m_3, m_4, m_5, m_6, k_1, k_2, k_3, k_4, k_5, k_6) and response variable (ω_1) was assumed to be quadratic. Linear or 2FI fit were not used because their R-Squared values were found to be lesser than that of a quadratic fit as is also clear from Table 3. Although cubic fit was having a higher R-Squared value than its quadratic

Table 4 Analysis of variance for natural frequency (after backward elimination)

Source	Sum of squares	Degrees of freedom	Mean square	F-value	Prob > F
Model	21.96	71	0.31	26,708	<0.0001
A-m1	0.01	1	0.01	517	<0.0001
B-m2	0.08	1	0.08	7,070	<0.0001
C-m3	0.34	1	0.34	29,130	<0.0001
D-m4	0.80	1	0.80	69,311	<0.0001
E-m5	1.34	1	1.34	116,051	<0.0001
F-m6	1.71	1	1.71	148,027	<0.0001
G-k1	6.92	1	6.92	597,549	<0.0001
H-k2	5.44	1	5.44	469,889	<0.0001
J-k3	3.28	1	3.28	283,328	<0.0001
K-k4	1.39	1	1.39	120,205	<0.0001
L-k5	0.34	1	0.34	29,191	<0.0001
M-k6	0.02	1	0.02	2,078	<0.0001
AG	0.00	1	0.00	36	<0.0001
AH	0.00	1	0.00	9	0.0032
AJ	0.00	1	0.00	6	0.0132
AK	0.00	1	0.00	3	0.0870
BD	0.00	1	0.00	7	0.0088
BE	0.00	1	0.00	16	<0.0001
BF	0.00	1	0.00	24	<0.0001
BG	0.00	1	0.00	30	<0.0001
BH	0.00	1	0.00	30	<0.0001
BJ	0.00	1	0.00	81	<0.0001
BK	0.00	1	0.00	39	<0.0001
BL	0.00	1	0.00	10	0.0016
CD	0.00	1	0.00	25	<0.0001
CE	0.00	1	0.00	59	<0.0001
CF	0.00	1	0.00	87	<0.0001
CJ	0.00	1	0.00	7	0.0073
CK	0.00	1	0.00	150	<0.0001
CL	0.00	1	0.00	39	<0.0001
CM	0.00	1	0.00	3	0.0945
DE	0.00	1	0.00	118	<0.0001
DF	0.00	1	0.00	176	<0.0001
DG	0.00	1	0.00	83	<0.0001
DH	0.00	1	0.00	40	<0.0001
DJ	0.00	1	0.00	3	0.0641
DK	0.00	1	0.00	7	0.0080
DL	0.00	1	0.00	84	<0.0001
DM	0.00	1	0.00	6	0.0113
EF	0.00	1	0.00	235	<0.0001

(continued)

Table 4 (continued)

Source	Sum of squares	Degrees of freedom	Mean square	F-value	Prob > F
EG	0.00	1	0.00	309	<0.0001
EH	0.00	1	0.00	178	<0.0001
EJ	0.00	1	0.00	42	<0.0001
EL	0.00	1	0.00	28	<0.0001
EM	0.00	1	0.00	9	0.0029
FG	0.01	1	0.01	520	<0.0001
FH	0.00	1	0.00	313	<0.0001
FJ	0.00	1	0.00	87	<0.0001
FL	0.00	1	0.00	29	<0.0001
FM	0.00	1	0.00	36	<0.0001
GH	0.04	1	0.04	3,647	<0.0001
GJ	0.03	1	0.03	2,873	<0.0001
GK	0.02	1	0.02	1,452	<0.0001
GL	0.00	1	0.00	395	<0.0001
GM	0.00	1	0.00	30	<0.0001
HJ	0.02	1	0.02	1,945	<0.0001
HK	0.01	1	0.01	996	<0.0001
HL	0.00	1	0.00	274	<0.0001
HM	0.00	1	0.00	21	<0.0001
JK	0.00	1	0.00	427	<0.0001
JL	0.00	1	0.00	120	<0.0001
JM	0.00	1	0.00	9	0.0024
KL	0.00	1	0.00	24	<0.0001
D^2	0.00	1	0.00	4	0.0402
E^2	0.00	1	0.00	7	0.0074
F^2	0.00	1	0.00	8	0.0044
G^2	0.01	1	0.01	949	<0.0001
H^2	0.01	1	0.01	743	<0.0001
J^2	0.01	1	0.01	501	<0.0001
K^2	0.00	1	0.00	247	<0.0001
L^2	0.00	1	0.00	61	<0.0001
Residual	0.01	474	0.00		
Lack of Fit	0.01	465	0.00		
Pure Error	0.00	9	0.00		
Cor Total	21.97	545			

counterpart, still the former was not chosen because of presence of aliasing effects in cubic fit. Central composite design method was used to develop the design matrix of actual physical variables by using Design-Expert software (2010). The design matrix consisted of a total of 546 test runs. The design matrix was then imported in Matlab and used as the input to the finite element model. Outputs of the finite element model along with the design matrix of actual physical variables were used in Design-Expert (2010) for generating the response models. Analysis of variance results for natural frequency parameter after backward elimination

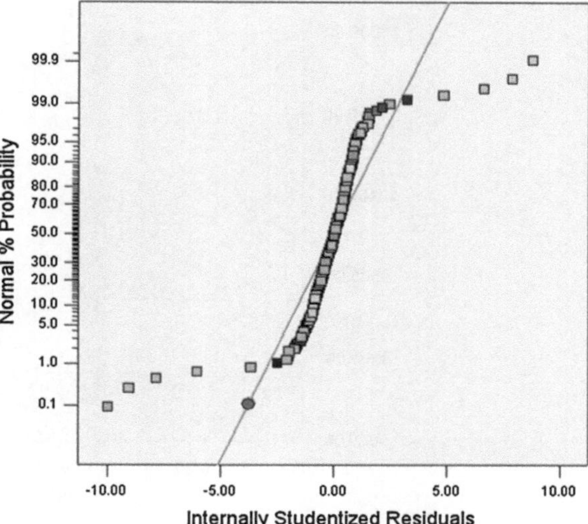

Fig. 3 Normal plot of residuals

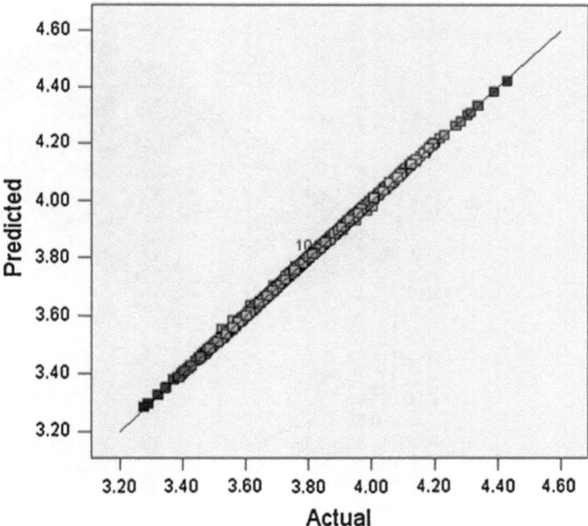

Fig. 4 Predicted versus actual values for natural frequency

process are presented in Table 4. It shows that the quadratic model is significant. Figure 3 shows the normal plot of residuals. As most of the points are falling along a straight line it means that the distribution is normal. Figure 4 shows predicted versus actual results for natural frequency. Here also the response model performs satisfactorily because the data points are falling along a straight line of unit slope. After checking the model for its significance, contour plots and three-dimensional surface plots of the mechanical system are drawn in Figs. 5 and 6 respectively.

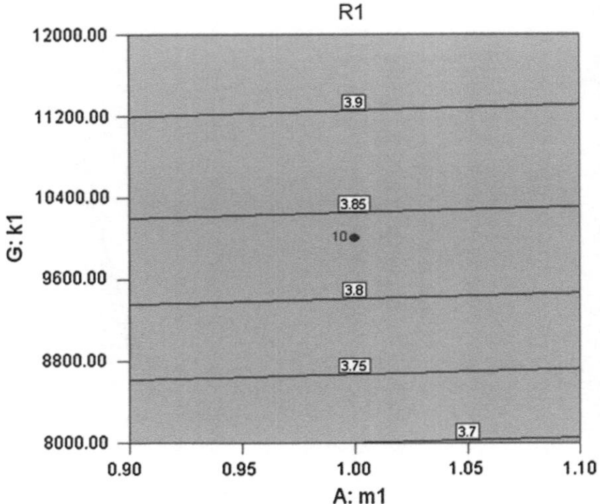

Fig. 5 Contour plots obtained from response model of mechanical system

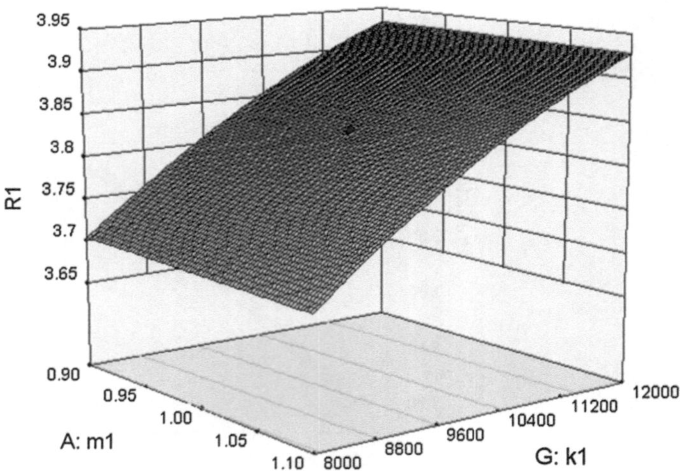

Fig. 6 Surface plot obtained from response model of mechanical system

4 Conclusion

Response surface based model of a mechanical spring-mass system has been developed. Response model is checked for its accuracy by using predicted versus actual data points for natural frequency. Significance of the model is also checked using analysis of variance method. Response model of mechanical spring-mass system can then be used in developing an efficient model updating technique.

Acknowledgments This work was supported by the university grants commission, New Delhi, India under major research project grant sanctioned via f-42-869/2013(SR).

References

Cochran G, Cox GM (1962) Experimental design. Asia Publishing House, New Delhi

Den Hartog JP (1934) Mechanical vibrations. McGraw-Hill Book Company Inc, US

Design-Expert (2010) User's guide for version 8 of design-expert. Minneapolis, MN: Stat-Ease Incorporation

Ewins DJ (2000) Modal testing: theory, practice and application. Research Studies Press Limited, England

Friswell MI, Mottershead JE (1995) Finite element model updating in structural dynamics. Kluwer Academic Publishers, The Netherlands

Kansal HK, Singh S, Kumar P (2005) Parametric optimization of powder mixed electrical discharge machining by response surface methodology. J Mater Process Technol 169:427–436

Matlab (2004) User's guide of matlab v7. The Mathworks Incorporation, New York

Montgomery DC (2004) Design and analysis of experiments. Wiley, Singapore

Petyt M (1998) Introduction to finite element vibration analysis. Cambridge University Press, Cambridge

Sehgal S, Kumar H (2012a) Structural dynamic analysis of cantilever beam structure. Int J Res Eng Appl Sci 2(2):1110–1114

Sehgal S, Kumar H (2012b) Modal testing of a cantilever beam structure. J Eng Technol Educ 6(1):25–29

Silva NMM, Maia JMM (1997) Theoretical and experimental modal analysis. Research Studies Press Limited, England

Vibration Analysis of Rotor Bearing Housings Considering Unbalancing in Two Planes

Sanjiv Kumar, Rajiv Kumar, Rakesh Sehgal and Sanjeev Bhandari

Abstract Unbalancing and misalignment are the most possible causes of machine vibrations. An unbalanced rotor always causes more vibration and generates excessive force in the bearing area and reduces the life of the machine. Understanding and practicing the fundamentals of rotating shaft parameters is the first step in reducing unnecessary vibration, reducing maintenance costs and increasing machine uptime. By the term two planes here, we mean that two rotors are used for the analysis of unbalanced vibrations. If only one rotor is used, then this system is called a single-plane system. In this paper, experimental studies were performed on a 2 rotor dynamic test apparatus to predict the vibration spectrum for rotor unbalance. Two rotor bearings were used in the experiments. The rotor shaft velocities were measured at rotor speed of 30 Hz using an accelerometer and a dual channel vibration analyzer (DCVA) under the balanced (baseline) and unbalanced conditions. The experimental frequency spectrum was also obtained for both baseline and unbalanced condition under different unbalanced forces. The experimental results of balanced and unbalanced rotors are compared at two different rotor locations.

S. Kumar (✉) · S. Bhandari
Department of Mechanical Engineering, Baba Banda Singh Bahadur Engineering College, Fatehgarh Sahib, Punjab, India
e-mail: sanjiv.kumar@bbsbec.ac.in

S. Bhandari
e-mail: sanjeevbhandari2007@gmail.com

R. Kumar
Department of Mechanical Engineering, Lovely Professional University, Jalandhar, Punjab, India
e-mail: rajivkumar267@yahoo.com

R. Sehgal
Department of Mechanical Engineering, National Institute of Technology, Hamirpur, Himachal Pradesh, India
e-mail: rakeshsehgal.nitham@gmail.com

S. S. Khangura et al. (eds.), *Proceedings of the International Conference on Research and Innovations in Mechanical Engineering*, Lecture Notes in Mechanical Engineering, DOI: 10.1007/978-81-322-1859-3_8, © Springer India 2014

Keywords DCVA · Flexible coupling · Rotor bearings housings · Unbalanced rotors

1 Introduction

Rotor unbalance is the most common reason in machine vibrations. Most of the rotating machinery problem can be solved by using the rotor balancing and misalignment. Mass unbalance in a rotating system often produces excessive synchronous forces that reduce the life span of various mechanical elements. A very small amount of unbalance may cause severe problem in high-speed rotating machines. Rotors are used in many engineering applications such as pumps, fans, propellers and turbo machinery. The vibration signature of the overhung rotors is totally different from the midway or intermediate rotors. The vibration caused by unbalance may destroy critical parts of the machine, such as bearings, seals, gears and couplings. Rotor unbalance is a condition in which the center of mass of a rotating assembly, typically the shaft and its fixed components such as disks and blades, is not coincident with the center of rotation. In practice, rotors can never be perfectly balanced because of manufacturing errors such as porosity in casting, non-uniform density of material, manufacturing tolerances and gain or loss of material during operation. As a result of mass unbalance, a centrifugal force is generated and must be reacted against by the bearings and support structures. Analytical methods of lateral response due to torsional excitation of geared rotors were reported by Rao et al. (1998). In continuation of this, Shiau et al. (1999) presented a dynamic behavior of geared rotors. Then, Lee et al. (2001) studied the rotor-dynamic characteristics of an APU gas turbine rotor–bearing system having a tie shaft. Unbalance response investigations of geared rotor bearing systems were carried out by Neriya et al. (1985) and Kahraman et al. (1992). Further, Iida et al. (1980) and Iwatsubo et al. (1984) reported on studies utilizing the usual procedure of solving simultaneous equations and Choi and Mau (1995) utilizing the frequency branching technique. Further, concerning unbalance response investigations of dual-shaft rotor bearing systems coupled by bearings, based on the transfer matrix modeling, Hibner (1975), Li et al. (1986) and Gupta et al. (1993) carried out investigations utilizing the usual procedure of solving simultaneous equations. Afterward Rao (1996) published a book on Rotor Dynamics. The study of critical speeds of a continuous rotor was reported by Eshleman and Eubanks (1969). Geared high-speed rotors were studied regarding their torsional lateral coupling by Mitchell and Mellen (1995). Further, the coupled lateral–torsional vibration characteristics of a varying speed geared rotor bearing system were studied by Lee and Ha (2003). Then, Rao et al. (1995) reported about the coupled bending–torsional vibrations of geared rotors. However, all the above investigations resulted in full numerical solutions of the unbalance responses of coupled shaft two rotor bearing systems.

In this work, a general method is presented for obtaining the unbalance response orbit based on the experimentation of a gear-coupled two-rotor shaft bearing system, where the shaft may rotate at different speeds. In this paper, balanced and unbalanced systems of 2 overhung rotors and 2 intermediate rotors are considered for the study. Experiments were conducted for two different positions of rotors (2 rotors overhung) and (2 rotors intermediate) for unbalanced weights at rotor speed of 30 Hz (1,800 rpm), and results are plotted. However, the setup can also be made to take results at different rotor speeds. The rotor unbalance can be detected by spectral analysis. The vibration frequency of rotor unbalance is synchronous that is one times the shaft rotational speed, since the unbalance can be reduced significantly by rotor balancing.

2 Experimentation

2.1 Description of Rotor Bearings System

A rotor bearing system (set of 2) is shown in Fig. 1. It has two bearing housings made up of cast iron. One housing contains one ball bearing and the other one the second ball bearing. The driver and driven shafts are mounted in between respective housings. Mild-steel is considered for the input and output shafts. Over these pins, a circular natural rubber bush is provided and its length is equal to the length of the hole. The diameter of the flange holes is equal to the diameter of pin plus the thickness of rubber bush.

The cast iron material is chosen for both left and right bearing housings. Figure 1 represents the three-dimensional model of rotor bearing system along with two intermediate rotors. Details of the rotor shaft and bearing housing and materials used are given in Tables 1 and 2, respectively.

2.2 Description of the Experimental Setup

The Experimental apparatus is shown in Fig. 1 (two rotors in intermediate position in two planes) and in Fig. 2 (two rotors in overhung position in two planes). It consists of a D.C. motor, a flexible coupling and a double-disk rotor. The rotor shaft is supported by two identical ball bearings and has a length of 660 mm with a bearing span of 460 mm. The diameter of the rotor shaft is 20 mm. Two disks of 128 mm in diameter and 10 mm in thickness are mounted on the rotor shaft non-drive end. The bearing pedestals are adjustable in vertical direction so that different misalignment conditions can be created. The rotor shaft is driven by 0.75 hp D.C. motor. The D.C. voltage controller known as control panel is used to adjust the power supply so that motor speed can be continuously increased or decreased

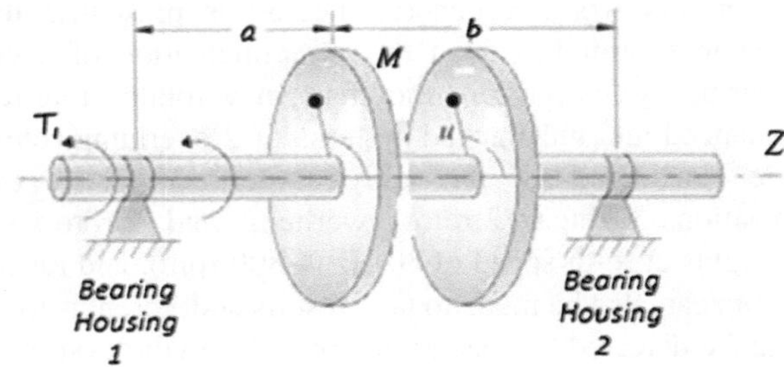

Fig. 1 Rotor bearing system along with two intermediate rotors

Table 1 Details of shaft and bearing housings

Sr.#	Description	Unit
1	Shaft diameter	20 mm
2	Length of the shaft	630 mm
3	Hub diameter	40 mm
4	Length of the hub	30 mm
5	No. of holes for pin	04 Nos.
6	Diameter of pin hole	4.2 mm
7	Diameter of pin	4 mm
8	Rubber bush outside diameter	11 mm
9	Rubber bush inside diameter	6 mm
10	*Keyway depth*	
	In shaft	4 mm
	In hub	3 mm
11	*Keyway X-section*	
	Height	6 mm
	Width	6 mm

Table 2 Material properties

Properties	Cast iron	M. steel	Rubber
Young's modulus (MPa)	1×10^5	2×10^5	30
Poisson's ratio	0.23	0.3	0.49
Density (Kg/mm^3)	$7{,}250 \times 10^9$	$7{,}850 \times 10^9$	$1{,}140 \times 10^9$

in the range from 0 to 3,000 rpm. The baseline signal has been measured at a rotor speed of 30 Hz to check the concentricity.

The instruments used in the experiments include accelerometers and a vibration analyzer, dual channel vibration analyzer (DCVA). The accelerometer directly measures the velocity of bearing housing vibrations and displays in the vibration analyzer.

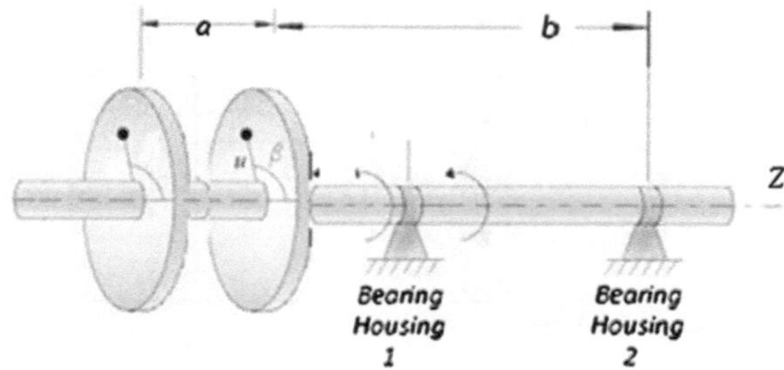

Fig. 2 Rotor bearing system along with two overhung rotors

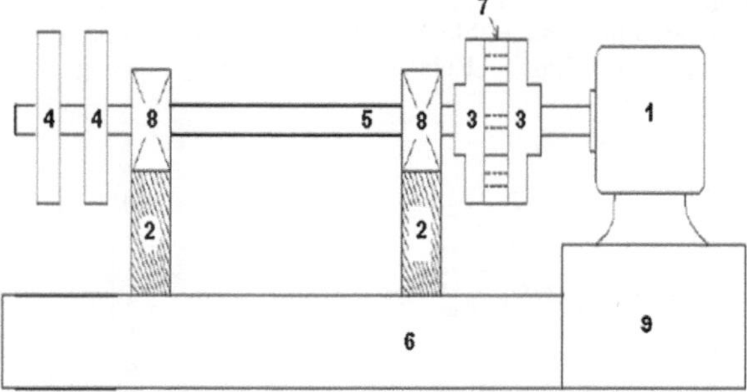

Fig. 3 Experimental setup with two rotors at overhung position. *1* D.C motor, *2* Bearings supported in plummer blocks, *3* Pin type flexible coupling, *4* Rotors (5–8 kg each, adjustable), *5* Rotor shaft, *6* Base, *7* Rubber, *8* Rotor bearing housing, *9* Control panel

2.3 Experimental Procedure

Experimental facility as shown in Figs. 2 (with two rotors in overhung position in two planes) and 3 (with two rotors in intermediate position in two planes) is used for both base line and unbalance tests. First, the setup is run for few minutes to settle down all minor vibrations. Before creating unbalancing, the shaft is checked for any misalignment and unbalance. After that an unbalance has been created by placing a mass of 18 g in the overhung rotor at a radius of 54 mm. Accelerometer along with the vibration analyzer is used to acquire the vibration signals. Figure 3 shows the setup for two rotors in overhung position in two planes, and Fig. 4 shows the setup for two rotors in intermediate position in two planes; the accelerometer is attached with the help of wires to take readings at three positions (Horizontal, vertical and axial) at Non-Drive End (NDE) and Drive End (DE) for both motor and rotor. Following are the three positions for motor and rotor (Fig. 4).

2.3.1 Motor

$[NDE\ (H)]_M$ Horizontal Non-Drive End of Motor
$[NDE\ (V)]_M$ Vertical Non-Drive End of Motor
$[NDE\ (A)]_M$ Axial Non-Drive End of Motor

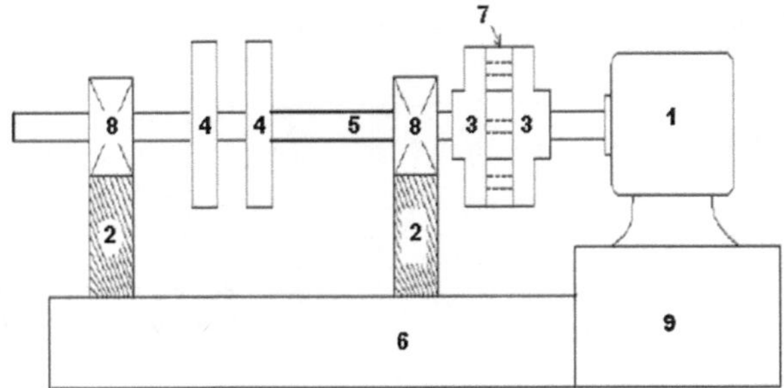

Fig. 4 Experimental setup with two rotors at intermediate position. *1* D.C motor, *2* Bearings supported in plummer blocks, *3* Pin type flexible coupling, *4* Rotors (5–8 kg each, adjustable), *5* Rotor shaft, *6* Base, *7* Rubber, *8* Rotor bearing housing, *9* Control panel

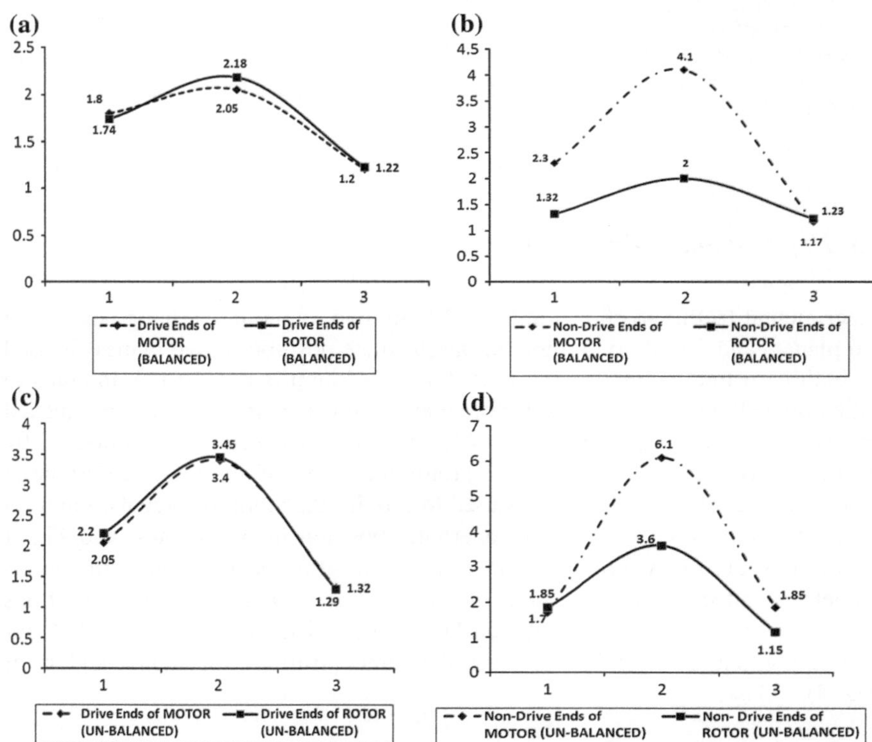

Fig. 5 Amplitude trends of NDE and DE of both motor and rotor in baseline and unbalanced conditions with two rotors in OVERHUNG POSITION. **a** Baseline amplitude trend of drive ends (motor and rotor) for 2 rotors overhung. **b** Baseline amplitude trend of non-drive ends (motor and rotor) for 2 rotors overhung. **c** Unbalanced amplitude trend of drive ends (motor and rotor) for 2 rotors overhung. **d** Unbalanced amplitude trend of non-drive ends (motor and rotor) for 2 rotors overhung

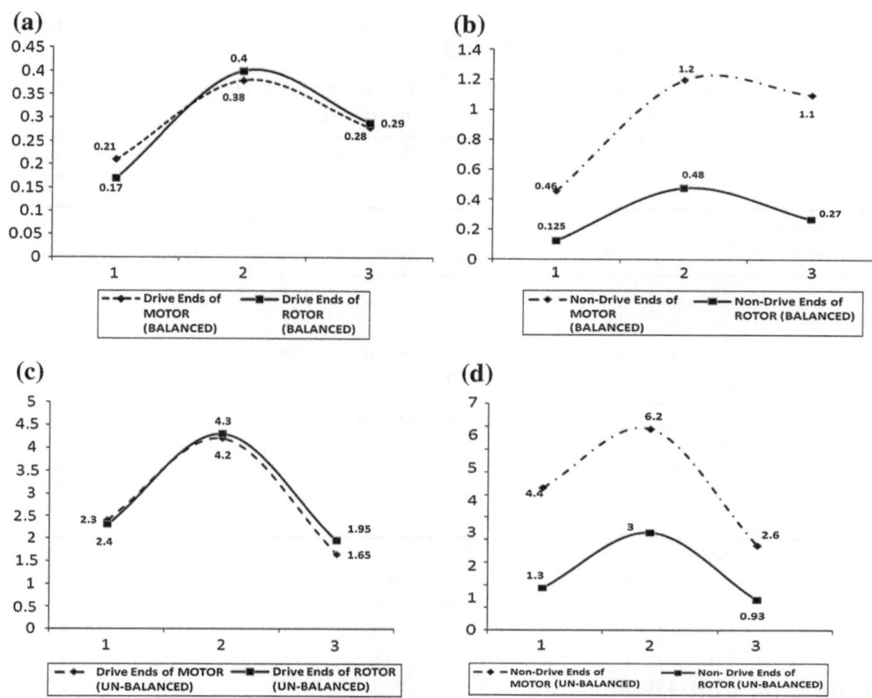

Fig. 6 Amplitude trends of NDE and DE of both motor and rotor in baseline and unbalanced conditions with two rotors in INTERMEDIATE POSITION. **a** Baseline amplitude trend of drive ends (motor and rotor) for 2 rotors intermediate. **b** Baseline amplitude trend of non-drive ends (motor and rotor) for 2 rotors intermediate. **c** unbalanced amplitude trend of drive ends (motor and rotor) for 2 rotors intermediate. **d** Unbalanced amplitude trend of non-drive ends (motor and rotor) for 2 rotors intermediate

$[DE\ (H)]_M$ Horizontal Drive End of Motor
$[DE\ (V)]_M$ Vertical Drive End of Motor
$[DE\ (A)]_M$ Axial Drive End of motor.

2.3.2 Rotor

$[NDE\ (H)]_R$ Horizontal Non-Drive End of Rotor
$[NDE\ (V)]_R$ Vertical Non-Drive End of Rotor
$[NDE\ (A)]_R$ Axial Non-Drive End of Rotor
$[DE\ (H)]_R$ Horizontal Drive End of Rotor
$[DE\ (V)]_R$ Vertical Drive End of Rotor
$[DE\ (A)]_R$ Axial Drive End of Rotor.

Table 3 Comparison of experimental vibration amplitudes when two rotors are overhung for baseline and unbalanced signals along MOTOR SIDE

Two rotors overhung baseline signals (motor side) (exp. values in mm/s)				Two rotors overhung unbalanced signals (motor side) (exp. values in mm/s)					
	H	V	A	H	V	A	Increase in amplitude		
							H	V	A
Non-drive end $(NDE)_M$	2.3	4.1	1.17	1.7	6.1	1.85	−0.6	2	0.05
Drive end $(DE)_M$	1.8	2.05	1.2	2.05	3.4	1.32	0.25	1.35	0.12

Table 4 Comparison of experimental vibration amplitudes when two rotors are overhung for baseline and unbalanced signals along ROTOR SIDE

Two rotors overhung baseline signals (rotor side) (exp. values in mm/s)				Two rotors overhung unbalanced signals (rotor side) (exp. values in mm/s)					
	H	V	A	H	V	A	Increase in amplitude		
							H	V	A
Non-drive end $(NDE)_R$	1.32	2	1.23	1.85	3.6	1.15	0.53	1.6	−0.08
Drive end $(DE)_R$	1.74	2.18	1.22	2.2	3.45	1.29	0.46	1.27	0.07

3 Results and Discussions

3.1 Amplitude Trend for Two Rotors with Overhung Position at Baseline and Unbalanced Conditions

The experimental frequency spectra were obtained to the baseline condition. The perfect alignment and balancing cannot be achieved in practice. Thus, a base line (well balanced and aligned) case is presented first to show the residual unbalance and misalignment. The measured amplitude trend of vibrations in the form of velocity of a baseline and an unbalanced system at Drive End (DE) and Non-Drive End (NDE) with rotor bearing housings at a frequency of 30 Hz (1,800 rpm) is shown in Fig. 3. The baseline spectrum is measured experimentally using DCVA. Tables 3 and 4 show the comparison of experimental vibration amplitudes when two rotors are in overhung position for both baseline and unbalanced signals on MOTOR SIDE and ROTOR SIDE, respectively. Figure 5a–d shows that the maximum amplitude 4.1 mm/s is observed at $[NDE (V)]_M$ in baseline condition and still higher amplitude of 6.1 mm/s is observed at same location in unbalanced condition. Peak amplitude values of small magnitudes are observed at other ends. This highest amplitude at this location is due to the centrifugal forces acting on the system. Small amount of peaks at harmonics of shaft speed are the indications of manufacturing errors of coupling and other elements like rubber.

Table 5 Comparison of experimental vibration amplitudes when two rotors are in intermediate for baseline and unbalanced signals along MOTOR SIDE

Two rotors intermediate baseline signals (motor side) (exp. values in mm/s)				Two rotors intermediate unbalanced signals (motor side) (exp. values in mm/s)					
	H	V	A	H	V	A	Increase in amplitude		
							H	V	A
Non-drive end $(NDE)_M$	0.46	1.2	1.1	4.4	6.2	2.6	3.94	5	1.5
Drive end $(DE)_M$	0.21	0.38	0.28	2.4	4.2	1.65	2.19	3.82	1.37

Table 6 Comparison of experimental vibration amplitudes when two rotors are in intermediate for baseline and unbalanced signals along ROTOR SIDE

Two rotors intermediate baseline signals (rotor side) (exp. values in mm/s)				Two rotors intermediate unbalanced signals (rotor side) (exp. values in mm/s)					
	H	V	A	H	V	A	Increase in amplitude		
							H	V	A
Non-drive end $(NDE)_R$	1.32	2	1.23	1.85	3.6	1.15	0.53	1.6	−0.08
Drive end $(DE)_R$	1.74	2.18	1.22	2.2	3.45	1.29	0.46	1.27	0.07

3.2 Amplitude Trend for Two Rotors with Intermediate Position at Baseline and Unbalanced Conditions

The experimental frequency spectra were obtained to the baseline condition. The perfect alignment and balancing cannot be achieved in practice. Thus, a baseline (well balanced and aligned) case is presented first to show the residual unbalance and misalignment. The measured amplitude trend of vibrations in the form of velocity of a *baseline* and *an unbalanced system* at Drive End (DE) and Non-Drive End (NDE) with bearing housing system at a frequency of 30 Hz (1,800 rpm) is shown in Fig. 4. Tables 5 and 6 show the comparison of experimental vibration amplitudes *when two rotors are in intermediate position* for both baseline and unbalanced signals on MOTOR SIDE and ROTOR SIDES, respectively. The baseline spectrum is measured experimentally using DCVA. Figure 6a–d shows that along motor side highest amplitude of 1.2 mm/s is observed at $[NDE\ (V)]_M$ in baseline condition and amplitude reaching to 6.2 mm/s is observed at the same location in unbalanced condition. Further along the rotor side, the maximum amplitude of 0.48 mm/s is observed at $[NDE\ (V)]_R$ in baseline condition and after unbalancing the amplitude increasing to 4.3 mm/s is observed at $[DE\ (V)]_R$.

As discussed earlier, the increase in amplitude at this location (vertical location) is due to increase in centrifugal forces acting on the system because of the created unbalance.

4 Conclusion

This paper presents mathematical analysis of the results taken from a DCVA both in base line and unbalanced conditions. Then, these results were compared and it was found that by creating an unbalance the amplitude of the system increases. This amplitude increase was noticed in all the positions (Horizontal, vertical and axial) of the unbalanced system, but severe boosting of the amplitudes was observed in the vertical position because of increase in centrifugal forces acting on the system. Based on this paper, further work can be initialized for fault analysis and diagnosis (condition monitoring) for different machines.

References

Choi ST, Mau SY (1995) Dynamic analysis of geared rotor-bearing systems by the transfer matrix method. In: ASME design engineering technical conference 3 (Part B), vol 84, pp 2967–2976

Eshleman R, Eubanks A (1969) On the critical speeds of a continuous rotor. J. Eng Indus 91:1180–1188

Gupta K, Gupta KD, Athre K (1993) Unbalance response of a dual rotor system: theory and experiment. Trans J Vib Acoust 115:427–435

Hibner DH (1975) Dynamic response of viscous-damped multi-shaft jet engines. J Aircraft 12(4):305–312

Iida H, Tamura A, Kikuchi K, Agata H (1980) Coupled torsional-flexural vibration of a shaft in a geared system of rotors: 1st report. Bull JSME 23(186):2111–2117

Iwatsubo T, Arii S, Kawai R (1984) Coupled lateral-torsional vibration of rotor system trained by gears: Part 1. Analysis by transfer matrix method. Bull JSME 27(224):271–277

Kahraman A, Ozguven HN, Houser DR, Zakrajsek JJ (1992) Dynamic analysis of geared rotors by finite elements. Trans ASME J Mech Des 114(03):507–514

Lee AS, Lee YS (2001) Rotor dynamic characteristics of an APU gas turbine rotor-bearing system having a tie shaft. KSME Intl J 15(2):152–159

Lee AS, Ha JW, Choi DH (2003) Coupled lateral and torsional vibration characteristics of a speed increasing geared rotor-bearing system. J Sound Vib 263(4):725–742

Li Q, Yan L, Hamilton JF (1986) Investigation of the steady-state response of a dual-rotor system with inter shaft squeeze film damper. J Eng Gas Turbines Power 108:605–612

Mitchell L, Mellen DDM (1995) Torsional- lateral coupling in a geared high-speed rotor system. In: ASME design engineering technical conferences 3 (Part B), vol 84(2), pp 977–989

Neriya SV, Bhat RB, Sankar TS (1985) Coupled torsional flexural vibration of a geared shaft system using finite element method. Shock Vib Bull 55(3):13–25

Rao JS (1996) Rotor dynamics, 3rd edn. New Age International Publishers, India
Rao JS, Chang JR, Shiau TN (1995) Coupled bending-torsion vibration of geared rotors. In: ASME design engineering technical conferences 3 (Part B), vol 84(2), pp 977–989
Rao JS, Shiau TN, Chang JR (1998) Theoretical analysis of lateral response due to torsional excitation of geared rotor. Mech Mach Theor 33(6):761–783
Shiau TN, Rao JS, Chang JR, Choi ST (1999) Dynamic behavior of geared rotors. J Eng Gas Turbines Power 121:494–503

Experimental Testing and Evaluation of a Prototype Magnetorheological Damper

Ashwani Kumar and S. K. Mangal

Abstract This paper presents experimental testing and evaluation of a prototype magnetorheological (MR) damper. An MR device fills the gap between purely passive and fully active control systems as it offers a reliability of passive systems, and yet it maintains the versatility and adaptability of the fully active devices. The MR devices are categorized as one of the semi-active control systems. Fast response, few moving parts, and low power requirements are one of the major factors that have generated special interest in MR devices. MR dampers are being used for control of vibrations in automobiles, for minimizing damage to civil engineering structures due to seismic motions, etc. In this paper, the basic theory behind the MR dampers and experimental testing is carried out, and its use in vibration control is also studied. For this purpose, an MR damper is fabricated in-house and tested in the newly setup laboratory of the department using an electrodynamic vibration shaker and associated data acquisition system. Its performance is then studied in the form of damping force, etc.

Keywords Magnetorheological damper · Passive and active control systems · Semi-active control system

1 Introduction

An MR damper is just like the standard viscous fluid damper; however, MR damper is filled with MR fluid and has one or more coils of electromagnetic wire wrapped around the grooves of a piston. The MR fluids are field responsive fluids

A. Kumar (✉) · S. K. Mangal
Department of Mechanical Engineering, PEC University of Technology,
Chandigarh 160012, India
e-mail: ashpatyal@rediffmail.com

S. K. Mangal
e-mail: skmangal_pec@rediffmail.com

S. S. Khangura et al. (eds.), *Proceedings of the International Conference on Research and Innovations in Mechanical Engineering*, Lecture Notes in Mechanical Engineering,
DOI: 10.1007/978-81-322-1859-3_9, © Springer India 2014

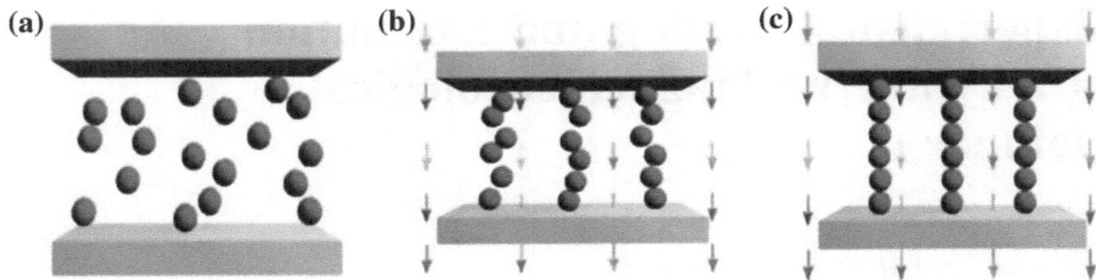

Fig. 1 Illustration of activation of the MR Fluid

as its properties are controlled by varying the external magnetic field. The discovery of MR fluids is credited to Jacob Rabinow at the US National Bureau of Standard in 1948 (Rabinow 1948).

The MR fluid contains micron-sized particles of polarizable material mostly carbonyl iron particles which are suspended in the fluid. When no current is applied to the solenoid/electromagnet in the piston, the particles remain suspended in the fluid, and the damper behaves like an ordinary viscous fluid damper. When an appropriate magnetic field is generated by the solenoid/electromagnet, these particles are arranged in a columnar fashion, thereby changing the viscosity of the fluid and increasing the resistance offered by it to the motion of the piston moving inside the cylinder (Carlson and Spencer 1996). Hence, by employing proper magnitude of current to the solenoid/electromagnet, the MR damper can be made to produce the desired damping force/damping coefficient.

When no magnetic field is applied on the MR fluid, the ferrous particles are randomly dispersed in the medium (Fig. 1a). In the presence of a magnetic field, the particles start to move in order to align themselves along the lines of magnetic flux, Fig. 1b. Figure 1c shows the formation of chains of ferrous particles, creating more viscosity and increased shear strength (Kumar and Mangal 2012). As this change occurs almost instantly, the MR fluids are attractive solutions for real-time control applications, for example, dampers, brakes, clutches, engine mounts, valves, etc. The changes of liquid–solid–liquid state or the consistency or yield strength of the MR fluid can be controlled precisely and proportionally by altering the strength of the applied magnetic field.

In the absence of an applied magnetic field, the MR fluids are reasonably well approximated as Newtonian liquids. For most engineering applications, a simple Bingham plastic model is effective to describe its essential, field-dependent fluid characteristics. A Bingham plastic is a non-Newtonian fluid whose yield strength must be exceeded to a desired value before the flow begins. Typical MR fluid can achieve yield strengths up to 50–100 kPa at the magnetic field strength of about 150–250 kA/m. These MR fluids are stable in temperature ranges from −50 °C to 150 °C (Kumar and Mangal 2013).

In this paper, the basic theory behind the MR damper, its fabrication, testing, and evaluation is carried out. For this purpose, an MR damper is fabricated and tested in the laboratory using an experimental setup.

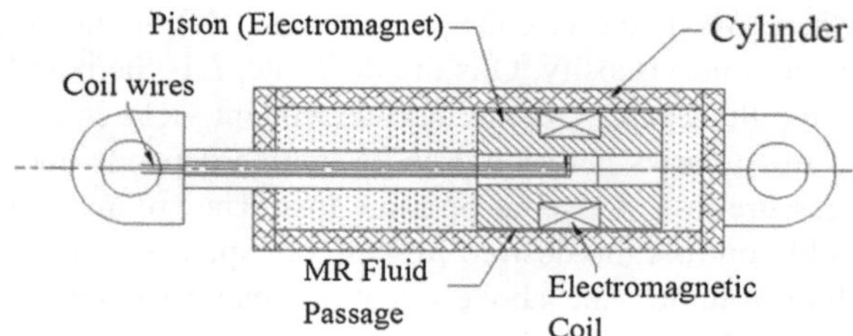

Fig. 2 Schematic diagram of MR damper

2 Design Parameters of MR Damper

An MR damper consists of a cylinder, piston, electromagnet, and the MR fluid which is enveloped in a cylinder. As a device acts on the magnetic effect, the MR damper should have optimized design to fulfill its magnetic performance and eventually to act as a satisfactory semi-active control device. The MR damper offers a highly reliable operation and can be viewed as fail-safe device if the control hardware malfunctions and the damper turns to be passive one. Figure 2 shows schematic diagram of the MR damper.

In the absence of an applied magnetic field, that is, off-state, the MR fluids are reasonably well approximated as Newtonian liquids. In the on state, that is, activated under magnetic field, the fluid behaves as a Bingham plastic with variable yield strength. Although the fluid does have the departures from this model, even then this gives a good reference for the behavior of the MR fluid (Carlson and Jolly 2000; Poyner 2001; Seval 2002). The shear stress associated with the flow of the MR fluid can, thus, be predicted by the Bingham equations (Spencer et al. 1996). In this model, the total fluid shear stress is given by

$$\tau = \tau_y(H) + \eta\dot{\gamma} \quad |\tau| > \tau_y \tag{1}$$

where τ_y is the yield stress (Pa), H is the magnetic field strength (A/m), η is the plastic viscosity (Pa · s), and $\dot{\gamma}$ is the shear strain rate (s^{-1}).

For the fluid shear stresses less than yield shear stress of the fluid (τ_y), the MR fluid behaves as visco-elastic material and is given as

$$\tau = G\gamma \quad |\tau| < \tau_y \tag{2}$$

where the G is the complex material modulus and γ is the fluid shear strain. The MR dampers generally use the flow mode of the fluid. In this mode, the pressure drop (loss) has two components, that is, pressure loss due to the viscous drag and pressure loss due to the field-dependent yield stress (Carlson and Spencer 1996). The total pressure drop (ΔP) in the fluid is given as (Spencer et al. 1996)

$$\Delta P = \Delta P_n + \Delta P_\tau = \frac{12\eta QL}{g^3 w} + \frac{c\tau_y L}{g} \tag{3}$$

where ΔP_η is the viscous pressure loss, ΔP_τ is the field-dependent pressure loss, η is the fluid viscosity, Q is the flow rate, L is the pole length, w is the pole width, g is the fluid gap, τ_y is the field-dependent yield stress, and c is constant. The value of the c varies between 2 and 3, which depends upon the ratio of field-dependent pressure loss to viscous pressure loss. The volume of fluid exposed to the magnetic field controls the desired MR effect (Spencer et al. 1996) and is known as active fluid volume. The above equations can be manipulated to determine the active fluid volume (V) and is given as

$$V = k\left(\frac{\eta}{\tau_y^2}\right)\lambda W_m \tag{4}$$

where k is a constant and λ is the desired control ratio required to achieve a specified mechanical power, W_m (Spencer et al. 1996). These variables can be defined as follows:

$$k = \left(\frac{12}{c^2}\right); \quad \lambda = \left(\frac{\Delta P_\tau}{\Delta P_\eta}\right); \quad W_m = Q\Delta P_\tau. \tag{5}$$

The Eq. (4) can be further solved to provide constraints and aspect ratios for an efficient use of the MR fluid as

$$wg^2 = \frac{12}{c}\left(\frac{\eta}{\tau_y}\right)\lambda Q. \tag{6}$$

3 Design of the Prototype MR Damper

A magnetic field in the flow path and most importantly in the space/gap between the piston and cylinder needs to be generated to use the MR damper effectively. The magnetic field is applied by a copper coil which is wounded around the grooves of the piston. The leads of the coil are taken out from the cylinder through a concentric hole in the piston rod to give the variable input current to the coil. It, thus, generates variable magnetic field which in turn produces variable damping effect.

The magnetic circuit concept is based on the phenomenon of magnetic conductor permeability which is much greater than the insulative material. It is assumed that the magnetic loop is formed only in the magnetic material, and the magnetic flux leakage is negligible (Zhang et al. 2006). The magnetic permeability of free space is assumed $\mu_o = 4\pi \times 10^{-7}$ H/m. The permeability of the steel is taken as $2{,}000\mu_o$, whereas the MR fluid permeability is taken as $6\mu_o$ (Dixon 2007). According to the structural design, a typical magnetic loop is shown in Fig. 3. In which, h is the clearance between piston and cylinder and is also known as MR

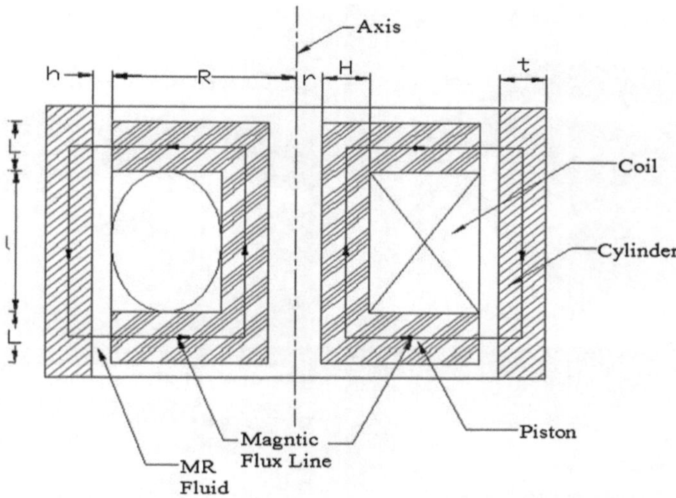

Fig. 3 Typical magnetic loop

Table 1 Various dimensions of the prototype MR damper

Serial no.	Parameter	Dimensions (mm)
1	Pole length (*L*)	28
2	Distance between the poles (ℓ)	22
3	Radius of the piston (*R*)	23
4	Piston rod radius (*r*)	0.6
5	Radial distance from piston rod to coil width (*H*)	0.9
6	Clearance between piston and cylinder (*h*)	0.8
7	Thickness of the cylinder (*t*)	0.8

fluid working clearance. The *R* is radius of the piston, *L* is the pole length, *t* is the thickness of the cylinder, and *r* is the piston rod radius.

Based on the literature survey, different dimensions selected for the MR damper prototype are given in Table 1.

To form an electromagnet, 26 AWG wire (diameter = 0.040386 mm) is wrapped around the grooves of the piston. The electromagnet is wrapped by the maximum possible turns which came out to be 350 turns.

4 Experimental Study of MR Damper Force

A low-carbon steel (EN1A) is selected for the manufacturing of various parts of the damper in order to get good and uniform magnetic field intensity. Based on the dimensions as listed in Table 1, an MR damper is fabricated and is shown in Fig. 4.

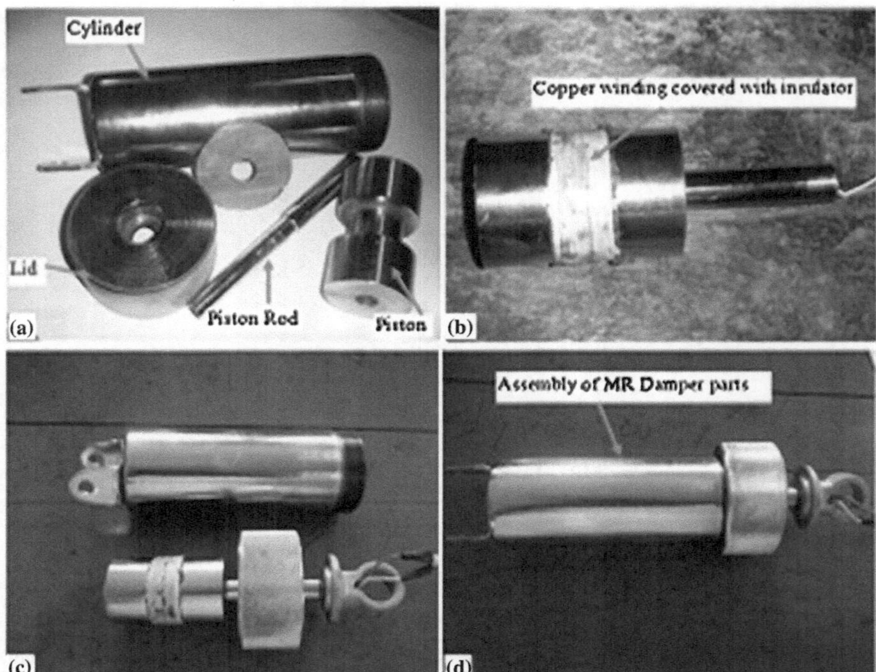

Fig. 4 Fabricated MR damper

The MR fluid (LORD-MRF-122EG) (www.lord.com) is used in the damper for testing its performance in the newly setup vibration control laboratory of the department (Fig. 5). The laboratory consists of an electrodynamic vibration (EDV) shaker, a compatible PC-based data acquisition and instrumentation system, which gives the data in the form of force, velocity and displacement in real-time manner.

The system also includes a PC-controlled mechanical loading frame to mount the dampers. For each testing of the damper, the shaker is driven with a sinusoidal signal at a fixed frequency of 1 Hz for number of cycles. The data are sampled at 256 ks/s. The input current is supplied and varied using Wonder Box kit provided by LORD® Corporation, USA (www.lord.com). The current is varied from 0.1 A to 0.6 A in a step of 0.1 A. The damping force experienced by the damper is sensed by a load cell fixed at the top of the MR damper. The data obtained are shown in Table 2 which presents the variation in damping force with respect to applied current. Figure 6 shows the same date in qualitative manner which shows that the value of current increases the damping force also increases.

Fig. 5 Experimental setup for the testing of MR damper (mounted)

Table 2 Experimental damping force versus applied current of prototype MR damper

Current (A)	Damping force (N)
0.1	350.81
0.2	481.84
0.3	524.31
0.4	564.98
0.5	601.35
0.6	621.92

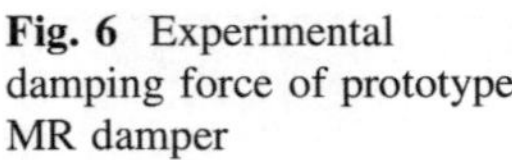

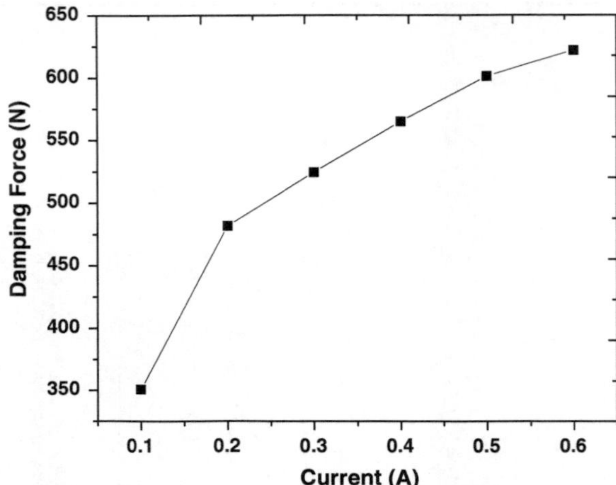

Fig. 6 Experimental damping force of prototype MR damper

5 Conclusion

In this paper, the basic theory behind the MR dampers, its testing, and evaluation procedure is summarized. A prototype of MR damper and its various components are fabricated and tested in the laboratory. The results presented in this paper show the good efficiency of vibration damping with variable damping. The above results show that the user can have a very good vibration control by the MR damper just by varying the input current supplied to the magnetic coil of the damper. In the MR damper, current at low voltage is sufficient to generate the desired magnetic field which can be easily obtainable from the battery of an automotive system. Thus, this type of damper is suitable and useful for an automotive industry.

References

Carlson JD, Jolly MR (2000) MR fluid foam and elastomer devices. Mechatron 10(4–5):555–569

Carlson JD, Spencer BF (1996) Magnetorheological fluid dampers for semi-active seismic control. In: 39th international conference on motion and vibration control, vol 3, pp 35–40

Dixon JC (2007) The shock absorber handbook, 2nd edn. Wiley, Great Britain

Kumar A, Mangal SK (2012) Properties and applications of controllable fluids: a review. Int J Mech Eng Res 2(1):57–66

Kumar A, Mangal SK (2013) Modeling testing and evaluation of magneto-rheological shock absorber. Int J Mech Eng Rob Res 2(1):71–80

Poyner JC (2001) Innovative designs for magneto-rheological dampers MS Thesis, Advanced Vehicle Dynamics Lab, Virginia Polytechnic Institute and State University. VA, Blacksburg

Rabinow J (1948) The magnetic fluid clutch. Trans AIEEE 67:1308–1315

Seval G (2002) Synthesis and properties of magneto rheological fluids Ph.D. thesis, University of Pittsburgh

Spencer BF, Dyke SJ, Sain MK, Carlson JD (1996) Phenomenological model of a magneto-rheological damper. ASCE J Eng Mech 123:1–23

Zhang HH, Liao CR, Chen WM, Huang SL (2006) A magnetic design method of mr fluid dampers and FEM analysis on magnetic saturation. J Intell Mater Syst Struct 17:813–818

Drilling of Glass Fiber-Reinforced Epoxy Laminates with Natural Fillers: Thrust Force Analysis

Vikas Dhawan, Kishore Debnath, Inderdeep Singh and Sehijpal Singh

Abstract An investigation has been carried out to study the effect of spindle speed and feed rate on thrust force generated in drilling glass fiber-reinforced plastics with natural fillers. An effort has been made to use abundantly available natural fibers, namely coconut coir, rice husk, and wheat husk as fillers along with synthetic glass fibers. The drilling experiments have been extensively conducted at six different levels of feed rate and spindle speed using carbide twist drills of 4 mm. Predictive model has been developed using Levenberg–Marquardt algorithm to predict the thrust force with material, spindle speed, and feed rate being the input parameters and thrust force being the output parameter. The results of the predictive model are in close agreement with the actual values. Coefficient of correlation between predicted and experimental values for training and testing data sets is 0.995 and 0.9849, respectively. The mean percentage error in training and testing is found to be 3.175 % and 5.31 %.

Keywords Glass–epoxy laminates · Natural fillers · Thrust force · Neural networks

V. Dhawan (✉) · S. Singh
Department of Mechanical Engineering, Guru Nanak Dev Engineering College, Ludhiana 141006, Punjab, India
e-mail: vikas251999@gmail.com

S. Singh
e-mail: mech@gndec.ac.in

V. Dhawan
Department of Mechanical Engineering, Maharaja Agrasen University, Baddi 174103, Himachal Pradesh, India

K. Debnath · I. Singh
Department of Mechanical and Industrial Engineering, Indian Institute of Technology, Roorkee 247667, Uttarakhand, India
e-mail: debnath.iitr@gmail.com

I. Singh
e-mail: dr.inderdeep@gmail.com

S. S. Khangura et al. (eds.), *Proceedings of the International Conference on Research and Innovations in Mechanical Engineering*, Lecture Notes in Mechanical Engineering, DOI: 10.1007/978-81-322-1859-3_10,

1 Introduction

Manufacturing of composites can be broadly classified as primary and secondary manufacturing. Although in most of the cases, primary manufacturing leads to near net-shaped products, sometimes the product has to be made in parts due to the design complexity involved. Thus, drilling of holes becomes unavoidable in order to enable fastening of the components for final assembly. Due to the anisotropic nature of FRPs and the constituent materials holding down to their individual properties, drilling of FRPs results in drilling-induced damage. It has been found by various researchers worldwide that thrust forces produced during drilling of FRPs directly affect the quality of the machined hole. Chen (1997) found a step linear relationship between the delamination factor and average thrust force for drilling unidirectional CFRP composite laminates. Ho-Cheng and Dharan (1990) performed the analysis of delamination and stressed the need of study of critical thrust force above which the damage occurs. Jain and Yang (1993) developed expressions for critical thrust and critical feed rate for unidirectional laminates. Thrust generated during the drilling operation was identified as the root cause for occurrence of delamination, and models for critical thrust force were developed using the concept of linear elastic fracture mechanics (Jain and Yang 1994). Mathew et al. (1999a, b) did an experimental investigation to study the effect of the geometry of a trepanning tool on thrust and torque generated during the drilling of unidirectional glass fiber-reinforced epoxy plastic (UD-GFREP) laminates. Various mathematical models to predict critical thrust force were developed by Zhang et al. (2001); Hocheng et al. (2003); Tsao and Hocheng (2005, 2008a), Lachaud et al. (2001), Langella et al. (2005) suggested a mechanistic model for predicting thrust and torque during drilling of GFRPs using traditional twist drills. Tsao and Hocheng (2008b) found a correlation between thrust force and cutting parameters using multi-variable linear regression and compared with the experimental results. Tsao (2008a) concluded that the feed rate and spindle speed were the main parameters that influenced the thrust force and delamination. The effect of diameter ratio was relatively insignificant. Fernandesa and Cook (2006) developed a mathematical model of the maximum thrust force and torque during drilling of carbon fiber using a 'one-shot' drill bit. Khashaba (2004) found that the delamination size was increased with increasing feed rate as a result of increasing thrust force. Khashaba et al. (2007) found that fiber volume fraction is directly proportional with thrust force and torque in drilling GFRP composites (Dharan and Won 2000). Dharan and Won(2000) conducted experimental studies on carbon fiber-reinforced composite laminates to obtain thrust force and torque responses for a wide range of feed rates in high-rate drilling using carbide-tipped twist drills. Durao et al. (2008a, b) stated that higher cutting speeds and high feed rates always caused higher values of maximum thrust force and delamination. Several researchers have used artificial neural networks to develop predictive models for drilling of FRPs which are more generic in nature. The benefit of using artificial neural networks is that there is no need to know the nature of the input parameters to solve complex and non-linear problems, and there is no constraint on number of input parameters being considered. A neural network control scheme was implemented by Stone and Krishnamurthy (1996) to minimize the delamination. An artificial neural network model was developed by Karnik et al. (2008) for delamination analysis in high-speed drilling with spindle speed, feed rate, and point angle as the affecting parameters. Latha and Senthilkumar (2010) and Mishra et al. (2010) used neural network based on back-propagation (BP) algorithm for predicting delamination factor. Radial basis function network (RBFN) was found to be more precise than response surface methodology (RSM) to predict thrust force for a core-center drill (Tsao 2008b). Athijayamani et al. (2010) found ANN models to be better than regression models.

In the present research endeavor, thrust force generated during drilling of GFRPs with and without natural fillers has been studied. Predictive model has been developed using artificial neural networks to predict thrust force generated during drilling of GFRPs with and without natural filler with spindle speed, feed rate, and material being the input parameters.

2 Levenberg–Marquardt Algorithm

The Levenberg–Marquardt algorithm uses batch learning process and is basically a Hessian-based algorithm for non-linear least squares optimization (RangaSuri et al. 2002). Levenberg–Marquardt algorithm allows the network to learn more subtle features of a complicated mapping. The training process converges quickly as the solution is approached, because the Hessian does not vanish at the solution. For neural network training, the objective function is the error function of the type shown in Eq. 11 in Chap. 5.

$$e(w) = E_P = \frac{1}{2}\sum_{p=1}^{P}\sum_{k=1}^{K}(d_{k,p} - O_{k,p})^2 \tag{1}$$

where, k is the number of outputs for P number of instances. $O_{k,p}$ is the actual output, and $d_{k,p}$ is the desired output at the output neuron k for the input p. P is the total number of training patterns, K represents the total number of neurons in the output layer of the network, and w represents the weights and biases of the network. Suppose there are I inputs in the input layer and Y neurons in the hidden layer, then N which is the total number of weights is given by equation.

$$N = I \times Y + Y \times K + Y + K \tag{2}$$

The steps involved in training of any artificial neural network in batch mode using Levenberg–Marquardt Algorithm are (Yu and Wilamowski 2011)

1. Present all inputs to the network and compute the corresponding network outputs and errors. Compute the mean square error over all inputs as in Eq. 11 in Chap. 5.
2. Compute the Jacobian matrix, $J(w)$, as shown in Eq. 4 where w represents the weights and biases of the network.
3. Calculate weight update as per Levenberg–Marquardt weight update Eq. 3.

$$\Delta w = [JT(w)J(w) + \mu I]^{-1} JTe(w) \tag{3}$$

where, e is a vector of errors of size PK, and μ is known as damping factor or training parameter.

4. Recompute the error using $w + \Delta w$. If error is smaller than that computed in step 1, then reduce the training parameter $\mu$by μ^{-1}, then update weights, $w = w + \Delta w$, and go to step 1. Typically, μ^{-1} is set to 0.1.

$$J(w) = \begin{bmatrix} \frac{\partial e_{1,1}}{\partial w_1} \frac{\partial e_{1,1}}{\partial w_2} \cdots & \frac{\partial e_{1,1}}{\partial w_N} \\ \frac{\partial e_{1,2}}{\partial w_1} \frac{\partial e_{1,2}}{\partial w_2} \cdots & \frac{\partial e_{1,2}}{\partial w_N} \\ \\ \frac{\partial e_{1,K}}{\partial w_1} \frac{\partial e_{1,K}}{\partial w_2} \cdots & \frac{\partial e_{1,K}}{\partial w_N} \\ \\ \frac{\partial e_{P,1}}{\partial w_1} \frac{\partial e_{P,1}}{\partial w_2} \cdots & \frac{\partial e_{P,1}}{\partial w_N} \\ \frac{\partial e_{P,2}}{\partial w_1} \frac{\partial e_{P,2}}{\partial w_2} \cdots & \frac{\partial e_{P,2}}{\partial w_N} \\ \frac{\partial e_{P,K}}{\partial w_1} \frac{\partial e_{P,K}}{\partial w_2} \cdots & \frac{\partial e_{P,K}}{\partial w_N} \end{bmatrix} \tag{4}$$

5. If the error is not reduced, do not change or update the weights but increase μ by μ^{+} and go back to step 3. Typically, μ^{+} is set to 10.
6. Iteration stops if error reduces to a particular value or training parameter reaches a certain value or numbers of iterations reach the maximum value.

3 Experimental Procedure

Composite laminates of 4 mm thickness were prepared using six layers of boron-free EC-R glass mats, epoxy resin, and natural fillers. Rice husk, wheat husk, and coconut coir were used as filler materials. Composite laminates were prepared by conventional hand lay-up technique in chrome-plated mild steel mold. Araldite epoxy resin LY556 (density 1.12 g/cm^3) and hardener HY 951 were mixed and stirred mechanically in a ratio of 10:1 by weight. The natural fillers were used in a proportion of 5 % of the weight of glass fibers. After the resin was thoroughly applied and fillers evenly spread between layers, the complete mold was placed in a press, and a compression load of 15 tones was applied. The mold was left for 10 h at room temperature to complete the curing process. The same technique was used to fabricate the other laminates. In all, four different specimens were fabricated. Three of the GFR epoxy composites (GFREC) consisted of different natural biofillers, that is, rice husk fillers, wheat husk fillers, and coconut coir fillers. The fourth laminate had a glass fiber alone as reinforcement (without filler).

Drilling experiments were conducted on all the four different composites using a carbide twist drill on a radial drilling machine at spindle speeds of 90, 224, 450, 900, 1,800, and 2,800 RPM and feed rates of 0.03, 0.05, 0.08, 0.12, 0.19, and 0.3 mm/rev. The drilling forces were recorded using the online Kistler make four-component drill dynamometer (type 9,272). The dynamometer was connected to the personal computer via analog to digital converter card. Hence, the force signals were recorded from the personal computer using standard data acquisition software, Dynoware. An average of three readings was taken to record the maximum thrust force. In all 144, thrust force readings were collected with 36 thrust force readings for each composite fabricated.

4 Predictive Modeling Using Levenberg–Marquardt Algorithm

An artificial neural network was used to develop a predictive model to predict thrust force. Material, spindle speed, and feed rate were taken as the input parameters and thrust force being the output parameter. A code based on Levenberg–Marquardt back-propagation algorithm was written in Matlab version R2008 b in order to develop the ANN model. A three-layer network with one input layer, one hidden layer, and one output layer was used for the present study. The number of input-layer neurons was three, the same as the number of input variables (material, spindle speed, and feed rate). The output layer consisted of one neuron corresponding to one output variable, that is, thrust force. It has to be pointed out here that the number of hidden neurons depends both on input vector size and number of input classifications. Too few neurons can lead to under-fitting, whereas too many neurons can contribute to over-fitting. Hence, the number of neurons in

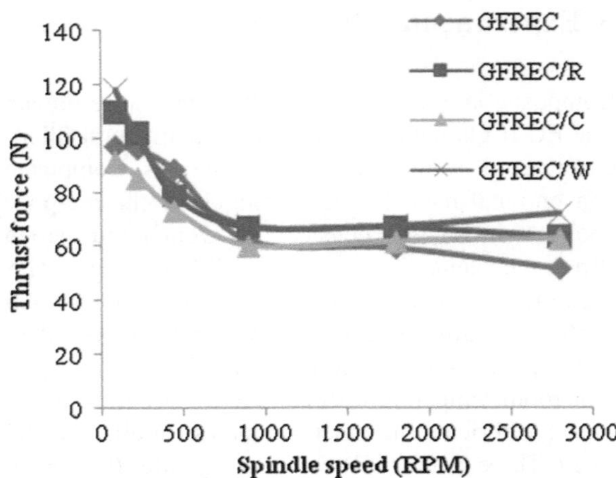

Fig. 1 Variation in thrust force with spindle speed at a feed rate of 0.03 mm/rev

hidden layer has to be varied to find an optimal solution. In the present case, number of neurons used in the hidden layer are 18. Tansigmoidal function was used in first layer, and pure linear function was used as an activation function for second layer to get output. The default values for damping factor were used. The default values of μ^{-1} is 0.1, μ^{+} is 10, initial value of μ is kept as 0.001, and maximum value of μ is set to 10^{10}. The number of iterations used was 25, and the initial weights were randomly chosen and were kept below one both in input and hidden layer. First, randomly 121 data sets were chosen and were used to train the algorithm, and the rest 23 data sets were used to test the program.

5 Results and Discussion

5.1 Thrust Force Response

Thrust force signals during drilling were acquired using a data acquisition system and were processed for analysis. The maximum thrust force was recorded. It was found that at low feed rates, the thrust force was found to decrease with spindle speed for all the composites as shown in Fig. 1. At high feed rates of 0.12 mm/rev and above, more abrupt variations in thrust force were recorded with increase in spindle speed for GFREC/R and GFREC/W as shown in Fig. 2, whereas for GFREC and GFREC/C, a decrease in thrust force was observed till 900 RPM, and then a gradual increase in thrust force was recorded with further increase in spindle speed. It was observed that thrust force recorded while drilling GFREC with coconut coir fillers (GFREC/C) was less or comparable to the thrust force recorded while drilling GFREC without fillers which suggests that delamination is less in GFREC with coconut fillers.

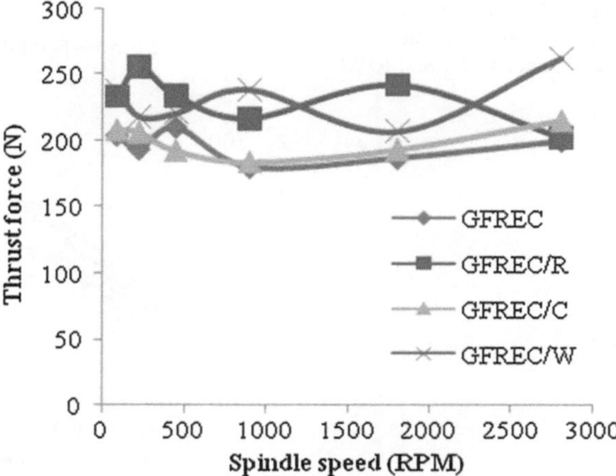

Fig. 2 Variation in thrust force with spindle speed at a feed rate of 0.3 mm/rev

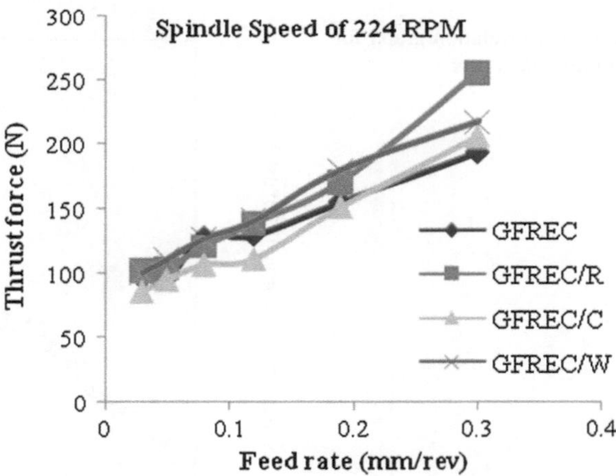

Fig. 3 Variation in thrust force with feed rate at a spindle speed of 224 RPM

It was observed that there was a steep increase in thrust force with increase in feed rate at all spindle speeds as shown in Figs. 3 and 4. The increase in thrust force with feed rate is almost linear, and the increase is such that the thrust force at 0.3 mm/rev feed rate is at least double or even more than that of the thrust force at 0.03 mm/rev for any spindle speed.

5.2 *Predictive Model for Thrust Force*

The data obtained for drilling-induced damage, thrust force, and torque in case of drilling of FRPs present a very complex input and output mapping. Hence, Levenberg–Marquardt algorithm was used. The coefficient of correlation between

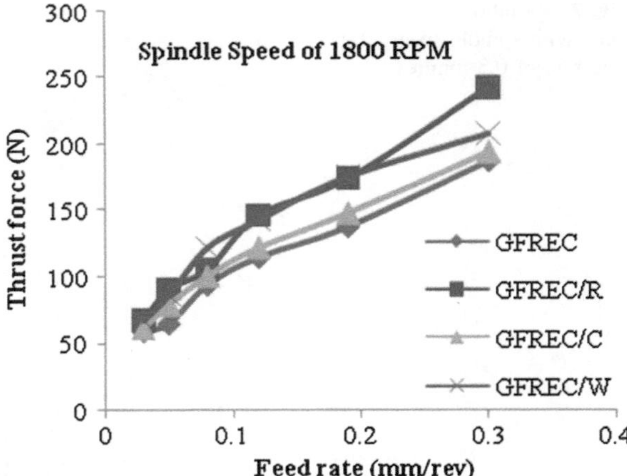

Fig. 4 Variation in thrust force with feed rate at a spindle speed of 1,800 RPM

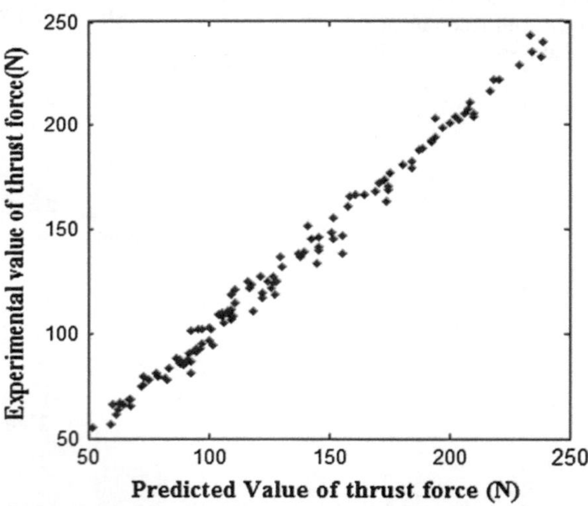

Fig. 5 Correlation graph for training data sets

predicted values and training data outputs was found to be 0.995, whereas the coefficient of correlation between predicted values and test data outputs was found to be 0.9849. The values of coefficient of correlation clearly indicate the fitness of the ANN models. The error was calculated as simple percentage error, and the mean percentage error during training was found to be 3.175 %, and the mean percentage error in testing was found to be 5.31 %. Figure 5 shows the comparison of the predictive outputs by the thrust force model with the experimental values for training, and the comparison of predicted values with the test data outputs during testing are shown in Fig. 6. The predicted values are in close agreement with the experimental values.

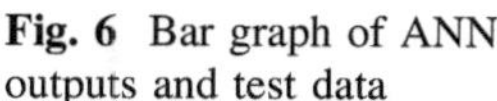

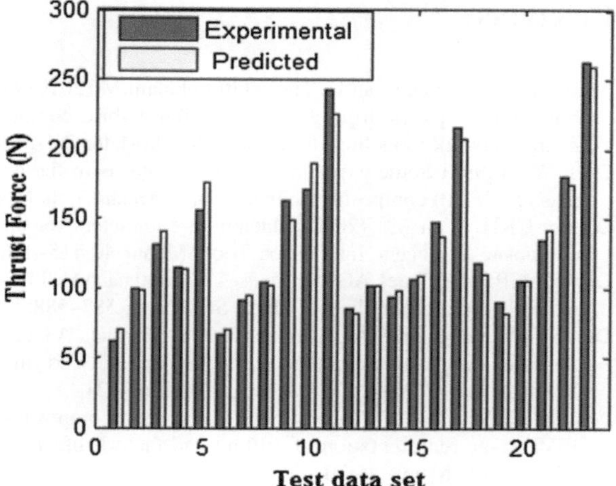

Fig. 6 Bar graph of ANN outputs and test data

6 Conclusions

Hybrid glass fiber-reinforced epoxy composites were fabricated using rice husk, coconut coir, and wheat husk as fillers along with plain glass fiber-reinforced composites, and drilling experiments were conducted using 4-mm carbide twist drill. Thrust forces generated during drilling were recorded, and a predictive model was developed using artificial neural networks to predict thrust force generated during drilling experiments. The following conclusions were drawn from the study.

1. At low feed rates, thrust force decreases with increase in spindle speed, but as such, there is no definite trend at high feed rates. The variation with increase in spindle speed at high feed rates (0.12 mm/rev or above) is more abrupt for composites with rice husk and wheat husk fillers.
2. Low feed rate and high spindle speed results in less thrust force for all the four composites fabricated.
3. Coconut coir fillers can be used to develop hybrid composites as the thrust force generated during drilling of glass fiber-reinforced composites with coconut coir as fillers is comparable or even less to the thrust force generated in drilling simple glass fiber-reinforced composites for all spindle speeds and feed rates.
4. Thrust force increases almost linearly with increase in feed rate.
5. A predictive model has been developed on the basis of artificial neural networks for predicting thrust force generated during drilling of GFREC, GFREC/R, GFREC/C, and GFREC/W.
6. The predicted values of thrust force for training data and testing data using the predictive model were close to the actual values. Coefficient of correlation between predicted and experimental values for training and testing data sets is 0.995 and 0.9849, respectively.
7. Similar models can be developed to predict damage and torque.

References

Athijayamani A, Natarajan U, Thiruchitrambalam M (2010) Prediction and comparison of thrust force and torque in drilling of natural fibre hybrid composite using regression and artificial neural network modeling. Int J Mach Mach Mater 8(1/2):131–145

Chen WC (1997) Some experimental investigations in the drilling of s carbon fibre-reinforced plastic (CFRP) composite laminates. Int J Mach Tools Manufact 37(8):1097–1108

Dharan CKH, Won MS (2000) Machining parameters for an intelligent machining system for composite laminates. Int J Mach Tools Manuf 40:415–426

Durao LMP, Magalhaes AG, Marques AT, Baptista AM, Figueiredo M (2008a) Drilling of fibre reinforced plastic laminates. Mater Sci Forum 587–588:706–710

Durao LMP, Magalhaes AG, Tavares JMRS, Marques AT (2008b) Analyzing objects in images for estimating the delamination influence on load carrying capacity of composite laminates. Electron Lett Comput Vis Image Anal 7(2):11–21

Fernandesa M, Cook C (2006) Drilling of carbon composites using a one shot drill bit. Part I: FIVE stage representation of drilling and factors affecting maximum force and torque. Int J Mach Tools Manuf 46:70–75

Ho-Cheng H, Dharan CKH (1990) Delamination during drilling in composite laminates. J Eng Ind 112:236–239

Hocheng H, Tsao CC (2003) Comprehensive analysis of delamination in drilling of composite materials with various drill bits. J Mater Process Technol 140(1–3):335–339

Jain S, Yang DCH (1993) Effects of feed rate and chisel edge on delamination in composite drilling. Trans ASME J Eng Ind 115:398–405

Jain S, Yang DCH (1994) Delamination-free drilling of composite laminates. J Eng Ind 116:475–481

Karnik SR, Gaitonde VN, Rubio JC, Correia AE, Abrao AM, Davim JP (2008) Delamination analysis in high speed drilling of carbon fibre reinforced plastics (CFRP) using artificial neural network model. Mater Des 29:1768–1776

Khashaba UA (2004) Delamination in drilling GFR-thermoset composites. Compos Struct 63:313–327

Khashaba UA, Seif MA, Elhamid MA (2007) Drilling analysis of chopped composites. Composites: Part A 38:61–70

Lachaud F, Piquet R, Collombet F, Surcin L (2001) Drilling of composite structures. Compos Struct 52:511–516

Langella A, Nele L, Maio A (2005) A torque and thrust prediction model for drilling of composite materials. Composites: Part A 36:83–93

Latha B, Senthilkumar VS (2010) Application of artificial intelligence for the prediction of delamination in drilling GFRP composites. Int J Precis Technol 1(3/4):314–330

Mathew J, Ramakrishnan N, Naik NK (1999a) Investigations into the effect of geometry of trepanning tool on thrust and torque during drilling of GFRP composites. J Mater Process Technol 91(1–3):1–11

Mathew J, Ramakrishnan N, Naik NK (1999b) Trepanning on uni-directional composites. Compos Part A Appl Sci Manuf 30(8):951–959

Mishra R, Malik J, Singh I (2010) Prediction of drilling-induced damage in unidirectional glass-fibre-reinforced plastic laminates using an artificial neural network. Proc Inst Mech Eng Part B J Eng Manuf 224(5):733–738

RangaSuri NNR, Deodhare D, Nagabhushan P (2002) Parallel Levenberg–Marquardt-based neural network training on Linux clusters—a case study. In: Linux clusters, ICVGIP 2002, 3rd Indian conference on computer vision, graphics and image processing, Ahmadabad

Stone R, Krishnamurthy K (1996) A neural network thrust force controller to minimize delamination during drilling of graphite-epoxy laminates. Int J Mach Tools Manure 36(9):985–1003

Tsao CC (2008a) Thrust force and delamination of core-saw drill during drilling of carbon fibre reinforced plastics (CFRP). The Int J Adv Manuf Technol 37(1–2):23–28

Tsao CC (2008b) Comparison between response surface methodology and radial basis function network for core-center drill in drilling composite materials. Int J Adv Manuf Technol 37:1061–1068

Tsao CC, Hocheng H (2005) Effects of exit back-up on delamination in drilling composite materials using a saw drill and a core drill. Int J Mach Tools Manuf 45(11):1261–1270

Tsao CC, Hocheng H (2008a) Analysis of delamination in drilling composite materials by core-saw drill. Int J Mater Prod Technol 32(2/3):188–201

Tsao CC, Hocheng H (2008b) Evaluation of thrust force and surface roughness in drilling composite material using Taguchi analysis and neural network. J Mater Process Technol 203:342–348

Yu H, Wilamowski BM (2011) Levenberg–Marquardt training, industrial electronics handbook, vol 5—Intelligent systems, 2nd edn. CRC Press, Taylor & Francis Group, Boca Raton, pp 12-1–12-16 (Chapter 12)

Zhang LB, Wang LJ, Liu XY (2001) A mechanical model for predicting critical thrust forces in drilling composite laminates. Proc Inst Mech Eng Part B J Eng Manuf 215(2):135–146

Assessment of Strength Evaluation and Structure Stability of Building Destroyed in Fire by Using Techniques of Non-destructive Testing

Inderpreet Kaur, Chander Sheikhar Singla and Amandeep Singh Gill

Abstract The two-storey industrial building with area of 41,000 sq. ft on each floor was framed structure. The machinery in the building was used for the manufacturing of cotton and synthetic fabric. Concrete of M20 grade and steel of Fe 415 grade were used for construction. A fire incident was occurred for 42–50 h after one year of construction. Visual examination of columns, slabs and beams of ground floor showed cracks on the concrete surface. At many places, concrete had fallen off, and reinforcement was exposed. At few places, colour of concrete had also been changed from its natural grey to reddish brown which indicated that the temperature during fire might had crossed 600 °C. In some of the circular columns, the thickness was reduced by about 25 % and fallen off concrete in between circular columns are a resulted in reducing the thickness of the slab by about 40 %. Due to inadequate circumstances for core extraction at some sections, decision of non-destructive testing like ultrasonic pulse velocity (UPV), rebound hammer (RH) were used to evaluate the elements. The quality of concrete was assessed by UPV. In situ strength of R.C.C. members, integrity and homogeneity of concrete help to evolve repair/strengthening/rehabilitation measures. Load test was also conducted to ascertain the actual load-carrying capacity of structural members. It was concluded that due to fire and excessive increase in temperature, the column concrete surface gets hardened, causing increase in value of rebound number. Further, results of concrete core test are more reliable when compared with rebound hammer test.

I. Kaur (✉) · C. S. Singla · A. S. Gill
Department of Civil Engineering, Guru Nanak Dev Engineering College, Ludhiana 141006 Punjab, India
e-mail: inderdhindsa@gmail.com

C. S. Singla
e-mail: singlagne@gmail.com

A. S. Gill
e-mail: engg.amangill@yahoo.com

S. S. Khangura et al. (eds.), *Proceedings of the International Conference on Research and Innovations in Mechanical Engineering*, Lecture Notes in Mechanical Engineering, DOI: 10.1007/978-81-322-1859-3_11,

Keywords Concrete core test · Fire damage · Load test · Rebound Hammer test · Strength · Ultrasonic testing

1 Introduction

Non-destructive testing (NDT) is a wide group of analysis techniques used in science and industry to evaluate the properties of a material, component or system without causing damage (Cartz 1995). The terms non-destructive examination (NDE), non-destructive inspection (NDI) and non-destructive evaluation (NDE) are also commonly used to describe this technology (Reis and Dilek 2012). NDT is commonly used in forensic, mechanical, electrical, civil, aeronautical engineering. As NDT does not permanently alter the article being inspected, it is a highly valuable technique that can save both money and time in product evaluation, troubleshooting and research. NDT is used in a variety of settings that covers a wide range of industrial activity, with new NDT methods and applications, being continuously developed. NDT is divided into various methods of non-destructive testing, each based on a particular scientific principle (http://en.wikipedia.org/wiki/nondestructive_testing) like Acoustic emission testing, Blue Etch Anodic, dye penetrant inspection, liquid penetrant testing, electromagnetic testing, ellipsometry, guided wave testing, hardness testing, impulsive excitation technique, laser testing, low coherence interferometry (Dufour et al. 2005; Losert 2009), metallographic replicas (BS ISO 3057 1998; ASTM 1351 2006), optical microscopy, rebound hammer testing, ultrasonic testing, vibration analysis and visual inspection.

The non-destructive and laboratory testing findings are used in determination of extent of heat damage to concrete and their potential effects. The in situ NDT phase consisted of locating affected areas using ultrasonic pulse velocity (UPV) through concrete members in the vicinity of the fire. Subsequently, cores are removed, and Young's modulus of 25-mm (1 in.)-thick concrete disk is determined by utilizing non-destructive measurement of resonant frequency (Hellier 2003). Determination of Young's modulus at small depth increments permitted assessment of the heat-induced damage front into the concrete. Based on the extent of damage identified, repairs are performed on the affected members.

Fire represents a transfer of energy from a stable condition to a transient condition as combustion occurs; examples are the burning of warehouse contents, office furniture, books, filing cabinet contents or other material (Tide 1998). The heat associated with fires may vaporize trapped concrete pore water. The lack of continuous voids for pressure relief creates internal tensile stresses that are relieved by cracks and spalls extending to the surface (Chiang and Tsai 2003). The spalling may be explosive in higher-strength concretes. Additionally, severe heat may cause chemical changes that lead to micro-cracking (visible only under magnification) and loss of strength and integrity.

All structures subjected to fire should be evaluated in a systematic manner to determine the extent, if any, of required repairs. The intensity and duration of the fire can be estimated by observing the collateral damage: a variety of testing methods and tools are available to evaluate the effects of the fire on both the materials and structural elements. These evaluations, combined with an engineering analysis, allow effective and economical repair details to be developed and installed as needed (Narendra et al. 2008).

2 Case Study

The data were collected from testing and consultancy cell of Guru Nanak Dev Engineering College, Ludhiana.

Industrial building under testing belonged to Jain Uday Hosiery Private Limited, Ludhiana. The main structural system of the fire-affected building was frame work of beams and columns as the main load-carrying elements. It was a two-storey industrial building with an area of 41,000 sq. ft of each floor. In plan, the building was divided into two parts with an expansion joint throughout the length. In plan, the ground floor and first floor were exactly same. The machinery in the building was used for the manufacturing of cotton and synthetic fabric and in the building, concrete of M20 grade and steel of Fe 415 grade were used. The fire incidence had occurred at night about a year after construction, and it continued for about 42 h. In some portions, the fire continued for more than 50 h. A visit to the site by engineers was aimed at deciding the location and types of tests to be conducted to assess the extent of damage caused by the fire. Following this, extensive testing work was carried out to check the strength and structural stability of the building.

2.1 Observations Based on Visual Inspection

2.1.1 Ground Floor

Visual examination of columns, slab and beams of ground floor showed cracks on the concrete surface. At many places, concrete fallen off and reinforcement was exposed. At few places, colour of concrete had also been changed from its natural grey to reddish brown which indicated that the temperature during fire might had crossed 600 °C. The building area was divided into two parts: (a) front portion with square/rectangular columns and (b) back portion with mostly circular columns. The damage during fire was more in the back area than in the front area. In some of the circular columns of the back area, the thickness was reduced by about 25 %. In the circular column area, concrete had fallen off at many places reducing the thickness of the slab by about 40 %. Steel was exposed in large portions of the slab. In some of the circular columns, cracks seem to be penetrating up to a large depth. Deflection in slab and beams was also excessive in the circular column area.

Concrete cores were taken out from some of the columns and slab to know the exact condition of concrete. Visual examination of cores and the places from where cores were taken out indicated the presence of cracks throughout the depth of the concrete members. At some places, it was not possible to take out cores as concrete was so soft that was breaking into small pieces during core extraction. Damage to the front portion of the building was less compared with back portion, although at some locations of the slab, the concrete had fallen off exposing the steel.

2.1.2 First Floor

On the first floor, damage due to fire was more in the front portion than in the back portion with circular columns. A portion of first floor adjoining the stair case and with nine columns was severely damaged during fire. The reason for this could be lesser roof height and storage of highly inflammable polyester material. Thickness of roof slab at certain portions was reduced by more than 50 %, and steel was exposed and had lost its elasticity. Visual examination of columns, slab and beams of first floor showed cracks on the surface. Cracks in the floor slab seemed to be penetrating up to a large depth. At many places on the roof slab, concrete had fallen off and reinforcement was exposed. The colour of concrete was also changed to reddish brown from its natural colour of grey. The back portion of the first floor was relatively less affected during fire. Columns, beams and roof slab of this portion were relatively in good condition compared with front portion. The concrete cores were taken out from some of the columns and slab of second floor to know the exact condition of concrete. Visual examination of cores and at places from where cores were taken out indicated the presence of cracks throughout the depth of the floor slab, while the columns and roof slab were in good condition of back portion. At some places of the floor slab, it was not possible to take out cores as concrete was so soft that was getting broken into small pieces during core extraction.

2.2 Non-destructive Testing of the Building

It is possible to determine various engineering properties of concrete like strength, quality, etc. by using NDT. The assessment of existing structure for damage, its repair, strengthening and rehabilitation areas which extensively postulates use of such test methods. In addition to check the strength and quality of concrete, NDT also gives an idea about the estimate of repairs, urgent and non-urgent repairs, etc. NDT provides invaluable means of both qualitative and quantitative determination of the extent and type of defects in R.C.C. members. It also helps in determining the mechanism of deterioration in the structure.

Purpose of NDT

1. To evaluate integrity and homogeneity of concrete for all accessible R.C.C. members.
2. To identify areas of concrete which were defective and not apparent on visual inspection.
3. To help evolve repair/strengthening/rehabilitation measures.

Types of NDT used in the building

1. Ultrasonic pulse velocity (UPV)
2. Rebound hammer (RH)
3. Concrete core test (partially destructive test method)

In addition to above, limited use of rebar locator was made. Basically, it was used to decide about the correct locations for UPV test points. For ground-floor columns, wherever possible or required, reinforcement was located using rebar locater, and UPV test points were selected such that reinforcement does not lie in or close to the direct path between transducers.

The UPV test was used with the following objectives:

(a) To judge homogeneity of concrete
(b) To judge the quality of concrete
(c) For qualitative comparison of one element of concrete in relation to another.

The RH test was used for:

(a) Assessing the most likely cube compressive strength of concrete.
(b) Qualitative comparison of one element of concrete in relation to another.

Examination of cores extracted from hardened concrete also enables visual inspection of interior regions of the structural member. Thus, core drilling and testing were considered to arrive at more accurate estimation of strength and to visually inspect the interior region of structural member. An attempt was made so that majority of structural elements are covered by any one of the test method, namely UPV, RH and concrete core test. Concrete core test shall give more exact idea about actual in situ condition of concrete, including compressive strength and thereby suggesting overall condition of the structure, due to certain restraints like accessibility, instrument handling, availability of proper surface for testing, time, finance, etc.

2.3 Ultrasonic Pulse Velocity Test

In this test, ultrasonic pulse is generated by electro-acoustical transducer. It undergoes multiple reflections at the boundaries of the different material phases within the concrete. A complex system of stress waves is developed which includes longitudinal (compressional), shear (transverse) and surface (rayleigh) waves. The receiving transducer detects the onset of the longitudinal waves, which is the fastest. Because the velocity of the pulses is almost independent of the geometry of the material through which they pass and depends only on its elastic properties, pulse velocity method is a convenient technique for investigating structural concrete. The underlying principle of assessing the quality of concrete is that comparatively higher velocities are obtained when the quality of concrete in terms of density, homogeneity and uniformity is good. In case of poorer quality, lower velocities are obtained. If there is a crack, void or flaw inside the concrete which comes in the way of transmission of the pulses, the pulse strength is attenuated and it passes around the discontinuity, thereby making the path length longer. Consequently, lower velocities are obtained. The actual pulse velocity obtained depends primarily upon the materials and mix proportions of concrete. Density and modulus of elasticity of aggregate also significantly affect the pulse velocity.

During testing, the ultrasonic pulse is produced by the transducer, which is held in contact with one surface of the concrete member under test. After traversing a known path length (*L*) in the concrete, the pulse of vibrations is converted into an electrical signal by the second transducer held in contact with the other surface of the concrete member, and an electronic timing circuit enables the transit time (*T*) of the pulse to be measured. The pulse velocity (*V*) is given by:

$$V = \frac{L}{T}. \tag{1}$$

Once the ultrasonic pulse impinges on the surface of the material, the maximum energy is propagated at right angles to the face of the transmitting transducer, and best results are, therefore, obtained when the receiving transducer is placed on the opposite face of the concrete member (direct transmission or cross-probing). The quality of concrete in terms of uniformity, incidence or absence of internal flaws, cracks and segregation, etc., indicative of the level of workmanship employed, can be assessed using the guidelines give in the Table 1, which have been evolved for characterizing the quality of concrete in structures in terms of the UPV.

UPV test point locations were decided as follows:

For columns: Within middle half of the height such that path length of ultrasonic wave was parallel to shorter dimension of the member. Wherever possible or required, due care was taken for locating reinforcement using rebar locator, and UPV test points were located such that the steel bar did not lie in or close to the direct path of wave propagation.

For beams: Within middle half of the span, such that path length of ultrasonic wave was parallel to width of the beam.

For slabs: All slabs could not be scanned for UPV due to non-accessibility and limiting length of instrument cable. Wherever possible, readings were taken near external boundary of the slab panel.

For all the elements mentioned above, readings were taken by direct method (cross-probing). Due care was taken for aligning transducers for shortest path length. Perfect contact between test surface and transducers was assured by use of grease.

Table 1 Velocity criterion for concrete quality grading (IS: 13311 Part I)

S. no.	Pulse velocity by cross probing (km/s)	Concrete quality grading
1	Above 4.5	Excellent
2	3.5–4.5	Good
3	3.0–3.5	Medium
4	Below 3.0	Doubtful

2.4 Rebound Hammer (RH) Test

The RH consists of a spring-controlled mass that slides on a plunger within a tabular housing. When the plunger of rebound hammer is pressed against the surface of concrete, the spring-controlled mass rebound depends upon the surface hardness of concrete, the surface hardness, and therefore, the rebound is taken to be related to the compressive strength of the concrete. The rebound is read off along a graduated scale and is designated as the rebound number. The rebound hammer method provides a convenient and rapid indication of most likely cube compressive strength of concrete.

For columns and beams, rebound hammer test locations were selected approximately within middle half of member height. Plaster was removed, and original concrete surface was exposed. The surface was rubbed and smoothened by carborundum stone, and readings were taken with horizontal position of hammer. Readings were taken on larger face of the column and beam. Similar procedure was followed for slabs, except that RH position was kept vertically upwards.

2.5 Concrete Core Test

The examination and compressive strength of cores drilled from hardened concrete enable the visual inspection of interior regions of the structural member along with estimation of actual in situ concrete strength. Use of rebar locator was made to decide correct location for drilling and extracting cores such that reinforcement bars were avoided. Columns—being prominent structural element—were selected randomly for this test, so as to represent both, good appearing and fire-affected concrete. Majority of samples were taken from fire-affected area. Cores were extracted from vertical.

2.6 Load Test

Load test is generally carried out to ascertain the actual load-carrying capacity of the structural members like beams and columns to verify the extent of direct or indirect damage caused by fire. There were certain portions of the building, where minor or no cracks were present on the surface, and direct affect of fire was negligible as compared to other parts. Other NDT tests also indicate negligible damage during fire.

Load test was conducted at four places on the slab, three on ground-floor slab and one on first-floor slab. The tests were conducted as specified in IS 456: 2006, in which the slab portion was subjected to a load equal to full dead load plus the 1.25 times the imposed load specified for industrial building for a period of 24 h. Deflection was recorded with full loading. Load was kept in position for 24 h, and deflection was again recorded. After recording the deflection, the imposed load was removed, and recovery in deflection was recorded.

3 Observations and Results

Data of the observations and results of all tests performed on 142 columns, 38 beams and 28 slab panels of Jain Uday Hosiery Private Lt., Ludhiana, are presented in Tables 2, 3, 4, 5, 6, 7, 8.

3.1 Load Test

Load test was conducted on three slab panels of ground-floor roof and one beam span of first-floor roof slab. All the slab panels were loaded with a uniformly distributed load of 12.5 kN/m^2 which was 1.25 times the imposed load specified in IS 456 for industrial buildings. A dial gauge was installed on the underside of the roof slab before applying load on the roof slab. Deflection was measured after the application of full load. Load was applied by putting sand bags on the specified area of the slab. Deflection was also measured after applying load for 24 h. After the removal of the load recovery in deflection was also measured (Table 9).

4 Discussion and Recommendation

1. Almost all possible columns within fire-affected area, and few columns outside this area were tested for both rebound number and UPV evaluation. Out of the 142 columns tested by rebound hammer, about 97 locations suggested strength less than 20 N/mm^2 (28 days cube compressive strength of M20 grade concrete). Evaluation of rebound member is directly related to hardness of concrete surface. Because of fire and great increase in temperature, sometimes the column concrete surface gets hardened, causing increase in value of rebound number. This resulted in conservative estimation of compressive strength of

Table 2 Rebound hammer and ultrasonic pulse velocity test of ground-floor columns (partial data)

Rebound hammer				Ultrasonic pulse velocity		
S. no.	Member	Avg. rebound hammer reading	Equivalent cube comp. strength (N/mm^2)[a]	Method applied	U.P.V. (Km/s)	Concrete quality grading
1	C22	32	26	Direct	4.7	Excellent
2	C71	30	24	Direct	4.2	Good
3	C82	30	24	Direct	4.1	Good
4	C11	30	24	Direct	4.0	Good
5	C65	28	20	Direct	3.4	Medium
6	C61	27	18	Direct	3.3	Medium
7	C47	23	14	Direct	2.3	Doubtful
8	C34	20	8	Direct	2	Doubtful
9	C118	20	8	Direct	1.1	Doubtful

[a] Required minimum 28-day cube strength for M20 grade concrete = 20 N/mm^2 , core diameter (mm) = 50

Table 3 Rebound hammer and ultrasonic pulse velocity test of first-floor columns (partial data)

Rebound hammer				Ultrasonic pulse velocity		
S. no.	Member	Avg. rebound hammer reading	Equivalent cube comp. strength (N/mm^2)[a]	Method applied	U.P.V. (Km/s)	Concrete quality grading
1	C91	29	22	Direct	3.9	Good
2	C72	28	20	Direct	3.5	Medium
3	C96	25	16	Direct	3.4	Medium
4	C88	26	17	Direct	3.2	Medium
5	C51	24	15	Direct	2.6	Doubtful
6	C11	24	15	Direct	2.5	Doubtful
7	C26	22	12	Direct	2.1	Doubtful

[a] Required minimum 28 day cube strength for M20 grade concrete = 20 N/mm^2 , Core diameter (mm) = 50

Table 4 Rebound hammer and ultrasonic pulse velocity test of ground-floor beams (partial data)

Rebound hammer				Ultrasonic pulse velocity		
S. no.	Member	Avg. rebound hammer reading	Equivalent cube comp. strength (N/mm^2)[a]	Method applied	U.P.V. (Km/s)	Concrete quality grading
1	B32-33	30	24	Direct	3.8	Good
2	B21-22	29	22	Direct	3.4	Medium

[a] Required minimum 28-day cube strength for M20 grade concrete = 20 N/mm^2 , core diameter (mm) = 50

Table 5 Rebound hammer and ultrasonic pulse velocity test of first-floor beams (partial data)

Rebound hammer				Ultrasonic pulse velocity		
S. no.	Member	Avg. rebound hammer reading	Equivalent cube comp. strength (N/mm^2)[a]	Method applied	U.P.V. (Km/s)	Concrete quality grading
1	B32-33	30	24	Direct	3.8	Good
2	B45-46	23	14	Direct	2.9	Doubtful

[a] Required minimum 28-day cube strength for M20 grade concrete = 20 N/mm^2 , core diameter (mm) = 50

Table 6 Rebound Hammer and ultrasonic pulse velocity test of ground-floor slab (partial data)

Rebound hammer				Ultrasonic pulse velocity		
S. no.	Member	Avg. rebound hammer reading	Equivalent cube comp. strength (N/mm^2)[a]	Method applied	U.P.V. (Km/s)	Concrete quality grading
1	SP 91-92-102-101	30	24	Direct	3.3	Good
2	SP 53-54-64-63	28	20	Direct	3.2	Medium
3	SP 95-96-106-105	18	05	Direct	1.7	Doubtful

[a] Required minimum 28-day cube strength for M20 grade concrete = 20 N/mm^2 , core diameter (mm) = 50

Table 7 Rebound hammer and ultrasonic pulse velocity test of first-floor slab (partial data)

Rebound hammer				Ultrasonic pulse velocity		
S. no.	Member	Avg. rebound hammer reading	Equivalent cube comp. strength (N/mm^2)[a]	Method applied	U.P.V. (Km/s)	Concrete quality grading
1	SP 75-76-86-87	30	24	Direct	3.3	Good
2	SP 32-33-42-43	21	10	Direct	2.0	Doubtful

[a] Required minimum 28-day cube strength for M20 grade concrete = 20 N/mm^2 , core diameter (mm) = 50

concrete, which was on higher side than the actual in situ strength. Such views are supported by the fact that concrete core test on selected columns gives much lower compressive strength when compared with rebound hammer test. Results of concrete core test are more reliable when compared with rebound hammer test. Analysis of core test results and relevant discussion is given in point No. 5.

2. Majority of columns in fire-affected area and few columns outside this area were tested for UPV test. Out of 142 columns tested, about 79 test locations gave doubtful concrete quality. There was large variation observed in the results, and no specific trend was observed. At few locations, large fluctuation was observed in readings. Such unsteady reading, when test is conducted according to standard procedure suggests cavity, voids, cracks and non-

Table 8 Concrete core test

S. no.	Member	Location	Crushing load (kN)	Equivalent cube strength (N/mm^2)[a]	Minimum required strength (N/mm^2)
1	Column C34	Ground floor	8.5	10	20
2	Column C36	Ground floor	10.0	12	20
3	Column C87	Ground floor	5.0	06	20
4	Column C95	Ground floor	4.20	05	20
5	Column C23	First floor	6.8	08	20
6	Column C46	First floor	7.6	09	20
7	Column C75	First floor	19.5	23	20
8	Slab panel 14-15-25-24	Ground-floor roof slab	8.5	10	20
9	Slab panel 22-23-33-32	Ground-floor roof slab	18.7	22	20
10	Slab panel 45-46-56-55	Ground-floor roof slab	9.3	11	20
11	Slab panel 95-96-106-105	Ground-floor roof slab	7.6	09	20
12	Slab panel 16-17-27-26	First-floor roof slab	12	14	20
13	Slab panel 36-37-47-46	First-floor roof slab	10.0	12	20
14	Slab panel 87-88-98-97	First-floor roof slab	17.0	20	20

[a] Required minimum 28-day cube strength for M20 grade concrete = 20 N/mm^2 , core diameter (mm) = 50

homogeneity within direct path length of the wave. Out of all the readings taken on columns, only 39 readings (27.5 %) showed good or excellent quality concrete.

3. RCC beams were selected randomly to represent the entire area of factory building and tested for RH and UPV tests. Out of 77 locations, 52 locations (67.5 %) had compressive strength evaluated using RH test less than 20 N/mm^2, whereas only seven locations (9.1 %) suggested "good" concrete quality.
4. Slab panels were selected randomly to represent the entire area of the building and tested for RH and UPV tests. Total 28 locations were tested for RH test and UPV test. Due to limitations of instrument handling and accessibility on opposite faces of slab, readings could be taken at about twelve locations. At 18 test points, RH test gave compressive strength that was much less than 20 N/mm^2; while, except for two test location, UPV measurements revealed "good" quality of concrete resulting in conclusion that all test locations did not comply with expected strength and quality criteria.
5. Concrete core test was performed on 14 locations (seven columns and seven slabs). Visual inspection of extracted concrete core invariably revealed the presence of voids on its surface. Few cores also revealed reddish brown colour

Table 9 Load test

S. no.	Slab	Deflection after application of full load (mm)	Deflection after application of full load for 24 h (mm)	Permissible deflection (mm)	Deflection after removal of full load (mm)	Remarks
1	Ground-floor roof slab panel 82-83-93-92	2.1	2.8	26.8	Not required as per IS 456	Deflection within permissible range
2	Ground-floor roof slab panel 22-23-33-32	2.3	3.0	26.8	Not required as per IS 456	Deflection within permissible range
3	Ground-floor roof slab panel 45-46-56-55	12.2	13.2	26.8	5.2	Recovery in deflection does not conform to IS:456
4	First-floor beam panel 73-83	–	–	–	–	Shear cracks start appearing in beam before the application of full load

of concrete. Change in colour of concrete takes place, when it is subjected to very high temperature. Wherever reddish brown colour of concrete was observed, the core disintegrated into finer pieces of basic ingredients, which reflected the effect of fire. The maximum six of coarse aggregate as observed in extracted core was not more than nominal size of 15 mm. At three locations, core was broken during extraction. It is mainly because of the presence of interconnected fractures and not due to mishandling of cores. To summary, seventeen attempts were made to extract cores, three cores got disintegrated while drilling and were broken during extraction, whereas remaining fourteen cores were taken to laboratory and tested for compressive strength strictly in accordance with IS 516 code of practice.

Categorization of area on the basis of intensity of damage: On the basis of intensity of damage, the whole area of two-storey building was divided into following four categories:

Severely damaged area during fire: This was the area where intensity of damage was maximum leading to major cracks penetrating to large depth of structure. Size reduction in columns and large-scale deflection could also be seen in some slabs and beams. Surface concrete had changed its colour to reddish brown and become brittle. In this area, the average compressive strength by rebound hammer was about 12 N/mm^2, while UPV indicated concrete of doubtful quality. Ground-floor area between few columns had concrete of doubtful quality. Steel exposure at many places and large-scale deflection in beams and slabs had also been noted. The intensity and duration must be very high which had turned a large thickness of surface concrete into ash. The area adjoining stair case and at level above first floor had also concrete of doubtful quality. In this area, the slab thickness was reduced by about 50 % due to spalling of concrete.

Keeping in view the severity of damage, it was recommended to dismantle this portion of the building.

Moderately damaged area during fire: In this area, damage caused by fire was not very severe. Minor cracks could be seen on the surface of concrete and plaster. In this area, the average compressive strength by RH test remained between 16 and 20 N/mm^2, while at most places, the UPV indicated concrete of "medium" quality.

Although this area was not so severely damaged to be recommended for dismantling but almost all of this area lied above the area which had been severely damaged by fire. Considering the above-mentioned facts, it was recommended to dismantle this portion of the building.

Area not affected by fire but cracks of adjoining areas penetrating in this area: There was no direct effect of fire on small portion of the ground-floor area due to the presence of partition walls, but cracks and deflection in the adjoining slab had damaged the slab of this area. In this area, the average compressive strength by RH test was more than 20 N/mm^2, while at most places the UPV indicated concrete of "good" quality. Load test was conducted on a portion of slab of this area, which indicated deflection within permissible range as per IS 456.

Although almost all the test results indicated that the direct damage to this portion of the building was negligible suggesting no need of dismantling, this area was of very small width and was situated along the periphery of severely damaged area by fire. Moreover, slab deflection and large-scale temperature variation in the adjoining portion had caused cracks in this area.

It was very difficult to isolate this area from the adjoining area during dismantling. Thus, this portion of the building was also recommended for dismantling due to its doubtful structural stability.

Area unaffected by fire: In this area, only top of slab had slightly been damaged, but its structural efficiency was not affected. Moreover, columns and beams of this area were also intact. The average compressive strength by RH test was more than 20 N/mm^2, while the UPV indicated concrete of "good" quality. Load test also indicated deflection within permissible range as per IS 456.

Although this portion did not require dismantling, it was to be isolated both vertically as well horizontally from the remaining building. Whenever some part of a framed building is required to be isolated from the part to be dismantled, the isolation/cutting is done at the bay line (along the line joining the columns). The isolation process is very laborious and requires skilled workmanship and costly equipments. Whole of the concrete slab, beams and reinforcement are to be cut skillfully along the bay line so that cracks do not penetrate the portion to be kept intact. If proper isolation can be possible, then there is no need to dismantle the unaffected area.

5 Conclusions

Field assessment of fire-damaged concrete members requires a systematic approach to determine their conditions. Visual observation provides the most practical means to assess the potential of damage in fire exposed members. NDT like UPV and RH provides the information regarding the strength and stability of members. Further, area can be categorized on the basis of intensity of damage.

Acknowledgments Authors are grateful to the testing and Consultancy Cell of Guru Nanak Dev Engineering College, Ludhiana, for their cooperation in assessing the data of fire-damaged building for technical support of this article.

References

ASTM E1351 (2006) Standard practice for production and evaluation of field metallographic replicas

BS ISO 3057 (1998) Non-destructive testing—metallographic replica techniques of surface examination

Cartz L (1995) Nondestructive testing. ASM International, ISBN 978-0-87170-517-4

Chiang C-H, Tsai C-L (2003) Time–temperature analysis of bond strength of a rebar after fire exposure. Cem Concr Res 33(10):1651–1654

Dufour ML, Lamouche G, Detalle V, Gauthier B, Sammut P (2005) Low-coherence interferometry, an advanced technique for optical metrology in industry. Insight Non-Destr Test Condition Monit 47(4):216–219

Gosain NK, Drexler RF, Choudhuri D (2008) Evaluation and repair of fire-damaged buildings. Structure Magazine, pp 18–22

Hellier C (2003) Handbook of nondestructive evaluation. McGraw-Hill, p 1.1. ISBN 0-07-028121-1

Losert R (2009) Solution for NDT inspection. NDT Magazine

Reis EM, Dilek U (2012) Non-destructive evaluation and laboratory testing of a concrete structure damaged by fire. Forensic Engineering ASCE, pp 1159–1166

Tide RHR (1998) Integrity of structural steel after exposure to fire. Eng J 35:26–38

Corrosion Monitoring and Service Life Prediction of Rebars in Structural Concrete: State-of-Review

Yuvraj Singh, Inderpreet Kaur and Amandeep Singh Gill

Abstract A vast governmental budget is spent annually, worldwide, to face corrosion problems of steel reinforcement in concrete bridges attributable to the extensive use of de-icing salts and is a worldwide durability problem. Corrosion controls the lifetime of a structural system, which has two distinct periods. During the first period, chlorides diffuse through the cover. When sufficient chlorides are formed at the rebars, corrosion initiates. This marks the start of the second period, during which rust with higher volume to bare steel is produced, and this happens mainly when the protective alkaline passive film, formed during the process of cement hydration around rebars, is broken. Several electrochemical methods are suggested in the published literature for the determination of corrosion rate in the rebar. All methods have some positive and negative points in their favour. This paper intends to present a summary of these research studies and outlines various preventive measures, methods of corrosion monitoring and service life prediction models that can be employed to minimize the rate of corrosion.

Keywords Corrosion · Rebar · Chlorides · Corrosion monitoring

---

Y. Singh (✉) · A. S. Gill
Master of Technology in Structural Engineering, Guru Nanak Dev Engineering College, Ludhiana 141006, Punjab, India
e-mail: yuvraj_23@yahoo.co.in

A. S. Gill
e-mail: engg.amangill@gmail.com

I. Kaur
Department of Civil Engineering, Guru Nanak Dev Engineering College, Ludhiana 141006, Punjab, India
e-mail: inderdhindsa@gmail.com

S. S. Khangura et al. (eds.), *Proceedings of the International Conference on Research and Innovations in Mechanical Engineering*, Lecture Notes in Mechanical Engineering, DOI: 10.1007/978-81-322-1859-3_12, © Springer India 2014

1 Introduction

Corrosion of steel in concrete has become a costly problem all over the world. Corrosion of rebars in reinforced concrete structure is a complex phenomenon. The problem of corrosion of rebars is caused primarily by inorganic salt in concrete. The salt, primarily chloride, penetrates the concrete from sources such as road-de-icing salts or sea exposure. It can also be built in through the use of salt-contaminated aggregate, seawater in the concrete or chloride-based admixtures. Due to the porous nature of concrete, corrosive species and chemical species supporting corrosion reactions can enter the concrete and lead to corrosion problems. Furthermore, corrosive species can enter the mix if "contaminated" mix ingredients are used (water, aggregates and additives). This mechanism is explained by the following reactions (Jones 1996):

$$FeOOH + H^+ + Cl^- \rightarrow FeOCl + H_2O \tag{1}$$

$$FeOCl + 2H^+ \rightarrow Fe^{3+} + Cl^- + H_2O \tag{2}$$

FeOOH represents the protective layer and reacts with hydrogen and chloride ions to form a solid of FeOCl, then, FeOCl further reacts with the hydrogen ion to produce ferric ion. These reactions are catalyzed by chloride ions.

Most of the studies on the corrosion of rebars embedded in cement have shown the corrosion damage on a structure depends on several factors such as interstitial solution chemistry and pH, water-to-cement ratio, concrete composition and cover thickness, pore and capillary structure and curing period and curing and exposure temperature (Montemor et al. 2003).

This paper addresses various preventive measures, methods of corrosion monitoring and service life prediction model that can be employed to minimize the rate of corrosion.

2 Mechanisms of Rebar Corrosion

Two important rebar corrosion mechanisms are as follows:

1. Chloride-induced rebar corrosion
2. Carbonation.

2.1 Chloride

Chlorides, if present in sufficiently high concentration, including sodium chloride, can promote the corrosion of embedded steel rebars. Chloride anions induce both localized corrosion and generalized corrosion of steel reinforcements. The use of de-icing salts on roadways, used to reduce the freezing point of water, is probably one of the primary causes of premature failure of reinforced or prestressed concrete bridge decks, roadways and parking garages. Another important source of chloride ions is from sea water. Sea water contains by weight approximately 3.5 wt. % salts. These salts include sodium chloride, magnesium sulphate, calcium sulphate and bicarbonates. In water, these salts dissociate into free ions (Na^+, Mg^{2+}, Cl^-, $SO_4{}^{2-}$ and HCO^{3-}) and migrate with the water into the capillaries of the concrete. Chloride ions are particularly aggressive for the corrosion of the carbon steel reinforcement bars and make up about 50 % of these ions.

2.2 Carbonation

Carbonation, or neutralization, is a chemical reaction between carbon dioxide in the air with calcium hydroxide and hydrated calcium silicate in the concrete. The water in the pores of Portland cement concrete is normally alkaline with a pH in the range of 12.5–13.5. This highly alkaline environment is one in which the embedded steel is passivated and is protected from corrosion. The carbon dioxide in the air reacts with the alkali in the cement and makes the pore water more acidic, thus lowering the pH. Carbon dioxide will start to carbonate the cement in the concrete from the moment the object is made. This carbonation process will start at the surface, then slowly move deeper and deeper into the concrete. If the object is cracked, the carbon dioxide in the air will be better able to penetrate into the concrete.

3 Preventive Measures of Corrosion

There are several corrosion protection strategies available. The most prominent approaches take advantages of properties of the corrosion cell to reduce corrosion damage to structure steel are as follows:

3.1 Hot-Dip Galvanized Steel

Galvanized reinforcing steel is effectively and economically used in concrete where unprotected reinforcement will not have adequate durability. The susceptibility of concrete structures to the intrusion of chlorides is the primary incentive for using galvanized steel reinforcement. American Galvanizers Association (2004) has explained that galvanized reinforcing steel is especially useful when the reinforcement will be exposed to the weather before construction begins. Galvanizing provides visible assurance that the steel has not rusted.

The reason for the extensive use of hot-dip galvanized steel is the twofold nature of the coating. As a barrier coating, galvanizing provides a tough, metallurgically bonded zinc coating that completely covers the steel surface and seals the steel from the environment's corrosive action. Additionally, zinc's sacrificial action (cathodic) protects the steel even where damage or a minor discontinuity occurs in the coating. It should be noted that the performance of hot-dip galvanized reinforcing steel in concrete is quite different than that of hot-dip galvanized steel in atmospheric conditions.

3.2 Sulphur Polymer Composite

Sulphur polymer composite can provide surface corrosion protection to the concrete and reinforcing steel in concrete. Corrosion rate for the steel has decreased by 2–3 orders of magnitude when covered with protective coating even though this latest became unseal at load exceeding 88.5 MPa (Ksiazek 2011). The favourable thickness of the polymer sulphur composite layer should be 1.5 mm, and concrete protected with this composite should not be used in the environment of water solutions of HNO_3 and hydroxides KOH and NaOH.

3.3 Inorganic Conversion Coatings

Corrosion resistance can be achieved using phosphate coating. Phosphate coatings increase cathodic areas on the rebar surface, resulting in a passivation and higher charge transfer resistance. Also by applying the phosphate layer on the rebar, cathodic area/anodic area ratio increases, thus the open-circuit potential (OCP) increases to a maximum point (Jalili et al. 2009).

3.4 Epoxy-Coated Reinforcement

Epoxy-coated reinforcement is used extensively in construction to protect steel from corrosion. Epoxy-coating works by preventing chloride and moisture from reaching the surface of the steel. Its greatest advantage lies in its applicability to existing designs without changes in load capacity or section size, the only change is in the modification of development length. FHWA's 2003 National Bridge Inventory showed more than 54,000 US bridges contained FBECR either in the top mat or in both the top and bottom mats (Lee et al. 1994).

4 Methods of Corrosion Monitoring

4.1 Embeddable Sensors

Durable, embeddable and inexpensive sensors for monitoring the corrosion process of new and existing reinforced concrete structures were developed. The sensor provides online information of the OCP and the corrosion rate of the rebars, the concrete electrical resistivity, the oxygen availability, the chloride content and the temperature inside the structure. Besides, the software generates a system of alarms that lead the specialist to make a decision concerning what mitigation strategies to apply. The geometric configuration and material composition of the sensor together with its electronic allow the system to produce accurate measurements over its specified operating range (Duffó and Farina 2009).

4.2 Piezoelectric-Based Non-destructive Monitoring

Piezoceramic (PZT) sensors were embedded in concrete by bonding the sensors on steel-reinforcing bars to perform non-destructive monitoring. The chosen PZT type was of "soft" materials with high permittivity, high coupling factor and high piezoelectric charge constant, and the size of $10 \times 10 \times 0.3$ mm was specifically ordered. A deformed steel rebar of 67 cm long was bonded with a PZT sensor at the mid-length. High-frequency excitation up to 400 kHz is generally used in the EM impedance or admittance measurement using PZT. In this study, an impedance analyzer (Agilent 4294A) was used for providing AC voltage sweep to vibrate the PZT sensor and measure its electrical response at the same time. The impedance analyzer can be used to measure simultaneously two complimentary admittance parameters in each measurement cycle. The act of bonding the PZT on the steel rebar caused the conductance values to drop significantly. The selected frequency range should include the resonance peak to facilitate the detection of small structural changes (Tawie and Lee 2010).

5 Service Life Prediction Model

5.1 Fuzzy Logic

The determination of building condition index (BCI) begins by subdividing a structure into its elements such as columns, beams, slabs. Though every element of a structure in consideration may have different structural importance but from corrosion point of view, all components are assumed to have equal importance.

Therefore, the determination of combined BCI is done assuming equal weights for condition index of individual elements. To facilitate the fuzzy calculations, a model in MS Excel has to be created using in-built functions and macros written in Visual Basic for Applications (VBA) (Mitra et al. 2010).

Following are the steps to calculate the BCI:

1. Input element condition codes.
2. Membership functions corresponding to the given condition codes are selected for various manifestations as discussed.
3. Selected Membership functions are then combined using fuzzy extension principle to obtain element membership function, corresponding area and "x" co-ordinate of area centroid (Cond. Rating technically).
4. Above step is repeated for all components (elements) of the building.
5. De-fuzzification using centre of sum method, of all individual element fuzzy subsets obtained above, followed by factorization for number of distress conditions used in fuzzy extension.
6. BCI.

References

American Galvanizers Association (2004) Hot-dip galvanizing for corrosion protection: a guide to specifying and inspecting. Hot-dip galvanized reinforcing steel

Duffó GS, Farina SB (2009) Development of an embeddable sensor to monitor the corrosion process of new and existing reinforced concrete structures. Constr Build Mater 23:2746–2751

Jalili MM, Moradian S, Hosseinpour D (2009) The use of inorganic conversion coatings to enhance the corrosion resistance of reinforcement and the bond strength at the rebar/concrete. Constr Build Mater 23:233–238

Jones DA (1996) Principles and prevention of corrosion, 2nd edn. Prentice-Hall, Upper Saddle River

Ksiazek M (2011) The experimental and innovative research on usability of sulphur polymer composite for corrosion protection of reinforcing steel and concrete. Compos B 42:1084–1096

Lee SK, McIntyre JF, Hartt WH (1994) New development in laboratory testing of epoxy coated reinforcing steel. NACE

Mitra Gopal, Jain KK, Bhattacharjee B (2010) Condition assessment of corrosion-distressed reinforced concrete buildings using fuzzy logic. J Perform Constructed Facil ASCE 24(6):562–570

Montemor MF, Simoes AMP, Ferreira MGS (2003) Chloride-induced corrosion on reinforcing steel: from the fundamentals to the monitoring techniques. J Cem Concr Compos 25:491–502

Tawie R, Lee HK (2010) Piezoelectric-based non-destructive monitoring of hydration of reinforced concrete as an indicator of bond development at the steel concrete interface. Cem Concr Res 40:1697–1703

Comparing Taguchi Method and RSM for Optimizing Flank Wear and Surface Roughness During Hard Turning of AISI D3 Steel

Rupinder Singh and J. S. Dureja

Abstract The present work attempts to investigate tool wear (flank wear) and surface roughness (Ra) during finish hard turning of AISI D3 steel (58HRC) with coated carbide (TiSiN–TiAlN PVD coated) cutting tool. Taguchi L9 (3)3 orthogonal array has been applied for experimental design. Signal-to-noise (S/N) ratio (Lower the best) and analysis of variance (ANOVA) analyses were performed to identify significant parameters influencing tool wear and Ra. The cutting speed and feed were the most significant factor influencing tool wear (flank wear), and feed is the most significant factor influencing Ra. Mathematical models for both response parameters, i.e., tool wear and Ra were obtained through regression analysis. The confirmation experiments carried out at optimal combination of parameters given by Taguchi analysis predicted the response factors with less than 5 % error. In addition to this desirability function module, response surface methodology (RSM) was applied to arrive at the optimal setting of input parameters to minimize tool wear and Ra and to compare this with optimal setting of parameters given by Taguchi analysis. The optimization results provided by desirability function optimization is quite close to the optimal solutions provided by Taguchi analysis.

Keywords Hard turning · Taguchi orthogonal array · Analysis of variance · S/N ratio · Flank wear · Surface roughness

R. Singh (✉)
University College of Engineering, Punjabi University, Patiala, Punjab, India
e-mail: rupinder_singh302@yahoo.com

J. S. Dureja
Department of Mechanical Engineering, University College of Engineering,
Punjabi University, Patiala, Punjab, India
e-mail: jsdureja73@yahoo.com

S. S. Khangura et al. (eds.), *Proceedings of the International Conference on Research and Innovations in Mechanical Engineering*, Lecture Notes in Mechanical Engineering, DOI: 10.1007/978-81-322-1859-3_13, © Springer India 2014

1 Introduction

The modern machining industries is mainly focused on the achievement of high quality, in terms of work piece dimensional accuracy, surface finish, high production rate, less wear on the cutting tools, economy of machining in terms of cost saving and increasing the performance of the product with reduced environmental impact. Carbides are the most prevalent tool materials, as they are tough and can be used for machining at higher cutting speed, feed rate, also applicable to difficult intermittent machining. Coated carbides consist of a hard carbide substrate with surface coating, which increases the thermo-chemical stability (carbides, nitrides, oxides and their combinations). These high-quality materials provide high rate of material removal even during intermittent machining. Conventional tool materials for hard turning applications are CBN, ceramics and PCD because of high hardness and toughness. Coated carbides on the other hand have been tried for these applications, and carbide tools are widely used in the metal-working industry and provide the best alternative for most turning operations (Mahesh 2010). Tungsten carbide (TiC) thin films act as heat barrier and protect the tools from being exposed to high cutting temperature in the cutting zone thereby enabling these to retain high hardness even at increased temperatures of the order of 400 °C (Sundgren and Hentzell 1986; Quinto and Willd 1998). Researchers are showing keen interest in developing WC–C composite coating, which have high heat resistance (Quinto and Willd 1998; Hoffman et al. 1992). The use of coating materials to enhance the performance of cutting tools is not a new concept. Coated carbides are the most popular and most common high production tool materials available today (Handbook 1980). The boost in wear resistance gave room for a significant increase in cutting speed and thereby improved productivity at the machine shop floor. Coated carbide tools account for nearly 70 % of the tools used in the industry (Abdullah 1996). The majority of carbide cutting tools in use today employ chemical vapor deposition (CVD) or physical vapor deposition (PVD) hard coatings. The high hardness, wear resistance and chemical stability of these coatings offer proved to be beneficial in terms of longer tool life and improved machining performance (Handbook 1980; Abdullah 1996; Lux et al. 1986; Layyous et al. 1992; Prengel et al. 1998). CBN and ceramics tools are used for hard turning applications owing to their higher hardness viz. clutch disks, gears, hubs, rock drills, bearings, shafts and axles, nozzles having hardness in the range of 40–70 HRC.

In present investigation, a newly developed coated carbide grade TH1000 (SECO-PVD coated) carbide has been employed as cutting tool for hard turning of AISI D3 steel. The superior edge toughness of this grade provides excellent performance in continuous and interrupted cuts in hardened steels. This coated carbide grade (TH1000) was selected for current investigation owing to its low cost compared to costlier CBN and ceramic grades to provide an economical alternative for the applications in industry predominantly machined by CBN and ceramic grades as mentioned above.

2 Taguchi Design

The DOE is a technique used to define what data, in what quantity and conditions should be collected during an experiment, trying to satisfy two major goals: the statistical accuracy of the response and achieving lower cost (Gunasegaram et al. 2009). Before starting the experimentation, it is important to establish the planning of experiments (Montgomery 1997). The traditional experimental design methods are too complex and difficult to use. Additionally, large numbers of experiments have to be carried out when the number of machining parameters increases. Therefore, the factors causing variations should be determined and checked under laboratory conditions. These studies are considered under the scope of off-line quality improvement (Hascalık and Caydas 2008). The Taguchi method is an experimental design technique, which is useful in reducing the number of experiments dramatically by using orthogonal arrays and also tries to minimize effects of the factors out of control. An orthogonal array means the design is balanced so that factor levels are weighted equally. Because of this, each factor can be evaluated independently of all the other factors, so that effect of one factor does not influence the estimation of another factor. In robust parameter design, we first choose control factors and their levels and choose an orthogonal array appropriate for these orthogonal arrays. The basic philosophy of the Taguchi method is to ensure quality in the design phase. The greatest advantages of the Taguchi method are to decrease the experimental time, to reduce the cost and to find out significant factors in a shorter time period (Chattopadhyay and Chattopadhyay 1982). The most reliable of Taguchi techniques is the use of parameter design, which is an engineering method for product or process design that focuses on determining the parameter (factor) settings producing the best levels of a quality characteristic (performance measure) with minimum variation. Taguchi analysis comprises of determining signal-to-noise (S/N) ratio as the quality characteristic of choice, i.e., the ratio of useful information to false or irrelevant data in a conversation or exchange (Anyılmaz 2006). Taguchi empirically found that the two stage optimization procedure involving S/N ratios indeed gives the parameter-level combination, where the standard deviation is minimum while keeping the mean on target.

To fit mathematical models of tool flank wear and surface roughness (Ra), multiple linear regression has been applied in terms of input parameters viz. cutting speed (A), feed rate (B), depth of cut (C). Thereafter, ANOVA has been applied (F-test and lack of fit tests) to establish the reliability of developed models for making future predictions.

3 Experimental Details

3.1 Equipment and Materials

The AISI D3 was selected as the work piece material in the form of round bars with 25 mm diameter and 190 mm axial cutting length. The specimens were

Table 1 Composition of Aisi D3 steel

C	Si	Mn	Cr	Mo	V
1.7	0.3	0.5	13.0	0.8	0.8

Table 2 Cutting conditions for the experiment

Level	Cutting speed (Vc) (m/min)	Feed rate (*f*) (mm/rev)	Depth of cut (ap) (mm)
1	130	155	180
2	0.05	0.10	0.15
3	0.1	0.25	0.4

through hardened to achieve hardness of 58HRC. The chemical composition of work material is given in Table 1. Specific applications of D3 steel are as follows: High-quality press tools, drawing and cutter dies, shear blades, thread rollers, punches, measuring tools, pressure casting molds, blanking, reamer etc. The turning tests were conducted under dry conditions on a CNC lathe (Make: Batliboi, Sprint 16 TC, India) having a maximum spindle speed of 5,000 rpm.

The cutting tool used was coated carbide TH1000 (SECO grade) TiSiN–TiAlN nano PVD coated which enhances edge toughness and results in improved surface finish. The inserts with ISO designation (DNMG150604-MF1) were clamped on 25/25M left-hand tool holder with ISO designation DDJNL2525M15-M. Three levels were specified for each of the factors as indicated in Table 2. These levels were selected based on the recommendations of manufacturer and are supported by previous work of various authors (Hascalık and Caydas 2008; Chattopadhyay and Chattopadhyay 1982; Anyılmaz 2006; Aslan et al. 2007).

Each test was performed by using a fresh cutting edge and over a constant helical cutting length of 900 m. The average values of tool flank wear (VB) and Ra were measured at the end of each test. The average flank wear was monitored with the help of a tool maker's microscope and Ra with a Ra analyzer.

The experiment was terminated when either of the following two conditions was reached:
Condition 1 VB $\geq$ 200 μm;
Condition 2 Ra $\geq$ 1.6 μm.

3.2 Design of Experiments

The cutting parameters are cutting speed (Vc), feed rate (*f*) and depth of cut (ap). The objective is to minimize Ra of machined specimen and tool wear (flank wear, VB).

Three levels were specified for each of the factors as indicated in Table 2. The orthogonal array chosen was L(OA)9 (3)3 comprising a total of 27 experiments (Table 3) where 9 means number of trials (3) means number of levels of each factor ( )3 means number of factors. Once the parameters are assigned to a

Table 3 Orthogonal array L_9 of Taguchi

Experiment	Speed (*A*)	Feed (*B*)	DOC (*C*)
1	1	1	1
2	1	2	2
3	1	3	3
4	2	1	2
5	2	2	3
6	2	3	1
7	3	1	3
8	3	2	1
9	3	3	2

particular column of the selected orthogonal array, the factors at different levels are assigned for each trial.

4 Analysis of Results (Taguchi Method)

Tables 4 and 5 illustrates the experimental results for VB and Ra. The main effects can be evaluated by analyzing raw data or of S/N ratios for flank wear and Ra as indicated in Tables 6 and 8. The analysis is done by averaging the raw or S/N data at each level of all parameters and plotting the values in graphical form. The level average responses from the raw data help in analyzing the trend of the performance characteristics with respect to the variation of the factor under consideration. The level average response plots based on the S/N data help in optimizing the objective function under consideration. The lower points of these plots correspond to the optimum condition for average tool wear (flank wear) and average Ra.

4.1 S/N Ratio and ANOVA Analysis for Flank Wear (VB)

It is clearly indicated from Fig. 1 that flank wear (VB) increases with increase cutting speed, where as S/N ratio decreases with increase in cutting speed. Corresponding to minimum cutting speed (130 m/min), minimum flank wear and high S/N ratio have been observed. Thus, cutting speed 130 m/min is the optimal level to attain minimum flank wear (VB).

Figure 2 shows the plot of S/N ratio and (VB) versus feed rate. The flank wear initially decreases with increase in feed rate up to 0.1 mm/rev followed by an increase in VB with feed rate to 0.15 mm/rev. Corresponding to minimum feed rate at 0.1 mm/rev, the S/N ratio is observed to be highest, thereby indicating that feed rate of 0.1 mm/rev to be the optimal level.

Figure 3 gives a plot of S/N ratio and (VB) versus depth of cut (DOC). The flank wear initially increases with increase in DOC followed by a sharp decreasing trend up to 0.4 mm depth of cut. The S/N ratio observed a decreasing trend initially followed by an increasing trend reporting maximum value at DOC = 0.4 mm, thereby indicating this to be optimal level.

Table 4 Experimental results for VB

Experiment	Factors			VB (μm)	S/N ratio
	Vc	*f*	ap		
1	130	0.05	0.10	121.67	−41.72
2	130	0.1	0.25	98.33	−39.87
3	130	0.15	0.40	106.67	−40.62
4	155	0.05	0.25	163.33	−44.27
5	155	0.1	0.40	123.33	−41.83
6	155	0.15	0.10	131.67	−42.40
7	180	0.05	0.40	151.67	−43.63
8	180	0.1	0.10	163.33	−44.29
9	180	0.15	0.25	201.67	−46.13

Table 5 Experimental results for Ra

Experiment	Factors			Ra (μm)	S/N ratio
	Vc	*f*	ap		
1	130	0.05	0.10	0.86	−41.72
2	130	0.1	0.25	0.62	−39.87
3	130	0.15	0.40	1.03	−40.62
4	155	0.05	0.25	0.51	−44.27
5	155	0.1	0.40	0.62	−41.83
6	155	0.15	0.10	0.98	−42.40
7	180	0.05	0.40	1.15	−43.63
8	180	0.1	0.10	0.62	−44.29
9	180	0.15	0.25	1.10	−46.13

Table 6 Response for S/N ratios (smaller is better)

Level	Speed (*A*)	Feed (*B*)	DOC (C)
1	−40.74	−43.20	−42.80
2	−42.83	−42.00	−43.42
3	−44.68	−43.05	−42.03
Max–min	3.94	1.2	1.39
Rank	1	3	2

4.2 S/N Ratio and ANOVA Analysis for Surface Roughness (Ra)

Figure 4 shows a plot of S/N ratio and Ra versus speed. Best surface finish and highest S/N ratio have been reported at cutting speed of 155 m/min, indicating this to be the optimal level. Figure 5 is a plot of S/N ratio and Ra versus feed rate. S/N ratio initially increases with increase in feed rate and is highest at 0.1 mm/rev, with best surface finish achieved at this level. S/N ratio and surface finish report

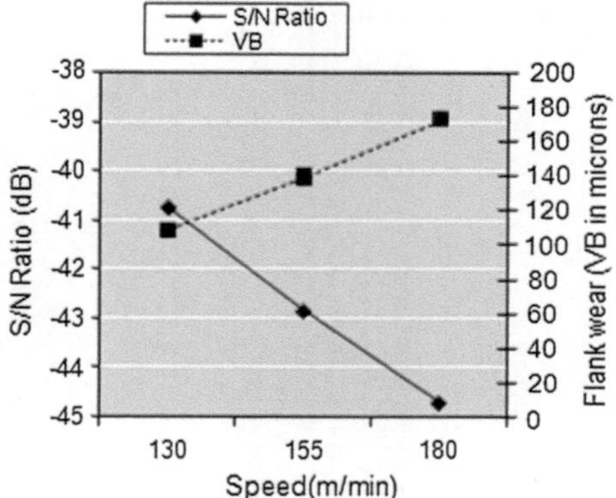

Fig. 1 S/N ratio values for speed

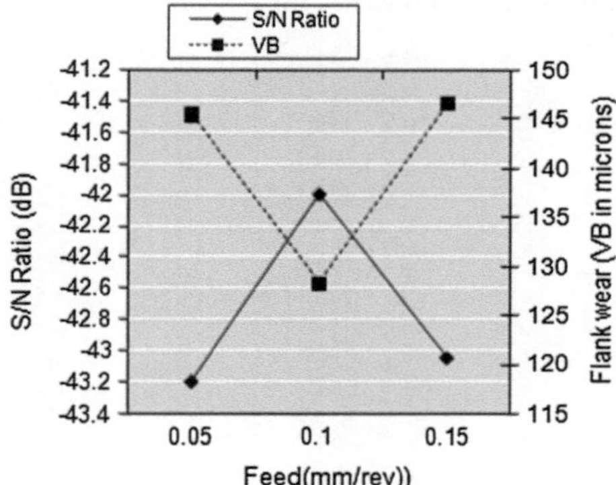

Fig. 2 S/N ratio values for feed

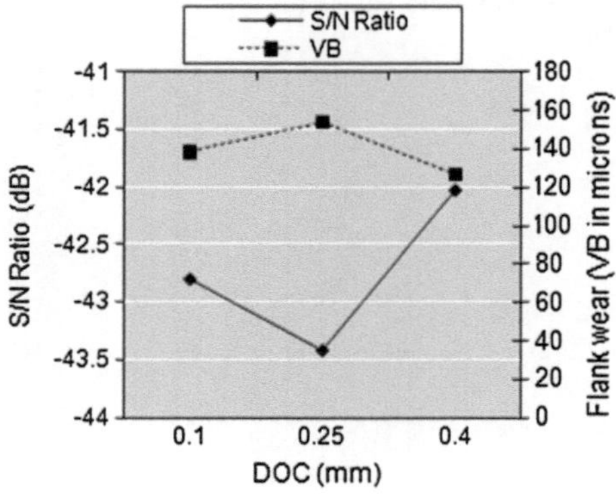

Fig. 3 S/N ratio values for DOC

decreasing trend with further increase in feed rate. Figure 6 gives a plot of S/N ratio and Ra versus DOC. Surface finish and S/N ratio improve with increase in DOC and report higher value at 0.25 mm DOC. Surface finish and S/N ratio decrease with further increase in DOC.

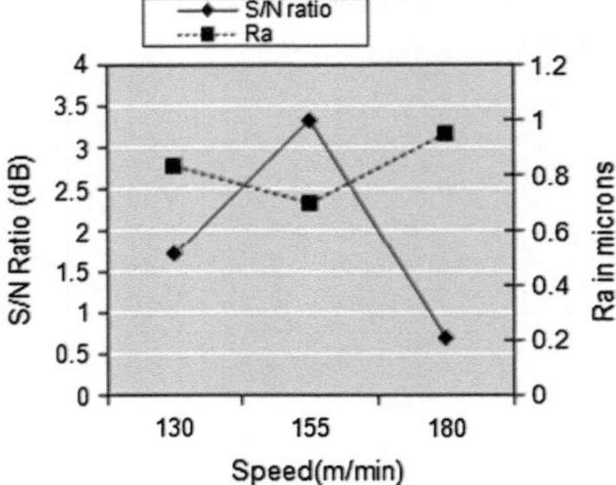

Fig. 4 S/N ratio values for speed

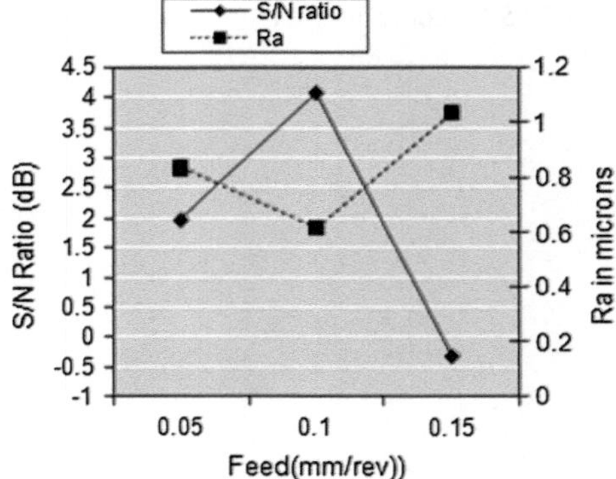

Fig. 5 S/N ratio values for feed

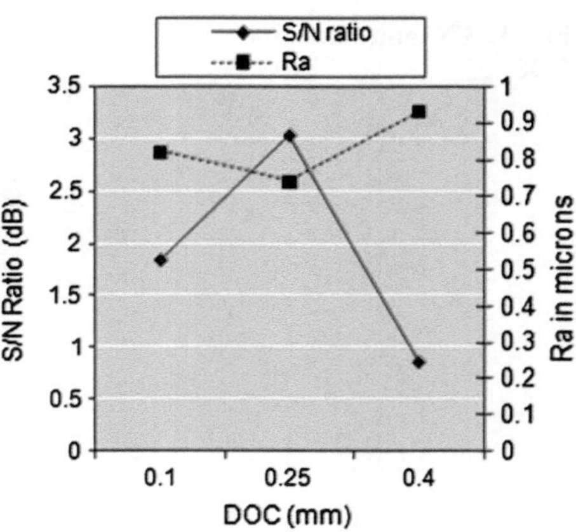

Fig. 6 S/N ratio values for DOC

The S/N ratios for VB versus speed and DOC are given in Table 6 and for Ra versus feed and DOC are given in Table 8. The speed and the depth of cut are two factors that have the highest difference between values (Max–Min), i.e., 63.33 and

Table 7 Optimal parameters for Ra and VB

Optimal parameters	Speed (Vc) m/min	Feed (*f*) mm/rev.	DOC (ap) mm
Flank wear (VB)	130	0.1	0.4
Surface roughness (Ra)	155	0.1	0.25

Table 8 Sequential model sum of squares for flank wear model (VB)

Source	Sum of source	DF	Mean square	*F* Value	Prob > *F*	
Mean	4.704E+005	1	4.704E+005			
Linear	19,360.74	3	6,453.58	17.47	<0.0001	
2FI	*3,610.04*	*3*	*1,203.35*	*5.41*	*0.0085*	*Suggested*
Quadratic	1,183.40	2	591.70	3.42	0.0598	Aliased
Cubic	0.000	0	173.06			Aliased
Residual	2,595.83	15				
Total	4.972E+005	24	20,714.58			

27.22 for flank wear, 4.4340 and 2.6704 for Ra, respectively. Based on the Taguchi prediction, larger difference will have a more significant effect on tool wear and Ra.

Thus, it can be concluded that increasing the speed will increase the tool wear (flank wear-VB) significantly and also the depth of cut. The Ra will be increased by increasing the feed rate significantly and also speed. The optimal setting of parameters obtained from main effects plot is in Table 7.

The optimal parameter-level combination obtained through Taguchi analysis to minimize flank wear (VB) is 1-2-3, i.e., 1st level for cutting speed, 2nd level for feed rate and the 3rd level for DOC. In order to validate this optimal solution provided by Taguchi analysis, conformation experiment was performed at this parameter-level combination, i.e., speed = 130 m/min, feed rate = 0.1 mm/rev., DOC = 0.4 mm. The flank wear (VB) observed at this combination is = 96 μm. The optimal setting of parameters given by Taguchi analysis 1-2-3 is closest to the 1-2-2 combination of parameters in Taguchi orthogonal array. Thus, the error between initial (*I*) and final (*F*) observed values of (VB) is the 3rd level for DOC. In order to validate this optimal solution provided by Taguchi analysis, conformation experiment was performed at this parameter-level combination, i.e., speed = 130 m/min, feed rate = 0.1 mm/rev., DOC = 0.4 mm. The flank wear (VB) observed at this combination is = 96 μm. The optimal setting of parameters given by Taguchi analysis 1-2-3 is closest to the 1-2-2 combination of parameters in Taguchi orthogonal array. Thus, the error between initial (*I*) and final (*F*) observed values of (VB) is calculated as:

$$\text{Error} = \text{Mode}\frac{I - F}{I} \times 100$$

Thus, the error between initial (*I*) and final (*F*) for VB is 2.33 %, i.e., within 95 % confidence interval. Similarly, the optimal parameter-level combination to obtain minimum Ra is 2-2-2. Thus, a conformation experiments was performed at this combination of parameters, i.e., speed = 155 m/min, feed rate = 0.1 mm/rev and DOC = 0.25 mm. The Ra observed at this combination is Ra = 0.57 μm. This optimal parameter-level combination is closest to the parameter-level combination in Taguchi orthogonal array of 2-2-3. Thus, the error between initial (*I*) and final (*F*) observed values of (Ra) was evaluated, and it came out to be 3.4 % again within 95 % confidence interval.

5 Modeling of Flank Wear and Surface Roughness

The Taguchi method enables designing experimental plan and arriving at optimal setting of parameters, but there is no module to develop models of responses in terms of input parameters. Therefore, response surface methodology (RSM) and multiple linear regression were applied to generate models of VB and Ra. The Taguchi orthogonal array $L(OA)_9$ $(3)^3$ design gave a total of 27 experiments as reported earlier. The experimental observations corresponding to theses 27 experiments were analyzed through historical data module of RSM using Design Expert software. The regression models developed are given as Eqs. 1 and 2. 2F1 models were fitted for flank wear and Ra because these were the only significant models as given in Table 8 for VB. The regression models of both responses, i.e., VB and Ra in terms of input parameters speed (*A*), feed (*B*) and depth of cut (*C*) are given:

$$\begin{aligned} \text{VB} &= +121.06 + 20.5 * A - 2.26 * B - 0.55 * C + 5.62 * A * B \\ &\quad - 0.080 * A * C + 15.00 * B * C \end{aligned} \tag{1}$$

$$\begin{aligned} \text{Ra} = &+ 0.73 + 0.12 * A + 0.55 * B - 0.38 * C - 0.18 * A * B \\ &+ 0.27 * A * C + 0.34 * B * C \end{aligned} \tag{2}$$

The ANOVA test was performed to evaluate statistical significance of the fitted 2F1 models and factors involved therein. In addition to this, the goodness of fit of the fitted 2F1 models was also evaluated through lack of fit test. The results obtained are summarized in Tables 10 and 12. Both the 2F1 models are found to be significant. The *p* value (Prob > *F*) for both the responses was observed to less than 0.0001 and clearly shows that the flank wear model and Ra model are statistically significant (Table 9). The lack of fit for the fitted quadratic models for VB and Ra was found to be insignificant, Tables 10 and 12. The "Lack of Fit F value" of 3.42 for VB implies that there is 5.98 % chance that a "Lack of Fit F value" larger than this could occur due to noise. On the other hand, "Lack of Fit F value" of 3.52 observed for Ra (Table 12) implies that there is a 5.56 % chance that a "Lack of Fit F value" larger than this could occur due to noise. The R^2 values for both VB and Ra 0.8587 and 0.9630 are approaching 1.0. The predicted and adjusted R^2 values for VB (Table 11) are in reasonable agreement, whereas for Ra these values are in excellent agreement (Table 13), which again signifies fitness of the developed models. The coefficient of variation, C.V. = (S.D./Mean) × 100, is a measure of error associated with the

Table 9 Sequential model sum of squares for surface roughness (Ra)

Source	Sum of source	DF	Mean square	*F* value	Prob > *F*	
Mean	18.15	1	18.15			
Linear	0.29	3	0.097	2.24	0.1145	
2FI	*0.83*	*3*	*0.28*	*151.42*	*<0.0001*	*Suggested*
Quadratic	9.953E–003	2	4.977E–003	3.52	0.0556	Aliased
Cubic	0.000	0	1.412E–003			Aliased
Residual	0.021	15				
Total	19.30	24	0.80			

Table 10 ANOVA results for flank wear (VB)

Source	Sum of square	DF	Mean source	*F* value	Prob > *F*	
Model	22,970.77	6	3,828.46	17.22	<0.0001	*Significant*
A	11,898.74	1	11,898.74	53.52	<0.0001	
B	14.46	1	14.46	0.065	0.8018	
C	0.86	1	0.86	3.856E–003	0.9512	
AB	385.34	1	385.34		0.2055	
AC	0.074	1	0.074	1.73	0.9856	
BC	720.42	1	720.42	3.341E–004	0.0896	
Residual	3,779.23	17	222.31	3.24		
Lack of fit	1,183.40		591.70		0.0598	*Not significant*
Pure error	2,595.83	2	173.06	3.42		
Cor total	26,750	23				

Table 11 Statistical summary of model for flank wear (VB)

Std. dev.	14.91	(R^2)	0.8587
Mean	140.00	Adjusted (R^2)	0.8089
C.V. (%)	10.65	Predicted (R^2)	0.7148
PRESS	7630.14	Adequate precision (AP)	12.0

model. The low value of C.V. obtained for both the models indicates improved precision and reliability of the experiments carried out.

The value of adequate precision (AP), defined as the signal-to-noise ratio, for both the models is significantly higher than 4 (Tables 10, 12), which indicates suitability of model, Eqs. (1) and (2) for making future predictions.

6 Optimization of Cutting Conditions (RSM)

In the present study, desirability function optimization has been employed for multi-response (VB and Ra) optimization. The optimization module searches for a combination of factor levels that simultaneously satisfy the requirements imposed

Table 12 ANOVA table for surface roughness (Ra)

Source	Sum of square	DF	Mean source	*F* value	Prob > *F*	
Model	1.12	6	0.19	102.15	<0.01	*Significant*
A	0.38	1	0.38	207.83	<0.01	
B	0.84	1	0.84	459.79	<0.01	
C	0.41	1	0.41	224.94	<0.01	
AB	0.39	1	0.39	213.86	<0.01	
AC	0.83	1	0.83	453.58	<0.01	
BC	0.36	1	0.36	197.92	<0.01	
Residual	0.031	17	1.832E–003			
Lack of fit	9.953E–003	2	4.977E–003	3.52	0.0556	*Not significant*
Pure error	0.021	15	1.412E–003			
Cor total	1.15	23				

Table 13 Statistical summary of model for surface roughness (Ra)

Std. dev.	0.043	(R^2)	0.9730
Mean	0.87	Adjusted (R^2)	0.9635
C.V. (%)	4.92	Predicted (R^2)	0.9503
PRESS	0.057	Adequate precision (AP)	28.974

Table 14 Constraints used for optimization

Name	Goal	Lower limit	Upper limit	Lower weight	Upper weight	Importance
Speed	Is in range	130	155	1	1	3
Feed	Is in range	0.05	0.15	1	1	3
DOC	Is in range	0.1	0.4	1	1	3
VB	Minimize	90	220	1	1	3
Ra	Is in range	0.47	1.18	1	1	3

Table 15 Optimization results

S. no.	Speed	Feed	DOC	VB	Ra	Desirability	
1	130.00	0.13	0.21	93.24	1.18	0.975	Selected
2	130.00	0.13	0.22	93.87	1.15	0.970	
3	130.44	0.13	0.22	94.10	1.18	0.968	
4	130.00	0.12	0.17	94.39	1.18	0.966	
5	130.00	0.12	0.16	94.75	1.18	0.963	
6	130.00	0.14	0.29	96.53	1.18	0.950	
7	130.00	0.12	0.25	97.09	0.93	0.945	
8	130.06	0.15	0.30	97.75	1.18	0.940	
9	130.00	0.11	0.29	100.02	0.53	0.923	
10	130.00	0.10	0.28	100.38	0.47	0.920	

on each of the response factor, in an attempt to establish the appropriate model. The objective of optimization is to find the optimal values of input parameters to minimize the value of flank wear (VB) and Ra during machining of AISI D3 tool steel with coated carbide tool. The constraints used for optimization are given in Table 14. The lower and upper limits of input parameters correspond to the range of input parameters selected in this study (Table 14). The optimal solutions obtained are reported in Table 15 in order of their decreasing desirability level.

It is clear from Table 15 that for simultaneously optimizing both the responses, i.e., VB and Ra, the most optimal solution is: Speed = 130 m/min., feed rate = 0.13 mm/rev. and DOC = 0.21 mm with VB and Ra approaching 93.24 and 1.18 μm, respectively. This solution provided by desirability function optimization is quite close to the optimal solutions provided by Taguchi analysis although for one response parameter (VB or Ra) at time.

7 Conclusions

- The optimum setting of input parameter-level combination suggested by both the techniques viz. Taguchi method as well as RSM is in close agreement, validating the use of both techniques in response factors optimization.
- The Taguchi analysis has suggested percentage contribution of input parameters influencing flank wear (VB) as: Speed = 73.65 %, Feed rate = 8.18 % and DOC = 9.28 %, signifying the cutting speed to be the most contributing factor influencing flank wear.
- The percentage contribution of input parameters influencing Ra is: Speed = 19.47 %, Feed rate = 52.98 % and DOC = 12.98 %, signifying the feed rate to be the most contributing factor influencing Ra.
- The optimal machining conditions for minimizing tool wear as per Taguchi analysis are approaching: cutting speed 130 m/min., feed 0.10 mm/rev., depth of cut 0.40 mm with an estimated flank wear 96 μm.
- The optimized machining conditions for minimizing Ra as per Taguchi analysis are approaching: cutting speed 155 m/min., feed 0.10 mm/rev., depth of cut 0.25 mm with an estimated Ra of 0.57 μm.
- The results of ANOVA test and validation experiments confirm that the mathematical models developed for tool flank wear and Ra excellently fit and predict the values of response factors close to experimentally achieved values with 95 % confidence interval.
- The desirability function optimization for simultaneously optimizing both the responses, i.e., VB and Ra gives the most optimal solution as: Speed = 130 m/min., feed rate = 0.13 mm/rev. and DOC = 0.21 mm and the VB is approaching 93.24 μm and Ra approaching 1.18 μm. This solution provided by desirability function optimization is quite close to the optimal solutions provided by Taguchi analysis although for one response parameter (VB or Ra) at time.

References

Abdullah A (1996) Machining of aluminum based metal matrix composite (MMC). Ph.D. thesis, University of Warwick, Warwick

Anyılmaz MS (2006) Design of experiment and an application for Taguchi method in quality improvement activity. M.S. thesis, Dumlupınar University, Turkey

Aslan E, Camus N, Bingoren B (2007) Design optimization of cutting parameters when turning hardened AISI 4140(63 HRC) with Al_2O_3 + TiCN mixed ceramic tool. J Mater Des 28:1618–1622

Chattopadhyay AK, Chattopadhyay AB (1982) Wear and performance of coated carbide and ceramic tools. Wear 80(2):239–258

Gunasegaram DR, Farnsworth DJ, Nguyena TT (2009) Identification of critical factors affecting shrinkage porosity in permanent mold casting using numerical simulations based on design of experiments. J Mater Process Technol 209(3):1209–1219

Handbook M (1980) American society for metals (ASM). Machining 7:773–783

Hascalık A, Caydas U (2008) Optimization of turning parameters for surface roughness, and tool life based on the Taguchi method. Int J Adv Manuf Technol 38:896–903

Hoffman A, Evans PJ, Choen DD, Paterson PJK (1992) J Appl Phys 15:72

Layyous AA, Frenkel DM, Israel R (1992) Al2O3-coated cemented carbides: optimization of structure, number of layers and type of interlayer. J Surf Coat Technol 56:89–95

Lux B, Columbier C, Atena H, Stemberg K (1986) Preparation of alumina coatings by chemical vapor deposition. J Thin Solid Films 138:49–64

Mahesh PJ (2010) Investigation in tool life of coated and uncoated carbide tools in turning. In: Proceedings of 2nd national conference TIME

Montgomery DC (1997) Design and analysis of experiments, 4th edn. Wiley, USA

Prengel HG, Pfouts WR, Santhanam AT (1998) State of the art in hard coatings for carbide cutting tools. J Surf Coat Technol 102:183–190

Quinto DT, Willd R (1998) Balzers Limited. In: 25th International, conference on metallurgical coatings and thin films invited talk, San Diego

Sundgren JE, Hentzell HTG (1986) A review of the present state of art in hard coatings grown from the vapor phase. J Vac Sci Technol 5:2259–2279

Fine Finishing of Metal Matrix Composite Plate with Magnetic Abrasives

Varun Sharma and Sehijpal Singh

Abstract In order to overcome the need of harder tools required for difficult-to-cut materials, the technology has advanced from traditional finishing processes to precision and ultra-precision processes based on non-traditional techniques. The limitations of traditional finishing processes need to be eliminated. This led to the development of loose abrasives-based finishing processes wherein low mechanical forces are used to finish the surfaces. Magnetic abrasive finishing (MAF) process is one of the non-traditional finishing processes. The diamond-based sintered magnetic abrasives are used to finish round plate of MMC. The MMC plate is prepared by mixing SiC abrasive and aluminium by stir casting method. Three process parameters (magnetic field density, rotational speed and abrasive/lubricant ratio) are taken as input variables. A well-known design of experiment approach "response surface methodology" is employed for the conduct and analysis of experimental work. The rotational speed taken in this work is bidirectional for ensuring proper dressing of abrasives. The combination of high magnetic flux density, average rotational speed and less lubricant caused better results, and the best surface finish of 1.12 μm is achieved.

Keywords Abrasives · Magnetic abrasive finishing · Response surface methodologies · Stir casting method

V. Sharma (✉)
Department of Mechanical Engineering, Lovely Professional University,
Phagwara, Punjab, India
e-mail: varun.gndec@gmail.com

S. Singh
Department of Mechanical Engineering, Guru Nanak Dev Engineering College,
Ludhiana 141006, Punjab, India
e-mail: Mech@gndec.ac.in

S. S. Khangura et al. (eds.), *Proceedings of the International Conference on Research and Innovations in Mechanical Engineering*, Lecture Notes in Mechanical Engineering,
DOI: 10.1007/978-81-322-1859-3_14, © Springer India 2014

1 Introduction

Abrasive machining is a machining process where material is removed from a work piece using a multitude of small abrasive particles. Abrasive processes are usually expensive, but capable of tighter tolerances and better surface finish than other machining processes. Abrasive machining works by forcing the abrasive particles, or grains, into the surface of the work piece so that each particle cuts away a small bit of material.

However, unlike conventional machining, the grains are much smaller than a cutting tool, and the geometry and orientation of individual grains are not well defined. For abrasive grains to effectively cut, they must be significantly harder than the work piece material.

The serious limitations of two body abrasions are the expenditure of huge specific energy and subsurface damage due to high temperature at tool and work interface. The need for high precision in manufacturing was felt by manufacturers worldwide to improve interchange ability of components, improve quality control and longer wear/fatigue life (Mc Keown 1987).

Magnetic abrasive finishing (MAF) is one such unconventional finishing process developed to produce good quality finish efficiently and economically on the internal and external surfaces of tubes as well as flat surfaces made of magnetic or non-magnetic materials. In this process, usually ferromagnetic particles are sintered with fine abrasive particles (Al_2O_3, SiC, CBN or diamond) and such particles are called ferromagnetic abrasive particles (or magnetic abrasive particles). The magnetic abrasive grains are combined with each other magnetically between magnetic poles along a line of magnetic force, forming a flexible magnetic abrasive brush, which acts as a multi-point cutting tool for finishing operations. MAF uses this magnetic abrasive brush for surface and edge finishing. The magnetic field retains the powder in the gap and acts as a binder causing the powder to be pressed against the surface to be finished (Kremen 1994). The important parameters influencing the surface quality generated during the MAF as: (1) voltage (DC) applied to the electromagnet, (2) working gap, (3) rotational speed of the magnet and (4) abrasive size (mesh number) (Jain et al. 2001). Using response surface methodology and analysis of variance, the performance of flexible magnetic abrasive brush is found to depend upon the current to the electromagnet, followed by machining gap, grain size and number of cycle (Singh et al. 2005).

The larger grain size of MAP is found to improve the surface finish significantly. The improvement in the surface finish is found to be directly proportional to the feed rate and current (Girma et al. 2007). Addition of machining fluid (such as stearic acid, straight oil type) to unbounded MAPs of alumina-based showed a remarkable effect on stock removal and surface finish (Shinmura et al. 1985).

The amount of powder has been found to have the largest impact on the process, while the rotational speed has the least effect on the surface finish (Kwak and Kwak 2010). In the drilling of Al_2O_3 aluminium-based metal matrix composites, it was found that polycrystalline diamond (PCD) drills outperformed all other drills in terms of drilled-hole quality and minimum-drilling forces induced (Ramulu et al. 2002). Coating of diamond by chemical vapour deposition (CVD) on tools is found to be suitable for machining of MMC. The high machining costs caused by the extreme abrasive tool wear can be reduced using CVD diamond technology (Teti 2002).The ductile matrix is the constituent that mostly influence the grind ability of MMCs, rather than the hard reinforcement (Ilio and Paoletti 2000).

With the change in the diameter of magnetic abrasive particle, both stock removal and surface finish was affected, whereas the finishing pressure depended only on the magnetic flux density and was independent of the size of the abrasive particle and magnetic abrasive particle size (Shinmura et al. 1987). When machining of the MMC A359 aluminium matrix composite reinforced with 20 % by volume fraction silicon carbide particles it was found that the reinforcement material, type of reinforcement (particle or whisker), volume fraction of the reinforcement and matrix properties as well as the distribution of these particles in the matrix are the factors that affect the overall machinability of these composites (Dandekar and Shin 2012).

2 Experimental Design and Procedure

2.1 Experimental Setup

The experimental set-up has been designed and developed to carry out the present research work, keeping in view the objectives and various design considerations and constraints. The experimental set-up for finishing of plane work pieces using magnetic abrasive finishing process consists of an electromagnet, AC-to-DC convertor, variac (to provide variable DC supply), nylon fixture, radial drilling machine and diamond-sintered magnetic abrasives. The electromagnet is installed on the machining table fitted with wooden board.

The electromagnet consists of enamelled copper wire of 26 standard wire gauge (SWG) wound over cylindrical section of 400 S stainless steel having 60 mm core diameter. The electromagnet is connected to the variable DC supply through the variac. The energised electromagnet produces magnetic field between the magnetic pole and the work piece. A nylon fixture is prepared to hold the work piece at a distance of 5 mm above the magnetic pole. The nylon fixture is rotated by holding it in the radial drilling machine spindle having eight different speed variation provisions by the belt drive. A changeover is installed beside the spindle in order to ensure the bidirectional rotation of the nylon fixture. The magnetic abrasive particles are placed on the top surface of the work piece, which are attracted against the rotating work piece by the stationary electromagnet. The magnetic field strength was varied for the experimentation with the help of the variac (Fig. 1).

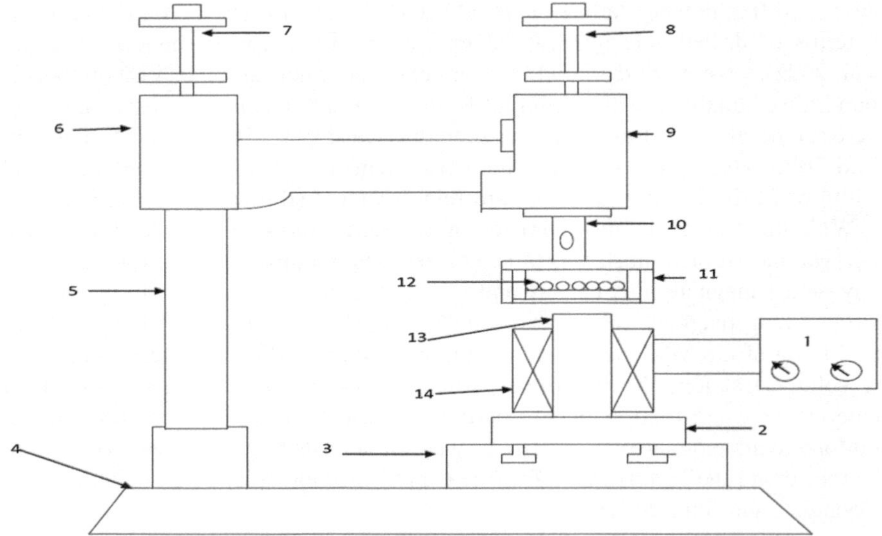

Fig. 1 Experimental Set-up. Parts name: *1* variac DC power supply, *2* wooden base, *3* machining table, *4* base, *5* column, *6* radial arm, *7* DC motor, *8* driving drill spindle, *9* drill head, *10* drill spindle, *11* fixture, *12* Magnetic abrasive particles, *13* core, *14* coil

2.2 Preparation of Magnetic Abrasive Particles and Work Piece Material

The magnetic abrasive powder was prepared by blending, compacting and sintering diamond and iron powders together with a binder in a furnace at 200 °C. A hard lump of particles formed after sintering is crushed into smaller size particles and sieved to obtain MAP grains of 44 mesh size (335 microns).

Aluminium matrix composite (AMC) having 10 % by volume SiC powder has been selected as a work piece for this experimentation. This material is widely used in different ranges of products. The AMCs provide significant benefits including performance benefits (component lifetime, improved productivity), economic benefits (energy savings or lower maintenance costs) and environmental benefits (lower noise levels and fewer airborne emissions).

The melting of the aluminium (98.41 %) scraps and silicon carbide powder (SiCp – 320 grit size) is carried out in the graphite crucible into the electric furnace 1 and 2, respectively. First, the scraps of aluminium were preheated for 3–4 hs at 450 °C and SiC powder was also heated at 900 °C and both the preheated mixtures were poured into the graphite crucible and kept into the electric furnace 1 at 760 °C temperature. The furnace temperature was first increased above the composites to completely melt the scraps of aluminium and then cooled down just below the components temperature and keep it in a semi-solid state.

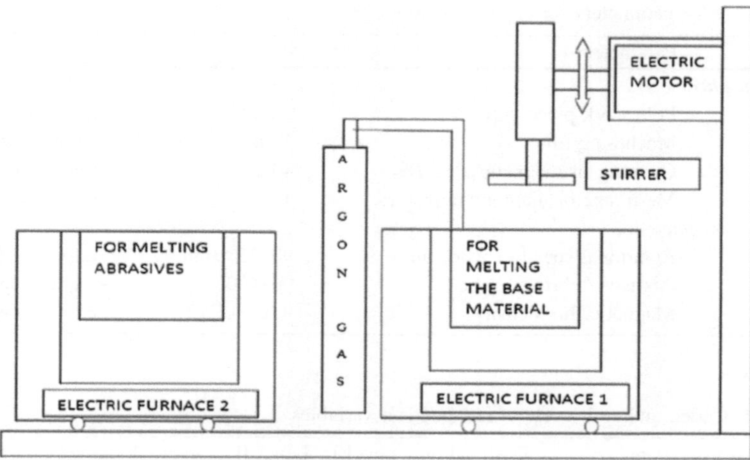

Fig. 2 Experimental set-up for manufacturing composite material

The supply of argon gas is ensured to the electric furnace 1 to hamper the process of oxidation of aluminium, which otherwise will form the aluminium oxide. When the manual mixing is complete, then automatic stirring is carried out for ten minutes with normal 400 rpm of stirring rate. The temperature rate of the electric furnace should be controlled at 760 ± 10 °C in final mixing process. After completing the process, the mould is allowed to solidify (Fig. 2).

2.3 Experimental Design

The experiment was designed on the basis of response surface methodology (RSM) using central composite design (CCD). The practical deployment of CCD often arises through the sequential experimentation.

In the present CCD, the total number of experimental runs came to be 20. So to complete the entire experimentation, 20 experiments were performed in order. The input process parameters and their levels are tabulated in Table 1.

2.4 Selection of Bidirectional Rotation

The preliminary experimentation done on the composite material reveals that there occurs a self-dressing of the abrasives by the bidirectional rotation of the work piece. With increase in the machining time, the dullness of the abrasives increases, and hence, there is a reduction in the percentage improvement of surface finish (PISF) (Table 2).

Table 1 MAF parameters for experimentation

S. No.	Parameter	Range	Mode of selection
Constant parameters			
1	Pole work piece gap	5 mm	Experimentation
2	Machining time	35 min	Experimentation
3	Quantity of magnetic abrasive	40 g	Literature
4	Mesh size of diamond abrasives	44	Literature
Variable parameters			
5	Rotational speed of work piece	90–350 rpm	Experimentation
6	Abrasive/lubricant ratio	90–400	Experimentation
7	Magnetic flux density	0.03–0.12 T	Experimentation

Table 2 Coded and real levels of independent variables

Input process parameter	Designation	Level I −1.68	Level II −1	Level III 0	Level IV 1	Level V 1.68
Rotational speed (RPM)	A	90	145	215	300	350
Magnetic flux density (tesla)	B	0.03	0.05	0.08	0.12	0.14
Abrasive/lubricant ratio	C	400	200	133	100	80

However, due to the bidirectional rotation the orientation of the abrasives changes, and hence, the amount of active grains coming in contact with the work surface increases and there is a noticeable improvement in the PISF (Fig. 3).

3 Experimental Procedure

In order to evaluate the performance of the developed set-up, 20 experiments are carried out at the above-mentioned levels according to the base-design matrix as shown in Table 3. Circular plates of aluminium-based composite material having 120 mm diameter are prepared. The work pieces are turned and faced on lathe to have initial roughness in certain range. Then, the initial surface roughness (Ra) values of the work pieces are measured.

The work piece is held in the nylon fixture, and 40 g of abrasive particles with desired amount of lubricant is placed over the work piece. The work piece is secured with the help of screws, and the nylon fixture is clamped in the radial drilling machine spindle (Table 4).

The electromagnet is placed beneath the nylon fixture and is connected to the variable DC supply. The speed variations are provided to the work piece with the help of belt drive of the radial drilling machine. The magnetic field density of the electromagnetic pole is controlled through the variac. The work piece is rotated for 35 min, having bidirectional rotation of 17.5 min each.

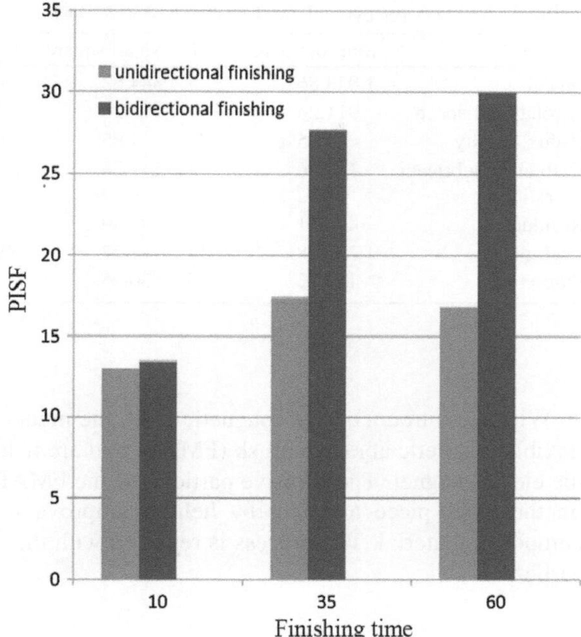

Fig. 3 Effect of unidirectional and bidirectional finishing on percentage improvement on surface finish

Table 3 Experimental data

Experiment No.	Independent parameters			Response
	A	B	C	PISF
1	0.00	−1.68	0.00	22.02
2	−1.00	−1.00	−1.00	27.49
3	0.00	0.00	0.00	19.03
4	0.00	1.68	0.00	31.12
5	0.00	0.00	0.00	26.63
6	0.00	0.00	1.68	18.44
7	0.00	0.00	0.00	25.00
8	0.00	0.00	−1.68	13.67
9	1.68	0.00	0.00	32.20
10	1.00	1.00	1.00	45.54
11	−1.00	−1.00	1.00	20.88
12	1.00	−1.00	−1.00	49.90
13	0.00	0.00	0.00	25.00
14	1.00	−1.00	1.00	51.01
15	0.00	0.00	0.00	26.00
16	0.00	0.00	0.00	25.00
17	−1.68	0.00	0.00	10.82
18	−1.00	1.00	−1.00	15.00
19	1.00	1.00	−1.00	28.91
20	−1.00	1.00	1.00	36.17

Table 4 ANOVA for percentage improvement in surface finish

Source	Sum of squares	df	Mean square	F value	P value	
Model	1,033.86	3	344.62	3.69	0.0342	Significant
A-rotational speed	911.26	1	911.26	9.76	0.0065	
B-flux density	4.85	1	4.85	0.052	0.8227	
C-abrasive/lubricant ratio	117.76	1	117.76	1.26	0.2780	
Residual	1494.21	16	93.39			
Lack of fit	1,339.48	11	121.77	3.93	0.0711	Not significant
Pure error	154.73	5	30.95			

With the introduction of magnetic field, the magnetic abrasive particles form a flexible magnetic abrasive brush (FMAB) and are held in their desired position by the electromagnet. The abrasive particles of the FMAB shear the asperities present on the work piece and thereby help to improve the surface roughness of the composite material. The process is repeated with the different parameters selected earlier.

4 Results and Discussions

Different tests were applied to select the adequate model that fits the performance characteristics namely PISF.

It was observed that PISF the linear model was appropriate. Addition of quadratic and cubic terms does not significantly improve the field because the terms are aliased and the analytical results of the model may be misleading, even though some of the terms were significant.

The interactive effect of parameters has been discussed with the help of surface plots obtained by design expert software. The inferences drawn from the surface plots have been discussed.

4.1 Statistical Inferences

1. The Model F value of 3.69 implies the model is significant. There is only a 3.42 % chance that a "Model F value" this large could occur due to noise.
2. The "Lack of Fit F value" of 3.93 implies there is a 7.11 % chance that a "Lack of Fit F value" this large could occur due to noise.

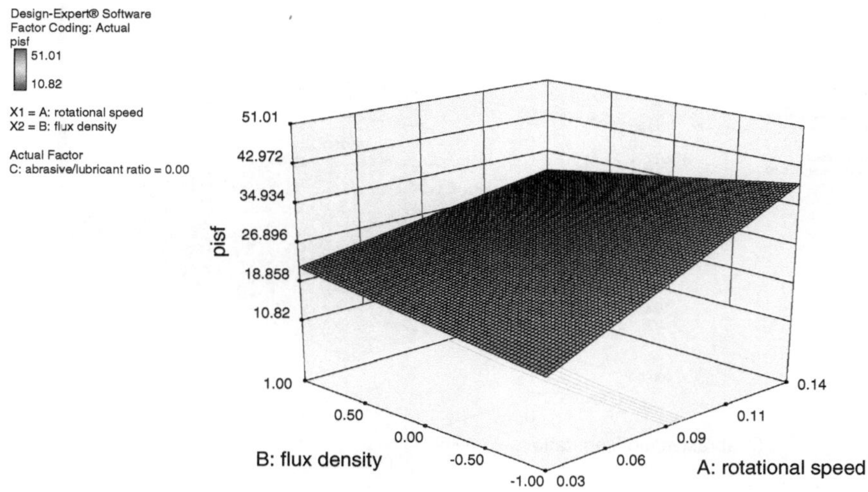

Fig. 4 Interaction effect of magnetic flux density and rotational speed on PISF

Final Equation in Terms of Coded Factors:

$$\text{PISF} = 27.15 + 8.17A - 0.60B + 2.94C.$$

4.2 Interaction Effect of Magnetic Flux Density and Rotational Speed on PISF

Figure 4 shows at higher value of magnetic flux density there is small variation in PISF with the variation of rotational speed. At lower value of magnetic flux density, the PISF increases with increase in rotational speed.

4.3 Interaction Effect of Abrasive/Lubricant Ratio and Rotational Speed on PISF

Figure 5 shows the interaction effect of abrasive/lubricant ratio and rotational speed on percentage improvement of surface finish. At lower values of rotational speed, the PISF is not much predominant with change in abrasive/lubricant ratio where as at higher values of rotational speed the abrasive /lubricant ratio change shows predominant effect on PISF.

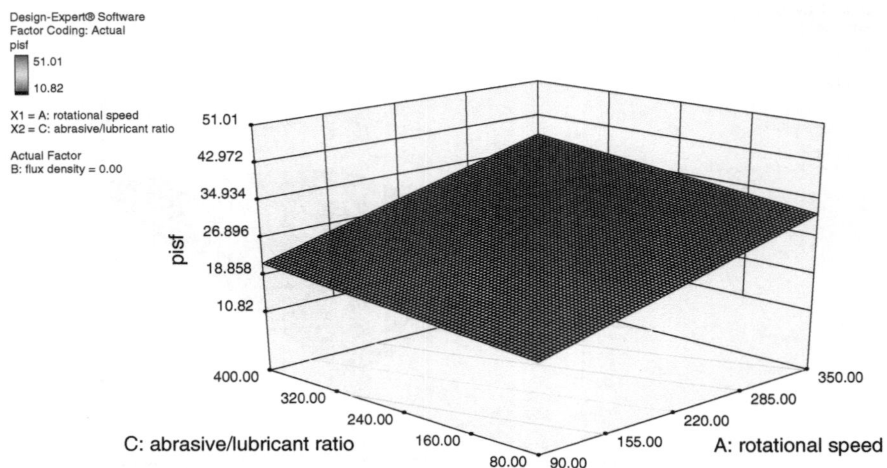

Fig. 5 Interaction effect of abrasive/lubricant ratio and rotational speed on PISF

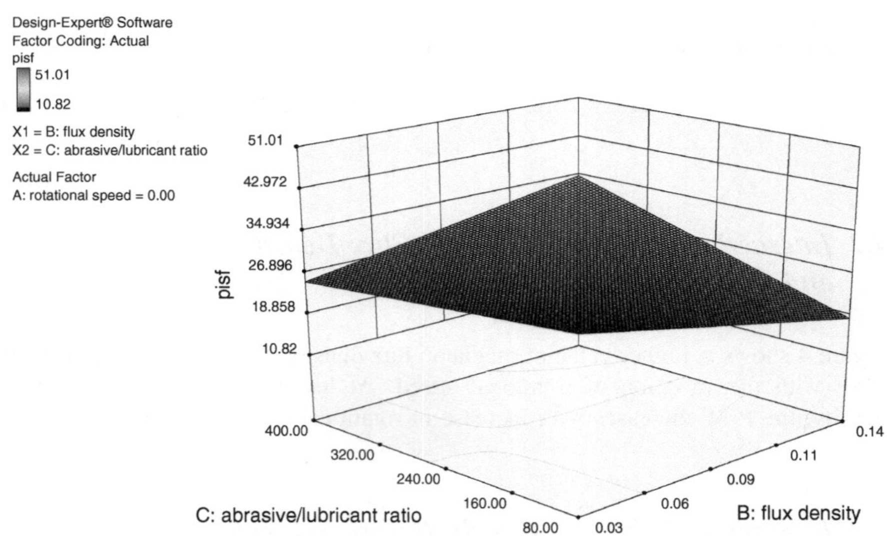

Fig. 6 Interaction effect of magnetic flux density and abrasive/lubricant ratio on PISF

4.4 Interaction Effect of Magnetic Flux Density and Abrasive/Lubricant Ratio on PISF

Figure 6 shows the interaction effect of magnetic flux density and abrasive/lubricant ratio on PISF by keeping rotational speed at a constant value. At lower values of magnetic flux density, there is a sharp decrease in PISF with change in abrasive/lubricant ratio. At higher values of flux density, there is not much significant effect of change in abrasive/lubricant ratio on PISF.

5 Conclusion

The experimental results obtained on the basis of present work indicate that MAF can be employed as a finishing technique for AMC plate. Following conclusion has been drawn in reference to finishing of AMC plate with MAF:

1. The selected process parameters namely magnetic flux density, rotational speed of work piece and abrasive/lubricant ratio significantly influence the percentage improvement in surface finish (PISF) of AMC plate.
2. For a given finishing time, the bidirectional rotation of the magnetic abrasives over the surface causes noticeable improvement in PISF as compared with that by conventional rotation in single direction.
3. There is significant effect of interaction of magnetic flux density (MFD) and rotational speed (RS) on PISF. Low MFD and higher RS gave best results in terms of PISF.
4. The empirical relation between PISF and input process parameters is as follows:

$$\mathrm{PISF} = 27.15 + 8.17A - 0.60B + 2.94C.$$

5. The best surface finish obtained was 1.12 μm in finishing time of 35 min.

References

Dandekar R, Shin C (2012) Modeling of machining of composite materials: a review. Int J Mach Tools Manuf 57:102–121

Girma B, Joshi SS, Raghuram MVGS, Balasubramaniam R (2007) An experimental analysis of magnetic abrasive finishing of plane surfaces. Mach Sci Technol 10(3):323–340

Ilio A, Paoletti A (2000) A comparison between conventional abrasives and superabrasives in grinding of SiC-aluminium composites. Int J Mach Tools Manuf 40:173–184

Jain VK, Kumar P, Behra PK, Jayswal SC (2001) Effect of working gap and circumferential speed on the performance of magnetic abrasive finishing process. Wear 250:384–390

Kremen GZ (1994) Machining time estimation for magnetic abrasive processes. Int J Prod Res 32(12):2817–2825

Kwak JS, Kwak TK (2010) Parameter optimization in magnetic abrasive polishing for magnesium plate. In: 2nd international conference on computer engineering and technology, vol 5, pp 544–547

Mc. Keown PA (1987) The role of precision engineering in manufacturing of the future. Ann CIRP 36/2:495–501

Ramulu M, Rao PN, Kao H (2002) Drilling of (Al_2O_3)p/6061 metal matrix composites. J Mater Process Technol 124:244–254
Shinmura T, Takazawa K, Hatano E (1985) Study of magnetic abrasive process—application to plane finishing. Bull Jpn Soc Precis Eng 19(4):289–291
Shinmura T, Takazawa K, Hatano E (1987) Study of magnetic abrasive process—effects of various types of abrasives on finishing characteristics. Bull Jpn Soc Precis Eng 20(2):79–84
Singh DK, Jain VK, Raghuram V (2005) On the performance analysis of flexible magnetic abrasive finishing brush. Mach Sci Technol 9:601–609
Teti R (2002) Machining of composite materials. Ann CIRP 51(1):611–634

Optimization of Dry Ball Burnishing Process Using Neuro-Fuzzy Interface System and Genetic Algorithm

Joginder Singh and Paramjit Singh Bilga

Abstract The present research paper deals with dry ball burnishing process undertaken to give significant improvements in both surface finish and surface hardness required for most of applications. Aluminum alloy (Al 6061) has been burnished using different burnishing parameters (number of revolution, feed, number of tool passes, and pressure force) with burnishing apparatus. A neuro-fuzzy inference model is generated from the experimental results, and genetic algorithm (GA) is employed to search the optimal solution on the response surfaces modeled by neuro-fuzzy inference system. The absolute average error between the experimental and predicted values from neuro-fuzzy inference model for surface roughness and surface hardness was calculated as 0.05 and 0.18 %. The optimum parameters found by GA in dry ball burnishing are feed 0.157 mm/rev, force 13.91 kg_f, rotational speed 145.09 rpm with two tool passes having response characteristic i.e., surface roughness 0.815 μm and surface hardness 71.3 HRB.

Keywords Burnishing process · Neural network · Adaptive neuro-fuzzy interface system · Genetic algorithm · Optimization

1 Introduction

Ball burnishing is one of the surface finishing processes used for producing surface with very high-quality texture (Fig. 1). Grinding, honing, and lapping are other processes, which are also used to produce fine surface, but these processes are chip

J. Singh (✉) · P. S. Bilga
Department of Mechanical Engineering, Guru Nanak Dev Engineering College, Ludhiana 141006 Punjab, India
e-mail: er.jogindersing@gmail.com

P. S. Bilga
e-mail: psbilga@gndec.ac.in

S. S. Khangura et al. (eds.), *Proceedings of the International Conference on Research and Innovations in Mechanical Engineering*, Lecture Notes in Mechanical Engineering, DOI: 10.1007/978-81-322-1859-3_15, © Springer India 2014

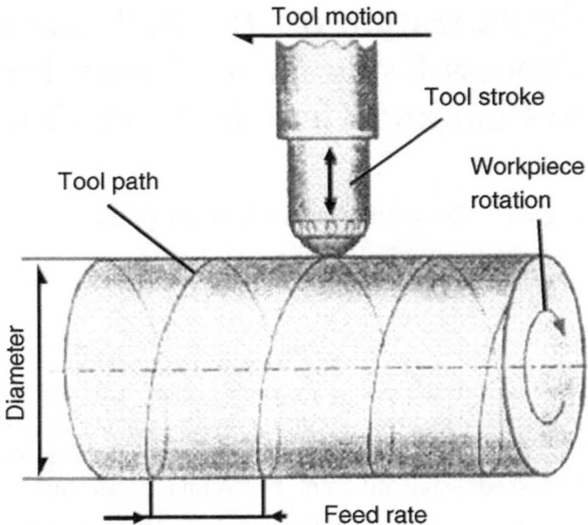

Fig. 1 Dry ball burnishing process

removal processes and always produce surface with geometrical defects in their topography and a metallurgically defective physical structure. Ball burnishing, on the other hand, is a chipless process that gives many advantages in comparison with chip removal processes (El-Axir et al. 2008; Jawalkar and Walia 2009; Singh 2001). The major advantages of burnishing process are decrease in surface roughness, increase in surface hardness, and increase in corrosion resistance, wear resistance, and fatigue life of components (Dabeer and Purohit 2010; El-Axir 2000; Hassan and Maqableh 2000; Ho et al. 2002).

It is cost-effective method of finishing of components as it can be performed on an easily available machine tools (lathe, vertical milling, etc.)

The dry ball burnishing here refers to the ball burnishing process carried out in the ambient air without any lubricants, coolants, etc. El-Axir and Ibrahim introduced a new burnishing tool (El-Axir and Ibrahim 2005). They used the center rest of a lathe as ball burnishing tool. The result of their investigation showed that the surface characteristics were improved with this burnishing tool. Rao et al. studied the performance of the ball and roller burnishing tools on lathe, along with the influence of number of burnishing tool passes on the surface roughness and surface hardness of brass specimens (Rao et al. 2011). Their results revealed that significant improvements in the surface finish and increase in the surface hardness are obtained by the increase in the number of burnishing tool passes in both ball burnishing and roller burnishing on the brass specimens. Fuzzy model is considered as the most effective model for optimization of machining parameters for economical machining processes (Liu 2004). Basak and Goktas (2009) used a fuzzy model to achieve the optimum burnishing parameters for nonferrous components.

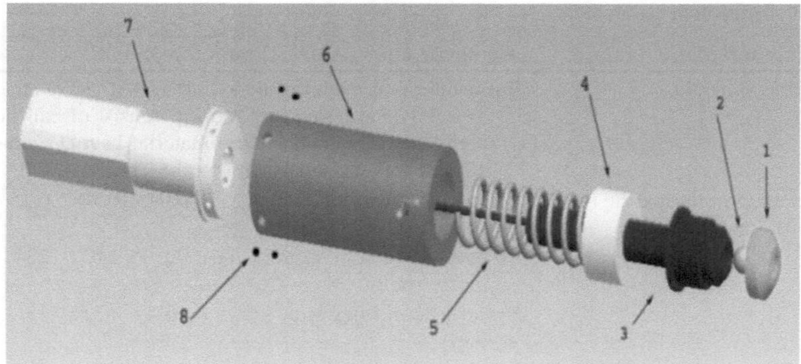

Fig. 2 Ball burnishing tool *1* adapter cover, *2* ball: diameter 10 mm, *3* adapter, *4* spring seat, *5* spring, *6* casing, *7* shank, *8* allen screw

The present research paper deals with the use of an adaptive neuro-fuzzy inference system (ANFIS) to model dry ball burnishing process and effectively employ genetic algorithm (GA) to search the optimal solution on the response surfaces modeled by neuro-fuzzy inference system.

2 Experimental Details

2.1 Ball Burnishing Tool

Figure 2 represents the 3-D drawing of ball burnishing tool that has been designed and fabricated for the present research work. A 10-mm-diameter ball made up of high carbon high chromium material has been used in the burnishing tool.

2.2 Ball Burnishing Process Parameters

Table 1 shows the ball burnishing process parameters considered in the present study based upon the results of pilot experiments and specifications of available lathe machine.

2.3 Measuring Instruments/Apparatus

Digital tachometer model DM 6234-P, Mitutoyo, has been used to measure the rpm. The feed of tool has been measured using a vernier depth gauge along with digital

Table 1 Burnishing parameters

Workpiece material	Aluminum alloy (Al 6061) rod of φ 38 mm					
Burnishing apparatus	Burnishing tool manufactured C3415, SAE 1050 etc. materials and burnishing tool tip consist of ball made of high carbon high chromium material having diameter 10 mm and surface roughness 0.15 μm					
Parameters used during the burnishing process	Longitudinal feed of tool (mm/rev)	x_1	0.045	0.104	0.15	0.208
	Rotational speed of work piece (rpm)	x_2	90	140	224	315
	Vertical burnishing force (kg_f)	x_3	10	12.5	15	18
	Number of tool passes	x_4	2	3	4	5
The machine used for the burnishing process	Lathe machine (model enterprise-1330), manufactured by Kirloskar India Ltd.					

stopwatch. The vertical burnishing force has been noted from deflection of dial gauge and use of spring constant values. Surface roughness values were measured using surface roughness Tester-Model SJ210, Mitutoyo, while surface hardness values were measured using Mitutoyo surface hardness measurement device. For each sample, three readings for input parameters viz. rpm, depth gauge, stop watch and force, and output parameters viz. surface roughness and surface hardness were measured and the arithmetic mean of the measured values with uncertainty (Holman 1994; Kline and McClintock 1953) was taken as the result.

3 Adaptive Neural Fuzzy Inference System (ANFIS)

Using a given input/output data set, the toolbox function ANFIS constructs a fuzzy inference system (FIS) whose membership function parameters are tuned (adjusted) using either a back propagation algorithm alone, or in combination with a least squares type of method. This allows your fuzzy systems to learn from the data they are modeling (Fig. 3).

3.1 ANFIS Editor GUI

These tools apply fuzzy inference techniques to data modeling. Using a given input/output data set, the toolbox function ANFIS constructs a FIS whose membership function parameters are tuned (adjusted) using either a back propagation algorithm alone, or in combination with a least squares type of method. In order to design an ANFIS model, the following steps must be performed (Chaturvedi 2010):

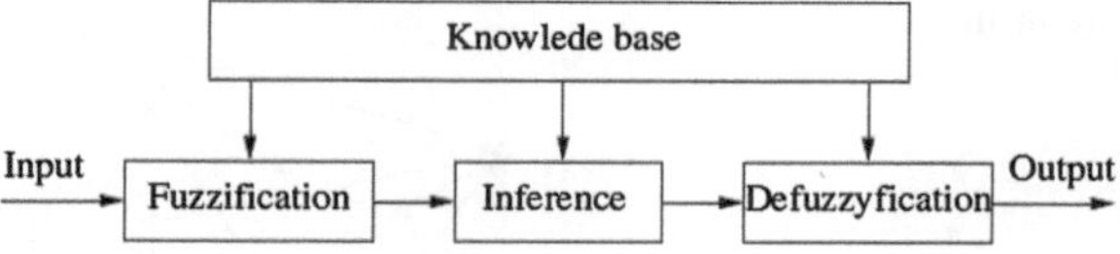

Fig. 3 Adaptive Neuro-Fuzzy Inference System

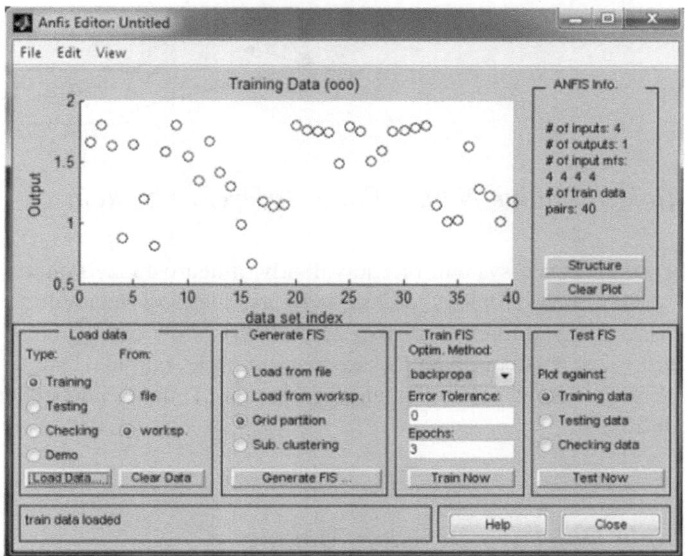

Fig. 4 ANFIS editor GUI

1. Load data (training, testing, and checking) by selecting appropriate radio buttons in the load data portion of the GUI and then clicking load data. The loaded data is plotted on the plot region (shown in Fig. 4).
2. Generate an initial FIS model or load an initial FIS model using the options in the generate FIS portion of the GUI.
3. View the FIS model structure once an initial FIS has been generated.
4. Choose the FIS model parameter optimization method: back propagation or a mixture of back propagation and least squares (hybrid method).
5. Choose the number of training epochs and the training error tolerance.
6. Train the FIS model by clicking the Train Now button.
7. This training adjusts the membership function parameters and plots the training (and/or checking data) error plot(s) in the plot region.

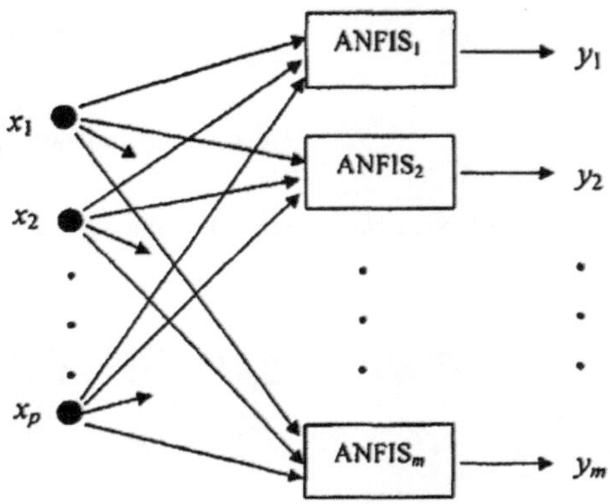

Fig. 5 Architecture of the MANFIS network

3.2 Multiple Adaptive Neuro-Fuzzy Inference System

A neuro-fuzzy inference system, or equivalently, a neuro-fuzzy system is a fuzzy inference system that employs neural network learning techniques. Multiple adaptive neuro-fuzzy inference system (MANFIS) is an extension of a single-output neuro-fuzzy system, ANFIS, so that multiple outputs can be handled. MANFIS (Fig. 5) can be viewed as an aggregation of many independent ANFIS (Chaturvedi 2010).

3.3 Multiple Objective Decision Making

To handle different objectives simultaneously, the multiple objective decision making (MODM) approach can be used. This MODM problem can be represented by (Cheng et al. 2002):,

$$\begin{aligned} &\max\ \lambda \\ \text{s.t}\quad &\lambda = \min\{\mu_{\hat{y}i}(\hat{y}i)\}, \quad i = 1,\ldots,m \\ & \qquad\qquad\qquad\qquad x \in B \end{aligned} \tag{1}$$

The membership function $\mu_{\hat{y}_i}(\hat{y}_i)$ can be defined based on the different categories. For the responses belonging to Category 1, or the-larger-the-better category, we assume that the maximum satisfaction is reached when the membership function reaches 1, or when $\hat{y}_{i(\max)} = \max_{x\in B}\{\hat{y}_i\}$. The membership function then decreases linearly to 0 at $\hat{y}_{i(\max)} = \min_{x\in B}\{\hat{y}_i\}$. The membership function, for $\hat{y}_i, I \in L$, can be represented as

$$\mu_{\hat{y}_i}(\hat{y}_i) = \begin{cases} 1, & \text{if } \hat{y}_i > \hat{y}_i^*, \\ \frac{\hat{y}_i - \hat{y}_i^-}{\hat{y}_i^* - \hat{y}_i^-}, & \text{if } \hat{y}_i^- < \hat{y}_i < \hat{y}_i^*, \quad \forall i \in L \\ 0 & \text{if } \hat{y}_i \leq \hat{y}_i^-, \end{cases} \tag{2}$$

For the responses belonging to Category 2, or the-smaller-the-better category, the membership function should be the reverse of Category 1. This membership function can be represented by

$$\mu_{\hat{y}_i}(\hat{y}_i) = \begin{cases} 1, & \text{if } \hat{y}_i < \hat{y}_i^-, \\ \frac{\hat{y}_i - \hat{y}_i^-}{\hat{y}_i^* - \hat{y}_i^-}, & \text{if } \hat{y}_i^- \leq \hat{y}_i \leq \hat{y}_i^*, \forall i \in S \\ 0, & \text{if } \hat{y}_i > \hat{y}_i^*, \end{cases} \tag{3}$$

Since the exact functional form of $\hat{y}_i$ is implicitly given through the trained MANFIS network, Eq. (1) cannot be solved easily in a straight forward fashion by the generally used optimization approaches. Derivative-free approaches such as GA and simulated annealing are ideally suited for solving problems where derivative information is unavailable. In this study, we will apply GA to solve Eq. (1).

4 Results

Burnishing process has been carried out in accordance with the parameters (Force, Feed, Number of revolution, and Number of passes) which are given in Table 1. After that MANFIS has been made that is an extension of a single-output neuro-fuzzy system, ANFIS, so that multiple outputs can be handled and optimized.

The experiments have four design variables and two responses. The design variables are Feed (x_1), Force (x_2), rpm (x_3), and number of passes (x_4) and the responses are surface roughness (y_1), and surface hardness (y_2). The problem is to minimize y_1 and maximize y_2. The experimental design data has been listed in Table 2.

The MANFIS network is used to model the response surfaces of this process. There are two output nodes corresponding to the two responses, which can be viewed as consisted of two independent ANFIS networks. Each ANFIS network has four input nodes and each input node associates with two nodes in Layer 1, which produces a combination of forty-eight nodes in Layers 2–4. The learning of the MANFIS network was carried out by training two separate ANFIS networks with the data listed in Table 2 as the training data. Using an ANFIS-constructing function in the fuzzy toolbox of MATLAB software, the training process of these two ANFIS networks can be carried out and the convergence rates are shown in Figs. 6 and 7 for responses y_1 and y_2, respectively. A comparison of the original data with the converged results obtained is shown in Fig. 8 for the percent conversion response (y_1). Similarly for y_2 is shown in Fig. 9.

Table 2 Experimental results of surface roughness and surface hardness for various values of the feed (mm/rev), number of revolutions (rev/min), number of passes (n), and force (kg_f)

S. N.	Feed (mm/sec) (%)	Force (Kgf) (%)	RPM (%)	No. of tool passes	Surface roughness (μm) (%)	Surface hardness (HRB) (%)
1	0.043 ± 0.9	10.07 ± 1.1	93.4 ± 0.1	2	1.66 ± 0.5	64 ± 1.8
2	0.043 ± 0.9	12.59 ± 1.1	93.4 ± 0.1	2	1.63 ± 0.4	64.3 ± 1.9
3	0.043 ± 0.9	15.07 ± 0.9	93.4 ± 0.1	2	1.51 ± 1.2	65.7 ± 1.8
4	0.043 ± 0.9	18.04 ± 1	93.4 ± 0.1	2	1.10 ± 5.5	67 ± 0.9
5	0.100 ± 1	10.07 ± 1.1	93.4 ± 0.1	2	1.64 ± 0.7	63 ± 1.8
6	0.100 ± 1	12.59 ± 1.1	93.4 ± 0.1	2	1.20 ± 0.3	64 ± 1.8
7	0.100 ± 1	15.07 ± 0.9	93.4 ± 0.1	2	1.10 ± 5.5	65.3 ± 1.3
8	0.100 ± 1	18.04 ± 1	93.4 ± 0.1	2	1.42 ± 0.6	66.3 ± 1.8
9	0.100 ± 1	10.07 ± 1.1	142.8 ± 0.2	2	1.62 ± 1.1	66 ± 1.7
10	0.100 ± 1	10.07 ± 1.1	221.8 ± 0.1	2	1.55 ± 1.1	67 ± 2.3
11	0.100 ± 1	10.07 ± 1.1	328.8 ± 0.1	2	1.35 ± 1.5	67.3 ± 1.8
12	0.100 ± 1	12.59 ± 1.1	142.8 ± 0.2	2	1.19 ± 1	65.3 ± 2.2
13	0.100 ± 1	12.59 ± 1.1	221.8 ± 0.1	2	1.15 ± 0.8	65.7 ± 2.7
14	0.100 ± 1	12.59 ± 1.1	328.8 ± 0.1	2	1.14 ± 1.3	66 ± 1.7
15	0.100 ± 1	15.07 ± 0.9	142.8 ± 0.2	2	0.99 ± 2.3	67 ± 1.7
16	0.100 ± 1	15.07 ± 0.9	221.8 ± 0.1	2	0.78 ± 2.4	67.3 ± 1.3
17	0.100 ± 1	15.07 ± 0.9	328.8 ± 0.1	2	1.06 ± 1.8	68 ± 0.8
18	0.100 ± 1	18.04 ± 1	142.8 ± 0.2	2	1.35 ± 1.3	68.3 ± 1.3
19	0.100 ± 1	18.04 ± 1	221.8 ± 0.1	2	1.48 ± 1	69 ± 3.6
20	0.100 ± 1	18.04 ± 1	328.8 ± 0.1	2	1.52 ± 1	69.3 ± 2.5
21	0.100 ± 1	10.07 ± 1.1	221.8 ± 0.1	3	1.54 ± 0.8	67.3 ± 1.8
22	0.100 ± 1	10.07 ± 1.1	221.8 ± 0.1	4	1.63 ± 0.5	68 ± 2.2
23	0.100 ± 1	10.07 ± 1.1	221.8 ± 0.1	5	1.68 ± 0.5	69 ± 0.8
24	0.100 ± 1	12.59 ± 1.1	221.8 ± 0.1	3	1.03 ± 2	67 ± 1.7
25	0.100 ± 1	12.59 ± 1.1	221.8 ± 0.1	4	1.48 ± 1.2	68 ± 0.8
26	0.100 ± 1	12.59 ± 1.1	221.8 ± 0.1	5	1.65 ± 0.5	70 ± 0.8
27	0.100 ± 1	15.07 ± 0.9	221.8 ± 0.1	3	1.12 ± 1.3	70 ± 1.6
28	0.100 ± 1	15.07 ± 0.9	221.8 ± 0.1	4	1.25 ± 0.7	72 ± 1.6
29	0.100 ± 1	15.07 ± 0.9	221.8 ± 0.1	5	1.32 ± 1.2	73 ± 0.8
30	0.100 ± 1	18.04 ± 1	221.8 ± 0.1	3	1.52 ± 0.6	71 ± 0.8
31	0.100 ± 1	18.04 ± 1	221.8 ± 0.1	4	1.72 ± 0.4	72.3 ± 0.5
32	0.100 ± 1	18.04 ± 1	221.8 ± 0.1	5	1.75 ± 0.9	74 ± 0.8
33	0.146 ± 1	10.07 ± 1.1	93.4 ± 0.1	2	1.32 ± 0.9	62 ± 0.9
34	0.146 ± 1	12.59 ± 1.1	93.4 ± 0.1	2	1.1 ± 1.1	63 ± 1.8
35	0.146 ± 1	15.07 ± 0.9	93.4 ± 0.1	2	1.03 ± 0.9	65 ± 1.8
36	0.146 ± 1	18.04 ± 1	93.4 ± 0.1	2	1.52 ± 1	66.3 ± 1.8
37	0.198 ± 1	10.07 ± 1.1	93.4 ± 0.1	2	1.2 ± 0.7	61 ± 0.9
38	0.198 ± 1	12.59 ± 1.1	93.4 ± 0.1	2	1.08 ± 0.8	63 ± 1.8
39	0.198 ± 1	15.07 ± 0.9	93.4 ± 0.1	2	0.92 ± 3.8	64 ± 1.8
40	0.198 ± 1	18.04 ± 1	93.4 ± 0.1	2	1.68 ± 0.5	65 ± 2.4

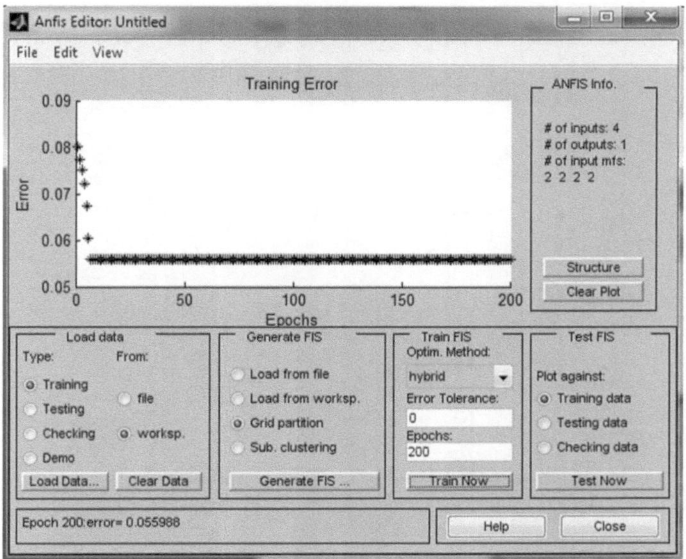

Fig. 6 Convergence rate of y_1

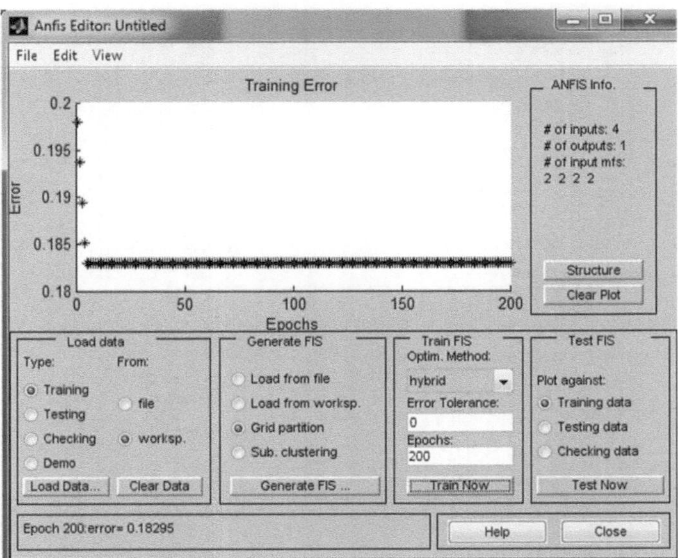

Fig. 7 Convergence rate of y_2

After the completion of the training of the MANFIS network, the multiple response problem can be optimized using the formulation of Eq. (1). The membership function for the percent conversion response, y_1, which belongs to the

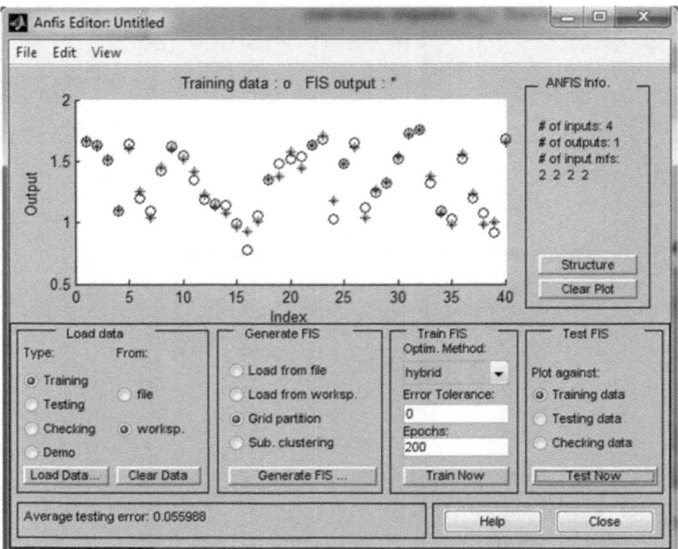

Fig. 8 Comparison of the original data with the converged results for y_1

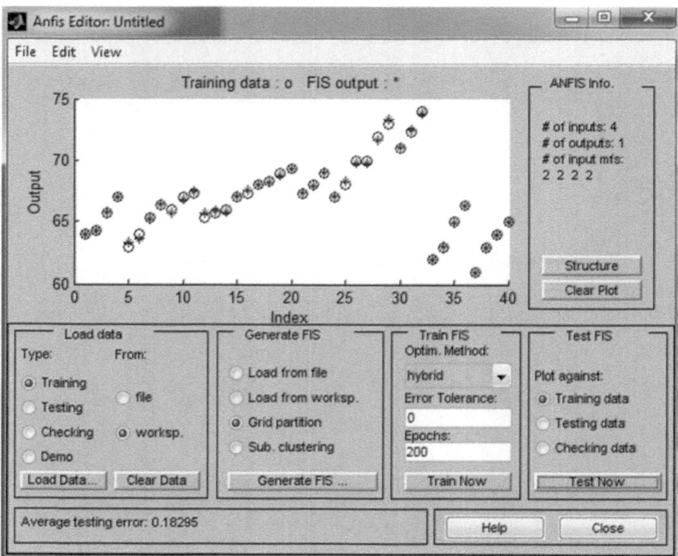

Fig. 9 Comparison of the original data with the converged results for y_2

category of the-smaller-the-better, should take the form of Eq. (3) and the membership function for the other response, y_2, which belongs to the category of the-larger-the-better, should take the form of Eq. (2). The limits, or the maxima and

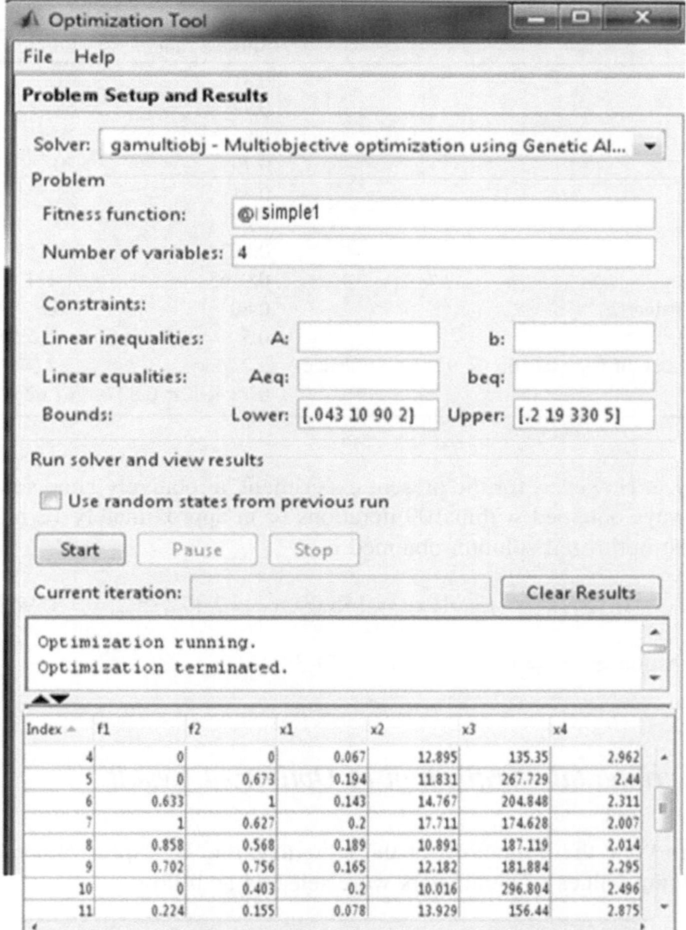

Fig. 10 Optimization tool box

the minima of these membership functions can be obtained either based on the given data or by solving the optimization problems for each response. For example, since the desired response of dry ball burnishing surface roughness must be between 0.66 and 1.8 μm, it is quite reasonable to set 0.66 and 1.8 μm as the minimum and maximum limits, respectively, for this response. Similarly, the minimum and maximum for surface hardness response are set as 55 and 73 HRB, respectively. The possible ranges for x_1, x_2, x_3 and x_4 are set within the interval [0.043, 10, 90, 2] to [0.2, 19, 330, 5]. The GA optimization algorithm was implemented on the MATLAB platform (as shown in Fig. 10) and run it. The GA algorithm is usually time-consuming and needs much iteration to obtain

Table 3 Experimental verification of optimized result

Experiment	Surface roughness	Surface hardness
1	0.8	71
2	0.78	68
3	0.82	70
4	0.79	70
5	0.81	67
6	0.82	69
Mean	0.803	69.16
Std. deviation	0.0163	1.471
Calculated value (μ)	0.80	70.5
t-value	0.5	−2.2188
Tabulated value of test statistic (at 95 % confidence)	±2.571	±2.571
Range	$0.78 < \mu < 0.82$	$67.62 < \mu < 70.71$

convergence. However, for the present experiment, a relatively good solution was almost always obtained within 100 iterations or in approximately 12 min.

The best optimized solution obtained is

$$x_1 = 0.157, x_2 = 13.91, x_3 = 145.09, x_4 = 2.172 \cong 2, \lambda = 0.964$$

with the responses $y_1 = 0.815$ and $y_2 = 71.39$.

4.1 Experimental Verification of Optimized Result

Keeping in view the limitations of the lathe machine in respect of rpm and feed rate, optimum values of parameters were selected as follows:

$$x_1 = 0.15, x_2 = 14, x_3 = 142.8, x_4 = 2$$

with the responses $y_1 = 0.80$ and $y_2 = 70.5$.with the selected optimum conditions, verification experiments were repeated six times. Surface roughness and surface hardness values obtained were measured and recorded in Table 3. The estimated value of surface roughness and surface hardness at the stated process parameters has been calculated using MANFIS model. The calculated value was compared with the mean observed value. For this comparison, student t-test has been applied with the test statistic and hypothesis as follows:

$$t = \frac{x_0 - x_c}{\mathrm{SD}/\sqrt{n}} \tag{4}$$

Hypothesis H_0: $x_0 = x_c$.

Where x_0 is the mean of the observed values, x_c is the calculated value, SD is the sample standard deviation, and n is total number of experiments.

It is known that if the calculated value of test statistic is less than the tabulated value, then hypothesis was checked at 95 % level of significance. For six experiments (degree of freedom = 5), the tabulated value of test statistic is ±2.571 (Montgomery 2001).

Since the t-values 0.5 and −2.21 are less than the tabulated value (±2.571), it can be concluded that x_0 and x_c are not statistically different. Therefore, within the specified range, the optimal values of process parameters and response characteristic (surface roughness and surface hardness) given by the developed response optimization model (MANFIS) are validated.

5 Conclusion

The experimental observations reveal that dry ball burnishing tool could produce surface roughness of the order of 0.78 ± 2.4 % μm and surface hardness of 74 ± 8 % HRB. The neuro-fuzzy inference system has been successfully employed to model the dry ball burnishing process using minimum number of experimental data. The absolute average error between the experimental and predicted values and neuro-fuzzy inference model for surface roughness and surface hardness obtained are 0.05 and 0.18 %.

The optimum parameter values suggested by genetic algorithm in dry ball burnishing process are feed 0.157 mm/rev, force 13.91 kg_f, rotational speed 145.09 rpm with two tool passes having response characteristic i.e., surface roughness 0.815 μm and surface hardness 71.3 HRB. The optimal values of process parameters and response characteristic (surface roughness and surface hardness) have been validated at 95 % confidence with experimental results within the specified range. The results of this research work reveal that percentage improvement in surface roughness using dry ball burnishing is 75.6 % and for hardness percentage improvement is 34.5 %.

References

Basak H, Goktas HH (2009) Burnishing process on Al-alloy and optimization of surface roughness and surface hardness by fuzzy logic. Mater Des 30:1275–1281

Chaturvedi DK (2010) Modeling and simulation of systems using MATLAB and Simulink. CRC Press, Boca Raton, p 709

Cheng CB, Cheng CJ, Lee ES (2002) Neuro-fuzzy and genetic algorithm in multiple response optimization. Comput Math Appl 44:1503–1514

Dabeer PS, Purohit GK (2010) Effect of ball burnishing parameters on surface roughness using surface roughness methodology. Adv Prod Eng Manage 5:111–116

El-Axir MH (2000) An investigation into roller burnishing. Int J Mach Tools Manuf 40:1603–1617

El-Axir MH, Ibrahim AA (2005) Some surface characteristics due to center rest ball burnishing. J Mater Process Technol 167:47–53

El-Axir MH, Othman OM, Abodiena AM (2008) Study on the inner surface finishing of aluminum alloy 2014 by ball burnishing process. J Mater Process Technol 202:435–442

Hassan AM, Maqableh AM (2000) The effects of initial burnishing parameters on non-ferrous components. J Mater Process Technol 102:115–121

Ho SY, Lee KC, Ho SJ (2002) Accurate modeling and prediction of surface roughness by computer vision in turning operations using an adaptive neuro fuzzy inference system. Int J Mach Tools Manuf 42:1441–1446

Holman JP (1994) Experimental methods for engineers, 6th edn. McGraw-Hill, New York

Jawalkar CS, Walia RS (2009) Study of roller burnishing process on En-8 specimens using design of experiments. J Mech Eng Res 1:38–45

Kline SJ, McClintock FA (1953) Describing uncertainties in single-sample experiments. Mech Eng 75:3

Liu ST (2004) Fuzzy geometric programming approach to a fuzzy machining economics model. J Prod Reach 42(16):3253–3269

Montgomery DC (2001) Design and analysis of experiments, 5th edn. Wiley, New York

Rao JNM, Reddy CK, Rao PV (2011) Experimental investigation of the influence of burnishing tool passes on surface roughness and hardness of brass specimens. Ind J Sci Technol 4:1113–1118

Singh R (2001) Some investigations into the burnishing process using different lubricants. M. Tech. Thesis, Mechanical and Production Engineering Department, G.N.D.E.C, Ludhiana

A Review on Advances in Wire Electrical Discharge Machining

Pratik A. Patil and C. A. Waghmare

Abstract Wire electrical discharge machining has become an important non-traditional machining process, as it provides an effective solution for producing components made of difficult-to-machine materials such as titanium, zirconium and intricate shapes, which are not possible by conventional machining methods. Due to large number of process parameters and responses, lots of researchers have attempted to optimize the process parameters. This paper reviews the advances in research of WEDM on relation between different process parameters, include pulse-on time, pulse-off time, servo voltage, peak current, dielectric flow rate, wire speed, wire tension on different process responses include material removal rate (MRR), cutting speed (Vc), surface roughness (Ra), and wire wear ratio (WWR) of the wire electrode. Effect of composition of material on the manufacturability of wire EDM has also been reviewed. And at last, wire failure analysis is discussed.

Keywords Material removal rate (MRR) · Taguchi method · Wire electrical discharge machining · Wire wear ratio (WWR)

1 Introduction

The electrical discharge machining (EDM) technology has developed rapidly in the recent years and has become important in precision manufacturing applications such as die and mold making, micro machining. Wire electrical discharge machining (WEDM) is a modified electrical discharge technique used for manufacturing components with intricate shapes and profiles, with the help of a numerically controlled traveling wire electrode. Material is eroded from the

P. A. Patil (✉) · C. A. Waghmare
Rajarambapu Institute of Technology, Rajaramnagar, Sangli, Maharastra, India
e-mail: papatil4@gmail.com

S. S. Khangura et al. (eds.), *Proceedings of the International Conference on Research and Innovations in Mechanical Engineering*, Lecture Notes in Mechanical Engineering,
DOI: 10.1007/978-81-322-1859-3_16, © Springer India 2014

workpiece by a series of discrete sparks between the workpiece and the wire electrode (tool) separated by a thin film of dielectric fluid (Janardhan and Samuel 2010). Whereas the wire does not touch the workpiece, so there is no physical pressure imparted on the workpiece and amount of clamping pressure required to hold the workpiece is minimal. Although electrical conductivity is an important factor in this type of machining, some techniques can be used to increase the efficiency in machining of low electrical conductive materials. The spark theory on a wire EDM is basically the same as that of the vertical EDM process. Many sparks can be observed at one time. This is because actual discharges can occur more than one hundred thousand times per second. The heat of each electricals park is around 15,000–21,000 °F. This process has been widely used in aerospace, nuclear and automotive industries, to machine precise, complex, and irregular shapes in various difficult-to-machine electrically conductive materials.

2 Parameters Affecting the Performance of WEDM

2.1 Pulse-on Time and Pulse-off Time

Electric discharge machining must occur (on time) and stop (off time) alternately during machining. During the on time, the voltage is applied to the gap between the workpiece and the electrode (wire), while no voltage is placed during the off time. Consequently, electric discharge occurs only for the duration of the on time. To have a long duration of electric discharge, it may be possible to select the great value for the on time; however, it may cause a short circuit to occur, resulting in wire breakage. To avoid such trouble, the off time must be inserted.

2.2 Peak Current and Gap Voltage

The peak current is basically a most important machining parameter in WEDM. It is the amount of power used in WEDM and measured in amperes. During each pulse-on time, the current increases until it reaches a preset level, which is expressed as the peak current. In both die-sinking and wire EDM processes, the maximum amount of amperage is governed by the surface area of the cut. Higher amperage is used in roughing operations and in cavities or details with large surface areas. Gap voltage or open-circuit voltage specifies the supply voltage to be placed on the gap. The electric discharge energy increases with gap voltage. However, normally these factors are not independent. In other words as the gap voltage increases, the peak current also automatically increases. In some WEDM machines, both of these factors show machining voltage.

2.3 Flushing Pressure

Electric discharge can occur in the air; however, it is not stable and cannot be used for rough cut machining. To obtain stable electric discharge, dielectric fluid is required. Within the dielectric fluid, electric discharge machining can be stabilized with efficient cooling and chip removal. The de-ionized water is typically used as a dielectric in wire EDM because it has environment-friendly characteristics. For example, due to low thermal conductivity in titanium alloy material, high flushing pressure is absolutely necessary for rough machining, otherwise the short-circuit phenomenon will cause to wire breakage.

2.4 Wire Speed and Wire Feed

Wire speed is an important parameter in WEDM that shows the speed of wire in WEDM. As the wire speed increases, the wire consumption also increases, which results in high cost of machining. While low wire speed can cause wire breakage at high cutting speed.

2.5 Wire Tension

Wire tension is the factor that can control the tension of wire in WEDM. If the wire tension is high enough the wire stays straight otherwise wire drags behind as shown in the Fig. 1.

3 An Overview of Advances in the Research of WEDM

Many researchers are working on performance improvement of wire electrical discharge machining. The term performance improvement implies the optimization of the process parameters, which affect the performance of WEDM. Lots of research papers are published that give a clear cut idea of optimization of process parameters. Some of the research papers are reviewed, and gap in the study is analyzed. Thus, the gap will be helpful for further investigation of WEDM.

Turning with WEDM is one of the emerging areas developed to generate cylindrical form on hard and difficult-to-machine materials by adding a rotary axis to WEDM. Several researchers have contributed to the development of the wire electrical discharge turning (WEDT). The optimization is carried out by analyzing the effect of machining parameters on the material removal rate (MRR), surface roughness and roundness error, using the pulse train data acquired at the spark gap.

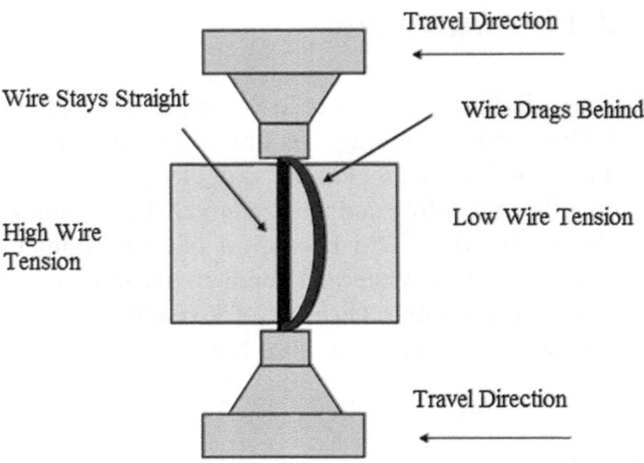

Fig. 1 Effect of wire tension

To achieve this objective, a simple and cost-effective spindle is developed for the WEDT process. Pulse train is acquired with a data acquisition system developed in the work. A pulse discrimination algorithm has been developed for classifying the discharge pulses into open circuit, normal, arc, and short-circuit pulses. With the help of algorithm, the number of arc regions, average ignition delay time, and the width of the normal and arc regions in the data acquired can also be obtained. The author has observed that the rotation of the workpiece has significant influence on the type of the discharges occurring at the spark gap. Preliminary experiments were conducted to compare the WEDM and WEDT processes disclosed that MRR is less in WEDT and the number of arcs and arc regions are more in WEDT. Also they have observed that the surface roughness and roundness error of the WEDT components are influenced by the occurrence of arc regions, width of arc, normal discharge regions, and average ignition delay time (Janardhan and Samuel 2010).

In die-sinking EDM and WEDM, all sparks occur in a dielectric that plays an important role in the process. In WEDM, the dielectric is flushed in the gap space, and its common material is liquid oil. Use of mineral oil-based dielectric liquids is the major cause of environmental concerns associated with the EDM process and its alternatives such as WEDM process. Dry WEDM is an environment-friendly modification of the oil WEDM process in which the liquid dielectric is replaced by a gaseous medium. Dielectric wastes generated during the oil WEDM process are very toxic and can not be recycled. Also, toxic fumes are generated during machining due to high-temperature chemical breakdown of mineral oils. The use of oil as the dielectric fluid also makes it necessary to take extra precaution to prevent fire hazards. Replacing liquid dielectric by gases is an emerging field in the environment-friendly EDM technology. The effect of pulse-on time, pulse-off time, gap set voltage, wire tension on cutting velocity, and surface roughness are analyzed. The experiments were conducted in dry conditions i.e., air was used as dielectric medium. Thus, it was concluded that air at inlet pressure of 1.5 bar leads to high MRR and low surface roughness. On the other hand, nitrogen gas has very low cutting velocity with respect to other gases. Further, it was concluded that cutting velocity and surface roughness increases by increasing the discharge current and pulse-on time. But increase in pulse-on time leads to low oversize while increasing discharge current firstly increases the oversize then decreases it (Shayan and Afza 2013).

In WEDM operations, MRR determines the economics of machining and rate of production but surface roughness is the measure of quality. Proper selection of process parameters is essential to obtain good surface finish and higher MRR. In setting the machining parameters, particularly in rough cutting operation, the goal is the maximization of MRR and minimization of SF. The machine tool builder provides machining parameter table to be used for setting optimal machining parameters, but in practice, it is very difficult to utilize the optimal functions of a machine owing to there being too many adjustable machining parameters. This process relies heavily on the experience of the operators. With a view to alleviate this difficulty, various investigations have been carried out by several researchers for improving selection of optimal parametric values for the MRR, surface finish. However, the problem of selection of machining parameters is not fully depending on machine controls rather material dependent. To improve manufacturing processes with single performance characteristic, the optimal selection of process parameters using Taguchi method has been extensively adopted. Traditional Taguchi method cannot solve multi-objective optimization problems. To overcome this, in Subrahmanyam and Sarcar (2013), the Taguchi method coupled with Grey relational analysis has a wide area of application in manufacturing processes. The work demonstrated the optimization of WEDM process parameters for the machining of H13 hot die steel, with multiple responses such as MRR, surface roughness (Ra) based on the Grey–Taguchi Method. Taguchi'sL27 (21 × 38) Orthogonal Array was used to conduct the experiments, which corresponds to randomly chosen different combinations of process parameter setting, with eight process parameters: TON, TOFF, IP, SV WF, WT, SF, and WP each to be varied in three different levels. Data related to each response, viz. MRR, surface roughness (Ra) have been measured for each experimental run. With Grey Relational Analysis, optimal levels of process parameters were identified. The relatively significant parameters were determined by analysis of variance. The variation of output responses with process parameters was mathematically modeled using nonlinear regression analysis. While applying the Grey–Taguchi method using L36 orthogonal array, it is observed that the MRR increased, surface roughness reduced, which are positive indicators of efficiency in the machining process. Thus, it was concluded that the Grey–Taguchi method is the most ideal and suitable for the parametric optimization of the wire-cut EDM process, when using the multiple performance characteristics such as MRR, surface roughness for machining the H13, or for the other material. A mathematical relation between the machining parameters and performance characteristics was established by the regression analysis method. The established mathematical models can be used in estimating the MRR, surface roughness without conducting experiments. Also in Kumar et al. (2010), the optimization of WEDM process parameters of Incoloy800 super alloy with multiple performance characteristics such as MRR, surface roughness, and Kerf is carried out based on Taguchi coupled with grey relational analysis. Thus, relation was established between the input parameters namely gap voltage, pulse-on time, pulse-off time, and wire feed and performance characteristics such as MRR, Ra, and kerf by the regression analysis method. The established mathematical models can be used in estimating the MRR, surface roughness, and kerf width without conducting experiments.

Machined surface produced by the wire EDM process consisted of many craters resulted from vaporization and melting of a small amount of material caused by each individual spark. The removal per pulse is related to the energy of each discharge pulse, which depends on the product of the discharge voltage, peak current, and the pulse duration of discharge current. The discharge voltage is kept almost constant during machining. Therefore, the discharge energy is almost dependent on the peak current and the pulse duration of the discharge current. Higher peak current and longer pulse duration result in larger craters and thus achieves higher metal removal rate and rough surface. After each spark, a quantity of molten material will be drawn back into the spark crater by surface tension and cooling effect. This resolidified material results in the recast layer of the wire EDMed surface. Since deionized water is used as a dielectric liquid and the wire electrode has a negative polarity in wire EDM, direct current passing through water causes ions to react chemically with the machined part, and an electrolytic effect increases the chemical erosion effect of the water dielectric. The recast layer accompanied by the electrolytic and corrosive effect can cause cracking, pitting, or flaking of the machined surface, a decrease of fatigue strength and an increase of surface roughness and hence contributing to the degradation of surface integrity and premature failure of the machined part. For overcoming this difficulty, in Yan and Lai (2007) a new fine-finish power supply in wire EDM is developed. The transistor-controlled power supply composed of a full-bridge circuit, two snubber circuits and a pulse control circuit was designed to provide the functions of anti-electrolysis, high frequency, and very low-energy pulse control. Test results indicated that the pulse duration of discharge current can be shortened through the adjustment of capacitance in parallel with the sparking gap. High value of capacitance contributes to longer discharge duration. A high current-limiting resistance results in the decrease of discharge current. The developed power supply using anti-electrolysis circuitry and CPLD-based pulse control circuit can provide low discharge energy pulses with a frequency of 500 kHz. Discharge duration as short as 150 ns and peak current as low as 0.7 A can be obtained through the adjustment of the capacitance and current-limiting resistance in the discharge circuit, respectively. Higher value of capacitance results in higher discharge energy and thus contributes to longer discharge duration. The peak current increases slightly with the increase in the pulse-on time. A higher current-limiting resistance results in a low peak current. Experimental results demonstrated that the developed fine-finish power supply can reduce the recast layer, eliminate rusting and bluing in titanium, and reduce cobalt depletion in tungsten carbide than a standard DC power supply. A fine-surface finish of 0.22 mm Ra can be achieved by means of four finish machining operations with proper machine settings.

DC53 is a newly developed cold die steel from Daido Steel, Japan. It is an improvement over the familiar cold die steel SKD11. Because DC53 is a new die steel, only little information is available in literature for its machining characteristics. In Kanlayasiri and Boonmung (2007), investigation is carried on the effects of machining parameters on surface roughness of wire EDMed DC53 die steel. The investigated machining parameters were pulse-on time, pulse-off time, pulse-peak current, and wire tension. Analysis of variance (ANOVA) technique was used to find out the parameters affecting the surface roughness. Assumptions of ANOVA were discussed and then examined through residual analysis. Quantitative testing methods on residual analysis were employed in place of the typical qualitative testing techniques. Results from ANOVA show that pulse-on time and pulse-peak current are significant variables to surface roughness of wire EDMed DC53 die steel. The surface roughness of test specimen increased as these two variables increased.

Since, Wire is used as electrode in WEDM; the percentage of copper in brass wire is one of the important factors affecting the performance of machining. In Parameswara and Sarcar (2009), the author has evaluated optimal parameters such as discharge current, voltage at rated wire speed, and tension for brass electrode of 5–80 mm size. Influence of optimal parameters on cutting speed, surface roughness, and spark gap is analyzed. Mathematical relations developed were more beneficial to estimate cutting time and cost of machining with the given range of machining. Effect of wire material on cutting criterion is also evaluated for workpiece (5 mm thick) using four wires of different copper percentages.

The demand for high-performance materials in modern industry has led to the development of cermets, such as WC-Co composites, combining high hardness and high strength with good wear resistance. About 70 % of the punches used within die for the production of high-quality sheet metal parts are made of carbide material. The two common processes to produce these punches are grinding and wire EDM. But wire EDM is more and more used because it can be fully automated and because complex geometrical shapes can be machined in one setup. In Lauwers et al. (2006), the influence of composition and grain size of WC-based cermet on manufacturability by wire EDM is described. It is shown that the cutting rate decreases with increasing WC-grain size, which can be mainly explained by the change in thermal conductivity of the material. Further, the toughness of the base material has an important influence on the bending strength of the wire EDM-machined samples. The lower the toughness of material, which is especially the case for carbide material with fine WC-grain size, the higher the reduction in the bending strengths of wire EDM-machined materials.

In wire EDM process, wire failure is a major problem and it increases machining processing time, decreases machining efficiency, and the quality of the machined surface. In Ranjanath et al. (2003), experimental and theoretical study of wire EDM process is carried. Wire erosion rate, which leads to wire failure, is analyzed using experimental results obtained by machining mild steel, OHN steel, and HCHCr steel work materials using bare brass and zinc-coated brass wires. Temperature and stress distribution on work and tool during machining and its effect on performance on wire EDM process are reported. SEM, EDAX, and ANN results are analyzed to study the failure characteristics of wire tool. This paper has concluded that wire failure is caused due to wire wear rate, which is a function of discharge current and discharge time. For the same MRR, the wire wear rate is observed to be lower with zinc-coated brass wire when compared with bare brass wire, resulting in higher wire erosion rate. Finite element analysis helps to study the temperature and stress distribution on tool and work in wire EDM process. This helps to select the optimum machining conditions, with a safe thermal and stressed condition, and minimum tool breakage.

Wires used in WEDM are the core of the system. Brass wire electrode is used extensively as a tool for WEDM. However, along with the recent diversification in applications of manufacturing fields, demand is expanding for a wire electrode with performance superior to the conventional brass wire electrode. These electrodes in use are generally zinc-coated wire with a copper/brass alloy or steel core, the brass containing either a small amount of Cr, or a high concentration of Zn. At present, the concern of EDM users is to shorten the machining time of products. A new high-performance EDM wire would be expected to provide both high cutting speed and improved accuracy. In Kapoor et al. (2010), evolution of EDM wire from copper to brass and from brass to various coated wires has been focused, which has helped make wire EDM machining, the method of choice for high-speed production applications, as well as applications requiring improved contour accuracy and improved surface finishes. Some of the characteristics of high-performance wire electrodes have been presented, which significantly increase the WEDM productivity.

Wire tension is one of the most dominant factors that affect wire bending and vibration. Conventionally, wire tension is open-loop controlled for a commercialized wire EDM machine. Since the open-loop control of wire tension does not account for the friction force between the wire electrode and the rollers of wire transportation mechanism, the inertia effect of wire reel and the disturbance load during machining process, the wire tension cannot be kept at a constant level. Accordingly, the machining accuracy is decreased due to the variation in wire tension. In Yan and Huang (2004), closed-loop wire tension control system for a wire EDM machine is used to improve the machining accuracy. Dynamic models of the wire feed control apparatus and wire tension control apparatus are derived to analyze and design the control system. PI controller and one-step-ahead adaptive controller are employed to investigate the dynamic performance of the closed-loop wire tension control system. In order to reduce the vibration of wire tension during wire feeding, dynamic absorbers are added to the idle rollers of wire transportation mechanism. Experimental results not only demonstrate that the developed control system with dynamic absorbers can obtain fast transient response and small steady-state error than an open-loop control system, but also indicate that the geometrical contour error of corner cutting is reduced with approximately 50 % and the vertical straightness of a workpiece can be improved significantly.

As surface roughness is one of the most important parameters in manufacturing, various investigations have been carried out by several researchers for improving the surface roughness of the WEDM process. These investigations show that the surface roughness of the process is closely dependent on the machining parameters. In Gokler and Ozanozgu (2000), the experimental study aims to select the most suitable cutting and offset parameter combination for the WEDM process in order to get the desired surface roughness value for the machined workpieces. A series of experiments have been performed on 1,040 steel material of thicknesses 30, 60, and 80 mm, and on 2,379 and 2,738 steel materials of thicknesses 30 and 60 mm. The test specimens have been cut using different cutting and offset parameter combinations of the "Sodick Mark XI A500 EDW" wire electrical discharge machine in the Middle East Technical University CAD/CAM/Robotics Center. The surface roughness of the test pieces has been measured using a surface roughness measuring device. The author concluded that the experimental results can be used in industry in order to select the best suitable parameter combination to get the required surface roughness values for the products. Also in Liao et al. (2004), in order to obtain good surface roughness, the traditional circuit using low power for ignition is modified for machining as well. With the assistance of Taguchi quality design, ANOVA and F-test, machining voltage, current-limiting resistance, type of pulse-generating circuit, and capacitance are identified as the significant parameters affecting the surface roughness in finishing process. In addition, it is found that a low conductivity of dielectric should be incorporated for the discharge spark to take place. After analyzing the effect of each relevant factor on surface roughness, appropriate values of all parameter are chosen and a fine surface of roughness $Ra = 0.22$ microns is achieved. The improvement is limited because finishing process becomes more difficult due to the occurrence of short circuit attributed to wire deflection and vibration when the energy is gradually lowered.

In wire EDM, it is absolutely essential to hold the wire in a designed position against the object because the wire repeats complex oscillations due to electro-discharge between the wire and workpiece. Normally, the wire is held by a pin guide at the upper and lower parts of the workpiece. In most cases, the wire will be discarded once used. However, there are problematic points that should be fully considered in order to enhance working accuracy. In Datta and Mahapatra (2010), quadratic mathematical model have been derived to represent the process behavior of WEDM operation. Experiments have been conducted with six process parameters: discharge current, pulse duration, pulse frequency, wire speed, wire tension, and dielectric flow rate; to be varied in three different levels. Data related to the process responses viz. MRR, roughness value of the worked surface (a measure of surface finish, SF), and kerf have been measured for each of the experimental runs; which correspond to randomly chosen different combinations of factor setting. These data have been utilized to fit a quadratic mathematical model (Response Surface Model) for each of the responses, which can be represented as a function of the aforesaid six process parameters. Predicted data have been utilized for identification of the parametric influence in the form of graphical representation for showing influence of the parameters on selected responses. Predicted data given by the models (as per Taguchi's L27 (3*6) Orthogonal Array (OA) design) have been used in search of an optimal parametric combination to achieve desired yield of the process: maximum MRR, good surface finish (minimum roughness value), and dimensional accuracy of the product. Grey relational analysis has been adopted to convert this multi-objective criterion into an equivalent single objective function; overall grey relational grade, which has been optimized (maximized) using Taguchi technique. Optimal setting, has been verified through confirmatory test, showed good agreement with the predicted value. This indicates utility of the Grey–Taguchi technique as multi-objective optimizer in the field of wire EDM.

4 Conclusion

WEDM is an advanced thermal machining process capable of accurately machining parts with complicated shapes, especially for the parts that are very difficult to be machine by traditional machining processes. It has been commonly applied for the machining and micro-machining of parts with in tricate shapes and varying hardness requiring high profile accuracy and tight dimensional tolerances.

- Optimization of the WEDM process parameters is essential because, WEDM is an expensive and widely used process. The ultimate goal of the WEDM process is to achieve an accurate and efficient machining operation. Several researchers have studied methods to improve the surface quality and increase the MRR of the WEDM process. However, the problem of selecting the process parameters in the WEDM process is not fully solved, even though the most up to date CNC-WEDM machines are presently available.
- Still more research can be extended to deal with wire breakage problem, which is the most crucial problem in machining with WEDM. Coated wire can be experimentally analyzed for reducing the wire rupture problems.
- There is limited work performed on electrode wear in WEDM. So, there is more scope to analyze Wire wear ratio (WWR), which is the most significant factor causing frequent wire breakages in WEDM.
- Also, further research can be carried out on the MRR, which enhances the productivity and plays an important role in economics of machining in terms of cost reduction.
- The modeling of the WEDM process by means of different approach like mathematical techniques such as response surface methodology and non-traditional methods such as particle swarm optimization, artificial neural network can be carried and thus more accurate results can be obtained.

Acknowledgment Pratik A. Patil thanks Prof. C.A. Waghmare (Project Guide) and Prof. M.V. Kavade (Head, Mechanical Engineering Department, RIT, Rajaramnagar) for their beneficial guidance, suggestion, support and constructive criticism.

References

Datta S, Mahapatra SS (2010) Modeling, simulation and parametric optimization of wire EDM process using response surface methodology coupled with Grey–Taguchi technique. Int J Eng Sci Technol 2:162–183

Gokler MI, Ozanozgu MO (2000) Experimental investigation of effects of cutting parameterson surface roughness in the WEDM process. Int J Mach Tools Manuf 40:1831–1848

Janardhan V, Samuel GL (2010) Pulse train data analysis to investigate the effect of machining parameters on the performance of wire electro discharge turning (WEDT) process. Int J Mach Tools Manuf 50:775–788

Kanlayasiri K, Boonmung S (2007) An investigation on effects of wire-EDM machining parameters on surface roughness of newly developed DC53 dies steel. J Mater Process Technol 187:26–29

Kapoor J, Singh S, Khamba JS (2010) Recent developments in wire electrodes for high performance WEDM. In: Proceedings of the world congress on engineering, vol 2, Jun 2010

Kumar M, Babu S, Venkatasamy R (2010) Optimization of the WEDM parameters on machining Incoloy800 super alloy with multiple quality characteristics. Int J Eng Sci Technol 2:1538–1547

Lauwers B, Lui W, Eeraerts W (2006) Influence of the composition of WC-Based cermets on manufacturability by WEDM. J Manuf Process 8:83–89

Liao YS, Huang JT, Chen YH (2004) A study to achieve a fine surface finish in wire-EDM. J Mater Process Technol 149:165–171

Parameswara CVS, Sarcar MMM (2009) Evaluation of optimal parameters for machining brass with wire cut EDM. J Sci Ind Reas 68:32–35

Ranjanath BJ, Sudhakar KG, Shrikantappa AS (2003) Wire failure analysis in wire-EDM process. In: Proceedings of the international conference on Mechanical Engineering, Dec 2003

Shayan AV, Afza RA (2013) Parametric study along with selection of optimal solutions in dry wire cut machining of cemented tungsten carbide (WC-Co). J Manuf Process 15:644

Subrahmanyam SV, Sarcar MMM (2013) Evaluation of optimal parameters for machining with wire cut EDM using Grey-Taguchi method. Int J Sci Res Pub. 3(3):1

Yan MT, Huang PH (2004) Accuracy improvement of wire-EDM by real-time wire tension control. Int J Mach Tools Manuf 44:807–814

Yan M, Lai Y (2007) Surface quality improvement of wire-EDM using a fine-finish power supply. Int J Mach Tools Manuf 47:1686–1694

References

[illegible]

Effect of Filler Wire Composition on Joining Properties of GTAW Stainless Steel 202

Gurmeet Kaur, Daljinder Singh and Jasmaninder Singh Grewal

Abstract This research intends to find out the effects of filler wire composition on joining properties of SS 202-welded with semi-automatic gas tungsten arc welding (GTAW) process. Hardness, tensile strength, surface roughness, and SEM/EDAX analysis of joints were evaluated with respect to different filler wires ER 308L, 316L, and 310. Single-V-butt joints were made on 6-mm-thick plates using semi-automatic GTAW process, while ER 308L exhibited high mechanical properties, and the ER 316L and 310 showed comparative less properties. SEM/EDAX analysis proved that high alloying elements are obtained in the joint made with ER 310, the ER 316L and ER 308L resulted in moderate and low amounts of elements that are responsible for resistance to corrosion and oxidation. It was concluded that ER 316L offered best compromise of mechanical properties and alloying elements from the view point of boiler application.

Keywords Welding processes · GTAW · EDAX · Surface roughness · Hardness · Tensile strength · SS 202

G. Kaur (✉)
Department of Mechanical Engineering, Guru Nanak Dev Engineering College,
Ludhiana 141006, Punjab, India
e-mail: gk1578@rediffmail.com

D. Singh
Department of Mechanical Engineering, Chandigarh Group of Colleges,
Landran, Mohali, Punjab, India
e-mail: ds.ubhi@yahoo.com

J. S. Grewal
Department of Production Engineering, Guru Nanak Dev Engineering College,
Ludhiana 141006, Punjab, India
e-mail: jsgrewal2000@yahoo.com

S. S. Khangura et al. (eds.), *Proceedings of the International Conference on Research and Innovations in Mechanical Engineering*, Lecture Notes in Mechanical Engineering,
DOI: 10.1007/978-81-322-1859-3_17, © Springer India 2014

1 Introduction

Gas tungsten arc welding (GTAW) is an arc welding process that produces coalescence of metals by heating them with an arc between a nonconsumable electrode and base metal. GTAW process is suitable for joining thin and medium thickness materials such as stainless steel sheets and for applications where metallurgical control of the weld metal in critical. SS 202 grade has wide applications in making seamless stainless tubes for boilers, heat exchanger tubes, super heater tubes, cookwares, etc. (Sudhakaran et al. 2010). It is worth noting that austenitic type is among the most easy to weld and allows fabrication of elevated toughness joints welded joints even in the as-welded conditions, without any further treatment (Aval et al. 2009). However, welding often leads to low mechanical properties owing metallurgical changes such as micro-segregation, precipitation of secondary phases, and solidification cracking (Yan et al. 2010; Lee and Jeng 2001). To overcome most of the short comings in welding, proper selection of filler wire is the key parameter for ensuring weld quality and microstructure (Yilmaz and Uzun 2002). So far very limited work had been carried out in assessing the properties of SS 202 with respect to the filler metals using semi-automatic GTAW process. In this research work, a study was made to investigate the effect of filler wires selection on the mechanical properties and SEM/EDAX analysis of the GTAW of SS 202.

2 Literature Review

Reddy et al. (1998) investigated the effect of filler metal composition on weldability of Al–Li alloy 1441. Hardness, tensile properties, and hot cracking susceptibility of Al–Li alloy 1441 were evaluated with respect to different filler alloys AA 2319, AA 4043, and AA 5356, as well as the parent alloys. The hardness in the as-welded condition was 70–90 HV and improved by 20, 30, 40, and 40 HV after heat treatment with AA 4043, AA 5356, and AA 2319, and 1441 fillers, respectively. Tensile strength showed similar trends as hardness in the as-welded and heat treated conditions. The hot cracking tendency was the maximum for welds deposited with 1441 filler and the minimum for welds deposited with AA 5356 filler, both values being less than the cracking tendency for autogenous welds. Hot cracking tendency was correlated with grain size, segregation distance, and distribution of the low melting phases.

Kuo and Lee (2002) studied the influence of filler metal composition on the corrosion resistance and mechanical properties of alloy 690 weldments. Alloy 690 was used as the base metal. Inconel I-52 and I-82 rods were used as filler metals. Manual GTAW was performed using four weld passes in three layers for a single-V-groove butt weld. Both had white particles dispersed in the fusion zone. Compositional analysis showed the I-52 weld's interdendritic region had higher

Al, Si, Ti, and N content than the dendritic core. The I-82 weld's interdendritic region had higher Al, Si, and Nb content than the dendritic core. The Ni and Cr content of interdendritic white particles of the I-52 and I-82 welds decreased to 38, 23 wt% and 11, 9 wt%, respectively, much lower than the base and filler metals. This situation causes high corrosion at the white particle sites during modified Huey testing.

Sireesha et al. (2000) studied the weld between 316LN and alloy 800. These welds were produced using three types of filler materials: austenitic stainless steels corresponding to 316, 16Cr–8Ni–2Mo, and the nickel-base Inconel 182. The weld fusion zones and the interfaces with the base materials were characterized in detail using light and transmission electron microscopy.

Yilmaz and Uzun (2002) studied the mechanical properties of austenitic stainless steels welded by gas metal arc welding (GMAW) and GTAW. In this study, AISI 304L and 316L types of austenitic stainless steels were welded by GMAW using only ER 316LSi filler metal and GTAW using ER 308L and 316L filler metals, respectively. Mechanical properties of 304L and 316L austenitic stainless steel weldments, such as tensile properties, hardness, and impact properties were determined. The results showed that the yield and tensile strength, hardness, and impact energy values of 304L and 316L stainless steels welded by GTAW are higher than that of GMAW.

Naffakh et al. (2007) carried out the work to characterize dissimilar welding of AISI 310 austenitic stainless steel to Inconel 657 nickel–chromium superalloy. The welds were produced using four types of filler materials; the nickel based corresponding to Inconel 82, Inconel A, Inconel 617, and austenitic stainless steels 310. The comparative evaluation was based on hot cracking and estimation of mechanical properties. While Inconel A exhibited highest resistance to solidification cracking, the Inconel 617 filler metal also showed moderate resistance and in addition, the later was superior from the mechanical properties viewpoints. It was concluded that for the joint between Inconel 657 and 310 SS, the Inconel A filler metal offered the best compromise.

Jang et al. (2008) studied the dissimilar metal welds composed of low alloy steel. Inconel 82/182 weld and stainless steel were prepared by GTAW and shielded metal arc welding techniques. Microstructures were observed using optical and electron microscopes. Typical dendrite structures were observed in Inconel 82/182 welds. Tensile tests using standard and mini-sized specimens and microhardness tests were conducted to measure the variation in strength along the thickness the weld as well as across the weld. In addition, fracture toughness specimens were taken at the bottom, middle, and top of the welds and tested to evaluate the spatial variation along the thickness. While the strength was about 50–70 Mpa greater at the bottom of the weld than the top of the weld, fracture toughness values at the top of the weld were about 70 % greater than those at the bottom of the weld.

Emamian et al. (2010) determined the effects of filler wire composition along with different pre- and post-heat treatment on mechanical properties of *AISI* 4130 steel welded by the GTAW process. Six test plates were joined by two types of filler wire with similar chemical composition to the base metal, and with lower carbon content and slightly higher alloy elements content compared with the first one. Test plates then exerted three different pre-heat and post-heat treatments on both groups. The three types of heat treatments were alternatively without pre-heat and post-heat, with pre-heat only, and finally with pre-heat and post-heat. Tensile, side bends, and impact tests (for weld zone and HAZ) have been conducted. Results show that using low-carbon filler wire along with pre-heat and post-heat resulted in outstanding mechanical properties.

Aziz et al. (2011) studied the effect of different fillers on microstructure and tensile properties of welded AA6061-T6 using GMAW. Butt-joint welds were made on 6-mm-thick plates using 21–22 V arc voltage. The results showed that yield strength of base metal were 330 MPa, while the yield strength of ER 5356 joints and ER 4043 joints were 200 and 235 MPa, respectively. The difference properties of strength in both weld metal was due to the difference major element in both filler composition. The amount of silicon content in ER 4043 (Al–Si5 %) filler was believed to play a role in the mechanical strength on weld metal. Microstructural examinations were carried out using a light optical and electron microscope.

Mohanraj et al. (2012) studied the influence of filler materials on mechanical and hot corrosion properties of gas tungsten arc welded AISI 304. This paper investigates the influence of filler materials such as E308L and ENiCu-7 utilized in the GTA welded AISI 304 with respect to mechanical and corrosion properties. It was found that E308L had shown better tensile strength and corrosion resistance as compared with ENiCu-7. A comparative analysis had been made on different regions of the weldment subjected to cyclic air oxidation at 700 °C. The corrosion products were examined using XRD and SEM/EDAX analysis.

Tatsukawa et al. (1987) carried out the work to study the influence of filler metal on weld bead penetration and shape in automatic TIG welding. The fundamental relationships during automatic TIG welding between the welding conditions, bead formation, and penetration shape were defined experimentally for mild steel and aluminum alloy. In addition, the effects filler rod feed upon this phenomenon were investigated from the viewpoint of penetration shape, bead formation, and temperature distribution in both the parent plate and filler rod.

The literature reveals that lot of work has been reported to study the effect of filler wire on properties of different alloys welded by GTAW. But, till now very less has been reported to study the effect of filler wire composition on properties of 202 stainless steels using semi-automatic GTAW process.

3 Experimental Procedure

3.1 Equipment

Automation of the arc welding processes is increasingly used with the aim of improving quality control of welded joints and workability. Gas shielded arc welding using consumable electrodes has already been developed to degree that it

Fig. 1 Welding machine

is practical with the use of welding robots. However, the applications for GTAW have increased, and the quality of joint materials and joint forms has been diversified. The automation of GTAW has become an important task (Tatsukawa et al. 1987). For this research work, experimental semi-automatic welding equipment was manufactured such that the travel of the workpiece, beneath the weld torch, and the feed of the filler rod could be separately adjusted. This ensures the uniform welding conditions across the weld. Figure 1 shows the semi-automatic GTAW equipment. In this system, the weld torch remains fixed, whereas the workpiece travels beneath the weld torch, placed on the workpiece carriage. The filler rod feeder device feeds the filler rod by two set of rollers; a wiper motor is used as the driving source and the feed rate is controlled by a driving circuit in the control box. The workpiece carriage is made of mild steel channel (12″ × 8″ × 3″), fitted with two bearings on each side to ensures its movement along two rails, driven by a wiper motor; its speed is controlled by a gear arrangement and a driving circuit in the control box. Rack and pinion gear arrangement is used to drive the carrier by the driving motor. The angle of the filler rod and weld torch is adjustable. The speed of welding carrier can be adjustable between 1.27 and 7.62 mm/s, and the range of speed of filler wire feeder is 1.48–9.1 mm/s. Limit switches are used to change the direction of the workpiece carriage.

3.2 Material

The base material selected for the research work was 6-mm-thick plates of SS 202. GTAW was carried out on SS 202 plates using ER 308L, 316L, and 310 filler wire having 2.4 mm diameter. The nominal chemical composition of the base material and filler wires are given in Table 1.

Table 1 Chemical composition of materials used (wt%)

Materials	C	Si	Cr	Mn	Ni	Mo	S	P
SS 202	0.13	0.09	17.25	8.33	5.84	–	0.015	0.052
ER 308	0.03	0.30–0.65	19–22	1–2.5	9–11	0.75	0.030	0.030
ER 316	0.03	0.30–0.65	18–20	1–2.5	11–14	2–3	0.030	0.030
ER 310	0.08–0.15	0.30–0.65	25–28	1–2.5	20–22.5	0.75	0.030	0.030

Welded joints were made between the base material plates of size (100 × 50 × 6 mm) using each of the two filler wires, employing a V-groove edge with an included angle of 60- and 2.4-mm root gap and 1.5-mm root face. Trial runs were performed to select the operating parameters. The welds were obtained in two passes to avoid distortion in the weld joints. Welding was carried out with 2.4-mm tungsten electrode. The arc length was 3 mm. Gas shielding for the welding was kept in the flowing rate of 17 l/min. Current intensity for first and second pass were selected as 150 and 170 A, respectively. The travel speed of the base metal beneath the weld torch was 120 mm/min, and the filler wire feeding speed was 216 mm/min. After the first pass of welding, weld was allowed to cool at room temperature. While, welding stainless steel the interpass temperature being restricted around 175° to avoid sensitization and distortion in the joints. After second pass and cooling of the joints, three samples were prepared for each filler wire that were ER 308L, 316L, and 310. The weld samples were cut into pieces for tensile test and SEM/EDAX analysis.

Transverse sections of the welds were metallographically characterized after etching in marble solution. The weld metals were examined by scanning electron microscopy, and chemical compositions were determined by energy dispersive X-ray spectrometer (EDAX) at IIT, Ropar. Tensile test was carried out on the welded samples as per ASTM E8, and test was conducted at MERADO, Ludhiana. Welded joints were investigated for surface roughness and hardness along the welding direction on the surface of the weld bead at Research and Development Centre for Bicycle and Sewing machines, Ludhiana.

4 Results and Discussion

4.1 SEM/EDAX Analysis

Figure 2 shows the micrograph of the weld metal of (a) ER 308L, (b) ER 316L, and (c) ER 310 filler wire, for base metal SS 202. Both white and black particles can be observed in the weld metals. Chemical analysis revealed that white particles contains less wt% of Chromium (Cr) and Nickel (Ni), but high Carbon (C), Oxygen (O), Silicon (Si), and Aluminum (Al) contents. However, black particles enriched with high Cr and Ni contents may provide high resistance to corrosion

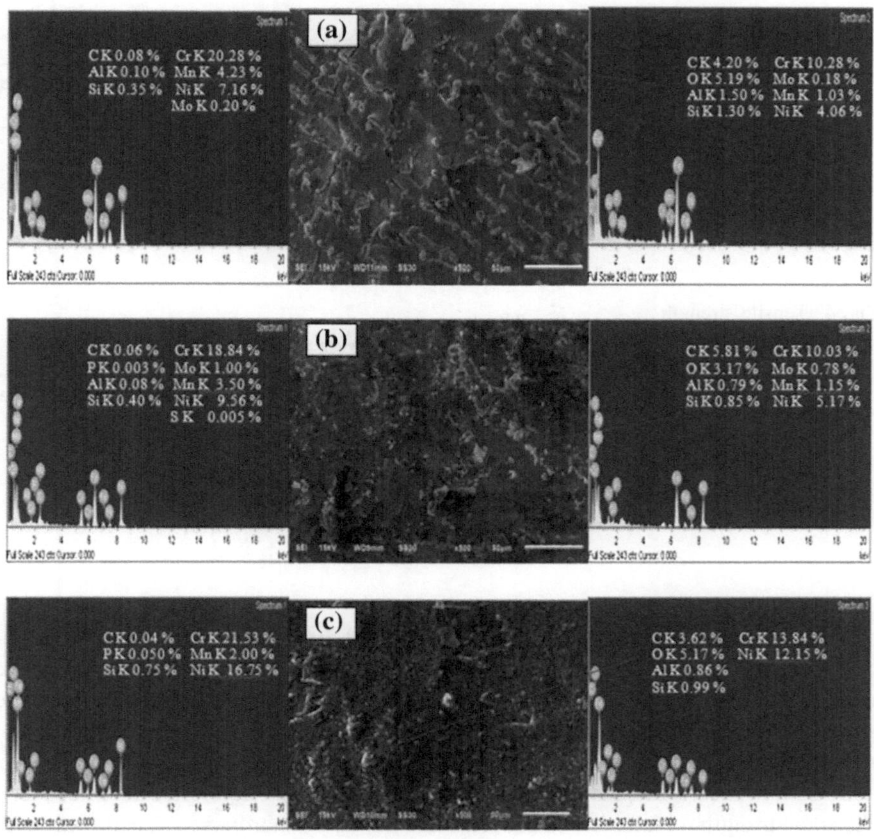

Fig. 2 Micrograph and EDAX patterns from *black* and *white* spots on weld metal of **a** ER 308L, **b** ER 316L, and **c** ER 310 filler wire, for base metal SS 202

than white particles (Kuo and Lee 2002). Weld metal (a) contains more numerous white particles as compared with weld metal (b) and (c). Flakes can be observed in weld metal (a). High manganese in the weld metals has tendency to remove the negative effects of impurities i.e., sulfur and phosphorous.

4.2 Mechanical Properties

Hardness measurements were carried out using Rockwell Hardness Tester. For each specimen, at least four hardness measurements were taken to ensure representative results. Figure 3 shows the variation in hardness with reference to filler wire for base metal SS 202. Filler wire ER 308L, ER 316L, and ER 310 gave the hardness values as 93HR B, 91HR B, and 90HR B, respectively. The hardness is

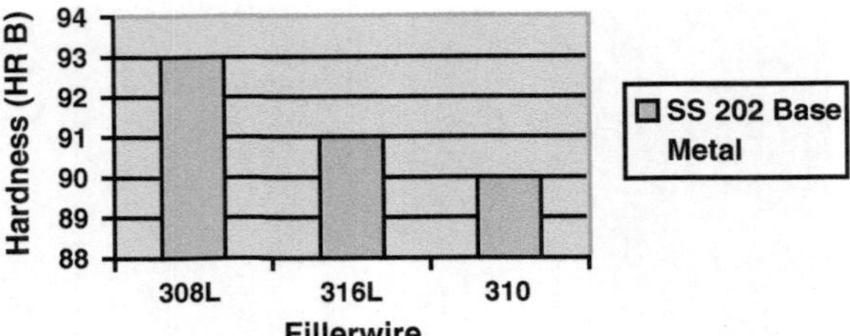

Fig. 3 Hardness across the welds produced with different filler wires

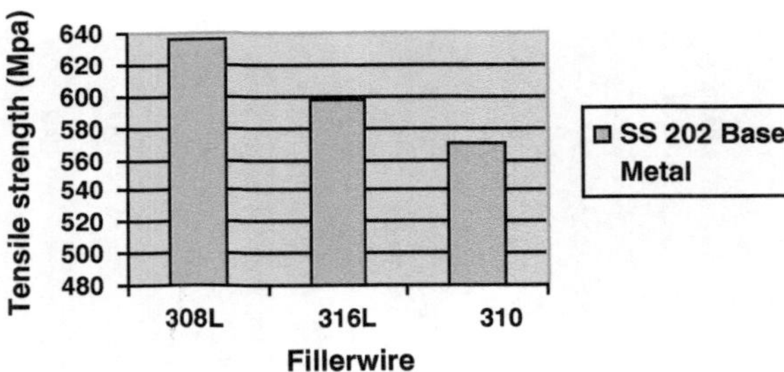

Fig. 4 Tensile strength obtained by different filler wires

attributed to the percentage of elements i.e., carbon (C), chromium (Cr), molybdenum (Mo), and Manganese (Mn) as observed in compositional analysis of the weldments. An increased amount of carbon increases the hardness and tensile strength, but it may affect the corrosion resistance. Mn, Cr, and Mo also promote strength and hardness.

Figure 4 shows the variation in tensile strength with reference to filler wire. Tensile strength results exhibited similar trends to hardness. Tensile strength results indicated that for base metal SS 202, filler wire ER 308L contributed to higher tensile strength (637 Mpa), while filler wire ER 316L displayed comparative less tensile strength (598 Mpa), and filler wire ER 310 resulted in tensile strength (570 Mpa) i.e., less than both the fillers.

The grain size of weld metal plays an important role for the mechanical properties i.e., hardness, tensile strength etc. of the joint. Fine-grained structure has higher strength than coarse-grained structure, and fine grains provide better resistance to cracking. Grain size depends upon cooling rate and amount of alloying elements involved. Welding speed and current conditions were same for all the weldments in this study. It is assumed that cooling conditions were same for the study. So, alloying elements must have some effects on the grain size of the weldments. As, aluminum (Al) restricts the grain growth. On other hand, silicon (Si) promotes large grain sizes. Surface roughness also contributes for the mechanical properties. Lower the surface roughness, more fine the grains. Thus, higher will be the mechanical properties.

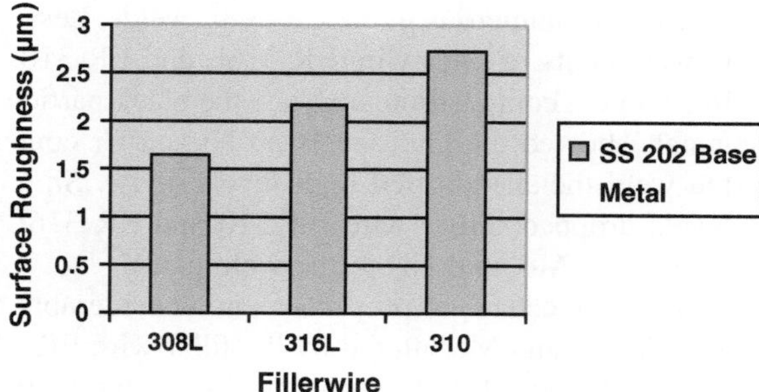

Fig. 5 Surface roughness produced with different filler wires

4.3 Surface Roughness

Figure 5 shows the variation in surface roughness with reference to filler wire. For base metal SS 202, filler wire ER 308L, ER 316L, and ER 310 displayed surface roughness as (Ra) 1.64, 2.18, and 2.75 µm, respectively.

Roughness plays an important role in determining how a real object, interact with its environment. Rough surfaces usually wear more quickly and have higher friction coefficients than smooth surfaces. Roughness is often a good predictor of the performance of a mechanical component, since irregularities in the surface may form nucleation sites for cracks or corrosion. On the other hand, roughness may promote adhesion. It has important relation with grain size. Coarse grains make the surface rough, while fine grains results in smooth surface. Also, during welding, it was observed that filler wire ER 308L melts comparatively easy than other filler wires. It may be attributed to comparative less molybdenum and less chromium than ER 316L and ER 310 filler wires, respectively.

5 Conclusions

- On the basis of mechanical properties and metallographic experimental observations accomplished and results obtained on the effect of filler wire composition on the semi-automatic GTAW process of SS 202, the following conclusions can be drawn:
- A surface roughness (*Ra*) of 1.5–3 µm and high-quality welds can be achieved with semi-automatic GTAW welding. Filler wire ER 308L provided best surface roughness results.
- In tensile test, filler wire ER 308L and 316L weldments had the better tensile results, whereas filler wire ER 310 weldments displayed the comparative less tensile strength. Hardness results exhibited similar trends to tensile properties.
- Metallographic inspection of the welds revealed that many white particles were distributed in the grain structure of all the welds, with white particles being

much more numerous in the ER 308L welds. Less white particles were observed in weldments of filler wire ER 316L and ER 310.
- In chemical composition analysis, the black particles in the micrographs of weld metals showed the high levels on Ni and Cr content. While, white particles of the weld metal indicated high levels of Al, Si, C, and O, whereas Ni and Cr levels dropped. Filler wire ER 310 and ER 316L provided high alloying elements (Cr, Mo, and Ni) to the weldments.
- Sufficient mechanical properties and considerable amount of alloying elements i.e., Cr, Ni, and Mo offered by the filler wire ER 316L, to boiler grade stainless steel 202 made it best compromise from the viewpoint of boiler application.

References

Aval HJ, Farzadi A, Serajzadeh S, Kokabi AH (2009) Theoretical and experimental study of microstructures and weld pool geometry during GTAW of 304 stainless steel. Int J Adv Manuf Technol 42:1043–1051

Aziz AAA, Ibrahim MFA, Jalar A, Junaidi S, Abdullah S, Rashdi N, Kornain Z (2011) Effects of different fillers on microstructure and tensile properties of welded AA6061-T6. Key Eng Mater 462–463:1189–1193

Emamian A, Emamian A, Kowkabi AH (2010) Effects of fillerwire composition along with different pre-and post-heat treatment on mechanical properties of AISI 4130 welded by the GTAW process. Mater Sci Appl 1:135–140

Jang C, Lee J, Kim JS, Jin TE (2008) Mechanical property variation within Inconel 82/182 dissimilar metal weld between low alloy steel and 316 stainless steel. Int J Press Vessels Pip 85:635–646

Kuo TY, Lee HT (2002) Effects of filler metal composition on joining properties of alloy 690 weldments. Mater Sci Eng A 338(1–2):202–212

Lee HT, Jeng SL (2001) Characteristics of dissimilar welding of alloy 690 to 304L stainless steel. Sci Technol Weld Joining 6(4):225–234

Mohanraj R, Kumar Rajesh D, Yeshwanth BK, Arivazhagan N, Devendranath Ramkumar K, Narayanan S (2012) Influence of filler materials on mechanical and hot corrosion properties of gas tungsten arc welded AISI 304. Solid State Phenom 185:113–115

Naffakh H, Shamanian M, Ashrafizadeh (2007) A comparative evaluation of welding consumables for dissimilar welds between 310 austenitic stainless steel and inconel 657. Int J ISSI 4(1–2):39–49

Reddy GM, Gokhale AA, Prasad Rao K (1998) Effect of filler metal composition on weldability of Al–Li alloy 1441. Sci Technol Weld Joining 3(3):151–158

Sireesha M, Albert SK, Shankar V, Sundaresan S (2000) A comparative evaluation of welding consumables for dissimilar welds between 316LN austenitic stainless steel and alloy 800. J Nucl Mater 279:65–76

Sudhakaran R, Vel Murugan V, Siva Sakthivel PS (2010) Optimization of process parameters to minimize angular distortion in gas tungsten arc welded stainless steel 202 grade plates using genetic algorithms. Int J Eng Sci Technol 2(5):731–738

Tatsukawa I, Satonaka S, Inada M (1987) Influence of filler metal on weld bend penetration and shape in automatic TIG arc welding. Q J Jpn Weld Soc 5(2):187–193

Yan J, Gao M, Zeng X (2010) Study on microstructure and mechanical properties of 304 stainless steel joints by TIG, laser and laser-TIG hybrid welding. Opt Lasers Eng 48:512–517

Yilmaz R, Uzun H (2002) Mechanical properties of Austenitic stainless steels Welded by GMAW and GTAW. J Marmara Pure Appl Sci 18:97–113

Effects of Machining Parameters on Performance of Electrical Discharge Surface Grinding of AISI D2 die Steel with Composite Tool Electrode

Rajesh Choudhary, Harmesh Kumar and Shankar Singh

Abstract Electrical Discharge Surface Grinding (EDSG) is much like electrical discharge machining except that the electrode is a rotating metal matrix composite (MMC) tool, processed by reinforcing abrasive particles in a conducting metal matrix by powder metallurgy route. The tool electrode is fed into the work material by a servo-controlled mechanism. The work material is machined and grinded by the combined action of electric sparks (thermal interaction) between the tool electrode and the work material, immersed in a dielectric fluid, and the abrasion (mechanical interaction) due to the abrasives mixed in the matrix. Each spark discharge melts or vaporizes a small amount of metal from the work surface, producing a small crater at the localized spot, followed by grinding by the abrasives, producing a ground surface. This paper focusses on understanding the mechanism of material removal in abrasion-assisted EDSG, for the machining of AISI D2 die steel. The rotary motion imparted to the metal matrix Cu/SiCp tool electrode using rotating spindle assembly mounted on electric discharge machine was employed to study the EDSG technology. Powder metallurgy route was adopted to fabricate the MMC electrode with different proportion of abrasives as SiC in copper powder. During the EDSG operation, the hump of material melted by the EDM is forcibly removed by the grinding/abrasion mechanism from the work surface, which enhances the material removal rate (MRR) of this hybrid process.

R. Choudhary (✉) · S. Singh
Department of Mechanical Engineering, Sant Longowal Institute of Engineering and Technology, Sangrur 148106 Punjab, India
e-mail: rajeshmimit@gmail.com

S. Singh
e-mail: singh_shankar@yahoo.com

H. Kumar
Department of Mechanical Engineering, University Institute of Engineering and Technology, Panjab University, Chandigarh 160014, India
e-mail: shaarut@yahoo.com

S. S. Khangura et al. (eds.), *Proceedings of the International Conference on Research and Innovations in Mechanical Engineering*, Lecture Notes in Mechanical Engineering, DOI: 10.1007/978-81-322-1859-3_18,

Keywords EDSG · Metal matrix composites · Powder metallurgy tool electrode · Material removal rate · Electrode wear rate

1 Introduction

The continuous development of technology in aircraft, automotives and manufacturing has led to researchers to explore advanced materials. These materials are having high heat- and wear-resistant properties along with lightweight characteristics to match the requirements of industries and space applications. Metal matrix composites (MMCs) find vast industrial applications because of their unique and outstanding physical and mechanical characteristics. Powder metallurgy processes are preferred to produce reinforced MMCs. The research on fabricating copper-based composites including Cu–Al_2O_3, Cu–Zr–Al_2O_3, Cu–TiO_2, Cu–Si_3N_4 and Cu–SiC is being extensively explored. Method of fabricating these composites includes casting, co-precipitation, internal oxidation and powder metallurgy (Denisenko et al. 1970, 1971; Fedorchenko et al. 1974; Kobayashi et al. 1990). The silicon carbide is characterized by its hardness, excellent high-temperature creep resistance, high thermal conductivity and excellent corrosion resistance; hence, it is mostly suitable as reinforcement in copper-based MMCs. Owing to these characteristics, SiC has been applied for service at high temperatures under corrosive conditions and in areas where wear must be prevented (Occhionero et al. 1999).

The production of the composite has many difficulties, for instance suitable bonding between SiC and copper, inhomogeneity of SiC particles in the matrix and the control of porosity. The possible solution of these problems is the coating of the SiC grains with suitable material to avoid interfacial reactions at interface. The most recommended plating material for SiC is copper.

This paper focusses on understanding the mechanism of material removal in abrasion-assisted Electrical Discharge Surface Grinding (EDSG), for the machining of AISI D2 die steel. The rotary motion imparted to the metal matrix Cu/SiCp tool electrode using rotating spindle assembly mounted on electric discharge machine (Fig. 1) was employed to study the EDSG technology. Powder metallurgy route was adopted to fabricate the MMC electrode with different proportion of abrasives as SiC in copper powder. During the EDSG operation, the hump of material melted by the EDM is forcibly removed by the grinding/abrasion mechanism from the work surface, which enhances the material removal rate (MRR) of this hybrid process. During EDM, the material in the form of resolidified layer (*white layer*) is deposited on the surface of work material, which requires supplementary processes such as polishing, lapping, which subsequently increase cost and production time. The white layer is often incapable of removal due to surface tension effect and is completely removed due to abrasion-assisted effect in

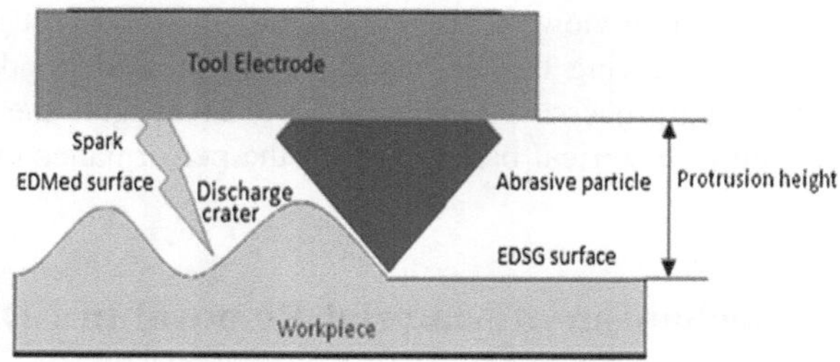

Fig. 1 Material removal mechanism of EDSG

EDSG technology, thus producing improved stock removal rate and better surface finish.

In the present investigation, the grinding effect was introduced with electrical discharge machining by selecting different levels of speeds and sizes of abrasives. Role of grinding and electrical discharge input parameters on the MRR, electrode wear rate (EWR) and surface roughness (SR) has been investigated to elucidate the mechanism of material removal during the process.

Most of the past research work has been conducted on development of various MMCs as workpiece machined with electrical discharge machining process. In the current study, authors have purposed a unique hybrid machining process in which grinding function can be performed with Cu/SiC$_P$ MMC electrode in addition to thermal erosion in electrical discharge machining process.

An investigation was made on the combined technology of electrical discharge machining and grinding (EDMG) by Shu and Tu (2003). A metal matrix (Cu/SiCp) electrode with a rotating device was made and employed to study the EDMG technology. It was reported that EDMG machining efficiency is three to seven times that of normal EDM operation, and the corresponding SR of work piece after EDMG operation is lower than that after EDM operation.

Recently, Yadav et al. (2008) investigated the machining parameters of electro-discharge diamond grinding for high-speed steel. From the study, it was found that most of the significant factors affecting the EDDG robustness are wheel speed and discharge current. The effect of current on metal removal rate (MRR) at constant speed was investigated; their results indicate that MRR increases with increase in current at constant RPM.

The study on abrasive electrical discharge grinding (AEDG) of Ti−6Al−4V titanium alloy was reported by Swiecik (2009). In this investigation, the effect of grinding conditions on the performance of AEDG process was studied. The results of the study were compared with that of convention grinding operation. It was observed from these investigations that application of AEDG process had a strong impact on the removal of machining allowance and surface machining texture (SGT). The effect of electric parameters, such as working voltage and intensity of current, on the tangential force was also investigated.

The hybrid technology like EDSG can also be successfully employed for drilling micro-holes in hard and brittle material such as glass and quartz. This technique was developed by Chen et al. (2011).

After the review of research trends, it is observed that MMCs have significant role in achieving higher material removal and good surface finish in advanced machining processes. A need is felt to investigate the combined effect of grinding as well as electrical parameters on the performance characteristics of the process.

2 Mechanism of Material Removal in EDSG

The principle of EDSG is illustrated in Fig. 1. When some suitable voltage is applied between both the electrodes of EDM, the spark is generated between them. The grinding action along with the electric sparks starts when the abrasive grits in the rotary electrode come in contact with the workpiece surface. This contact does not result in short circuit between tool and workpiece because of non-conductive nature of abrasive grits. As a result, the removal of material by combined action of erosive discharges and grinding from surface of workpiece starts. The mechanical interaction due to abrasion helps in speeding up the removal of workpiece material. With this kind of unconventional setup, the removal of non-conductive particles with conductive electrodes is also possible (Yan et al. 2000; Kozak 2002).

3 Materials

3.1 Tool Electrode

Fabrication of composite tool was carried out by powder metallurgy route. The reinforcements used for the fabrication of composites are in 60, 120 and 220 mesh sizes, respectively. All the abrasives of different sizes were free from foreign particles and impurities. Electroless plating process was adopted to electroplate the surface of SiC abrasives. This process includes three steps: surface cleaning, sensitization and activation and finally copper plating.

After electroless plating, powders of copper-coated SiC were mixed in copper powder with different proportions by wt. of 5, 10 and 15 %, respectively. After milling the powders in ball mill, these powders were compressed under direct loading of 55 KN on compression testing machine with the help of die steel die with suitable designed dimensions. After removing the green compacts, sintering was carried out at 800 °C in inert gas (N_2) in the tube furnace to prevent oxidation of green compacts. These compacts were used as tool electrode on a ZNC EDM machine with attached rotary spindle attachment.

Fig. 2 Experimental setup of EDSG

3.2 Workpiece

Hardened AISI D2 die steel (60–61 BHN) plate having size 100 × 100 × 12 mm (2.25 % C, 12 % Cr, 0.25 % Cu, 1 % V, 1 % Mo, 0.6 % Si, 0.6 % Mn, 0.03 % P, 0.03 % S, 0.3 % Ni and balance Fe) was used as a work piece. Due to extreme demand of manufacturing and automobile industries for the same material, it was selected as candidate work material for the current investigation. EDSG of workpiece was conducted up to 30 min for each run of the experiment to measure the performance characteristics.

4 Experimental Procedure

Experiments were performed using a ZNC EDM Sparkonix S50 die sinking machine mounted with fabricated rotary spindle assembly. Experimental setup is shown in Fig. 2. Range of electrical and grinding parameters such as gap parameters current and rotational speed was selected to determine MRR, EWR and SR as performance measures of the process. During the experiments, the pulse on time and pulse off time were kept constant at 150 μs, whereas the gap current varied from 1.5 to 9 A. The gap voltage remained between 55 and 60 V. Polarity of the tool is negative. Composite tool electrode with 60 mesh APS and 10 % APC was selected to perform the experiments.

5 Results and Discussions

The following discussion focuses mainly on the essential parameters which significantly contribute to the process performance. The performance of the combined electrical discharge machining and conventional grinding was verified by evaluating MRR, EWR and SR.

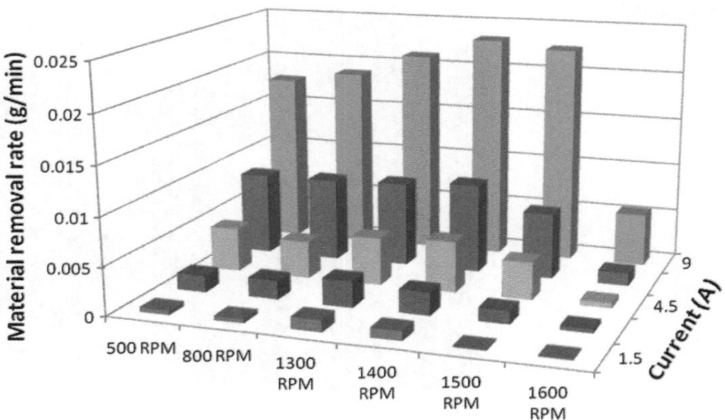

Fig. 3 Effect of gap current on metal removal rate at variable speed

5.1 Analysis of MRR

Comparison of 3D graph in Fig. 3 shows variation of MRR with gap current at different rotational speeds. MRR observed to be increase with gap current. Increase in rotational speeds also increases MRR upto 1,400 rpm. As the peak current has more impact on discharge energy, hence it removes more material with large-size craters from workpiece surface. Increase in speed enhances grinding effect contribution with spark erosion process. Increase in rotational speed of the tool electrode facilitates in removal of debris from the inter-electrode gap and increases machining stability which improves machining efficiency. The material that gets melted due to spark erosion is removed by the abrasive grains with rotational effect. Increase in rotational speed of electrode increases MRR up to 1,500 rpm and then reduces considerably. After 1,500 rpm, the effect of high speed dominates the process of EDM thermal erosion due to reduction in strength of plasma channel and contamination of gap with more debris within the discharge gap which reduces MRR.

5.2 Analysis of EWR

Figure 4 shows the comparison of 3D graph for EWR at different settings of rotational speeds. Electrode removal rate (EWR) increases with increase in gap current. Large peak current generates more discharge energy and hence causes large EWR. Sharp rise in EWR has been observed after current of 6 A. EWR also increases with increase in rotational speed of the tool electrode. Maximum values of EWR are observed at 1,600 rpm. At higher rotational speeds, grinding effect becomes more pronounced and hence contributes to increase in wear rate of tool electrode as

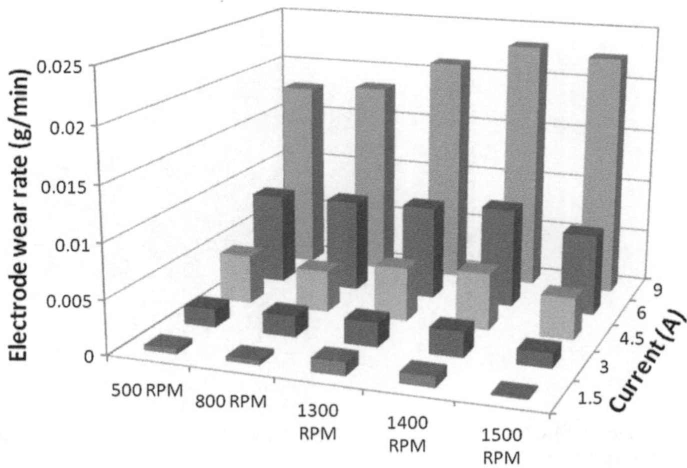

Fig. 4 Effect of gap current on EWR at variable speed

observed by Yan et al. in their investigations (Shu and Tu 2006). At higher values of rotational speeds, EWR is high because the grinding effect increases with increase in rotational speed of the electrode. This effect is attributed to the fact that at higher rotational speeds, more material is removed by abrasion which causes excessive wear and pull out of abrasives from the face of the tool electrode.

5.3 Analysis of SR

Figure 5 indicates the increase in SR with increase in gap current at selected range of grinding speed. At higher amperage, more discharge energy is released due to which more material is removed per spark and hence increases SR of the workpiece. SR decreases with increase in rotational speed. This is due to the fact that higher grinding energy is available at higher speeds due to which more material is removed uniformly from the surface of workpiece. Moreover, resolidified layer is also thin due to fast removal of the material from workpiece surface before solidification.

5.4 Comparison of Material Removal Rate in EDM and EDSG

The electrical discharge machining of AISI D2 die steel was also carried out by restricting rotary motion of tool holder assembly to measure MRR. Figure 6 shows the comparison of MRR during EDM and EDSG processes. It has been clearly

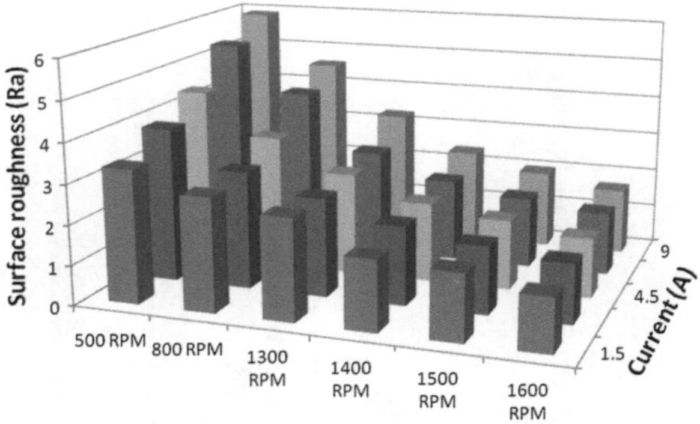

Fig. 5 Effect of gap current on SR at variable speed

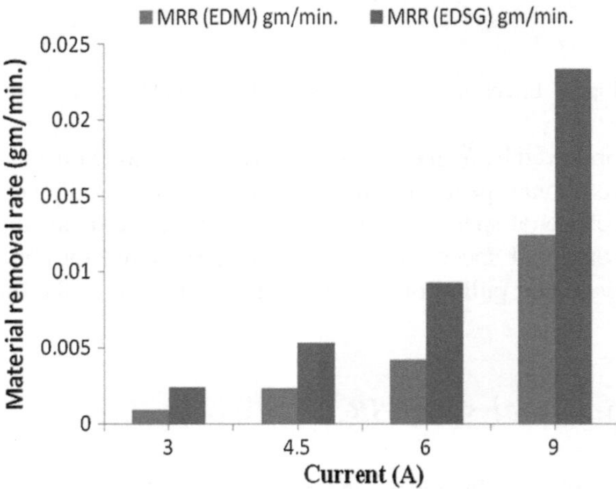

Fig. 6 Comparison of MRR of workpiece during EDM and EDSG

observed that MRR during EDSG is considerably increased from 2 to 3 times as compared to EDM process. Hybridization of specific energies of EDM and grinding causes increased and controlled removal of material from the workpiece surface. These energies assist to each other in such way so that the limitation of one is overcome by another.

5.5 Topography of Machined Surface

Workpiece surface machined by EDSG process was observed under scanning electron microscope (SEM). Figure 7a and b shows the surface topography of machined samples of AISI D2 die steel at 2,200 and 500 magnifications. Figure 7c shows the EDSG machined surface with pronounced effect of grinding at higher rotational speed. The surface was machined with 60-mesh (abrasive size)

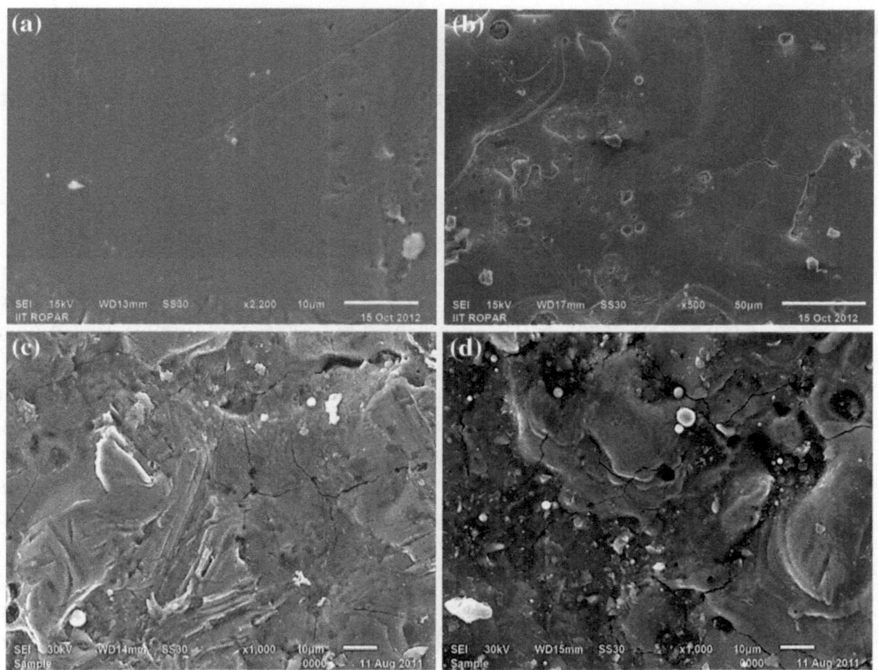

Fig. 7 Surface topography of samples machined by EDSG process, (**a**) 2,200×, (**b**) 500×, (**c**) grinding marks due to pronounced grinding effect (60 mesh SiC) at 1,500 rpm, (**d**) EDMed surface machined with die sinking at 8 A, 1,000×

composite electrode, which causes more abrasion at higher range of grinding speed at low values of gap current. Electrical Discharged Machined (EDMed) surface of AISI D2 die steel machined with composite tool electrode is shown in Fig. 7d. Micrograph of EDMed surface reveals the presence of debris and micro-cracks on the uneven surface. However, the surface machined with noble EDSG process is free from foreign particles and surface irregularities (Fig. 7b). Due to improved flushing and rotational effort, eroded material from workpiece and electrode is completely removed from the surface to reduce thickness of resolidified layer. Densities of surface cracks are minimum, which shows the uniform flow of heat to the workpiece surface at faster rate during the machining process which increases the life of the tool. Surface finish obtained is better than electrical discharge machining process which is clearly observed from Fig. 7b.

Scope of the work can be enhanced by optimizing the control parameters of EDSG process to improve MRR, EWR and surface finish.

6 Conclusions

The experimental EDSG setup was developed to investigate the effect of gap current and rotational speed on MRR and EWR. Based on experimental results, following conclusions are drawn:

1. EDSG experiments reveal that MRR increases with increase in gap current and rotational speed of the electrode.
2. At higher values of gap current 6–9 A, MRR sharply increases with increase in rotational speed. Grinding effect is more pronounced at higher rotational speeds. Above 1,500 rpm, MRR decreases due to gap contamination at higher speeds.
3. EWR increases with increase in gap current and rotational speed. Higher grinding speed leads to increased wear of electrodes at higher amperage. To yield optimum results of EWR, gap current and rotational speed should be low.
4. SR increases with increase in gap current and decreases with increase in rotational speed of the electrode. At low values of current and higher rotational speed, better surface finish can be attained by this process.
5. Grit protrusion height contributes major role in EDSG process. MRR increases with increase in size of the abrasives in composite electrode due to increased protrusion height of abrasives in composite electrode, as more protrusion height removes more material from workpiece surface at higher speed.

References

Chen ST, Jiang ZH, Wu YY, Yang HY (2011) Development of a grinding–drilling technique for holing optical grade glass. Int J Mach Tools Manuf 51:95–103

Denisenko ET, Voinitskii AG, Polushko AP, Kostenko AD, Filatova NA (1970) Some deformation characteristics of copper containing oxide inclusions. Probl Prochn 11:54–57

Denisenko ET, Polushko AP, Filatova NA, Kostenko AD (1971) Effect of various factors on the properties of copper hardened with dispersed oxides. Poroshk Metall 10:49–54

Fedorchenko IM, Denisenko ET, Sakhnenko AV, Polushko AP, Perepelkin AV (1974) Bulk forging technology and properties of dispersion-strengthened copper electrode materials. Transactions of the fourth international conference on powder metallurgy in Czechoslovakia, pp 171–187

Kobayashi KF, Tachibana N, Shingu PH (1990) Formation of amorphous Al–Cr alloys by mechanical alloying of elemental aluminium and chromium powders. J Mater Sci 25:3149–3154

Kozak J (2002) Abrasive electrical discharge grinding of advanced materials. Archives of civil and mechanical engineering, 2(1–2):1–19

Occhionero M, Richard A, Kevin F (1999) A new substrate for electronics packaging: aluminum–silicon carbide (AlSiC) composites. In: Proceedings of the 4th annual portable by design conference, electronics design, 24–27 March, pp 398–403

Shu KM, Tu GC (2003) Study of electrical discharge grinding using metal matrix composite electrodes. Int J Mach Tools Manuf 43:845–854

Shu KM, Tu GC (2006) Fabrication and characterization of Cu–SiCp composites for electrical discharge machining applications. Mater Manuf Process 20:483–502

Swiecik R (2009) Experimental investigation of abrasive electro-discharge grinding of Ti–6Al–4V titanium alloy. J Achievement Mater Manuf Eng 37:706–711

Yadav SKS, Yadava V, Narayana VL (2008) Experimental study and parameter design of electro-discharge diamond grinding. J Adv Manuf Technol 36:34–42

Yan BH, Wang CC, Liu WD (2000) Machining characteristics of Al_2O_3 /6061Al composite using rotary EDM with a disk like electrode. J Adv Manuf Technol 16:322–333

Optimization of Single Point Machining Parameters for Minimum Energy Utilization

Ardamanbir Singh Sidhu, Sehijpal Singh and Paramjit Singh Bilga

Abstract In the present research, a tractor part (king pin) was considered for single point machining. Keeping in view the product dimensions and capability of existing machine in the given industry, optimum cutting parameters for minimum energy consumption were recommended by using process window optimization technique. The recommended parameters were depth of cut 2.5 mm, cutting velocity 172 m/min, spindle rpm 1,273, feed rate 0.4 mm/rev, and cutting time 0.52 min resulting in energy consumed per volume of 4.9 Ws/mm^3. All these parameters were available on existing machine. This enabled the manufacturer to easily upgrade the changes without incurring additional cost like replacement of existing machine by more energy efficient machine like CNC. This method implemented in industry proved that energy consumed per volume of material removed decreased by 12.5 % and production rate increased by 48.5 %. The research therefore presents a simple methodology which can be easily implemented on single point machining process in any industry to obtain optimum parameters not only for minimizing energy consumption but also for increasing production rate and enhancing Green machining.

Keywords Energy · Green machining · Optimization · Turning

A. S. Sidhu (✉)
Department of Mechanical Engineering, Ludhiana College of Engineering and Technology, Katani Kalan, Ludhiana, Punjab, India
e-mail: rajacyber001@sify.com

S. Singh · P. S. Bilga
Department of Mechanical Engineering, Guru Nanak Dev Engineering College, Ludhiana 141006, Punjab, India

S. S. Khangura et al. (eds.), *Proceedings of the International Conference on Research and Innovations in Mechanical Engineering*, Lecture Notes in Mechanical Engineering, DOI: 10.1007/978-81-322-1859-3_19,

1 Introduction

The term green manufacturing reflects new manufacturing techniques that employ various strategies to become more eco-efficient. The present trend is to move from less green into a greener and eco-efficient manufacturing (Deif 2011). Human population is constantly increasing. Currently, it is 7.05 billion. India's population stands at 121,093,422 as per Registrar general and census commissioner (2012) as in 2012.

According to United Nations report "World population to 2,300" as in (2004), the world population is projected to peak at 9.22 billion in the year 2075, and Indian population is projected to be 1,531.4 million in 2050. It will surpass China's population. Such a huge population will increase load on manufacturing sector. In Indian context, manufacturing sector is one of the major consumers of electricity. This is verified by the data supplied by Ministry of Power, Government of India, as reported in (2010), the percentage utilization of electrical energy by industrial sector was highest, that is, 35.7 %. Further, generation of electricity is one of the major contributors to the green house gases (GHG). This is illustrated by factual data from Indiastat.com as in (2007), highlighting that green house gas emission of carbon dioxide equivalent was highest, that is, 719.3 million tons by electricity production in India. Therefore, there is intense need to reduce energy consumption in manufacturing sector. This can be brought about by conserving the energy being utilized in machining processes.

1.1 Literature Review

Most of the research in machining has been focused on maximization of production rate and cost minimization. Little work has been done in the field of optimization of energy in the machining process, keeping environmental factor in view. Some of the works done in the field of energy conservation in machining process are:

The energy used per volume of material removed in milling process was determined in Dhamus and Gutowski (2004). The reference data were provided for energy foot print of tooling (Y_E). According to them, the main contributor to the energy account and CO_2 emissions is the energy used in the machining process. The authors identified that energy requirements of actual material removal are quite small when compared to the total energy associated with machine tool operation.

The authors in Rajemi (2010; Rajemi et al. 2010; Mativenga and Rajemi 2011) derived theoretical criterion for optimal tool life $T_{\text{opt}-C}$ for minimum cost and minimum energy $T_{\text{opt}-E}$. They observed that optimum machining conditions for minimum costs do not necessarily satisfy the minimum energy criteria. According to author's energy, foot print of tooling plays important role in the tool life. The

Authors later develop methodology to select a feasible combination of cutting parameters which satisfies minimum energy criteria for turning process. According to them, optimum cutting conditions for minimum energy have to be selected or evaluated by the industrialists to suit their need for manufacturing.

The energy efficiency of machining process was reported in Neugebauer et al. (2011). First being drilling into solid material and second being cylindrical hard turning process on precision lathe machine, their work highlights the importance of tool selection and selection of process parameters for energy efficiency of manufacturing process. According to authors, the most efficient means for achieving energy savings is to reduce the machining time by proper selection of cutting tool, while maintaining the specified work piece characteristics.

The research work in Mori et al. (2011) was focused on identifying optimal cutting conditions for minimum energy consumption in milling and drilling process. Four machining parameters were considered with three levels of each, and Taguchi method was applied for optimization. Best results for ideal cutting conditions which fall within tool manufacturer's recommendation were considered.

2 Problem Formulation

2.1 Single Point Machining Process

It is a machining process using cutting tool having single cutting edge. The basic machining parameters which are involved in single point machining process under this study are cutting speed (V_c) in m/min, rpm (N), depth of cut (a_p) in mm, material removal rate (MRR) (v) in mm^3/s, feed rate (f) in mm/rev, and specific cutting energy requirement of work piece material (k) in Ws/mm^3.

2.2 Evaluation of Total Energy Consumed in Turning Process

As in Mativenga and Rajemi (2011), the total energy E used in turning operations can be evaluated from the energy consumed by the machine during setup operation E_1, energy consumed during cutting operations E_2, energy consumed during tool change E_3, energy consumed to produce a cutting tool and normalized per cutting edge E_4, and energy consumed to produce work piece material E_5. The energy of the work piece material E_5 was not considered as it is independent of the machining strategy and does not affect the optimization of production parameters. Therefore, the energy in single pass turning operation can be calculated as

$$E = E_1 + E_2 + E_3 + E_4 \tag{1}$$

If P_0 is power consumed in [W] when machine is running idle, and t_1 time taken when machine is running but cutting has not started. Then,

$$E_1 = P_0 t_1 \tag{2}$$

If t_2 is the time taken for cutting, then

$$E_2 = (P_0 + k)t_2 \tag{3}$$

If t_3 is tool change time and T is the tool life, then

$$E_3 = P_0 t_3 \left(\frac{t_2}{T}\right) \tag{4}$$

In present case, E_3 is not considered as in the given industry, the machine operator always changed the tool when the machine was powered off. Y_E is the energy embodied in the tool per cutting edge. Then,

$$E_4 = Y_E \left(\frac{t_2}{T}\right) \tag{5}$$

Total energy consumed in single pass turning process under the given case study

$$E = P_0 t_1 + (P_0 + kv)t_2 + Y_E \left(\frac{t_2}{T}\right) \tag{6}$$

According to Taylor tool life equation, T is tool life in minutes, the constant n is tool life constant, and the parameter C_t is an empirical tool life constant. Stephenson and Agapiou as in (2006)

$$V_c T^n = C_t \tag{7}$$

Extended Taylor tool life equation includes effects of feed rate f in mm/rev and depth of cut d in mm.

$$V_c T^n f^a d^b = K_t \tag{8}$$

If $\frac{1}{n}$ is cutting velocity exponent, $\frac{a}{n}$ is the feed exponent, and $\frac{b}{n}$ is the depth of cut exponent, then

$$T = \frac{K}{V_c^{\frac{1}{n}} f^{\frac{a}{n}} d^{\frac{b}{n}}} \tag{9}$$

Since cutting velocity V_c and feed f are the first and second most dominating criteria affecting the tool life, respectively,

$$T = \frac{K}{V_c^{\frac{1}{n}} f^{\frac{a}{n}}} \tag{10}$$

The cutting time t_2 for a single pass turning operation is given as

$$t_2 = \frac{\pi D_{\text{avg}} l}{f V_c} \tag{11}$$

Material removal rate (MRR), v

$$v = \frac{V_c \cdot a_p \cdot f \times 1{,}000}{60} \text{ mm}^3/\text{s} \tag{12}$$

$$v = \frac{\pi}{4}\left(D_i^2 - D_f^2\right) \frac{f V_c}{\pi D_{\text{avg}}} \tag{13}$$

Substituting the values k, t_2, v, and T in (6), we get Eq. (14) as

$$E = P_0 t_1 + P_0 \cdot \left(\frac{\pi D_{\text{avg}} l}{f V_c}\right) + k\left(\frac{\pi l}{4}\left(D_i^2 - D_f^2\right)\right) + \frac{Y_E \pi D_{\text{avg}} l V_c^{\left(\frac{1}{n}-1\right)} f^{\left(\frac{a}{n}-1\right)}}{K} \tag{14}$$

2.3 Optimum Tool-Life for Minimum Energy Consumption

The optimum tool life for minimum energy is obtained by differentiating total energy (14) with respect to cutting velocity and equating it equal to zero, $\frac{\partial E}{\partial V_c} = 0$. Derivative is taken with respect to V_c because it is the most dominating criteria which affect the tool life.

$$T_{\text{opt}-E} = \left(\frac{1}{n} - 1\right)\left(\frac{Y_E}{P_0}\right) \tag{15}$$

Equation 8 can be written as

$$V_c f^a d^b (T_{\text{opt}-E})^n = K_t \tag{16}$$

3 Experimentation

3.1 Details Regarding Industry, Workpiece, Tool, and Machine

This research was conducted in Auto International Pvt. Ltd., an industry situated near Ludhiana (India). The work piece considered was king pin of a tractor. The work piece material was alloy steel EN 42 CrMo 4 (AISI/ASTM Designation 4140). Total length of cylindrical work piece was 313 mm. The machining length was 266 mm. The initial diameter was 45.3 mm, and the final diameter was

40.4 mm. Numbers of work pieces were taken and were machined at different cutting speeds, feed, and depth. The cutting tool used was CVD-coated carbide insert of SECO Company having designation WNMG 060408-M3 (1000T), having six cutting edges. The tool holder used was of SECO Company designated as PWLNL2525MO6. The machine used was copying lathe (5 H.P) using three-phase motor. It was manufactured by Pars Co., Ludhiana.

3.2 Method for Measurement of Electrical Power

The electrical power consumption was measured using a digital clamp meter (MECO 2250F). The power relation used

$$P = V \cdot I\sqrt{3}(\mathrm{W}) \tag{18}$$

The voltage being supplied by three-phase transformer was 415 V. Average idle power P_0 noted during the experiment was 3,243 W.

3.3 Evaluation of Specific Cutting Energy Constant (k) for the Given Work Piece Material

Specific cutting energy k (Ws/mm^3) is the obtained as the ratio of net power consumed (W) to the MRR in (mm^3/s). This was obtained from slope of the graph as shown in Fig. 1.

Therefore, specific cutting energy evaluated for the material alloy steel 42 Cr Mo_4 was $k = 2.54$ Ws/mm^3. This value lies within the range for steels given by Kalpakjian and Schmid as in (2009) providing credibility to the result.

3.4 Determination of Taylor Tool Life Equation Constants and Exponents

Taylor tool life equation is $VT^n = C_t$ (7). Different work pieces were machined at different cutting speeds; the tool life was calculated by observing the standards ISO 3685. Microscopic examination of the tool was carried out at regular intervals. Whenever the flank wear was 0.3 mm, the tool was considered to be worn out and tool life time obtained. The exponent n in Eq. (7) was determined from slope of a graph plotted between Log V and Log T as in Fig. 2.

$$\text{The value of } \frac{1}{n} = 2.63 \tag{19}$$

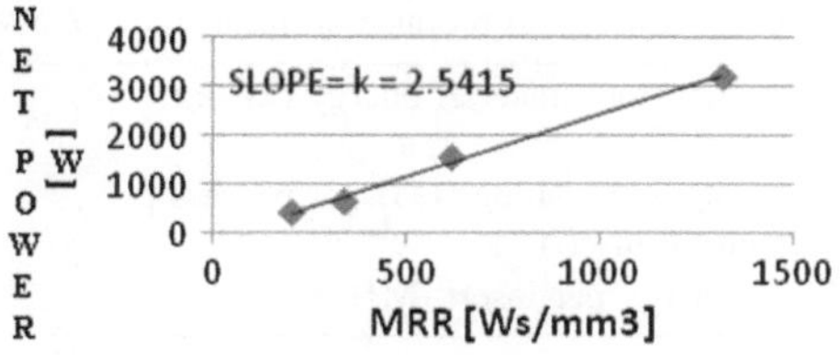

Fig. 1 Variation in power with MRR of alloy steel 42 Cr Mo4

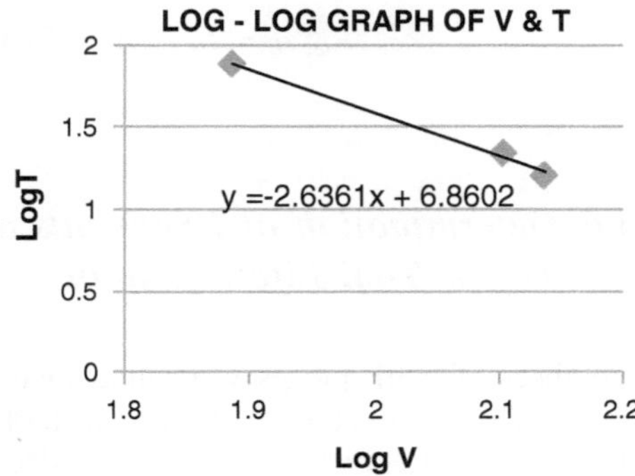

Fig. 2 Log—log graph between Log V and Log T

$$\text{Value of } n = 0.38 \tag{20}$$

The value of n lies within the range recommended by Kalpakjian (2009), giving credibility to result.

3.5 Determination of Optimal Tool Life for Minimum Energy Consumption for Given Work Piece Tool Material

As per Eq. (14) $Y_{E,}$ that is, energy embodied in the cutting tool used in the industry. Work of Dhamus and Gutowski as in (2004) presents reference data for energy footprint of tooling. Average weight of insert was 5.7 g, and numbers of cutting sides were 6.

Y_E from Table 1.

$$Y_E = \frac{\text{Energy}}{\text{Cutting..Edges}} = \frac{3.78 \text{ MJ}}{6} \quad \text{from} \quad Y_E = 63 \times 10^4 \text{ J} \tag{21}$$

$$P_0 = 3{,}243 \text{ W}, \quad \frac{1}{n} = 2.63$$

Table 1 Energy to produce an insert tool [WNMG 060408-M3 (1000T)]

Embodied tool material energy (MJ/kg)	400	2.28 (for the given insert of weight 5.70 g)
Sintering and coatings (MJ per process per cutting insert)	1–2 (avg. 1.5)	1.5
Total energy per insert (MJ)		3.78

Putting these values in (14)

$$T_{\text{opt}-E} = 5.3\,\text{min} = 5.3\text{min per cutting edge.} \tag{22}$$

3.6 Determination of Exponents and Constant in Extended Taylor Tool Life Equation

Numbers of work pieces were machined, and tool life tests were conducted for each set of parameters. The average tool life period was recorded, for determination of feed exponent "*a*" in extended Taylor tool life equation. Keeping V_c, RPM N, and depth of cut a_p same but altering feed rate *f* to values say f_3 and f_4, corresponding tool life values T_3 and T_4 were obtained. Substituting the values in "(8)" and dividing, we obtain

$$\left(\frac{f_3}{f_4}\right)^a = \left(\frac{T_4}{T_3}\right)^n \tag{23}$$

$$\text{From which value of "}a\text{"} = 0.26 \tag{24}$$

For determination of depth of cut exponent "*b*," V_c, RPM N, and feed rate *f* were kept same but depth of cut a_p was altered to values say a_{p3} and a_{p5}, and corresponding tool life values T_3 and T_5 were obtained. Substituting the values in "(8)" and dividing, we obtain.

$$\left(\frac{a_{p3}}{a_{p5}}\right)^b = \left(\frac{T_5}{T_3}\right)^n \tag{25}$$

$$\text{From which "}b\text{"} = 0.16 \tag{26}$$

Therefore, $VT^{0.38}f^{0.26}d^{0.16} = K_t$. Putting the values of *V*, *f*, and *d* in the given equation, the value of $K_t = 295$. The extended Taylor tool life equation for given work piece material and the insert tool is

$$VT^{0.38}f^{0.26}d^{0.16} = 295 \tag{27}$$

4 Results

4.1 Process of Optimization of Machining Parameters for Minimum Energy Consumption

This was carried out by using process window technique. A feed rate and depth of cut process window were constructed on basis of tool supplier data as in (Catalogue and Technical Guide 2012), and the range of values was further limited according to the requirements of industry. Therefore, maximum depth of cut "a_p" and feed rate "f" selected for process window were 2.5 mm and 0.4 mm/rev. The process window was converted to grid points, and values of corresponding parameters evaluated. The node with minimum energy consumed per unit volume gave optimum set of cutting conditions.

The stepwise optimization procedure followed was

1. Construction of process window for cutting speed and noting values of V_c at various nodes.
2. Construction of process window for cutting time t_2 and noting cutting time at various nodes.
3. Construction of process window for MRR.
4. Construction of process window for total energy consumed at various nodes.
5. Construction of process window for volume of material removed at various nodes.
6. Construction of process window for energy consumed per volume of material removed as in Table 2 and evaluation of minimum energy per unit volume and corresponding cutting parameters (Table 3).

4.2 Optimized Machining Parameters

The minimum energy consumed per volume of material removed was considered as criterion for optimization. Corresponding minimum energy consumed per volume, from the process window as in Table 1, was 4.90 Ws/mm^3, and corresponding optimum cutting parameters were V_c 172 m/min, f 0.4 mm/rev, a_p equal to 2.5 mm, and rpm equal to 1,273.

5 Discussion

Adopting these parameters, the energy consumed per volume of material removed decreased by 12.5 %, and the production rate increased by 48.5 %. Measurements of electric power consumed in the experiment highlight that only 12–48 % of total

Table 2 Process window for energy consumed per volume of material removed (Ws/mm^3)

Depth a_p (mm)	2.5	6.94	6.14	5.70	5.39	5.15	4.90
	2	7.71	6.84	6.26	5.88	5.56	5.34
	1.5	9.09	7.93	7.25	6.83	6.40	6.10
	1	11.65	10.10	9.19	8.51	8.02	7.56
	0	0.15	0.2	0.25	0.3	0.35	0.40
					Feed (mm/rev)		

Table 3 Comparison between existing and optimized machining parameters

Existing parameters		Recommended parameters	Remarks
Cutting speed V_c	127 m/min	172 m/min	Available on the machine
RPM N	945	1,273	Available on the machine
Depth of cut a_p	2.5 mm	2.5 mm	Available on the machine
Feed f	0.27 mm/rev	0.4 mm/rev	Available on the machine
Cutting time t_2	1.01 min	0.52 min	48.5 % increase in production rate
Energy consumed per volume	5.6 Ws/mm^3	4.9 Ws/mm^3	12.5 % decrease in energy/vol

power drawn is used for actual machining. The bulk power was spent for the non-cutting operations and for running the machine in the idle mode. Extended Taylor tool life equation for the given work piece material, that is, alloy steel EN 42 Cr Mo4 and the cemented carbide insert (WNMG 060408-M3 [1000T] of Seco Indian grade) comes out to be $VT^{0.38}f^{0.26}d^{0.16} = 402$, indicating maximum effect of cutting speed on tool life followed by effect of feed and then depth of cut, respectively. The criteria of optimal tool life for minimum energy consumption reveal that energy embodied in the tool Y_E has great effect on tool life and optimum cutting parameters. The results from the process window of energy consumed per volume of material removed highlight that greater is the MRR or greater is the feed and depth of cut, lower is the energy consumed per volume.

6 Conclusions

1. A simplified approach based upon the techniques introduced by Rajemi (Neugebauer et al. 2011) for determining minimum energy consumption in single point machining operation has been developed and applied to a small-scale industry.
2. The developed approach may prove to be cost-effective way of reducing machining costs and energy consumption for small-scale industry that cannot afford costs of CNC-based technology.

References

Deif MA (2011) A system model for green manufacturing. Adv Prod Eng Manag 6:27–36

Dhamus J, Gutowski T (2004) an environmental analysis of machining. In: ASME international mechanical engineering congress (IMECE 2004-62600) and R&D expo, pp 1–10, Anaheim, California

Indiastat.com. Greenhouse gas emissions of carbon dioxide equivalent from various sectors in India 2007. http://www.indiastat.com/table/environmentandpollution/11/environmentalstandards/384

Kalpakjian S, Schmid SR (2009) Manufacturing engineering and technology, 4th edn. Pearson Education, inc. and Dorling Kindersley Publishing Inc. Delhi, pp 571–578

Mativenga PT, Rajemi MF (2011) Calculation of optimum cutting parameters based on minimum energy footprint. CIRP Ann Manuf Technol 60:149–152

Ministry of Power, Govt. of India (2010). Region/State/Category-wise percentage of utilisation of electrical energy (Utilities) in India (2009–2010). http://www.indiastat.com/table/power/26/consumptionandsale/70/652717/454090/data.aspx

Mori M, Fujishima M, Inamasu Y, Oda Y (2011) A study on energy efficiency improvement for machine tools. CIRP Ann Manuf Technol 60:145–148

Neugebauer R, Schubert A, Reichmann B, Dix M (2011) Influence exerted by tool properties on the energy efficiency during drilling and turning operations. CIRP J Manufact Sci Technol 4:161–169

Office of Registrar General and Census Commissioner, India (2012) State wise provisional population by sex in India (As per 2011 Census). http://www.indiastat.com/table/demographics/7/population2011census/527359/527360/da

Rajemi MF (2010) Energy analysis in turning and milling. Ph.D. dissertation, School of Mechanical, Aerospace and Civil Engineering, University of Manchester, Manchester, UK

Rajemi MF, Mativenga PT, Aramcharoen A (2010) Sustainable machining: selection of optimum turning conditions based on minimum energy considerations. J Clean Prod 18:1059–1065

SECO (2012) Turning. Catalogue and technical guide. Vyoma Graphics, Pune

Stephenson DA, Agapiou JS (2006) Metal cutting theory and practice, 2nd edn. CRC Press Taylor and Francis group, Boca Raton, FL, pp 523

United Nations Department of Economic and Social Affairs, Population Division (2004) World Population to 2300.pdf. http://www.un.org/esa/population/publications/longrange2/WorldPop2300final.pdf

Development of Biomedical Implant (Hip Joint) by Combining Fused Deposition Modelling and Investment Casting

Rupinder Singh, Sunpreet Singh and Prince Kapoor

Abstract This research work is aimed at the development of biomedical implant (hip joint) by combination of fused deposition modelling (FDM) and investment casting. Three controllable factors of the investment casting process (namely slurry layer combination, viscosity, dry time of primary coating) were studied at three levels each by Taguchi's parametric approach, and single-response optimization was conducted to identify the main factors controlling dimensional accuracy (Δd). ABS patterns to be used for investment casting were made with FDM process, and castings were produced using stainless steel at recommended parameters through ceramic shell investment casting process. The results of the study suggest that for Δd, contribution of viscosity of primary dip and dry time of primary coating is 23 and 72 %, respectively.

Keywords Hip joint · Fused deposition modelling · Investment casting · Slurry viscosity · Abs · Dimensional accuracy

1 Introduction

Biomedical implants are manufactured at high rate in today's world. So, there is critical need of development of such implant in order to make these implants highly compatible physically with human body. However, chemical compatibility is also desired, and after impartment, they can serve for longer time without

This work was supported in part by the AICTE, New Delhi, under NCP Grant.

R. Singh (✉) · S. Singh · P. Kapoor
Department of Production Engineering, Guru Nanak Dev Engineering College, Ludhiana 141006, Punjab, India
e-mail: rupindersingh78@yahoo.com

S. S. Khangura et al. (eds.), *Proceedings of the International Conference on Research and Innovations in Mechanical Engineering*, Lecture Notes in Mechanical Engineering, DOI: 10.1007/978-81-322-1859-3_20, © Springer India 2014

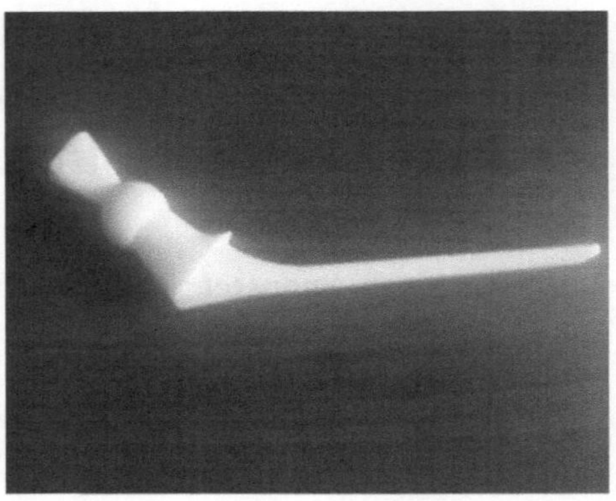

Fig. 1 Schematic of hip joint (ABS)

causing critical problem (Hanemann et al. 2006). Investment casting has come to occupy a key position in the range of modern metal casting techniques (Dong et al. 2011). Investment casting consists of manufacturing a wide range of industrial components with net clear shape (Hooker et al. 1993). It has been investigated that Δd of investment cast components can be controlled by altering input parameters (Hung et al. 2003). Viscosity of primary slurry layer affects the final mechanical properties in IC (Ibrahim et al. 2011). Viscosity of the primary slurry is an important measurement of the flow characteristics of the slurry, which determines the quality of the moulds (Jiang and Liu 2007). The quality of shell is dependent on slurry and slurry material as well as the method by which they built (Jiann 2005). Rapid prototyping polymer patterns are one of the major advancements in investment casting process (Omar et al. 2013). The direct conversion of 3-D CAD data into rapid prototyping pattern decreases development time, chances of costly mistakes, minimizes sustaining engineering changes and extend product lifetime by adding necessary features and eliminating redundant features early in the design (Liu et al. 2002). FDM is one of the rapid prototyping technologies by which physical objects are created directly from CAD data (Kenned et al. 2011; Singh 2010, 2012). For present research work, hip joint (Fig. 1) was selected as benchmark. The major technical challenge of the proposed system is to extract some typical 3-D geometry parameters with respect to the patient's 3-D bone anatomy and then creates a custom-made hip implant based upon the extracted parameters with hybridization of FDM and investment casting.

CAD model of the hip joint ABS pattern is shown in Fig. 2.

2 Design of Experiments

Before conducting final experimentation, pilot experiments were conducted for controlling the levels of input parameters. For the present study, different number of layers, viscosity of the primary slurry and drying time of primary coating have

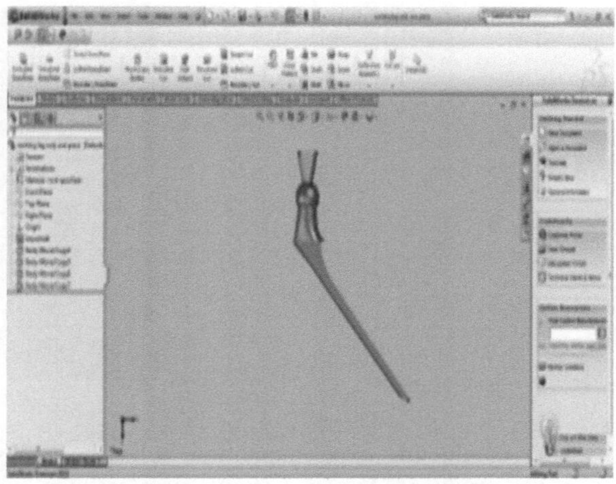

Fig. 2 CAD model of hip joint

Table 1 Input parameter

Parameter 1			Parameter 2			Parameter 3		
No. of slurry layers			Viscosity–efflux time (s)			Drying time of primary coating (min)		
L1	L2	L3	L1	L2	L3	L1	L2	L3
11	12	13	80	90	100	300	360	420

Table 2 Control log for experimentation

S. No.	No. of slurry layers	Viscosity–efflux time (s)	Drying time of primary coating (min)
1	11	80	300
2	11	90	360
3	11	100	420
4	12	80	360
5	12	90	420
6	12	100	300
7	13	80	420
8	13	90	300
9	13	100	360

been optimized. It has been observed that by increasing the viscosity level between 100 and 110 s (I.S. no. 3944 B.4 cup), shell built was of heavy weight. Further, decreasing it to level of 65–75 (I.S. no. 3944 B.4 cup), shell was cracked. So for present research work, viscosity level was varied from 80 to 100 (I.S. no. 3944 B.4 cup). Number of slurry layers and drying time of primary coat along with viscosity are shown in Table 1. Taguchi optimization technique has been used to optimize Δd of hip joint. Twenty-seven numbers of experiments were conducted, and control log experimentation is shown in Table 2.

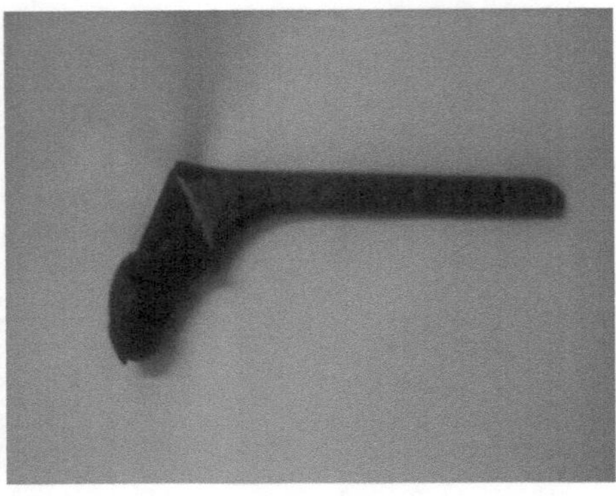

Fig. 3 Final casted hip joint

After conducting experimentation, it has been observed that very close tolerance was produced. Figure 3 shows stainless steel casted hip joint.

3 Result and Discussion

The Δd test of casted hip joint was performed on coordinate measuring machine (CMM). It should be noted that Δd was measured as difference in dimensions of the cast component from the nominal dimension (based upon actual drawing). Hip joint, as already discussed, is a biomedical implant; testing was performed precisely. Based on control log of experimentation (refer Table 2), three repetitions of experiment were made as R1, R2 and R3 for measurement of Δd. The results for Δd are shown in Table 3. Further, Table 4 shows sum of squares and signal to noise (S/N) ratio for present set of experiments. As per the values of Table 4, Figs. 4, 5 and 6 show the highest value of the S/N ratio and Δd with respect to input parameters.

Figure 7 shows pie chart for percentage contribution of input parameters for Δd.These results are valid at 95 % confidence level. The confirmatory experiment was conducted which shows percentage improvement in Δd is 2.62 % at proposed parametric settings. The results are in line with the observations made by other investigators (Jiann 2005; Singh 2010, 2012; Singh and Singh 2013; Yadav and Karunakar 2011).

Table 3 Observations of final experimentation for Δd

S. No.	Δd (mm)		
	R1	R2	R3
1	0.3	0.27	0.31
2	0.2	0.18	0.21
3	0.27	0.29	0.26
4	0.18	0.2	0.17
5	0.25	0.26	0.24
6	0.35	0.33	0.37
7	0.26	0.24	0.26
8	0.27	0.29	0.26
9	0.26	0.23	0.25

Table 4 S/N ratio for Δd

S. No.	Sum square	S/N ratio	Average
1	0.08633	10.6382	0.29
2	0.03883	14.108	0.2
3	0.07487	11.2571	0.27
4	0.03377	14.7151	0.18
5	0.06257	12.0366	0.25
6	0.12277	9.1092	0.35
7	0.06427	11.9201	0.25
8	0.07487	11.2571	0.27
9	0.061	12.1467	0.0666

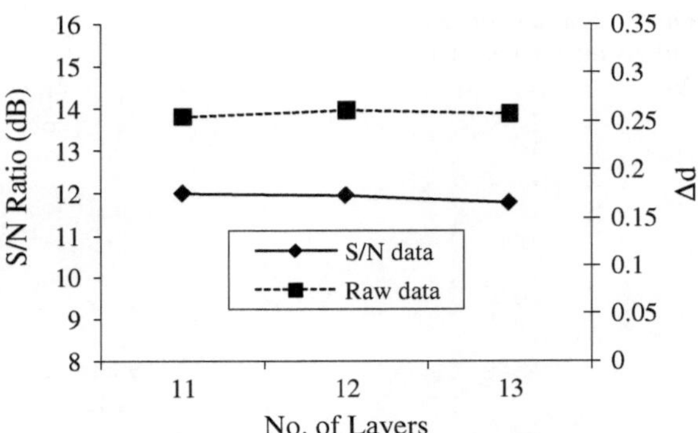

Fig. 4 S/N ratio and Δd with respect to number of layers

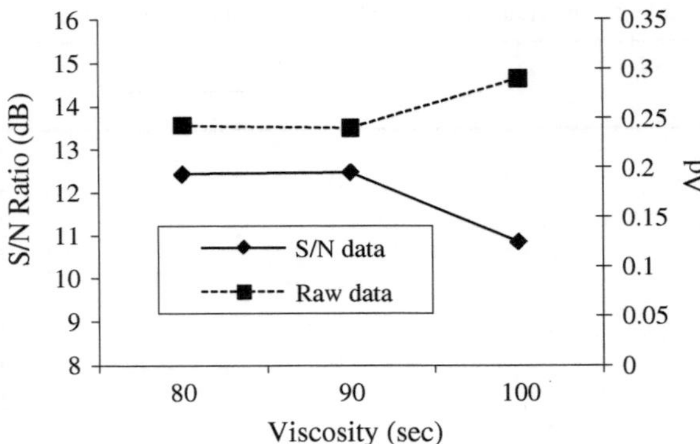

Fig. 5 S/N ratio and Δd with respect to viscosity

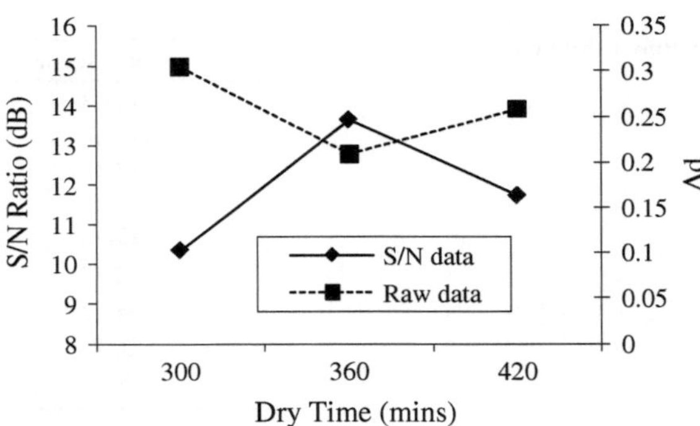

Fig. 6 S/N ratio and Δd with respect to drying time (min)

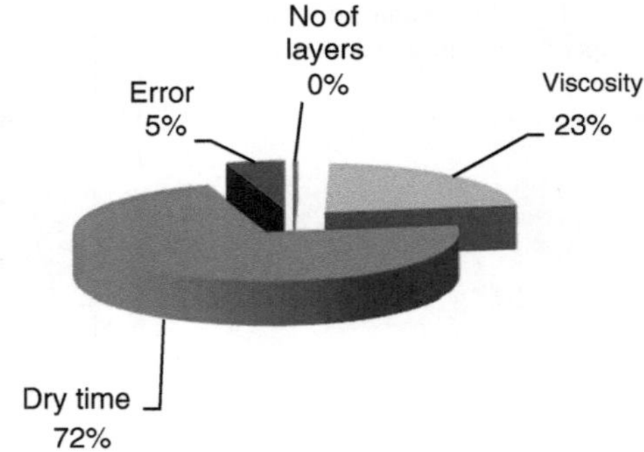

Fig. 7 Contribution of various input parameters for Δd

4 Conclusions

Drying time is one of the major parameters affecting Δd of parts. Longer as well as shorter drying time leads to shrinkage of casting. Mould with minimum layers can perform good job. Viscosity of the slurry plays significant role in investment casting.

- From present research work, it has been concluded that number of layers had no affect (0 %) on dimensional accuracy of hip joint. Number of layers had no effect on Δd of part, but it might have an effect on other mechanical properties of cast component.
- On the other hand, viscosity of primary slurry layer has affected 23 % for Δd. At high viscosity, mould built with thicker shell decreased solidification rate and hence affected metal shrinkage.
- Drying time had maximum impact (72 %) on dimensional accuracy. From Fig. 6, it was observed that shorter drying time leads to improper coat deposition, whereas longer dying time generates stresses in constituent parts of mould.
- At viscosity, 80–90 s and drying time of 360 min, high accuracy of 0.17–0.21 mm was achieved.

Acknowledgement The authors are thankful to AICTE New Delhi for financial support.

References

Dong Y, Bu K, Dou Y, Zhang D (2011) Determination of interfacial heat-transfer coefficient during investment-casting process of single-crystal blades. J Mater Process Technol 211(12):1231–2123

Hanemann T, Bauer W, Knitter R, Woias P (2006) Rapid prototyping and rapid tooling techniques for manufacturing of silicon, polymer, metal and ceramic microdevice. Handbook techniques and applications, pp 801–869. ISBN 978-0-387-24520-1

Hooker MW, Taylor TD, Leigh HD (1993) Non-aqueous slip casting of $YBa_2Cu_3O_7$-x ceramics. Appl Super Conductivity J 1(7–9):1303–1311

Hung CC, Hou GL, Tsai CC, Huang CC (2003) Effect of zirconia-modified magnesia investment on the casting of pure Titanium. Kaohsiung J Med Sci 19(3):121–125

Ibrahim KM, Mhaede M, Wanger L (2011) Mechanical characterization of cp-Ti produced by investment casting. Trans Non-ferrous Met Soc Chin 21(8):1735–1740

Jiang J, Liu XY (2007) Dimensional variations of castings and moulds in the ceramic mould casting process. J Mat Process Technol 189(1–3):247–255

Jiann PY (2005) Fe evaluation of thermal property of mould wall material for investment casting and the effect of layers on the hardness of the casting product. University of Technology Malaysia, Malaysia, pp 13–42

Kenned DM, Startford YPW, Robarts S, Gollish JD (2011) Using outcome measure results to facilitate clinical decisions the first year after total hip arthroplasty. J Orthop Sports Phys Ther 41(4):232–239

Liu Q, Sui G, Leu MC (2002) Experimental study on the ice pattern fabrication for the investment casting by rapid freeze prototyping (RFP). J Comput Ind 48(3):181–197

Omar MFM, Sharif S, Ibrahim M, Idris MH, Fadzil ASA, Mohd A, Azriszul (2013) Differential ceramic shell thickness evaluation for direct rapid investment casting. Appl Mech Mater 315:418–423

Singh R (2010) Three dimensional printing for casting applications: a state of art review and future perspectives. Adv Mater Res 83–86:342–349

Singh R (2012) Effect of work piece volume on statistically controlled rapid casting solution of aluminum alloys using three dimensional printing. Mater Manuf Process 27(4):377–382

Singh R, Singh S (2013) Effect of process parameters on surface hardness, dimensional accuracy and surface roughness of investment cast components. J Mech Sci Technol 27(1):191–197

Yadav N, Karunakar DB (2011) Effect of process parameters on mechanical properties of the investment castings produced by using expandable polystyrene pattern. Int J Adv Eng Technol 1(3):128–137

Multi-response Optimization of WEDM Using Utility-Based Taguchi Approach

Jatinder Kapoor

Abstract This paper presents a utility-based Taguchi loss function strategy for the multi-response optimization of wire electrical discharge machining (WEDM) process. The utility-based Taguchi loss function strategy has been used for the multi-response optimization of WEDM with the use of cryogenically treated wire electrode. The approach utilizes a composite utility function using the Taguchi loss functions of WEDM responses in selecting the optimal parametric settings such that the overall functional utility of the product is maximized. The use of the Taguchi loss functions ensures the process is robust against the random variations. Actual experiments confirm the feasibility of the strategy over a wide range of machining conditions employed in WEDM.

Keywords WEDM · Multi-response optimization · Taguchi loss function · Utility function

1 Introduction

In today's highly competitive environment, companies are incited to constantly improve the quality of their products, because the market has raised the customer's expectation to unparalleled levels. To stay in competition, a company must supply the high-quality product and services according to the needs of the customers. The overall cost of the product is ascertained with respect to a number of quality characteristics. These quality characteristics are often interrelated and need to be considered simultaneously. A single setting of process parameters may be optimal for one quality characteristics, but it may yield adverse results for other quality

J. Kapoor (✉)
Department of Mechanical engineering, Guru Nanak Dev Engineering College, Ludhiana 141006, Punjab, India
e-mail: Jatinder_kapur@yahoo.com

S. S. Khangura et al. (eds.), *Proceedings of the International Conference on Research and Innovations in Mechanical Engineering*, Lecture Notes in Mechanical Engineering,
DOI: 10.1007/978-81-322-1859-3_21, © Springer India 2014

characteristics. Therefore, optimization strategy is required to provide an integrated criterion so as to represent the overall optimal setting of process parameters with respect to all the responses. The multi-response optimization techniques can handle these types of optimization problems. For obtaining the multi-characteristic optimization of product quality, various techniques have been developed. Taguchi method is one of the most accurate and effective techniques to solve the multi-characteristics optimization problems. Antony (2000) and Tong and Su (1997) have reported few applications of multi-response problems in their study.

The Taguchi design has established to be an effective tool to produce high-quality products at low cost. The objective of Taguchi robust design is to decide the optimal process parameter settings and allow the process optimization with a minimum number of experiments at reduced time and lowest cost of experimentation. Byrne and Taguchi (1987) presented a case study, where the quality characteristics were independently optimized and then results were compared to select the best levels in terms of the quality of interest. Logothetis (2000) utilized linear programming and multiple regression techniques for multi-response optimization of five responses using Taguchi design. The objective function was optimized successively taking into consideration one of the responses as principal while ignoring the possible relationship among the responses. Shiau (1990) employed weighted S/N ratio approach to signify overall performance of a process. Phadke (1989) used multi-characteristic optimization technique for the products and assigned weights to different characteristics from practical experience. Elsayed and Chen (1993) devised a model by employing loss function approach to determine the optimal setting of parameters of a production process for multiple characteristic products. Pignatiello (1993) employed squared deviation and variance to form a probable loss function for optimizing multiple responses. The procedure for three simultaneous functions—the weight loss function, desirability function, and distance function—was presented by Layne (1995). Tong and Su (1997) used Taguchi method to propose multi-response optimization by fuzzy multiple attribute decision-making technique using multi-response S/N ratio decided through assimilation of quality loss for all the responses. Graphical approach for multi-response optimization was used by Floros and Chinnan (1998) in order to examine for overlapping response contours. The condition was suitable for all the responses, but the method was not suitable for more than three responses. Principal component analysis (PCA) was introduced by Antony (2000) for multi-response optimization of submerged arc welding parameters. In PCA, uncorrelated principal component characteristics are identified through transformation of set of correlated response characteristics.

In order to demonstrate the potential of the proposed methodology, a case study of maximizing material removal rate (MRR), minimizing surface roughness (SR), and improving wire wear ratio (WWR) in wire electrical discharge machining (WEDM) with cryogenic treated wire electrodes and untreated wire electrodes has been analyzed. Actual experiments are carried out to verify the feasibility of the strategy over a wide range of machining conditions employed in WEDM.

2 Experimentation

2.1 Experimental Material and Equipment

A Robofil-290 CNC wire-cut electrical discharge machine and brass wire as electrode was used for conducting the experiments. EN-31 plate of thickness 11 mm was selected as workpiece material. The MRR is calculated as,

$$\text{MRR} = K.t.V_c$$

Here, K is the kerf width, t is the thickness of workpiece (11 mm), and V_c is the cutting speed. The kerf width was measured using the Carl Zeiss tool maker microscope (X 100). The kerf width value is the average of three measurements made from each cut length of workpiece.

3 Utility-Based Multi-Response Optimization and Taguchi Method

In today's globally competitive business circumstances, a product is evaluated by a customer on a number of diverse quality characteristics. In present investigation, a simplified methodology based on Utility concept and Taguchi method is developed for determining the optimal setting of process parameters for multi-response characteristics of WEDM process.

3.1 Multi-Characteristic Optimization Algorithm

The stepwise algorithm for carrying out multi-response optimization with utility-based Taguchi approach is given in following steps:

1. Find the optimal values of each of the selected performance characteristic using Taguchi experimental design and analysis.
2. Construct a preference scale for each performance characteristic based on the optimal values and minimum acceptable levels.
3. Assign weights W_i, i = 1, 2 ….n, based on the experience and application of the components to various output characteristics.
4. Find the utility values for each experimental trial conditions for all the response characteristics involved in multi-response optimization.
5. Use the values obtained in step 4 as a response of the different trial conditions of the selected experimental matrix.
6. Analyze the results as per the procedure proposed by Taguchi method.

Table 1 Process parameters and their values at different levels

Parameters	Designation	Levels		
		L1	L2	L3
Type of wire	*A*	Untreated brass wire electrode	Deep cryogenic treated (−184°C) brass wire electrode	–
Pulse width (μs)	*B*	0.4	0.8	1.2
Time between two pulses (μs)	*C*	4	10	16
Wire tension (daN(Kg))	*D*	0.6	1.3	2.0
Servo reference voltage (V)	*E*	20	35	50

7. Find the optimal setting of the process parameters for mean and minimum deviation (optimum unity) based on analysis executed in step 6.
8. Predict the individual characteristic values that maximize the overall utility as determined in step 7.
9. Conduct confirmation experiments at the optimal settings to verify the optimal results.

4 Multi-Response Optimization for Deep Cryogenic Treated Wire Electrode

The optimal setting of the process parameters for WEDM with deep cryogenic treated wire electrode for different response characteristics shows the conflicting behavior. For example, minimum surface roughness (SR) was obtained at the cost of material removal rate (MRR). Therefore, it becomes imperative to obtain a single set of optimal levels of cryogenic treated wire electrode. Table 1 shows the process parameters and their values at different levels for conducting the experiment.

4.1 Determination of Optimal Values of Performance Characteristics

Step. 1 Determination of optimal results

The summary of results of optimal values (optimized individually) of MRR, SR, and WWR (performance characteristics) is given in Table 2.

Step. 2 Construction of preference scale

a. Preference scale for MRR

Y^* = Optimal value of MRR = 66.85 mm^3/min

Table 2 Optimal settings of process parameters and predicted optimal quality characteristics

Performance characteristic	Optimal setting of process parameters	Significant process parameters	Predicted optimal value of quality characteristics
MRR	A_2, B_3, C_1, D_1, E_3	*A*, *B*, *C*, *D*	66.85 mm^3/min
SR	A_2, B_1, C_3, D_2, E_2	*A*, *B*, *C*	1.93 μm
WWR	A_2, B_1, C_3, D_2, *E*2	*A*, *B*, *C*	0.021

$Y_i^{/}$ = Minimum acceptable value of MRR = 12.38 mm^3/min (All the observed values of MRR are greater than 12.38 mm^3/min).

The preference scale for MRR was constructed by using these values in equation.

$$P_{\mathrm{MRR}} = 12.28\ \log(Y_{\mathrm{MRR}}/12.38) \tag{1}$$

b. Preference scale for SR

Y^* = Optimal value of SR = 1.93 μm

$Y_i^{/}$ = Minimum acceptable value of SR = 3.30 μm (all the observed values of SR are less than 3.30 μm.

The preference scale for SR was constructed by using these values in equation.

$$P_{\mathrm{SR}} = -38.63\ \log(Y_{\mathrm{SR}}/3.30) \tag{2}$$

c. Preference scale for WWR

Y^* = Optimal value of WWR = 0.021

$Y_i^{/}$ = Minimum acceptable value of SR = 0.075 (all the observed values of WWR are less than 0.075

The preference scale for SR was constructed by using these values in equation

$$P_{\mathrm{WWR}} = -16.52\ \log(Y_{\mathrm{WWR}}/3.30) \tag{3}$$

Step. 3 Weight assignment

The weights to the response characteristics have been assigned on the basis of consideration that MRR contributes more to the machining of workpiece than SR and WWR.

The weight assignments to all the three quality characteristics are given as follows:

Weight assignment to MRR = W_{MRR} = 0.55

Weight assignment to SR = W_{SR} = 0.25

Weight assignment to WWR = W_{WWR} = 0.20

Table 3 Calculated utility data based on reponses (MRR, SR, WWR)

Expt. No	Utility values			Mean value
	*R*1	*R*2	*R*3	
1	4.59	4.55	4.60	4.58
2	4.29	4.34	4.30	4.31
3	4.18	4.17	3.99	4.11
4	4.99	5.15	5.10	5.08
5	5.22	4.95	4.93	5.03
6	4.64	4.44	4.53	4.54
7	4.71	4.55	4.55	4.60
8	4.86	5.04	5.09	4.99
9	5.01	4.99	5.02	5.01
10	5.62	5.44	5.65	5.57
11	5.66	5.75	5.58	5.66
12	4.99	5.34	5.02	5.12
13	6.51	6.44	6.42	6.46
14	5.91	5.97	5.74	5.87
15	7.10	7.07	7.08	7.08
16	5.92	5.93	5.72	5.86
17	5.46	5.47	5.61	5.51
18	5.77	5.88	5.67	5.77

*R*1, *R*2, *R*3 = Repetitions of experiments against each trial conditions

Step. 4 Calculation of utility value

The utility value of each casting has been calculated by using the following equation:

$$U(n,R) = P_{\mathrm{MRR}}(n,R)\times W_{\mathrm{MRR}} + P_{\mathrm{SR}}(n,R)\times W_{\mathrm{SR}} + P_{\mathrm{WWR}}(n,R)\times W_{\mathrm{WRR}} \quad (4)$$

Where n is the trial number, n = 1, 2,…18, and R is the repletion number, R = 1,2,3…n.

The calculated utility values are given in Table 3.

Step. 5 Utility data analysis for optimal stetting of process parameters

The utility values plotted are shown in Fig. 1. It can be observed from figure that the second level of type of wire (*A*2), second level of pulse width (*B*2), first level of time between two pulses (*C*1), first level of wire tension (*D*1), and second level of servo reference voltage (*E*2) are expected to yield a maximum value of the utility and S/N ratio within the experimental space. The analysis of variance for utility data is given in Table 4. It can be observed from Table 4 that parameter *A* (type of wire), factor *B* (pulse width), and factor *D* (wire tension) have significant effect (at 95 % confidence level) on the utility function.

The type of wire is the most dominant parameter, whereas the time between two pulses and servo reference mean voltage is statistically insignificant in affecting the utility function. However, the optimal levels of these parameters are very much

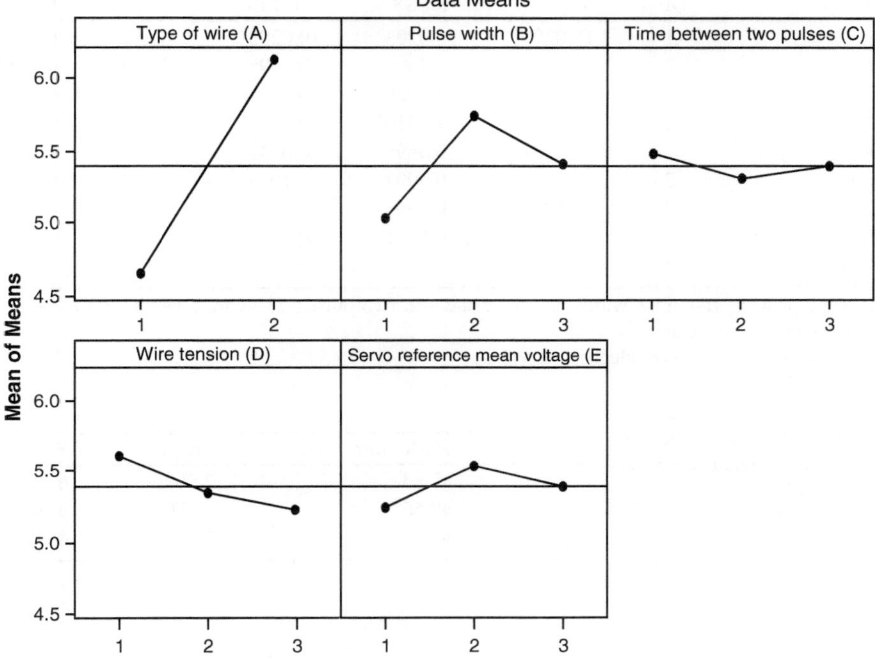

Fig. 1 Main effects plot for means (utility data)

necessary to be decided. On the basis of the results, for a particular setting of parameters (as per machine settings), the minimum values of the time between two pulses (first level) and middle value of servo reference mean voltage (second level) seem to be better from MRR and SR point of view. The optimal values of utility and thus the optimal values of response characteristics in consideration are predicted at above levels of significant parameters.

Step. 6 Predicted means (optimal values of performance characteristics)

The average values of all the response characteristics at the optimal levels of significant parameters with respect to utility function are recorded in Table 5.

The optimal values of the predicted means (μ) of different response characteristics can be obtained from the following equation:

$$\mu = A_2 + B_2 + D_1 - 2T \tag{5}$$

where A_2—second level of type of wire, B_2—second level of pulse width, and D_1—first level of wire tension

The 95 % confidence interval of confirmation experiments (CI_{CE}) was calculated by using the following expression:

Table 4 Analysis of variance for means (utility data: MRR, SR, WWR)

Source	DF	Seq SS	Adj SS	Adj MS	F	P
A	1	10.0203	10.0203	10.0203	360.66	0.000*
B	2	1.4990	1.4990	0.7495	26.98	0.005*
C	2	0.1029	0.1029	0.0514	1.85	0.270
D	2	0.4476	1.0310	0.5155	18.55	0.009*
E	2	0.2701	0.0694	0.0347	1.25	0.379
*A***B*	2	0.3909	0.3909	0.1954	7.03	0.049*
*A***C*	2	0.7963	0.7963	0.3982	14.33	0.015*
Residual error	4	0.1111	0.1111	0.0278		
Total	17	13.6383				

A—type of wire, *B*—pulse width, *C*—time between two pulses, *D*—wire tension, and *E*—servo reference mean voltage
* Significant at 95 % confidence interval

Table 5 Average values of performance characteristics at optimum levels

Levels	MRR (mm^3/min.)	SR(μm)	WWR
A_2	47.83	2.509	0.04222
B_2	49.56	2.670	0.04767
D_1	46.98	2.658	0.04767

$$\mathrm{CI}_{\mathrm{CE}} = \pm\sqrt{F_x(1,f_e)\times V_e\left[\frac{1}{r_{\mathrm{eff}}}+\frac{1}{R}\right]} \tag{6}$$

where $F_\alpha\,(1, f_e)$ is the F-ratio at a confidence level of $(1-\ \alpha)$ against DOF one and error degree of freedom $f_{e.}$

$$N_{\mathrm{eff}} = \frac{\text{Total number of trials}}{1+\text{DOF of all factors used for estimation mean}}$$

1. Material Removal Rate (MRR)

   The predicted optimal range for MRR is 51.40 mm^3/min < MRR < 62.18 mm^3/min

2. Surface Roughness (SR)

   The predicted optimal range for SR is 2.28 μm < SR < 2.80 μm

3. WWR.

   The predicted optimal range for WWR is 0.037 < WWR < 0.051.

Step. 7 Confirmation experiments

Table 6 Results of confirmation experiment for MRR

Exp. No.	MRR (mm3/min)			Mean MRR	SR(μm)			Mean SR	WWR			Mean WWR
	*R*1	*R*2	*R*3		*R*1	*R*2	*R*3		*R*1	*R*2	*R*3	
1	57.10	56.60	57.50	57.07	2.40	2.50	2.30	2.40	0.040	0.042	0.040	0.041
2	57.30	56.70	57.10	57.04	2.30	2.20	2.50	2.33	0.042	0.044	0.039	0.042
3	56.20	56.80	56.90	56.64	2.50	2.60	2.40	2.50	0.041	0.042	0.044	0.042
	Overall Mean			56.92	Overall Mean			2.41	Overall Mean			0.042

* Not significant hence selected as per machine setting

Three experiments were conducted at optimal settings as suggested by Taguchi analysis of utility data, and results have been depicted in Table 6. It can be seen that confirmatory results fall within the 95 % of the CI_{CE}.

5 Conclusions

Following conclusions have been drawn from the research work:

1. The experimental results indicate that the behavior of response characteristics (MRR, SR, WWR) in WEDM depends upon the wire electrode (type of wire) and machine-based parameters.
2. In the present research, the parametric optimization of process parameters with deep cryogenic treated wire electrodes in WEDM indicates that the optimal settings of experimental results suggested by Taguchi method lies well within the predicted range.
3. The multi-response optimization model developed based on "utility concept" provides a simplified method for investigating the combined effect of any number of process parameters on the quality characteristics of the process. The predicted results agree with the experimental results within the specified range of parameters.

References

Antony J (2000) Multi-response optimization in industrial experiments using Taguchi quality loss function and principal component analysis. Qual Reliab Eng Anal 16(1):3–8

Byrne DM, Taguchi G (1987) The Taguchi approach to parameter design. Qual Prog 20(12):19–26

Elsayed EA, Chen A (1993) Optimal levels of process parameters for product with multiple characteristics. Int J Prod Res 31(5):1117–1132

Floros JD, Chinnan MS (1998) Computer graphic–assisted optimization for product and process development. Food Technol 42(1):72–78

Layne KL (1995) Methods to determine optimum factor levels for multiple responses in the designed experiments. Qual Eng 7(4):649–656

Logothestis N (2000) Managing for total quality, from Deming to Taguchi and SPC. PHI, New Delhi

Pahadke MS (1989) Quality engineering using robust design. Prentice Hall, USA

Pignatiello JJ (1993) Strategies for robust multi-response quality engineering. IIE Trans 25(3):5–15

Shiau GH (1990) A study of the sintering properties of iron ore using Taguchi's parameter design. J Stat Assoc 28:253–275

Tong LI, Su CT (1997a) Optimizing multi-response problems in the Taguchi method by fuzzy multiple attribute decision making. Qual Reliab Eng Int 13(1):25–34

Tong LI, Su CT (1997b) Optimizing multi-response problems in the Taguchi method by fuzzy multiple attribute decision making. Qual Reliab Eng Int 13(1):25–34

Finishing of Micro-channels Using Abrasive Flow Machining

G. Venkatesh, Tarlochan Singh, Apurbba Kumar Sharma and Akshay Dvivedi

Abstract Abrasive flow machining (AFM) is one of the nontraditional finishing processes. In AFM process, tool used for finishing have some special characteristics; it is pliable in nature and is generally known as media. Medium is a key element in the process; it contains a mixture of polymer, abrasives, and processing oil. It has a capability of machining micro-channels, micro-bores, blind holes, and tiny holes due to its flexibility and rheological properties. Facilitating a micro-channel workpiece through proper fixture will be a difficult task. However, in the present paper, a special flexible tooling was designed and fabricated to hold micro-channels. Micro-channels of Si wafers were initially prepared using ultrasonic machining (USM) and further finished using AFM process. A set of 20 experiments were designed using response surface methodology for finishing of micro-channels. Extrusion pressure, wt.% of processing oil, and processing time were taken as varying process parameters. It was observed that the surface finish improved more than 50 %, whereas improvement in material removal was marginal.

Keywords AFM · Micro-channels · Surface finish · USM

G. Venkatesh (✉) · T. Singh · A. K. Sharma · A. Dvivedi
Research Scholar in Mechanical and Industrial Department, Indian Institute of Technology Roorkee, Roorkee 247667, Uttarakhand, India
e-mail: venkatesh8054@gmail.com

T. Singh
e-mail: er.tarlochansingh@gmail.com

A. K. Sharma
e-mail: akshafme@gmail.com

A. Dvivedi
e-mail: akshaydvivedi@gmail.com

S. S. Khangura et al. (eds.), *Proceedings of the International Conference on Research and Innovations in Mechanical Engineering*, Lecture Notes in Mechanical Engineering,
DOI: 10.1007/978-81-322-1859-3_22,

1 Introduction

Fabrication of micro-channels has gained importance over the last few years because of their outstanding modern industrial applications, especially in areas such as electronic cooling (Weisberg and Bhu 1992), biomedical engineering (Beebe et al. 2002), biochemistry (Zhao et al. 2002), and electrophoretic applications (Xu et al. 2001). As per the application requirements advanced machining techniques have been developed to fabricate micro-channels which include laser micro-machining (Cheng et al. 2003), focused ion beam micro-machining (Reyntjens and Puers 2001), micro-electro discharge milling (Hung et al. 2011), micro-EDM (Jahan et al. 2009) and micro-ultrasonic machining (Masuzawa and Fujino 1996). Micro-ultrasonic machining is one of the best techniques to fabricate the micro-channels on hard and brittle materials (Jain 2012).

After the fabrication of micro-channels through micro-ultrasonic machining, surface gets partially damaged due to pitting action that strongly influence the quality of the micro-channels. Surface quality and shape precision of micro-channels are greatly affected by machining quality. The present paper focuses on the development of appropriate process to increase the quality of the surface of micro-channel.

AFM is innovative finishing technique to finish intricate and complex surfaces with close tolerances by removing micro- to nano-level material. Layer by layer material is removed by the flow of pressurized abrasive media over or through the surface (Rhodes 1991). Efficiency of the AFM process depends upon process parameters which are further controlled by three major elements, namely the machine, workpiece fixture (tooling), and media. Recently, diesel injectors nozzles (Jung et al. 2008), spring collects (Kim and Kim 2004), micro-bores (Yin et al. 2004), and EDM fabricated micro-holes (Lin et al. 2007) have been finished by AFM efficiently. The present work highlights on the finishing of micro-channels fabricated by micro-ultrasonic machining.

2 Experimental Details

The experimental investigation on finishing of micro-channel was carried out in an indigenously developed experimental setup of AFM. The three-dimensional image of the developed AFM setup was presented in Fig. 1. The developed setup is a horizontal type and consists of two hydraulic cylinders, two media cylinders, and workpiece tooling is at middle. The machine is associated with power supply, hydraulic unit, and DC motor to control the media flow rate. In the present work, natural polymer media (Rajesha et al. 2011) was used to carry out the experimental trials for finishing of micro-channels. The workpiece tooling also play an important role to facilitate the media in desired direction. Owing to this, a flexible tooling fixture for holding micro-channels was designed and developed.

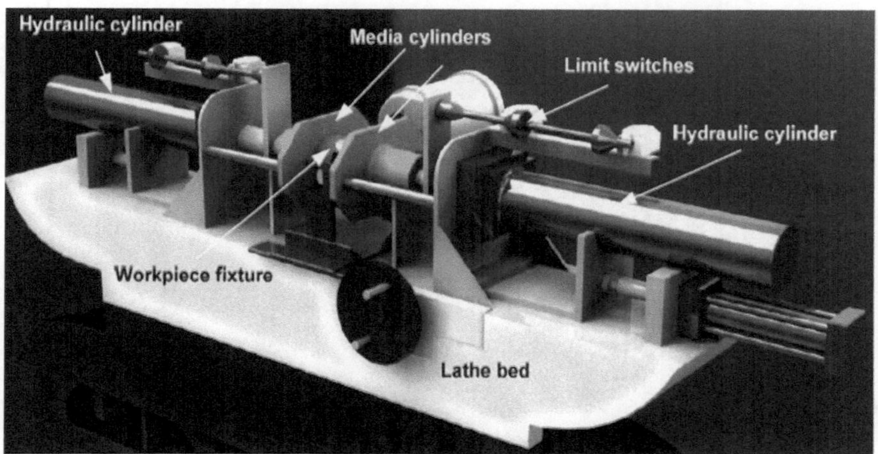

Fig. 1 A two-way horizontal type abrasive flow machine

Table 1 Chemical composition of Si wafer used as a workpiece

Element	O	Na	Al	Si	S	K
Wt %	42.25	0.75	19.74	2.96	33.61	0.32

2.1 Workpiece and Media Preparation

Selection of workpiece material is basically driven by practical requirement. The Si grade of wafers is one of the most widely used engineering materials, especially used in pressure sensor, acoustic wave resonator, and optical chopper. Accordingly, Si wafer was selected as the workpiece material. The chemical composition of the workpiece was confirmed using an EDS/facility.

The major elements present in the workpiece are tabulated in Table 1. Figure 2 shows the peaks of the spectrum for silicon grade wafer. The dimensions of flat rectangular workpiece having 8 mm width × 15 mm length were used for the experimentation. A channel of 1.5 mm width × 0.5 mm depth was machined on the workpiece flat surface using ultrasonic machining. The width and depth of the channel were maintained marginally coarse owing to measurement probe requirement. The size of the micro-channels was limited by the characterization requirements; however, micro-channels in the range of width = 1,500 μm and depth = 500 μm were fabricated on a regular basis using the same procedure.

As discussed in the above section, a natural polymer-based medium was used to conduct the experimental trials. The major elements of medium are natural organic polymer mixed with SiC particles and naphthenic based processing oil. The 50:35:±15 ratio (abrasive: carrier: processing oil) was maintained to achieve the percentage medium concentration by weight (wt.%) initially using a mechanical

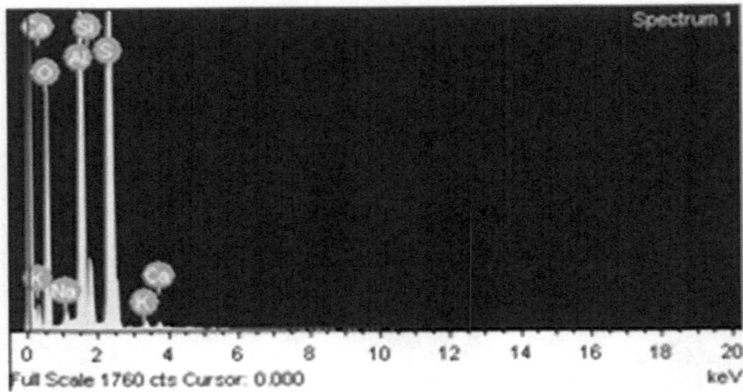

Fig. 2 EDX spectrum of Si wafer workpiece

stirrer. The wt.% of abrasive particles (50 %) and natural polymer (35 %) was maintained constant throughout the experiment.

Weight percentage of an ingredient was calculated as

$$= \frac{\text{weight of ingredient}}{\text{total weight of the media}} \times 100. \quad (1)$$

2.2 *Special Flexible Fixture for Holding a Micro-Channel Workpiece*

In AFM process, fixture or tooling plays a major role. A design of a flexible fixture will always be a cheer work in AFM area. Owing to this, a flexible fixture was designed and fabricated in the laboratory. Figure 3 shows a developed fixture. It mainly consists of Teflon holder, resting spool, locking rings, etc. It facilitates the media flow direction from one cylinder to the other cylinder by passing through the micro-channel workpiece and causes abrasion.

3 Experimental Procedure

Twenty experiments were conducted according to the central composite rotatable design (CCRD) of response surface methods (Montgomery 2004). Half replications for three variables were considered. The process parameters and their levels in the experimentation are shown in the Table 2. The percentage improvement in surface finish (ΔR_a) and material removal (MR) was calculated by using the relations (2) and (3), respectively.

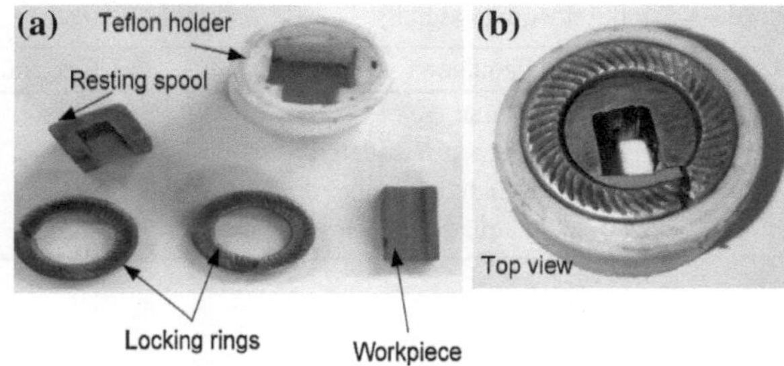

Fig. 3 **a** Special tooling to hold micro-channels. **b** Assembled top view of tooling

Table 2 Process parameters and their levels

S. no.	Process parameters	Units	Levels
1	Extrusion pressure	Bar	15, 20, 25
2	Wt.% of processing oil	ml	10, 12, 15
3	Processing time	Min	5, 7, 9

$$\Delta Ra = \frac{\text{Intial}\, R_a - \text{final}\, R_a}{\text{intial}\, R_a} \times 100 \qquad (2)$$

$$\text{MR} = [\text{Intialweight} - \text{finalweight}] \qquad (3)$$

In the present experimental data, the ANOVA was carried out at 95 % confidence level. The results show that the model is significant. The adjusted and predicted R-squared values are well within the range (Table 3), and hence, the model is acceptable. The results indicated that the model is significant and also there is no significant lack of fit.

4 Results and Discussions

The data obtained for finishing of micro-channels was well fitted for CCRD of RSM. In this paper, only the individual effect of process parameters on process response had been discussed. Further, parametric optimization of the factors and its contribution on the responses were plotted and discussed.

4.1 Observations on Surface Finish

The surface finish improvement of a micro-channel increase with the increasing extrusion pressure up to 20 bar. The improvement of ΔR_a increases from 15 to 20 bar. As the extrusion pressure increases, it leads to increase in axial force

Table 3 Model adequate signals

S. no.	Adequate signals	Surface finish	Material removal
1	*R*-squared	0.979	0.984
2	Adjusted *R*-squared	0.961	0.969
3	Predicted *R*-squared	0.843	0.969
4	Adequate precision	29.053	31.818

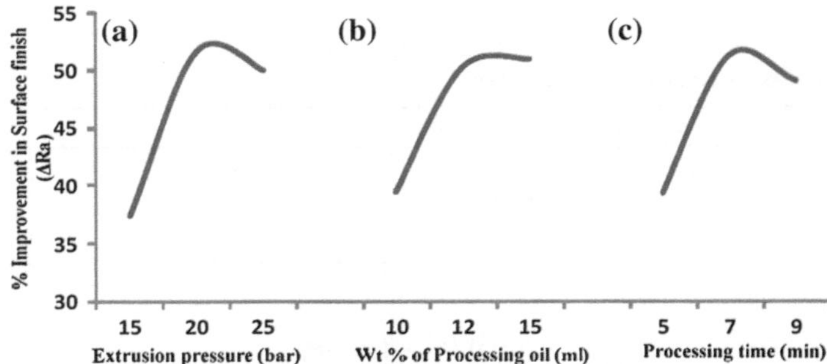

Fig. 4 Response curves showing the individual effect of process parameters on surface finish

resulting in higher ΔR_a improvements. Further increase in extrusion pressure from 20 to 25 bar begins to decrease in ΔR_a improvement. This may due to the fact that at high pressures, number of active abrasive particles becomes less as the highly viscous medium becomes stiffer (Fig. 4a).

Processing oil is used for proper mixing and providing better hold between SiC abrasives and natural polymer. It is observed from the Fig. 4b that at low processing oil content, the bonding among the ingredients (natural polymer and abrasives) in the medium is poor. The improvement of ΔR_a increases from 10 to 12 ml of processing oil. As the Wt.% of processing oil increases, it leads to increase in viscosity of the media, resulting in higher rate of ΔR_a improvements. Further increase in oil content from 12 to 15 ml begins to decrease in ΔR_a improvement. This may be due to the fact that at more addition of processing oil, the medium will lose its stiffness and it becomes less effective. Processing time is the total time of the workpiece surface exposed to abrasive action during the AFM process. It is observed that (Fig. 4c) during the initial stages of processing time, surface finish improves sharply and further drop in percentage surface finish improvement can be attributed to the fact that the sharp abrasives edges get blunt due to initial machining process which results in only marginal improvements in surface finish continued machining.

4.2 Observations on Material Removal of Micro-Channel Surface

Figure 5 illustrates the effect of the parameters such as extrusion pressure, wt.% of oil, and processing time on effect of material removal using AFM process. It is observed from the trend curves that there is an increase in MR with an increase in extrusion pressure. In each trend curve, the process parameter of interest was varied from its low to high level, whereas the values of the rest of the parameters were maintained at middle level (Table 2). The material removal increases with the increasing extrusion pressure up to 20 bar. The improvement of MR highly increases from 15 to 20 bar. As the extrusion pressure increases, it leads to increase in axial force resulting in higher rate of MR improvements.

Further increase in extrusion pressure from 20 to 25 bar, however, does not cause any further visible improvements in MR. This may due to the fact that at high pressures, number of active abrasive particles becomes less as the highly viscous medium becomes stiffer (Fig. 5a).

It is observed from Fig. 5b that when Wt.% of processing oil increases, MR decreases. This is due to the fact that at higher % of processing oil mixed with the media will leads to decrease in the media viscosity. Thus, at lower viscosity of the media, the MR observed was marginal.

Processing time is the total time of the workpiece surface exposed to abrasive action during the AFM process. It is observed that (Fig. 5c) during the initial stages of processing time, a significant improvement is observed in MR. Further drop in improvement of MR can be attributed to the fact that the sharp abrasives edges get blunt due to initial machining process which results in only marginal improvements in MR.

4.3 SEM Analysis on Micro-Channel

Scanning electron microscopy was used to observe the topography of the machined surface. Figure 6 shows a typical surface of a Si wafer workpiece prior to machining through AFM and after AFM. The pit surface was observed clearly on the machined surface before finishing process. After finishing with AFM process, glazed surface was observed (Fig. 6b).

4.4 Parametric Optimization

Figure 7 shows the percentage contribution of processing parameters on the process performance as observed from the ANOVA results. It was found that the extrusion pressure is the most influencing factor in case of surface finish

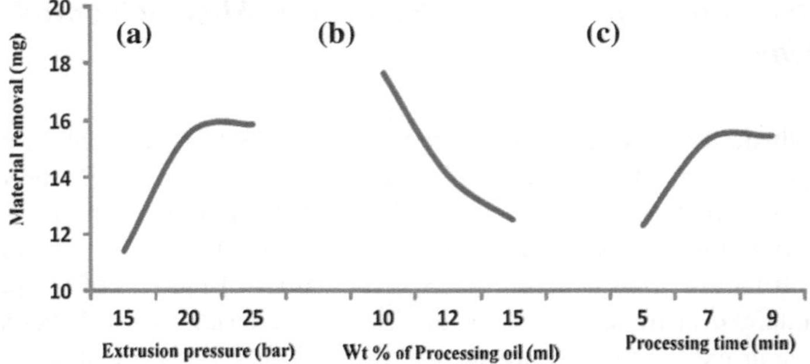

Fig. 5 Response curves showing the individual effect of process parameters on MR

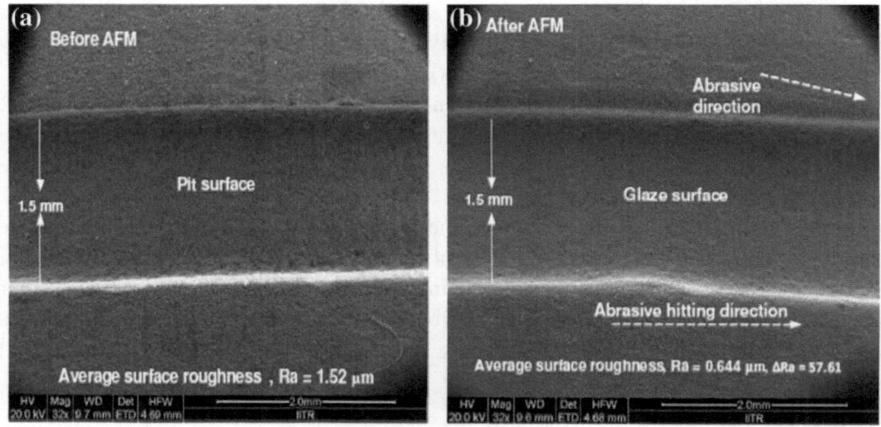

Fig. 6 SEM micrographs of Si wafer micro-channel specimen. **a** Micro-channel machined by USM. **b** Micro-channel finished by AFM (200#, 15 ml, 20 bar, 7 min)

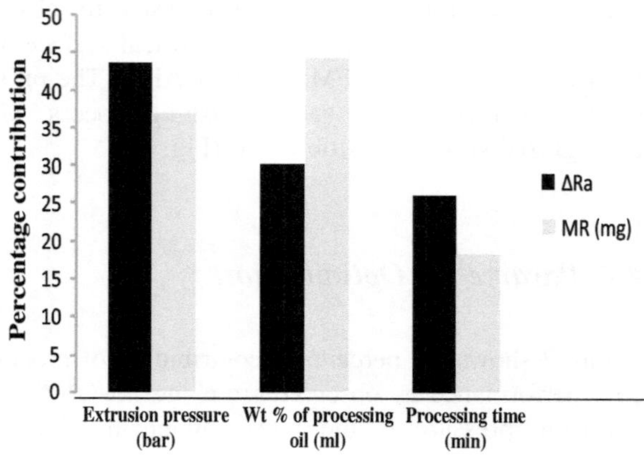

Fig. 7 Contribution of process parameters on ΔR_a and MR

improvement, whereas processing oil is the most influencing parameter for MR. Extrusion pressure with the highest percentage contribution (43.78 %) and wt.% of processing oil which is contributing (30.16 %) have the major influence on SF. Further, the wt.% of processing oil (44.49 %) and extrusion pressure (37.12 %) are major contributing parameters significantly affecting the removal of Si wafer. However, in both the outputs, the less contribution was observed by processing time.

5 Conclusion

The present study was carried out to evaluate the performance of a newly developed natural polymer-based AFM media in terms of the improvement in surface quality of a micro-channel surface of Si wafer workpiece. Experimentation was carried out using RSM to finish the micro-channel at different levels of important AFM process parameters. Following conclusions were drawn.

- The newly developed tooling was flexible enough to hold micro-channel workpieces.
- Extrusion pressure has the highest contribution in finishing of micro-channels, whereas wt.% of processing oil had obtained highest contribution in material removal aspect of a micro-channel.
- SEM analysis on finishing of micro-channels before and after AFM results shows neat agreement with the surface finish improvement results.

References

Beebe DJ, Mensing GA, Walker GM (2002) Physics and applications of micro-fludics in biology. Annu Rev Biomed Eng 4:261–286

Cheng Y, Sugioka K, Midorikawa K (2003) Control of the cross-sectional shape of a hollow micro-channel embedded in photostructurable glass by use of a femtosecond laser. OPT LETT 28(1):55–56

Hung JC, Yang TC, Li K (2011) Studies on the fabrication of metallic bipolar plates Using micro electrical discharge machining milling. J Power Sources 196:2070–2074

Jahan MP, Wong YS, Rahman M (2009) A study on the quality micro-hole machining of tungsten carbide by micro-EDM process using transistor and RC-type pulse generator. J Mater Process Technol 209:1706–1716

Jain VK (2012) Some studies on development of micro-channels using micro-ultrasonic machining. Ph.D. Thesis, Indian Institute of Technology Roorkee, Mechanical and Industrial Engineering Department, Roorkee, India

Jung D, Wang WL, Knafl A, Jacobs TJ, Hu SJ, Assanis DN (2008) Experimental investigation of abrasive flow machining effects on injector nozzle, Engine performance and Emission in a diesel engine. Int J Automot Technol 9(1):9–15

Kim JD, Kim KD (2004) Deburring of burrs in spring collets by abrasive flow machining. Int J Adv Manuf Technol 24:469–473

Lin YC, Chow HM, Yan BH, Tzeng HJ (2007) Effects of finishing in abrasive fluid machining on micro-holes fabricated by EDM. Int J Adv Manuf Tech 33:489–497

Masuzawa T, Fujino M (1996) Micro ultrasonic machining and its applications in MEMS. Sens Actuators, A 57:159–164

Montgomery DC (2004) Design and analysis of experiment. Wiley, New York, pp 427–500

Rajesha S, Venkatesh G, Sharma AK, Kumar P (2011) A natural polymer media for abrasive flow machining process and a process for preparation. Indian patent 1349/DEL/2011

Reyntjens S, Puers R (2001) A review of focused ion beam applications in micro-system technology. J Micromech Microeng 11:287–300

Rhodes LJ (1991) Abrasive flow machining: a case study. J Mater Process Technol 28:107–116

Weisberg A, Bhu H (1992) Analysis of micro-channels for integrated cooling. Int J Heat Mass Transf 35(10):2465–2474

Xu W, Uchiyama K, Shimosaka T, Hobo T (2001) Fabrication of polyester microchannels and their applications to capillary electrophoresis. J Chromatogr 907:279–289

Yin L, Ramesh K, Wan S, Liu XD, Huang H, Liu YC (2004) Abrasive flow polishing of micro bores. Mater Manuf Process 19(2):187–207

Zhao B, Viernes NOL, Moore JS, Beebe DJ (2002) Control and applications of immiscible liquids in microchannels. J American Chem Soc 124(19):5284–5285

An Ultrasonic Micromachining Setup for Machining of 3D Geometries

Manjot Singh Cheema, Akshay Dvivedi and Apurbba Kumar Sharma

Abstract Ultrasonic machining technique is best suitable for machining of hard and brittle materials. Miniaturization is the need of hour. Microultrasonic machining is a feasible solution for development of complex shapes at microlevels. This paper presents a simple setup which can be used for 3D micromachining purposes. Further, some key issues have also been discussed in this tool-based micromachining technique.

Keywords Ultrasonic machining · Micromachining · Serpentine channel · 3D geometries

1 Introduction

Tool-based micromachining has provided an excellent solution for eliminating costly setups of lithography and chemical etching. Chemical techniques of micromachining need special preparations for every material. These techniques shall be suitable for mass production, but for developing prototypes, tool-based micromachining comes to be an excellent solution (Boy et al. 2010). Further, the health hazards involved in tool-based micromachining are minimal in comparison to chemical techniques. One such tool-based micromachining process is

M. S. Cheema (✉) · A. Dvivedi · A. K. Sharma
Department of Mechanical and Industrial Engineering, Indian Institute of Technology Roorkee, Roorkee 247667, UT, India
e-mail: manjotcheema@hotmail.com

A. Dvivedi
e-mail: akshaydvivedi@gmail.com

A. K. Sharma
e-mail: akshafme@iitr.ernet.in

S. S. Khangura et al. (eds.), *Proceedings of the International Conference on Research and Innovations in Mechanical Engineering*, Lecture Notes in Mechanical Engineering,
DOI: 10.1007/978-81-322-1859-3_23, © Springer India 2014

microultrasonic machining (micro-USM). Micro-USM proved to be an excellent solution for machining of hard and brittle materials such as silicon, glass, zirconia. (Jain 2012; Sun et al. 1996a, b). The micro-USM is derived from conventional ultrasonic machining (USM), which is an established industrial process for machining hard and brittle materials. The micro-USM, however, has higher frequency (above 20 kHz) and lower amplitude (0.5–5 μm) than conventional USM. The size of abrasive particles in micro-USM is in few microns (0.5–5 μm). Literature reveals that the micro-USM technology was reported in the 90s. With technological advancements, lot of improvements were done in the conventional USM processes (Boy et al. 2010; Kuriyagawa et al. 2001). It was observed that in stationary tools, asymmetrical forces are applied on tool during machining. Due to this reason, a non-uniform wear was observed. This non-uniform wear exerted a bending force on the tool and finally it broke. So to increase the machining rate and reduce the tool wear, researchers provided vibrations to the workpiece and rotation to the tool. To resolve the problem of tool eccentricity during rotation, a high-precision spindle mechanism was also developed. Holes of diameter 9 μm were developed with less out of roundness (Egashira and Masuzawa 1999). Further researchers developed aerostatic spindle head and integrated it with a 3-axis table (Kuriyagawa et al. 2002). The microtools for micro-USM were machined online through wire electric discharge grinding (Egashira et al. 1997). On machine tool, fabrication provided better tolerances and the dimensional accuracies while machining; moreover, chucking of microtools was not so easy due to their very small size. To reduce the tool-handling problem, on machine tool, fabrication was done. The flexibility of the process can be very well observed as holes of range of 5 μm were developed on silicon wafers (Egashira and Masuzawa 1999). Another main advantage of micro-USM is the high aspect ratio. Similarly, layer-to-layer machining was done for development of 3D cavities in silicon (Yu et al. 2004). Tool compensation was provided in the CAD/CAM system. Reverse die sinking was also attempted for transferring pattern on ceramics (Li and Gianchandani 2006). As the tool shape gets replicated on the workpiece surface, the selection of the right tool is very essential. The shape, size and material of the tool are very important in deciding the form accuracy. The tools mainly used for micro-USM were tungsten, stainless steel (SS), tungsten carbide, cemented carbide, etc. (Jain et al. 2011). Single-point cutting tool acts as a very good solution for layer-by-layer machining. The single-point cutting tool can be treated similarly to an end mill in micromilling. Complex and intricate shapes can be developed easily. In this paper, we present a new ultrasonic micromachining setup which was used to fabricate serpentine microchannels. Layer-to-layer machining was used for the development of microchannels. A simple method for fabrication of microchannels using single-point cutting tool is presented. Some technical problems have also been discussed.

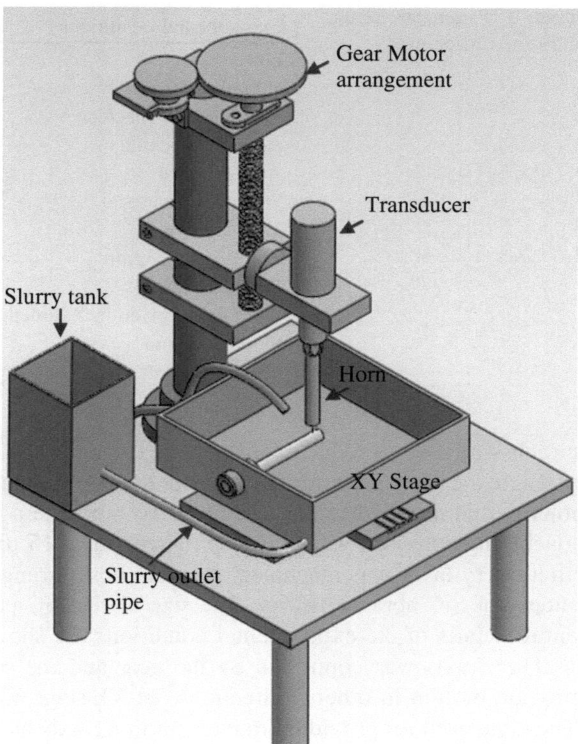

Fig. 1 Ultrasonic micromachining setup

2 Setup

The most important part of USM is the transducer. The transducer consists of a piezoelectric material. When current is passed through this piezoelectric material, it expands and contracts. This expansion and contraction is so quick that it makes the tool vibrate at high frequencies. The energy is amplified from the booster and later on passed to the horn. The output is finally received at the tool end which is brazed to the horn. These vibrations can be used for ultrasonic drilling, welding, soldering, stirring, etc. In this case, machining was to be done so abrasive slurry was flowed through it. The tool strikes the abrasive particles at a very high frequency. Abrasive particles later on striking the workpiece material lead to propagation of microcracks on the glass surface. Finally, material is removed in the form of microchips. Similarly in this setup, the ultrasonic transducer and booster assembly was tightened with an EN-8 horn brazed with a SS tool. Figure 1 shows the entire assembly. The power of transducer was 800 W and a frequency of 25 kHz was generated. The entire assembly was mounted on a bracket which was connected to the pillar. The up-and-down movement to the entire assembly was given through a gear arrangement. A high torque and low r.p.m motor was used to precisely control the movement of motor. A pulse-width modulation (PWM)

Table 1 Parameters of the ultrasonic machining

Experimental conditions	
Power	800 W
Vibration frequency	25 kHz
Amplitude of machine	10–15 μm
Abrasive material	Silicon carbide
Abrasive size	12 μm
Workpiece material	Borosilicate glass
Tool material	Stainless steel
Feed in *X* direction	100 mm/min
Depth of cut	15 μm
Total depth given in *Z* direction	300 μm
Slurry medium	Water
Slurry concentration	15 % abrasive by volume

circuit was used to reduce the speed of the motor. A feed of 15 μm was obtained through the gear system. The feed and depth of cut indication was given by a micron dial attached to the assembly. To fabricate a microchannel, layer-by-layer machining approach was used. An increment of 15 μm was given to the assembly after every layer was machined. A pump hose arrangement was used for constant supply of SiC abrasive slurry. The size of the SiC particle used was 12 μm. The entire details of the experimental conditions are shown in Table 1.

The *Z* axis was controlled by the gear and the servo motor arrangement. To provide motion in other 2 directions, an *XY* stage was integrated with the setup. The table used for providing movement in *XY* axis had a resolution of 0.1 μm. So a very precise control of the workpiece motion was obtained through the integrated *XY* stage. The programming was done through dedicated software of the machine. Coordinates were entered in the program. The *Z* axis was carefully brought near to the glass workpiece. An initial gap of 15 μm was obtained through point of contact method. A conductive gauge was kept between tool and glass workpiece. When the tool came in contact with the gauge, the continuity was checked with continuity tester. After this, the width of gauge was compensated and a constant gap was maintained between tool and workpiece.

The transducer was connected to the generator supply through which the settings and frequency tuning of the horn was done. To adjust the power rating, a variator was used. The experiments were done at 80 % power rating. After this, the machine was switched on and the program was run simultaneously. The final profile obtained on glass workpiece is shown in Fig. 2a.

3 Results and Discussion

Serpentine channel is one of the most common geometries used in electronic cooling devices. The feed rate for development of this channel was 100 mm/min. The diameter of the semicircular portion was 2.5 mm, and the distance between

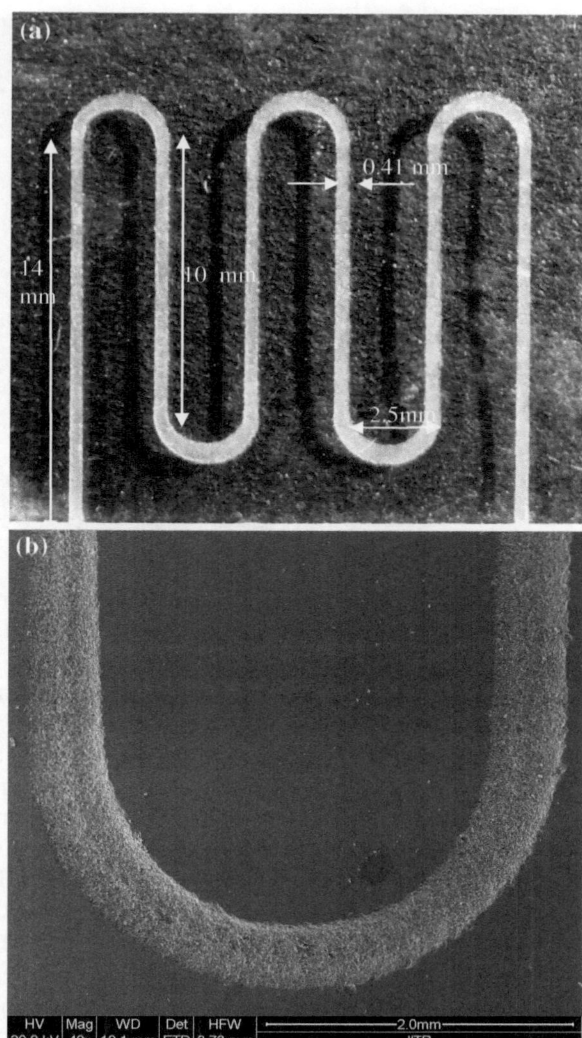

Fig. 2 **a** Serpentine channel of width 410 μm and depth 250 μm. **b** SEM view of the curved periphery

end points of two opposite arcs was 10 mm. At last, the width of microchannel obtained was 410 μm and depth was of 250 μm. Figure 2b shows the SEM image of the curved periphery. The tool wear in the development of microchannel was in a non-uniform fashion. The SEM image of the tool has been shown in Fig. 3.

The gap between the microchannel wall and the tool is known as the lateral gap. More wear was involved in those sides where lateral gap was involved. A plowing effect was observed on the tool face and a built-up edge was formed on the face of the tool. The abrasives after striking the tool need an exit. The intake of fresh slurry and vibratory motion of the tool forces the abrasive particles to enter the

Fig. 3 SS tool and the material plowed along

lateral zone. These particles entering the lateral zone plow some tool material along with it and a built-up edge was formed. The built-up edge was also responsible for the increased diameter of the tool.

Proper selection of the parameters is very essential in micromachining. Abrasive slurry concentration, feed rate, power rating, abrasive size, etc. are the major factors on which quality of a developed microchannel depends. The effect of feed rate can be seen in the SEM image (Fig. 4). The top semicircular microchannel was developed at a feed of 400 mm/min. The feed marks can be clearly seen on the surface of this channel. The feed marks are visible due to non-uniform machining of the microchannel. Below is a semicircular microchannel with a feed of 200 mm/min which had comparatively lesser feed marks than upper semicircular microchannel. Similarly, other problem that exists in the microchannels is the surface damage. Surface damage on the top layer is decided by the abrasive slurry concentration. More the concentration, more will be the number of particles coming from the lateral gap. These abrasive particles shear some material along with it and damage the top surface in form of lateral cracks. The damage is shown in the Fig. 5. The top surface damage in the microchannels is to be minimized. This is possible only when the appropriate slurry concentration is used. The surface finish of the microchannel depended mainly upon the power rating. At high power ratings, high material removal and low surface finish were observed. At lower power ratings, low material removal and high surface finish were observed. The surface finish is an important parameter with respect to fluid flow in the microchannels. A surface finish of .42 R_a was obtained at reduced power ratings.

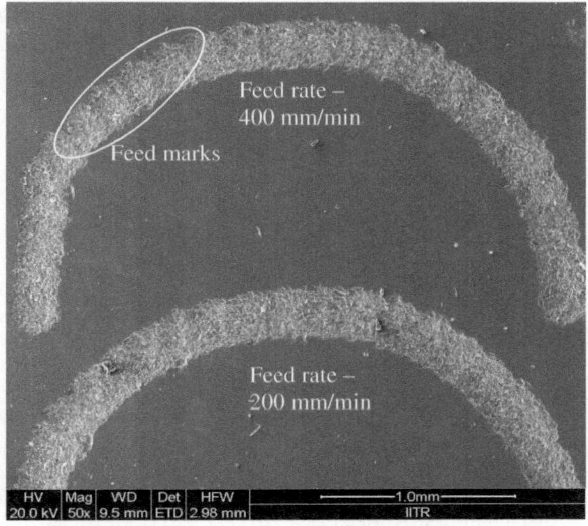

Fig. 4 The effect of feed rate on circular microchannels

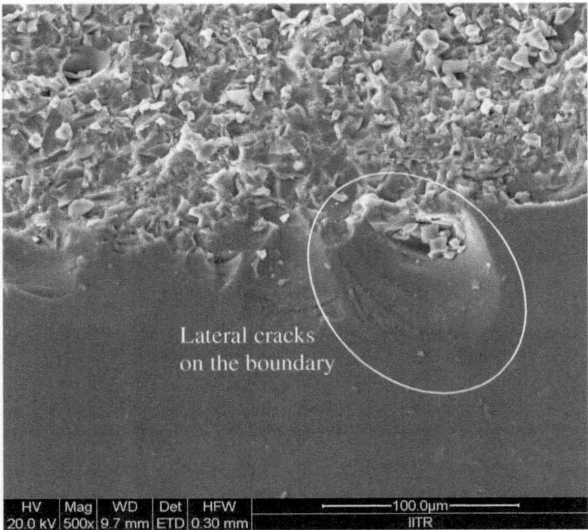

Fig. 5 Outer edge of the channel showing lateral cracks

4 Conclusions

The setup shows the potential of developing complex and intricate profiles. This was proved by developing a serpentine microchannel of width 410 μm and depth of 250 μm. Micro-USM machining can prove to be an effective solution for making prototypes and small-scale productions. Extensive care has to be given in handling the microtools. Appropriate selection of process parameters is essential to get good surface finish, form accuracy, less surface damage and less tool wear. A layer-by-layer machining approach can be used for fabrication of a number of complex profiles.

References

Boy JJ, Andrey E, Boulouize A, Khan-Malek C (2010) Developments in microultrasonic machining (MUSM) at FEMTO-ST. Int J Adv Manuf Technol s47:37–45

Egashira K, Masuzawa T (1999) Microultrasonic machining by the application of workpiece vibration. Annals CIRP 48(1):131–134

Egashira K, Masuzawa T, Fujino M, Sun XQ (1997) Application of USM to micromachining by on-the-machine tool fabrication. Int J Electr Mach 2: 31–36

Jain V (2012) Investigations on development of microchannels using micro-ultrasonic machining. Ph.D dissertation, Mechanical and Industrial Engineering Department, IIT Roorkee

Jain V, Sharma AK, Kumar P (2011) Recent developments and research issues in microultrasonic machining. ISRN Mech Eng 2011(413231). doi:10.5402/2011/413231

Kuriyagawa T, Shirosawa T, Saitohand O, Syoji K (2001) Micro ultrasonic abrasive machining for three-dimensional milli-structures of hard–brittle materials. In: Proceedings of 16th annual meeting of the ASPE, Crystal-City, Virginia, USA, 2001, pp 525–528

Kuriyagawa T, Shirosawa T, Saitoh O, Syoji K (2002) Development of micro ultrasonic abrasive machining system (1st report, studies in micro ultrasonic abrasive machining). JSME Int J Ser C 45(2):593–600

Li T, Gianchandani YB (2006) A micromachining process for die-scale pattern transfer in ceramics and its application to bulk piezoelectric actuators. J Microelectromech Syst 15(3):605–612

Sun XQ, Masuzawa T, Fujino M (1996a) Micro ultrasonic machining and self-aligned multilayer machining/assembly technologies for 3d micromachines. In: Proceedings of 9th annual international workshop on micro electro mechanical systems (MEMS'96), pp 312–317

Sun XQ, Masuzawa T, Fujino M (1996b) Micro ultrasonic machining and its applications in MEMS. Sens Actuators, A 57(2):159–164

Yu Z, Rajurkar KP, Tandon A (2004) Study of 3D micro ultrasonic machining. ASME J Manuf Sci Eng 126:727–732

An Experimental Investigation During Wire Electrical Discharge Machining of Al/SiC-MMC

Manish Kumar, Alakesh Manna, S. K. Mangal and Anup Malik

Abstract This paper presents the effects of various parameters of wire electrical discharge machining (WEDM) on material removal rate (MRR) and surface finish during machining of Al/SiC-MMC. The Taguchi method, a powerful tool in the design of experiments, is utilized to optimize the parametric combination of WEDM for effective machining of such composite. The mathematical models for different response characteristics are developed to investigate the influence of cutting parameters during machining. Confirmation test results proved that the developed mathematical models are appropriate for machining of such composite by WEDM.

Keywords Al/SiC-MMC · WEDM · Material removal rate · Surface finish

1 Introduction

The advanced materials have attractive properties, i.e., high strength, high bending stiffness, good damping capacity, low thermal expansion, better fatigue characteristics which make them potential material for modern day industrial application. Present manufacturing industries are facing challenges from these advanced materials viz. super alloys, ceramics and composites that are hard and difficult to machine, requiring high precision, surface quality which increases machining cost. To meet these challenges, new processes with advanced methodology and tooling need to be developed. The conventional machining of such advanced materials is often difficult due to the improved thermal, chemical and mechanical properties of new advanced materials. Conventional machining such as turning, milling and

M. Kumar · A. Manna (✉) · S. K. Mangal · A. Malik
Department of Mechanical Engineering, PEC University of Technology, Scctor-12, Chandigarh 160012, India
e-mail: kgpmanna@rediffmail.com

S. S. Khangura et al. (eds.), *Proceedings of the International Conference on Research and Innovations in Mechanical Engineering*, Lecture Notes in Mechanical Engineering, DOI: 10.1007/978-81-322-1859-3_24, © Springer India 2014

drilling shows ineffectiveness in machining of advanced materials, since it results in poor materials removal rate, excessive tool wear and increased surface roughness. The parts are usually made of composite materials by casting process which is not enough to make a part with desire complicate shape, size, contour and surface finish. Non-conventional machining processes such as ultrasonic machining (USM), water jet machining (WJM), abrasive jet machining (AJM), electrical discharge machining (EDM), electron beam machining (EBM), laser beam machining (LBM), photo chemical machining (PCM) can be used to machine such materials. Aerospace and nuclear industries have developed a large range of super alloys and heat resistant materials such as ceramics and composite materials. Such materials can only be effectively machined if manufacturing engineers understand the reasons for their poor machinability and take effective measures to counter them. Keeping in view an attempt has been made to machine Al/SiC-MMC materials by wire electrical discharge machine (WEDM) and reported the results and thereafter analysis in the paper.

From the review of literature, it is evident that many research works on WEDM have been carried out for processing of different materials by the different researchers, but still a lot of applied research in the above fields is required so as to explore the successful utilizations of the process in the area of machining of conductive metal matrix composite materials. Some of the published research work reviewed in brief and is listed as follows. Boujelbene and Bayraktar (2009) studied the influence of machining parameters on the surface integrity in EDM. Authors concluded that increasing energy discharge increases instability and thereby quality of the workpiece surface becomes rougher and the white layer thickness increases. Daneshmand et al. (2012) studied on the WEDM during machining of nickel–titanium shape memory alloys and concluded that the surface roughness increases by increasing peak current, pulse on time and wire speed. Gatto and Luliano (1997) studied on the cutting mechanism and surface features of WEDM of metal matrix composites with 15 % whiskers and 20 % particles reinforcement and SiC/Al alloy. Authors concluded that the machined surface finish produced was rough without proper setting parameters. Huang et al. (1999) studied on the finish cutting utilized WEDM and concluded that a better surface quality and accurate dimension value can be obtained in less machining time by optimizing the machining parameters. Manna and Bhattacharyya (1998) carried out a comparative study on various inserts during conventional machining of Al/10 Vol% and 20 Vol% SiC-MMC. Authors performed the different sets of experiments to study the effect of the main machining parameter, e.g., cutting speed on the machining performance criteria such as surface finish and tool wear. The effect of tool materials and geometry of different inserts on the surface finish and tool wear is also investigated during machining experimentations. The effect of use of cutting fluid during machining on the surface finish and tool wear is also investigated. Authors concluded that the coolants as well as the SiC particles presence in Al/SiC-MMC are resisting the cutting action during machining of Al/SiC-MMC. Manna and Bhattacharyya (2006) studied on a dual response approach for parametric optimization of CNC wire cut EDM of particulate reinforced

Table 1 Cutting parameters and their levels

S. no.	Symbols	Machining parameters	Levels			Units
			1	2	3	
1	A	T_{ON} (X_1)	0.5	0.9	1.3	μs
2	B	T_{OFF} (X_2)	18	22	26	μs
3	C	Peak current (X_3)	100	120	140	A
4	D	Wire feed (X_4)	8	9	10	m/min
5	E	Wire tension (X_5)	900	1,140	1,380	g

aluminum silicon carbide metal matrix composite (PRAl/SiC-MMC). Authors used Taguchi method for experimental design and optimize the machining parameters for better responses. Different authors are also worked on WEDM for processing of different materials (Ho and Newman 2003; Hargrove and Ding 2007; Liao and Woo 1997).

2 Planning for Experimentation

The workpiece samples have been fabricated through liquid stir casting process. Samples of different sizes Al/10 wt% SiC-MMC are prepared for experimental investigation on WEDM. The SiC particles with 45 μm average particle size are used as abrasive reinforced particles. Different sets of experiments are performed using Electronica Supercut-734 CNC wire cut EDM machine. According to the Taguchi robust design, a L_{18} ($2^1 \times 3^7$) mixed orthogonal array is employed for the experimentation. A total of five WEDM parameters with three levels for each parameter such as control factor, parameter *A* (pulse on time), parameter *B* (pulse off time), parameter *C* (peak current), parameter *D* (wire feed rate) and parameter *E* (wire tension) are considered as the controlling factors for optimally analysis during machining of Al/SiC-MMC. Table 1 represents planning for the experimental design is considered for the investigation in the WEDM process during machining of Al/SiC-MMC. Surface roughness (R_a, μm) is measured using Surfcom 130 A surface roughness measuring instrument. The material removal rate (MRR) is calculated from the following relationship:

$$Y_{\mathrm{MRR}} = Vc.b.h\,(\mathrm{mm}^3/\mathrm{min})$$

where Vc = Cutting speed (mm/min), which is calculated based on the mean of cutting speed data recorded directly from the machine during cutting and data recorded for actual length of machining with time required for continuous machining. b = Width of cut (mm), it is determined from the relation of 2 × spark gap width plus wire diameter. h = depth of the workpiece (mm).

3 Results and Discussions

Different sets of experiments are performed to identify the effect of various parameter of WEDM on response characteristics. The experiments are performed in constant voltage mode of the WEDM. In the first set of experiments, pulse on time (T_{ON}) is varied from 0.5 to 2.1 μs with increment of 0.4 μs. All other input parameters such as wire feed, wire tension, peak current and pulse off time are kept constant. Table 2 shows L_{18} ($2^1 \times 3^7$) mixed orthogonal array, experimental results and S/N ratio (dB) (See Appendix for Table 2).

Figure 1 shows the effects of T_{ON} on MRR during machining of Al/SiC-MMC on CNC wire cut EDM. From Fig. 1, it is clear that the MRR (mm^3/min) increases with increase in T_{ON}(μs). Maximum MRR is observed at 2.1 μs T_{ON}. It may be due to the increase in duration of sparking time and enhances formation of crater and vaporization of metal and thereby increases MRR. Figure 2 shows a cure represents the effects of wire tension on MRR during machining of Al/SiC-MMC on CNC wire cut EDM. From Fig. 2, it is clear that the MRR (mm^3/min) increases with increase in wire tension (g). Maximum MRR is observed at 1,620 g wire tension. The results are obtained during machining of Al/SiC-MMC with 250-μm-diameter brass wire. It may be due to more straightening the traveling wire by increasing wire tension which increases striking rate of sparking per unit area and thereby increases MRR.

Figures 3 and 4 show the actual machined surface roughness profile generated during WEDM of Al/SiC-MMC. During this experiment, the cutting speed observed is 2.95 m/min. At this particular parametric setting, i.e., *A*1 *B*1 *C*1 *D*1 *E*1, the average machined surface roughness height, R_a, (μm) is measured and found 2.9859 μm (Fig. 3). Similarly, Fig. 4 shows another actual machined surface profile generated during WEDM of Al/SiC-MMC at parametric setting, i.e., *A*2 *B*1 *C*2 *D*3 *E*1 and found that the average machined surface roughness height, R_a, (μm) is 2.9226 μm (See Appendix for Figs. 3, 4).

Figure 5 shows the average signal-to-noise ratio (S/N, dB) and grand mean of signal-to-noise (S/N) ratio (dB) graphs for material removal (MR, mm^3/min). From Fig. 5, it is clear that the optimal parametric combination for maximum MRR (mm^3/min) is *A*2 *B*1 *C*2 *D*1 *E*2. Figure 6 shows the average signal-to-noise ratio (S/N, dB) and grand mean of signal-to-noise(S/N) ratio (dB) graphs for surface roughness (R_a, μm). From Fig. 6, it is clear that the optimal parametric combination for minimum surface roughness height, R_a, (μm) is *A*1 *B*1 *C*2 *D*1 *E*1.

Table 3 represents ANOVA and "F" test values with percentage of contribution, i.e., effectiveness of the individual machining parameter on MRR (mm^3/min). From ANOVA (Table 3), it is clear that the T_{ON} (X_1, μs) and T_{OFF} (X_2, μs) are the most significant and significant parameters on MRR with 46.36 and 19.54 % contributions, respectively. Similarly, ANOVA and "F" test analysis was done for surface roughness height, R_a (not shown in the paper), and it is found that the T_{ON} (X_1, μs) and T_{OFF} (X_2,μs) are the most significant and significant parameters on surface roughness height, R_a with 59.91 and 20.09 % contributions, respectively.

Table 2 L_{18} ($2^1 \times 3^7$) mixed orthogonal array, experimental results and S/N ratio (dB)

Exp. No.	WEDM parameters					MRR during WEDM of Al/SiC-MMC					SR during WEDM of Al/SiC-MMC	
	A T_{ON}	*B* T_{OFF}	*C* IP	*D* WF	*E* WT	MRR1	MRR2	MRR3	Average MRR	S/N ratio (dB)	Average R_a (μm)	S/N ratio (dB)
1	1	1	1	1	1	12.483	12.854	12.676	12.671	22.056	2.9712	−9.5015
2	1	2	2	2	2	12.379	12.452	12.254	12.362	21.842	3.1250	−9.9164
3	1	3	3	3	3	10.098	10.186	10.112	10.132	20.114	3.8147	−11.6305
4	2	1	1	2	2	17.364	17.541	17.412	17.439	24.827	3.4040	−11.5906
5	2	2	2	3	3	18.113	18.282	18.198	18.917	25.537	3.3184	−10.4389
6	2	3	3	1	1	15.115	15.277	15.186	15.193	23.633	3.2171	−10.1511
7	3	1	2	1	3	16.783	16.912	16.850	16.848	24.531	3.4079	−10.6571
8	3	2	3	2	1	17.917	17.998	17.956	17.957	25.085	3.8111	−11.6305
9	3	3	1	3	2	17.281	17.376	17.318	17.325	24.773	3.5696	−11.0660
10	1	1	3	3	2	7.438	7.566	7.488	7.497	17.498	2.9856	−9.5079
11	1	2	1	1	3	7.043	7.214	7.156	7.137	17.070	3.0768	−9.7585
12	1	3	2	2	1	6.886	6.950	6.914	6.916	16.797	2.9541	−9.3796
13	2	1	2	3	1	8.312	8.416	8.378	8.369	18.453	2.9187	−9.3153
14	2	2	3	1	2	7.784	7.998	7.948	7.910	17.964	3.1668	−10.0088
15	2	3	1	2	3	6.913	6.980	6.948	6.947	16.836	3.5493	−10.9986
16	3	1	3	2	3	7.973	7.996	7.984	7.984	18.044	3.4240	−10.7038
17	3	2	1	3	1	5.884	5.664	5.542	5.697	15.113	3.5466	−10.9974
18	3	3	2	1	2	8.146	8.220	8.168	8.178	18.253	3.7235	−11.4325

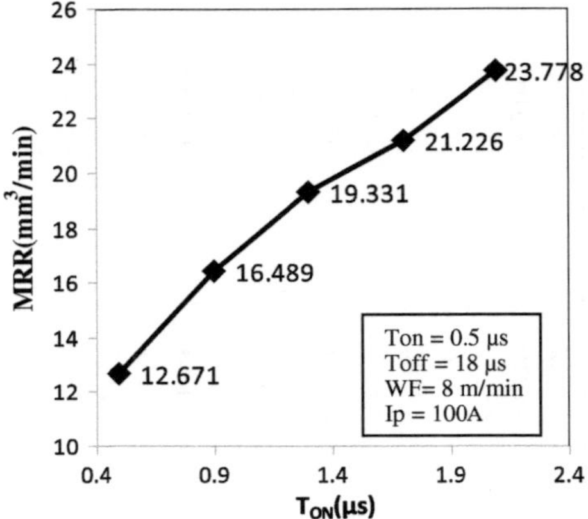

Fig. 1 Effect of T_{ON} on MRR

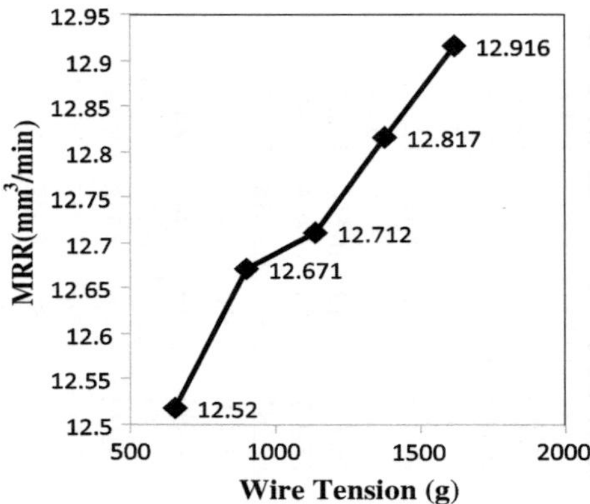

Fig. 2 Effect of wire tension on MRR

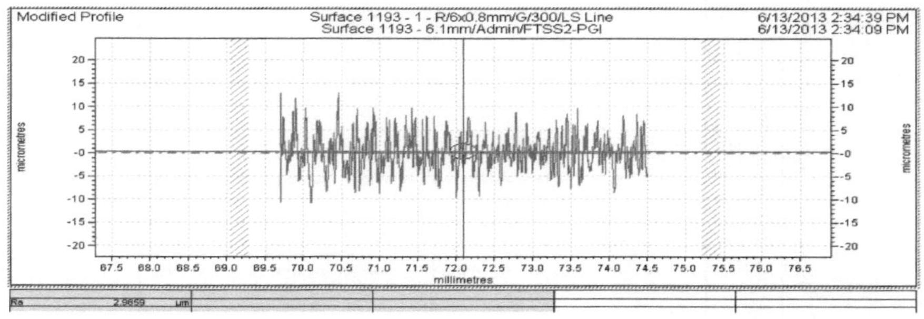

Fig. 3 Machined surface roughness height, R_a, (μm) at parametric setting $A1$ $B1$ $C1$ $D1$ $E1$

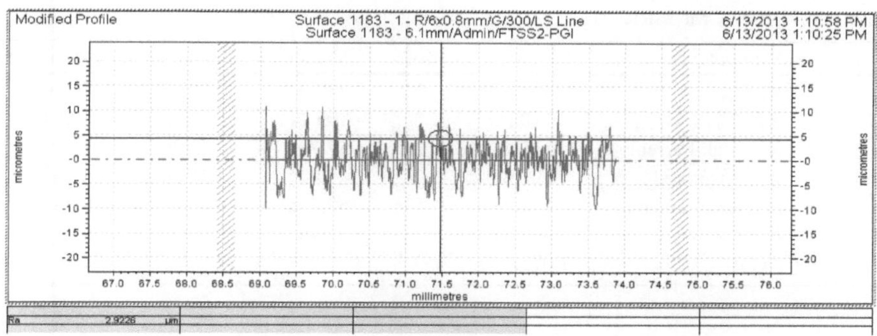

Fig. 4 Machined surface roughness height, R_a, (µm) at parametric setting *A*2 *B*1 *C*2 *D*3 *E*1

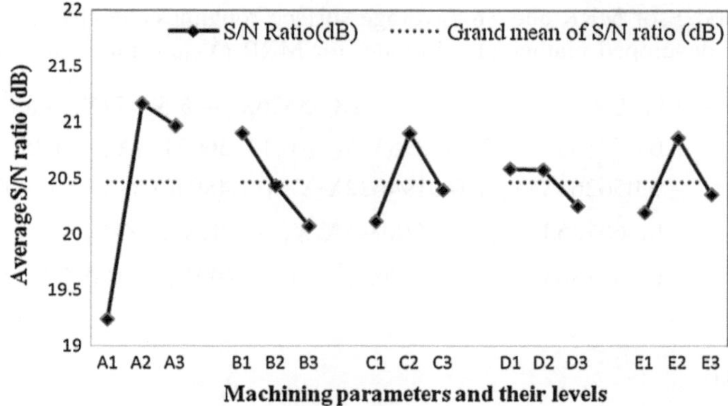

Fig. 5 S/N ratio for MRR (mm^3/min)

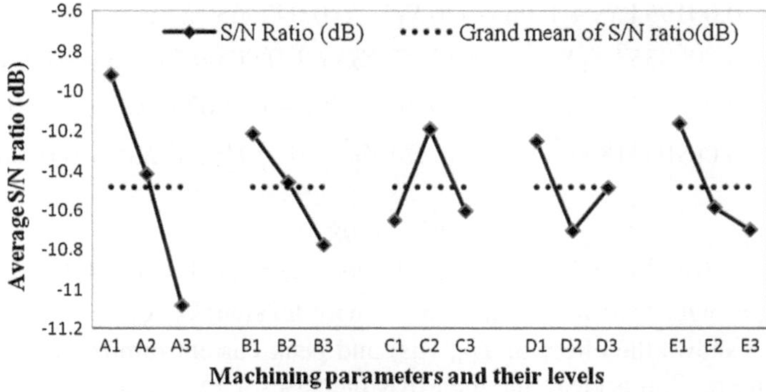

Fig. 6 S/N ratio for surface roughness

Table 3 ANOVA for MRR

S. no	Parameters	SS	dof	Variance	F_0	% of Cont
1	X_1: T_{ON}	498.68	2	249.34	195.56	46.36
2	X_2: T_{OFF}	210.22	2	105.11	82.43	19.54
3	X_3: Peak current	115.48	2	57.74	45.29	10.74
4	X_4: Wire feed	91.64	2	45.82	35.94	8.56
5	X_5: Wire tension	104.83	2	52.415	41.11	9.75
6	Error	54.81	43	1.275	–	5.05
7	Total	1,075.66	53	–	–	100

4 Development of Mathematical Models

Considering all parameters selected for experimental investigation, the mathematical model for MRR and SR (average surface roughness height, R_a) is developed. The developed mathematical model for MRR (Y_{MRR}, mm^3/min) is

$$\begin{aligned} Y_{MRR} = & -0.1126498 + 0.012679X_1 - 28.35326X_2 + 5.57977X_3 - 8.83526X_4 \\ & + 0.057535X_5 + 2.048298X_1X_2 + 0.14156051X_1X_3 + 0.35740089X_1X_4 \\ & - 0.056264X_1X_5 + 0.0194022X_2X_3 + 1.489401X_2X_4 - 0.003361X_2X_5 \\ & - 0.1667853X_3X_4 - 0.000861X_3X_5 + 0.018423X_4X_5 + 0.124986{X_1}^2 \\ & + 0.306186{X_2}^2 - 0.015126{X_3}^2 - 1.325409{X_4}^2 + 0.000003149{X_5}^2 \end{aligned} \tag{1}$$

$R^2 = 0.94$

The developed mathematical Model for surface roughness height (Y_{Ra}, μm) is

$$\begin{aligned} Y_{Ra} = & -0.196255 + 0.19675X_1 - 0.15620009X_2 - 0.2145712X_3 + 3.926717X_4 \\ & - 0.001954X_5 + 0.0651526X_1X_2 - 0.0424639X_1X_3 + 0.4323894X_1X_4 \\ & + 0.0008353X_1X_5 + 0.0032357X_2X_3 + 0.0034657X_2X_4 + 0.000106X_2X_5 \\ & + 0.018248X_3X_4 - 0.000003679X_3X_5 - 0.00023103X_4X_5 + 0.023679{X_1}^2 \\ & - 0.00905318{X_2}^2 + 0.0001274{X_3}^2 - 0.35419345{X_4}^2 + 0.000000861{X_5}^2 \end{aligned} \tag{2}$$

$$R^2 = 0.98$$

where $X_1 = T_{ON}(X_1, \mu s)$; $X_2 = T_{OFF}(X_2, \mu s)$, $X_3 = \text{peak current}(X_3, A)$,
$X_4 = \text{wire feed}(X_4, m/\text{min}), X_5 = \text{wire tension}(X_5, g)$.

Figure 7 shows the effect of T_{ON} (μs) and peak current (amp) on MRR (mm^3/min). From Fig. 7, it is clear that the MRR increases with increase in T_{ON}. It is also clear that the MRR increases with increase in peak current up to a certain value and then it starts decreases. It is because of after certain parametric value of peak current supply, the current density increased and rate of vaporization also

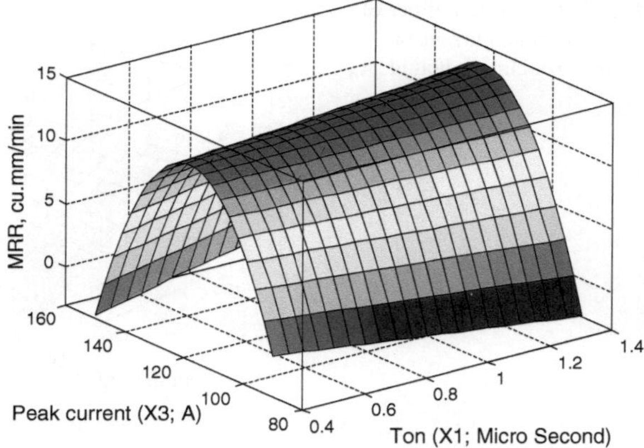

Fig. 7 Interaction effect of T_{ON} and peak current on MRR (mm^3/min)

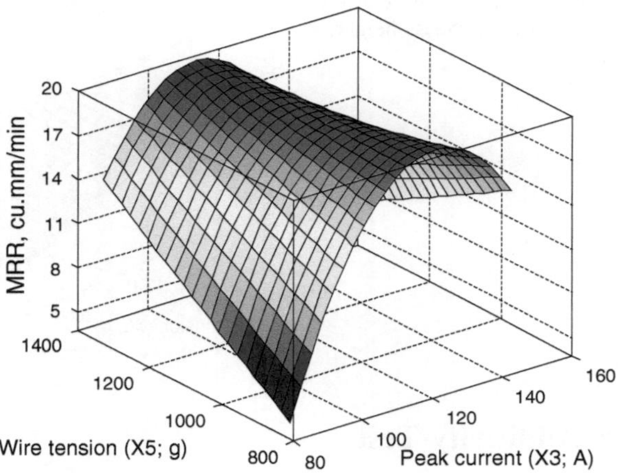

Fig. 8 Interaction effect of peak current and wire tension on MRR (mm^3/min)

increased; some of the particles already cratered and vaporized are not removed, but rather adhere with the workpiece parent metal and thereby reduce MRR after certain limiting value of current supply. Figure 8 shows the effect of peak current (amp) and wire tension on MRR (mm^3/min). From Fig. 8, it is clear that the MRR increases with increase in wire tension. It is also observed that MRR increases with increase in peak current up to a particular point and then it starts decreases.

Figure 9 shows the effect of T_{ON} (μs) and peak current (amp) on surface roughness (R_a, μs). From Fig. 9, it is observed that the surface roughness height (R_a, μs) increases with increase in T_{ON}. It is also observed that surface roughness (R_a, μs) increases with increase in peak current. Figure 10 shows the effect of peak current (amp) and wire tension on surface roughness (R_a, μs). From Fig. 10, it is clear that the surface roughness height (R_a, μs) increases with increase in wire tension. It is also clear that the surface roughness height (R_a, μs) increases with increase in peak current.

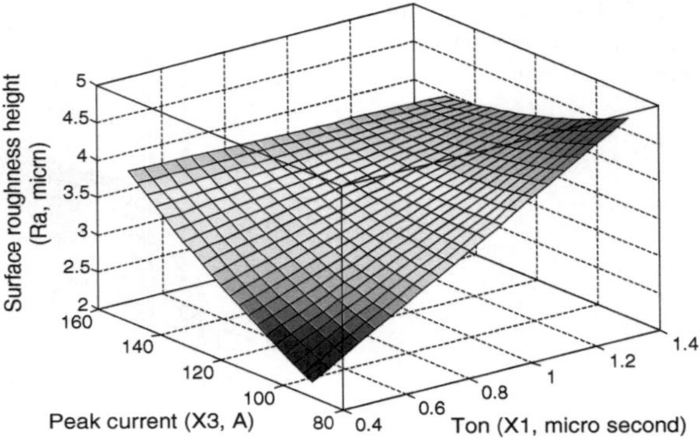

Fig. 9 Interaction effect of T_{ON} and peak current on surface roughness height, R_a

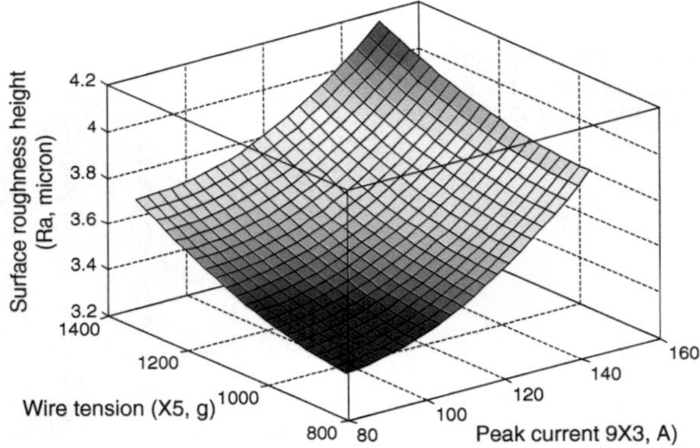

Fig. 10 Interaction effect of peak current and wire tension on surface roughness height, R_a

5 Additivity Test

The parametric conditions were used for additivity test to validate the developed mathematical models for MRR and surface roughness height, $R_{a,}$ during WEDM of Al/SiC-MMC. The % of errors between the calculated values from the developed mathematical models with respect to the experimental values are not more than 6.72 %. Hence, it is concluded that the developed mathematical models Eqs. (1) and (2) for MRR and surface roughness height, $R_{a,}$ respectively, for effective machining of Al/10 wt% SiC-MMC by WEDM bear a good agreements with the experimental test lines.

6 Conclusions

Based on the experimental results, during machining of Al/10 wt% SiC-MMC by WEDM, the following conclusions are drawn as listed below.

1. Pulse on Time (T_{ON}) and Pulse off Time (T_{OFF}) are the most significant and significant machining parameter for controlling the metal removal rate (MRR).
2. Pulse on Time (T_{ON}) and Pulse off Time (T_{OFF}) are the most significant and significant influencing machining parameters, respectively, for the surface roughness height (R_a, μm).
3. For high cutting speed and for maximum MRR, the recommended parametric combination is *A*2 *B*1 *C*2 *D*1 *E*2.
4. For smooth cutting and better surface finish, the recommended parametric combination is *A*1 *B*1 *C*2 *D*1 *E*1.
5. The developed mathematical models for MRR and surface roughness height, $R_{a,}$ are successfully proposed for proper selection of machining parameters and for evolution of surface roughness values and MRR under various combinations during machining of Al/10 wt% SiC-MMC by WEDM.

References

Boujelbene M, Bayraktar E (2009) Influence of machining parameters on the surface integrity in electrical discharge machining. Arch Mater Sci Eng 37:110–116

Daneshmand S, Hessami R, Esfandiar H (2012) Investigation of wire electro discharge machining of Nickel-Titanium shape memory alloys on surface roughness and MRR. Life Sci J 426:2904–2909

Gatto A, Luliano L (1997) Cutting mechanisms and surface features of WEDM metal matrix composite. J Mater Process Technol 65:209–214

Hargrove SK, Ding D (2007) Determining cutting parameters in wire EDM based on workpiece surface temperature distribution. Int J Adv Manuf Technol 34:295–299

Ho KH, Newman ST (2003) State of the art electrical discharge machining EDM. Int J Mach Tools Manuf 43:1287–1300

Huang JT, Liao YS, Hsue WJ (1999) Determination of finish-cutting operation number and machining-parameters setting in wire electrical discharge machining. J Mater Process Technol 87:69–81

Liao YS, Woo JC (1997) The effects of machining setting on the behavior of pulse train in the WEDM process. J Mater Process Technol 71:433–439

Manna A, Bhattacharyya B (1998) An investigation on Al/SiC metal matrix composite during machining. The Institution of Engineers (Bangladesh). "Proceedings", Fifth Annual Paper Meet. 5–7 Nov 1998, Paper no-46, pp 368–375

Manna A, Bhattacharyya B (2006) Taguchi and Gauss elimination method: a dual response approach for parametric optimization of CNC wire cut EDM of PRAlSiCMMC. Int J Adv Manuf Technol 28(1–2):67–75. ISSN:0268-3768

Optimization of Magnetic Abrasive Finishing Parameters with Response Surface Methodology

Palwinder Singh and Lakhvir Singh

Abstract Fine surface finish is in high demand in modern time, with the development of industry manufacturing technology, in a wide range of industrial applications. Now a days, in medical instruments and aerospace components, fine surface finish is highly desirable. The inner surfaces of workpieces used in critical applications are finished using magnetic abrasive finishing (MAF) process. The process principle and the finishing characteristics of MAF of cylindrical pipes using sintered magnetic abrasives are described in this research work. The surface roughness measurements and material removal rate (MRR) measurements resulting from finishing experiments are described in this research work. Response surface methodology (RSM) technique is used for optimization of process parameters for the analysis of surface roughness in terms of percent improvement in surface finish (PISF) and MRR. The obtained maximum PISF was 92 %, and minimum surface roughness was 0.04 μm. The surface was microscopically examined using scanning electron microscopy (SEM) to further study the improvement in surface finish.

Keywords Magnetic abrasive finishing (MAF) · Sintered · Scanning electron microscopy (SEM)

P. Singh (✉) · L. Singh
Mechanical Engineering Department, B.B.S.B. Engineering College, Fatehgarh Sahib 140407, Punjab, India
e-mail: palwindergill13@gmail.com

L. Singh
e-mail: lakhvir1972@gmail.com

S. S. Khangura et al. (eds.), *Proceedings of the International Conference on Research and Innovations in Mechanical Engineering*, Lecture Notes in Mechanical Engineering, DOI: 10.1007/978-81-322-1859-3_25,

1 Introduction

Magnetic abrasive finishing (MAF) is one of the super polishing processes involving extremely small amount of material removal even to the extent of an atomic cluster. This process uses magnetic force for material removal. In this process, the cutting forces of extremely small magnitude are applied on the work piece surface. The forces are uniformly distributed on the work surface and are easily controllable. Hence, they minimize the damage to it. The process is capable of achieving surface roughness of the order of nanometric level. The finish, accuracy, and surface integrity of parts produced by conventional machining (i.e., tool harder than the work piece) depend on the machine tool system which comprises of machine tool, cutting tool, work piece, cutting conditions, and cutting fluid.

In conventional super-finishing processes, tools are mechanically pressed against the work piece surface, which results in surface defects like microcracks, geometrical and dimensional errors, and distortions. Thus, MAF has been taken as an alternative super-finishing process for efficient and precision finishing of surfaces. In MAF, material removal is carried out using magnetic force applied on abrasive particles, and the cutting process becomes virtually resilient. This in turn results in minimizing the damage to the machined surface. The MAF is found to be a versatile process for finishing internal, external, flat, and complex shaped surfaces.

Harry Coats described one of the non-conventional machining processes, that is, MAF in 1938 in a patent. The countries which are involved in the study and development of this process are USA, CIS, France, England, Bulgaria, Japan, and Germany. A relatively new finishing advanced machining process in which cutting force is primarily controlled by the magnetic field. MAF is a fine finishing technique which can be employed to produce optical, mechanical, and electronic components with micrometer or submicrometer form accuracy and surface roughness within nanometer range with hardly any surface defects. Finishing of bearings, precision automotive components, shafts, and artificial hip joints made of oxide ceramic and cobalt alloy are some of the products for which this process can be applied. The stainless steel workpiece material (non-ferromagnetic) was finished by Jain et al. (2001), and the results showed that working gap and circumferential speed are the influential parameters affecting the material removal and surface roughness value. This process can be used to produce efficiently good surface quality on at surfaces as well as internal and external surfaces of tube type work pieces (Shinmura et al. 1985; Yamaguchi and Shinmura 1999). The method is used for both ferromagnetic materials as well as non-ferromagnetic materials such as stainless steel, aluminum, and brass.

The application of MAF for finishing of the inner surfaces of alumina ceramic components using diamond-based magnetic abrasives is studied by Yamaguchi and Shinmura (2004). The effects of different machining parameters like magnetic flux density (MFD), vibration frequency and amplitude, machining time, and pole-work gap on finishing characteristics using sintered magnetic abrasives are studied by Shinmura et al. (1984). They described that the two parameters, vibration and MFD, remarkably affects the finishing efficiency. Shinmura et al. (1994) concluded that with the increase in rotational speed of magnetic pole, the metal removal rate increases. They almost keep a linear relationship under given experimental conditions. The metal removal of brass work piece was highest with the artificial abrasives (alloyed Titanium Carbide and Iron). Shinmura and Aizawa (1989) mixed the diamond abrasives with iron to form diamond magnetic abrasives. The two types of magnetic abrasives were prepared by sintering by Shinmura et al. (1987). The diameter of iron particle was varied in first sample, and in second, diameter of abrasive particle varied. They reported that diameter of iron particle effects both stock removal and surface finish. The finishing of ceramics was examined. The results showed that finishing efficiency increased with diamond magnetic abrasives with increase in speed of tool. The magnetic abrasives were prepared by blending of Al_2O_3 (15 %) and iron powders (85 %), compacting them by a bench press, sintering the mixture in a furnace at 1,400 °C in an inert environment, crushing the compacts into small particles and then sieving to different ranges of sizes (Khairy 2001). The silver steel bars were finished with these sintered magnetic abrasives for various combinations of finishing parameters. The process mechanism was studied by using sintered magnetic abrasive particles by Mori (2003). A new type of magnetic abrasives composed of WC/Co sintered particles were developed for cleaning the tubes by Kim (2003) and found the optimal finishing parameters. The finishing of non-ferromagnetic material, SUS304, was examined by Lin et al. (2007). The sintered magnetic abrasives were prepared by typically mixing iron powder and Al_2O_3 powder with composition of 60:40 of wt% and compressing mixture into the cylindrical shape. These compacts were sintered into a vacuum furnace. After sintering process, these cylinders were crushed to produce magnetic abrasives. Singh et al. (2010) concluded that the sintered magnetic abrasives give highest surface finish on most of the work materials among all the available types of magnetic abrasives, and he highlighted major existing technologies that are used to manufacture magnetic abrasives. The internal finishing of three kinds of work materials, that is, Ly12 aluminum alloy, 316L stainless steel, and H62 brass, was studied by Wang and Hu (2005). They concluded that material removal rate (MRR) of brass is highest among all three materials. The finishing of SUS304 stainless steel bent tubes using aluminum oxide composite magnetic abrasive was examined by Yamaguchi et al. (2005).

2 Process Principle and Experimental Setup

Figure 1 shows a schematic of internal MAF process using a stationary pole system. Magnetic abrasives introduced into the pipe are conglomerated at the finishing zone by a magnetic field, generating the finishing force against the inner

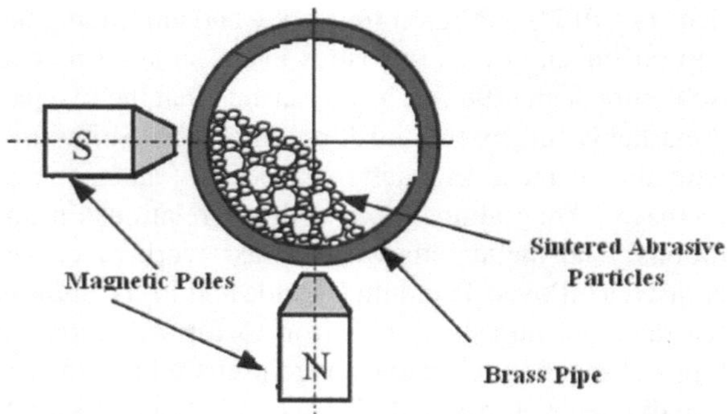

Fig. 1 Schematic of internal magnetic abrasive finishing

surface of the tube. In the process, magnetic abrasive particles introduced into the workpiece are attracted by the magnetic field and bear on the inner surface of the workpiece. These particles join each other along the lines of magnetic force due to dipole–dipole interaction and form a flexible magnetic abrasive brush (FMAB) which pushes against the work piece surface and develops finishing pressure. This finishing pressure originates microindentations in the workpiece surface. The tangential force developed by FMAB is the major cutting force responsible for microchipping. Abrasives generally rely upon a difference in hardness between the abrasive and the material being worked upon, the abrasive being the harder of the two substances.

The magnetic poles N and S were placed face to face with their axes crossing at right angle with a brass pipe in the configuration as shown in Fig. 1. The magnetic field extends in the inner region of the pipe without regard to the presence and absence of the pipe to the machined and actuates magnetic force to magnetic abrasive particles packed inside pipe. Work piece is kept between the two magnets in MAF operation. Abrasive particles can be used as unbonded, loosely bonded, or bonded. Bonded magnetic abrasive particles are prepared by sintering of ferromagnetic powder (iron) and abrasive powder (Al_2O_3) at a very high pressure and temperature in H_2 gas atmosphere. Loosely bonded MAPs are prepared by mechanical mixing of ferromagnetic powder and abrasive powder with a small amount of lubricant to give some holding strength between the abrasive and ferromagnetic particles. Unbonded magnetic abrasive particles are mechanical mixture of ferromagnetic and abrasive particles without any lubricant. The magnetic abrasive particles join each other along the lines of magnetic force and form a FMAB between each magnetic pole and the workpiece. This brush behaves like a multipoint cutting tool for finishing operation. The magnetic force on the abrasive particles provides the necessary machining force. This force is responsible for the abrasion of the pipe by magnetic abrasive particles.

An external photograph of experimental set up is shown in Fig. 2, which shows the principles of internal finishing. The major components are electromagnet (12k Gauss), control unit, D.C. motor, and variable D.C. supply. The cylindrical work

Fig. 2 External photograph of experimental setup

Table 1 Experimental conditions

Workpiece material	Brass
Machining time	45 min
Pole vibration-amplitude	5 mm
Frequency	0.8 Hz
Lubricant	Light oil (5 % of quantity of abrasives)
Workpiece-pole gap	1 mm

piece, that is, brass pipe was held in the chuck attached to D.C. motor, and abrasives were packed inside the pipe. Magnetic field was applied to the abrasives by electromagnet. Magnetic field strength is varied for experimentation with the help of variable D.C. supply. Electromagnet plays an important role in present experimentation. The space between workpiece and electromagnet is kept constant. The magnetic field strength depends upon weight percentage of the magnetic particles, present in the magnetic abrasive powder. Both the working gap and size of the workpiece are taken into consideration, while designing. The objective of the design is to give rotational motion to the cylindrical workpiece. A D.C. motor is chosen for providing rotational motion to the work piece. A schematic view of the setup is shown in the figure. Magnetic abrasive particles through magnetic pressure finish the workpiece. Al_2O_3-based sintered magnetic abrasives are used as magnetic abrasives in this work.

3 Experimental Process Variables

Alumina (Al_2O_3)-based sintered magnetic abrasives were used in this work for internal finishing of cylindrical brass pipes. The Alumina (Al_2O_3)-based sintered magnetic abrasives were prepared by blending of Al_2O_3 (10 %) of 200 mesh size

Table 2 Coded and real levels of independent variables

Input process parameter	Coded	Level I −2	Level II −1	Level III 0	Level IV 1	Level V 2
Rotational speed (rpm)	A	250	500	750	1,000	1,250
Magnetic flux density (T)	B	0.4	0.6	0.8	1	1.2
Grit size (μm)	C	85	120	200	300	420
Quantity of abrasives (g)	D	4	8	12	16	20

Table 3 Experimental data

Experimental no.	Independent parameters				Responses	
	Speed (rpm)	MFD (T)	Grit size (μm)	Quantity (g)	PISF (%)	MRR (mg/min)
1	500	1.0	300	8	92.06	3.44
2	750	0.4	200	12	79.70	6.48
3	1,000	0.6	300	8	77.13	2.98
4	1,250	0.8	200	12	87.04	6.97
5	750	0.8	200	12	67.24	5.85
6	750	0.8	200	12	64.67	7.06
7	250	0.8	200	12	88.03	7.86
8	1,000	1.0	300	8	75.59	3.67
9	500	0.6	120	8	69.60	4.35
10	1,000	1.0	300	16	52.80	5.09
11	750	0.8	420	12	80.16	6.43
12	750	0.8	200	12	65.63	7.16
13	750	0.8	85	12	70.57	5.31
14	500	1.0	120	16	57.41	4.44
15	500	0.6	300	16	81.71	5.64
16	750	1.2	200	12	71.84	4.13
17	750	0.8	200	12	65.15	7.35
18	750	0.8	200	12	61.20	6.53
19	500	1.0	120	8	75.34	4.98
20	1,000	0.6	120	8	87.03	2.85
21	1,000	1.0	120	16	52.58	2.94
22	500	1.0	300	16	80.32	4.51
23	500	0.6	300	8	69.78	2.41
24	750	0.8	200	12	66.23	7.20
25	500	0.6	120	16	64.98	5.73
26	1,000	0.6	120	16	87.45	4.74
27	1,000	0.6	300	16	63.23	6.19
28	1,000	1.0	120	8	84.13	3.10
29	750	0.8	200	4	53.45	0.48
30	750	0.8	200	20	28.01	1.47

Table 4 ANOVA results for PISF

Source	Sum of squares	Df	Mean square	*F*-value	*p* value Prob > *F*	
Model	5,257.52	14	375.54	25.96	<0.0001	Significant
A—speed	14.59	1	14.59	1.01	0.3312	
B—magnetic flux density	79.46	1	79.46	5.49	0.0333	Significant
C—grit size	5.22	1	5.22	0.36	0.5572	
D—quantity	816	1	816	56.41	<0.0001	Significant
AB	295.84	1	295.84	20.45	0.0004	Significant
AC	613.14	1	613.14	42.38	<0.0001	Significant
AD	129.16	1	129.16	8.93	0.0092	Significant
BC	146.7	1	146.7	10.14	0.0062	Significant
BD	378.69	1	378.69	26.18	0.0001	Significant
CD	19.3	1	19.3	1.33	0.2661	
Residual	217	15	14.47			
Lack of fit	195.5	10	19.55	4.55	0.0542	Not significant
Pure error	21.5	5	4.3			
Cor total	5,474.51	29				
Standard deviation	3.80			*R*-squared		0.9604
Mean	70.67			Adj *R*-squared		0.9234
C. V.	5.38			Pred *R*-squared		0.7508
PRESS	1,364.12			Adequate precision		22.612

(74 μm) and iron powders (90 %) of 300 mesh size (51.4 μm), compacting them by a universal testing machine (UTM), sintering the mixture in a sintering set up at 1,100 °C in H_2 gas environment, crushing the compacts into small particles, and then sieving to different ranges of sizes. The obtained sizes are 85, 120, 200, 300, and 420 μm. The experimental conditions are shown in Table 1 and coded, and real levels of independent variables are shown in Table 2. Cylindrical brass pipes (Ø36 mm × Ø33 mm × 77 mm) were used for the experiments as workpieces. The experimental variables such as abrasive grit size, circumferential speed, quantity of abrasives, and MFD were considered in this work. The finishing characteristics of magnetic abrasives were analyzed by measuring the surface roughness, using a Mitutoyo surface roughness tester (SJ-210P) having a least count of 0.001 μm (cutoff length = 0.8 mm). Also surface finish was analyzed using response surface methodology (RSM). Therefore, finishing characteristics in terms of percentage improvement in surface finish (PISF) and MRR were analyzed.

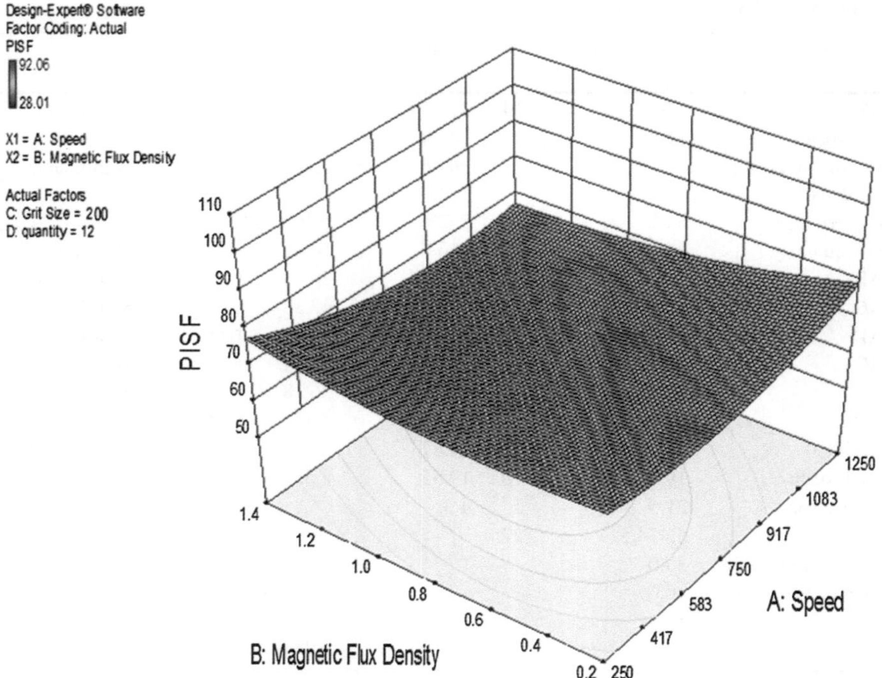

Fig. 3 Effect of interactions between speed (A) and magnetic flux density (B) on PISF

4 Experimental Results and Discussions

The results of different experimental investigations carried out under the study are tabulated in Table 3.

4.1 Analysis of Percentage Improvement in Surface Finish

ANOVA results for PISF are shown in Table 4.

The Model *F*-value of 25.96 implies the model is significant. There is only a 0.01 % chance that a "Model *F*-value" this large could occur due to noise. Values of "Prob > *F*" less than 0.0500 indicate model terms are significant. In this case, MFD, quantity of abrasives, interactions between (a) rotational speed of work piece and MFD, (b) rotational speed and grit size, (c) rotational speed and quantity of abrasives, (d) MFD and grit size, (e) MFD and quantity of abrasives are significant model terms. Values greater than 0.1000 indicate the model terms are not significant. The "Lack of Fit *F*-value" of 4.55 implies there is a 5.42 % chance

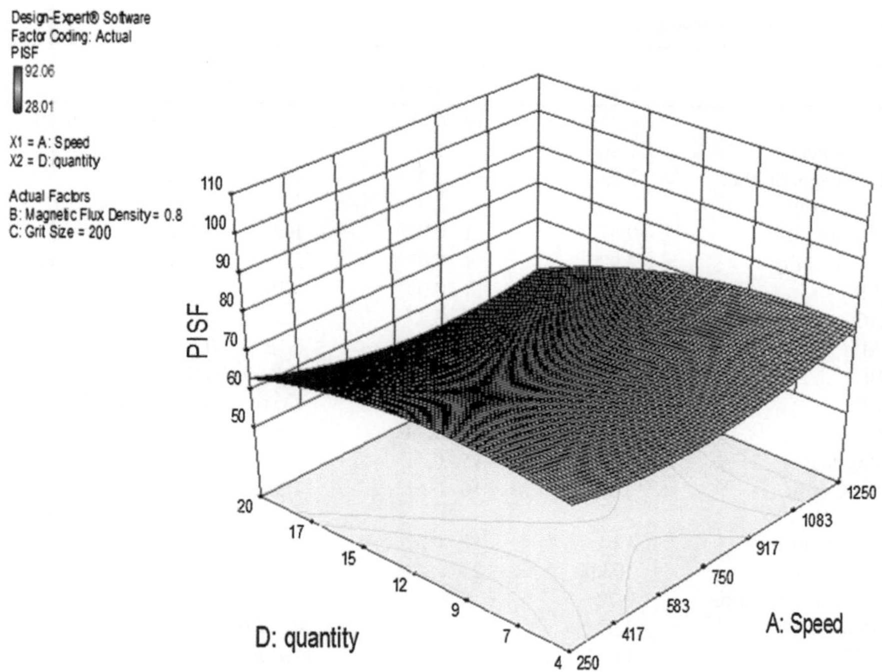

Fig. 4 Effect of simultaneous variation of speed (A) and quantity (D) on PISF

that a "Lack of Fit *F*-value" this large could occur due to noise. Lack of fit is bad; we want the model to fit.

The effects of interactions of different process parameters such as circumferential speed of the work piece, MFD, and abrasive grit size on PISF and MRR were analyzed using RSM. The maximum PISF of 90 % and minimum surface roughness of 0.04 μm were obtained.

Figure 3 shows the effect of simultaneous variation of speed (A) and MFD (B) on PISF. At lower level of speed, with increase in MFD, PISF increases but at higher speeds when MFD increases surface finish decreases. At lower level of MFD with increase in speed, PISF first decreases then starts increasing but at higher level of MFD, with increase in speed PISF goes on decreasing.

Figure 4 shows the effects of speed (A) and quantity (D) on the PISF. It can be seen that as quantity of abrasives increases, PISF increases up to certain extent but after 12 g, PISF starts decreasing. The PISF is better at lower level of quantity and higher level of speed.

Table 5 ANOVA results for MRR

Source	Sum of squares	Df	Mean square	F-value	p-value Prob > F	
Model	95.34	14	6.81	13.3	<0.0001	Significant
A—speed	1.16	1	1.16	2.27	0.1525	
B—magnetic flux density	2.22	1	2.22	4.34	0.0547	
C—grit size	0.87	1	0.87	1.69	0.2127	
D—quantity	7.97	1	7.97	15.58	0.0013	Significant
AB	0.09	1	0.09	0.18	0.6809	
AC	3.77	1	3.77	7.37	0.016	Significant
AD	0.093	1	0.093	0.18	0.6759	
BC	0.24	1	0.24	0.46	0.5074	
BD	3.92	1	3.92	7.66	0.0144	Significant
CD	2.74	1	2.74	5.36	0.0352	Significant
Residual	7.68	15	0.51			
Lack of fit	6.06	10	0.61	1.88	0.2524	Not significant
Pure error	1.61	5	0.32			
Cor. total	103.02	29				
Standard deviation	0.72			R-squared		0.9255
Mean	4.91			Adj R-squared		0.8559
C.V.	14.57			Pred R-squared		0.5059
PRESS	50.90			Adequate precision		9.922

4.2 Analysis of Material Removal Rate

ANOVA results of MRR are shown in Table 5.

The Model F-value of 13.3 implies the model is significant. There is only a 0.01 % chance that a "Model F-value" this large could occur due to noise. Values of "Prob > F" less than 0.0500 indicate model terms are significant. In this case, quantity of abrasives, interactions between (a) rotational speed of workpiece and grit size, (b) MFD and quantity of abrasives, and (c) grit size and quantity of abrasives are significant model terms. Values greater than 0.1000 indicate the model terms are not significant. The "Lack of Fit F-value" of 1.88 implies the lack of fit is not significant relative to the pure error. There is a 25.24 % chance that a "Lack of Fit F-value" this large could occur due to noise. Non-significant lack of fit is good.

Figure 5 shows the relationship between the speed (A) and MFD (B) keeping the value of grit size and quantity to constant level. At all levels of speed, with increase in MFD, MRR increases before it starts decreasing. The MRR starts decreasing after MFD of 1.0 T.

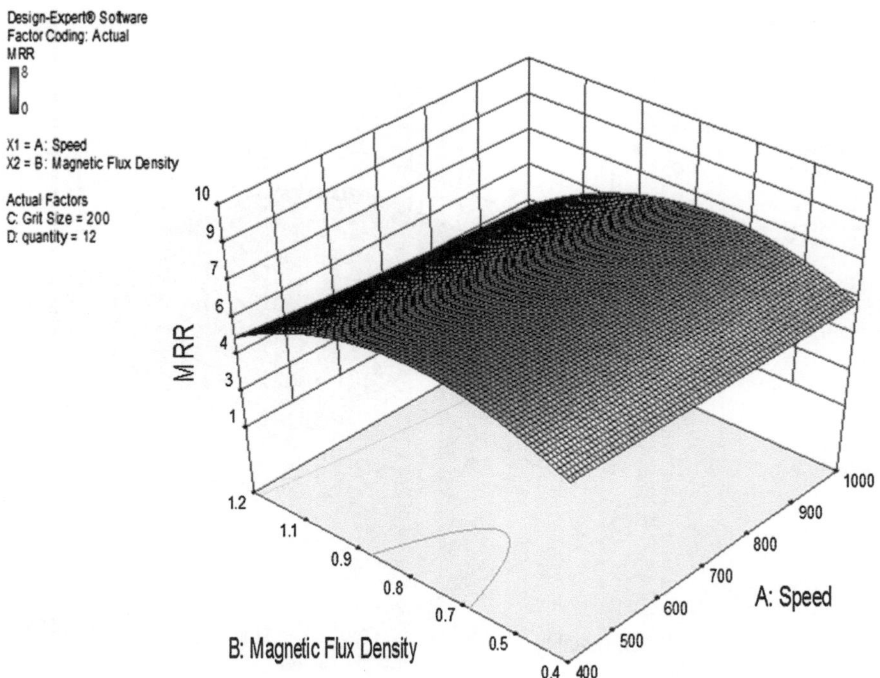

Fig. 5 Effect of interactions between speed (A) and magnetic flux density on MRR

Figure 6 shows the relationship between the speed (A) and quantity (D) keeping the value of MFD and grit size to constant level. At all levels of speed, with increase in quantity, MRR increases before it starts decreasing. The MRR starts decreasing after quantity of 16 g.

5 Microstructure Examination

The surface structure was analyzed by SEM before and after finishing of brass tube. The typical SEM micrographs of as received after boring the workpiece and magnetic abrasive finished surfaces are shown in the figures. The initial surface profile has periodic peaks and valleys generated by boring. It is quite clear from Fig. 7a that deep cutting marks left by the boring operation have been removed and replaced by the new texture generated during the MAF process as shown in Fig. 7b.

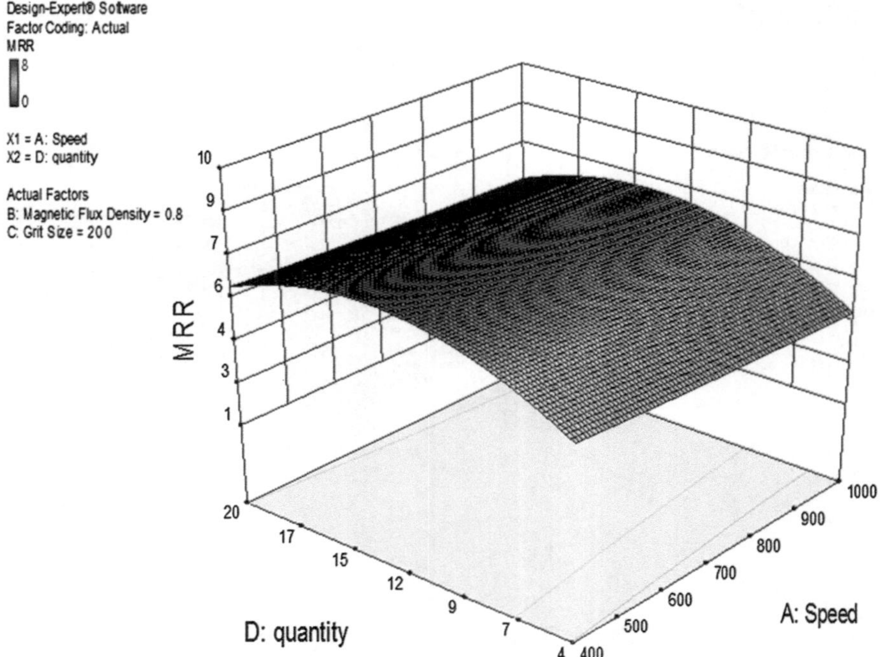

Fig. 6 Effect of simultaneous variation speed (A) and quantity (D) on MRR

6 Conclusions

The conclusions of this research work are as follows:

1. PISF was significantly affected by MFD, interactions between rotational speed of workpiece and MFD, rotational speed and grit size, MFD, and grit size.
2. MRR was significantly affected by quantity of abrasives, interactions between rotational speed of workpiece and grit size, MFD, and quantity of abrasives and grit size and quantity of abrasives.
3. The SEM analysis shows that the tool marks and scratches are either removed or reduced resulting in fine surface after MAF.
4. Maximum PISF obtained was 92 % and Ra 0.04 μm.

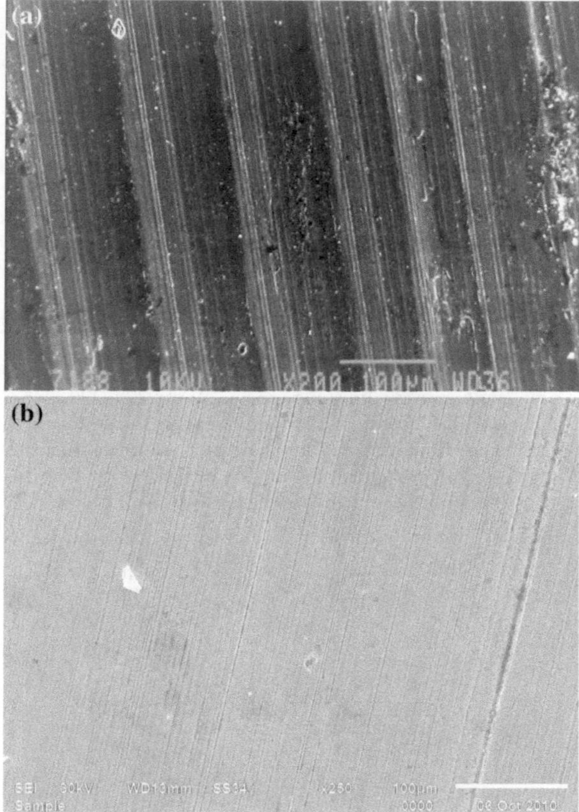

Fig. 7 SEM micrographs of inner surface of pipe before (**a**) and after (**b**) finishing for 45 min

Acknowledgments The authors acknowledge the support of Baba Banda Singh Bahadur Engineering College, Fatehgarh Sahib, for carrying out experimentation work.

References

Jain VK, Prashant K, Behra PK, Jayswal SC (2001) Effect of working gap and circumferential speed on the performance of magnetic abrasive finishing process. Wear 250:384–390

Khairy AB (2001) Aspects of surface and edge finish by magnetoabrasive particles. J Mater Process Technol 116:77–83

Kim J-D (2003) Polishing of ultra-clean inner surfaces using magnetic force. Int J Adv Manuf Technol 21:91–97

Lin C-T, Yang L-D, Chow H-M (2007) Study of magnetic abrasive finishing in free-form surface operations using the Taguchi method. Int J Adv Manuf Technol 34:122–130

Mori T, Hirota K, Kawashima Y (2003) Clarification of magnetic abrasive finishing mechanism. J Mater Possessing Technol 143–144

Shinmura T, Aizawa T (1989) Study on internal finishing of non-ferromagnetic tubing by magnetic abrasive machining process. Bull Jpn Soc Precis Eng 23(1):37–41

Shinmura T, Takazawa K, Hatano E, Aizawa T (1984) Study on magnetic abrasive process. Bull Jpn Soc Precis Eng 18(4):347–348

Shinmura T, Takajava K, Hatano E (1985) Study on magnetic abrasive process—application to plane finishing. Bull Jpn Soc Precis Eng 19(4):289–291

Shinmura T, Takazawa K, Hatano E (1987) Study on magnetic abrasive finishing-effects of various types of magnetic abrasives on finishing characteristics. Bull Jpn Soc Precis Eng 21(2):139–141

Shinmura T, Wang F, Aizawa T (1994) Study on a new finishing process of fine ceramics by magnetic abrasive machining. Jpn Soc Precis Eng 28(2):99–104

Singh L, Singh S, Mishra PS (2010) Performance of abrasives used in magnetically assisted finishing: a state of the art review. Int J Abras Technol 3(3):215–227

Wang Y, Hu D (2005) Study on inner surface finishing of tubing by magnetic abrasive finishing. Int J Machine Tools Manuf 45:43–49

Yamaguchi H, Shinmura T (1999) Study of the surface modification resulting from an internal magnetic abrasive finishing process. Wear 225–229, 246–255

Yamaguchi H, Shinmura T (2004) Internal finishing process for alumina ceramic components by a magnetic field assisted finishing process. Precis Eng 28:135–142

Yamaguchi H, Shinmura T, Sekine M (2005) Uniform internal finishing of SUS304 stainless steel bent tube using a magnetic abrasive finishing process. ASME J Manuf Sci Eng 127:605–611

Nanofinishing of Steel Rollers (SUS 304) Using Diamond-Based Loosely Bonded Magnetic Abrasives

Partap Singh Samra and Lakhvir Singh

Abstract Nanofinishing of steel rollers (SUS304) has been done using loosely bonded diamond-based magnetic abrasives prepared by homogeneous mixing of magnetic powder [Fe powder of 300 mesh size (51.4 μm)], abrasive powder [diamond particles of 200 mesh size (74 μm)], and lubricant. A series of experiments have been conducted using in-house fabricated setup. It has been found that magnetic flux density, quantity of magnetic abrasives, rotational speed of workpiece, and percentage of abrasives in magnetic abrasives have significant effect on PISF. Scanning electron microscope (SEM) photographs show that the surface generated by turning on lathe consists of deep scratches (pits and digs). These scratches have been removed by MAF resulting in improved surface finish, but fine scratching marks produced by MAF appear on the surface. Atomic force microscopy (AFM) micrographs show that valley to peak height is reduced considerably by MAF, and there is no bad effect of MAF on surface of finished workpiece. The surface roughness is reduced to 10 nm Ra as measured by AFM. Diamond-based loosely bonded magnetic abrasives have nanofinishing capability when used in MAF.

Keywords Magnetic abrasive finishing (MAF) · Scanning electron microscope (SEM) · Atomic force microscopy (AFM)

P. S. Samra (✉)
Department of Mechanical Engineering, Sri Guru Granth Sahib World University, Fatehgarh Sahib 140407, Punjab, India
e-mail: Partap_20@yahoo.com

L. Singh
Department of Mechanical Engineering, B. B. S. B. Engineering College, Fatehgarh Sahib 140407, Punjab, India

S. S. Khangura et al. (eds.), *Proceedings of the International Conference on Research and Innovations in Mechanical Engineering*, Lecture Notes in Mechanical Engineering, DOI: 10.1007/978-81-322-1859-3_26, © Springer India 2014

1 Introduction

Finish, accuracy, and surface integrity of parts produced by conventional machining (i.e., tool harder than the workpiece) depend on the machine tool system which comprises of machine tool, cutting tool, workpiece, cutting conditions, and cutting fluid. In conventional super finishing processes, tools are mechanically pressed against the workpiece surface; this results in surface defects like microcracks, geometrical and dimensional errors, and distortions. There is also a high machine–tool–workpiece interactions taking place, and it is difficult to achieve requisite tolerances and surface quality. To minimize the damage due to machining, it is necessary to process such materials under gentle conditions. Thus, magnetic abrasive finishing (MAF) has been taken as an alternative super finishing process for efficient and precision finishing of surfaces. MAF is found to be a versatile process for finishing internal, external, flat, and complex shaped surfaces.

MAF is one of the super polishing processes involving extremely small amount of material removal even to the extent of an atomic cluster. This process uses magnetic force for material removal. These forces are uniformly distributed on the work surface and are easily controllable. Hence, there is negligible damage to the surface of workpiece. The process is capable of achieving surface roughness of the order of nanometric level.

Literature reveals that different techniques have been used for preparing magnetic abrasives and may be categorized as follows:

a. Sintered abrasives
b. Adhesive Based (Glued)
c. Plasma Based (Powder Melting/Plasma Spraying)
d. Mixing (Loosely bonded/Unbonded)
e. Mechanical alloying.

An exhaustive review of various available methods has been presented by Singh et al. (2010).

Kim (2003) used sintered magnetic abrasives for internal finishing of SUS304 stainless steel tubes and found that surface finish and material removal were affected by grain size, weight of magnetic abrasives, flux density, speed of workpiece, machining fluid, and machining time. The optimal working conditions reported were 90 %, 12 gm, 0.4 T, 112 m/min, 1 ml, and 15 min, respectively, and a surface finish of 0.05 μm (Ra) was obtained. Lin et al. (2007) prepared the magnetic abrasives by typically mixing iron powder (60 wt%) and Al_2O_3 (40 wt%) with average size of 50 μm and compressing mixture into the cylindrical shape. These compacts were sintered into a vacuum furnace. After sintering process, these cylinders were crushed to produce magnetic abrasives of average size 150 μm. The ball-shaped magnetic pole with special grooves was used with these magnetic abrasives. It was found that the design increased the finishing efficiency and created a good surface finish for the non-ferromagnetic material, SUS304. The best surface finish was obtained at a working gap of 2.5 mm, a feed rate of 10 mm/min, and an abrasive mass of two grams. In sintering technique, the magnetic abrasives were prepared under high temperature and pressure conditions in inert medium. The solid mass of iron and abrasives formed during sintering was crushed into desired size. Due to this complex production process, the sintered magnetic abrasive is expensive.

Feygin et al. (1998) prepared magnetic abrasives by mixing iron powder, Al_2O_3, and glue as adhesive (commercially known as industrial crazy glue). Iron and abrasive particles were strongly bonded with each other by the glue. They reported that this method was simple as compared to the other methods for preparation of the magnetic abrasives. MRR was higher as compared to magnetic abrasives prepared by other methods. Kremen et al. (1999) also developed magnetic abrasives using an adhesive to bind magnetic component (iron powder) with abrasive component (diamond powder). All the three components were mixed thoroughly, dried, and crushed into small particles of desired size for machining. Then, using these glued magnetic abrasive powder and keeping magnetic flux density 0.4 T, machining time 5 min and 4 % boric acid in water used as cooling fluid, investigated the effect of powder grain size on the surface roughness and MRR of a silicon wafer and tube.

Handa et al. (2008); Yamaguchi and Handa (2008) also developed spherical iron-based composite powder with carried diamond particles using a plasma-spraying technique. Spherical carbonyl iron powder with 7.2 μm in average and diamond particles with 0.3 μm in average were mechanically mixed and then were plasma-sprayed at various plasma currents, and the spherical iron-based composite powders with carried diamond particles which have a particle size less than 10 μm were obtained. These fabricated spherical magnetic abrasives were used for finishing of SUS304 plate on setup in dry conditions.

Yin and Shinmura (2004a, b) finished three different materials with loosely bonded magnetic abrasives (4 gm iron particles of size 330 μm mixed with 1 gm of Al_2O_3 of size 80 μm and 2 ml of straight oil type grinding fluid). It was reported that volume removal rate of magnesium alloy was more as compared to stainless steel and brass. Wang and Hu (2005) also used loosely bonded magnetic abrasives (mixer of 1.5 gm iron of size 510 μm, 1.5 gm Cr_2O_3 of 3 μm size, and 0.5 ml distilled water). They reported that wet finishing gives better surface finish as compared to dry finishing. Singh et al. (2005) used UMA prepared by homogeneous mixing of 25 % SiC abrasives (mesh no. 400) and 75 % iron particles (mesh no. 300) in 3 % oil (SAE30) for finishing of alloy steel tubes. The surface roughness of the workpiece was decreased from 0.58 to 0.11 μm Ra.

Sran and Khangura (2012) used the bonded magnetic abrasives prepared by a new technique called mechanical alloying for the internal finishing of the brass tubes. After rough boring operation, the inner surface of the tubes is finely finished by newly developed magnetic abrasives; best surface finish obtained is of the order of 3 nm.

Irrespective of type of magnetic abrasive used, the percentage improvement in surface finish over original finish of the surface varies in 75–99 %. Only few researchers have used diamond as abrasive component in loosely bonded/

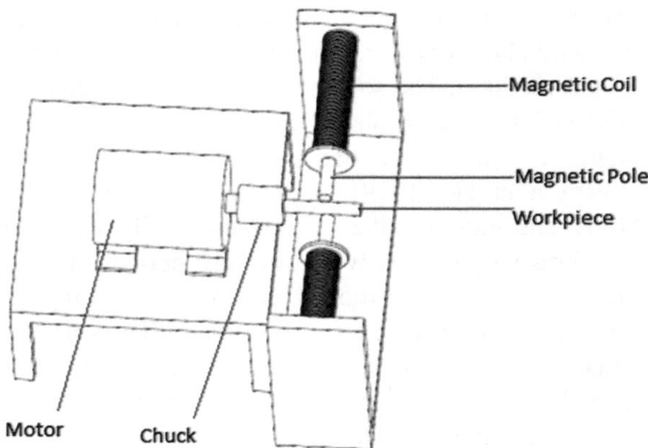

Fig. 1 Block diagram of the experimental setup

unbonded magnetic abrasives. Most of the abrasive manufacturing methods are either proprietary or difficult. In case of unbounded/loosely bonded magnetic abrasives, few researchers have used diamond as abrasive component.

The present work is aimed at studying the performance of loosely bonded diamond-based magnetic abrasives for external finishing of steel rollers.

2 Experimental Setup

A schematic view of the setup is shown in the Fig. 1. The fabricated setup has major components like electromagnet, control unit, dc motor with three jaw chuck, and variable dc supply. Magnetic flux density is variable from zero to 1 T by varying the current supplied to the magnetic coils. To rotate the workpiece, 1 HP DC shunt motor was used and the workpiece was mounted in three jaw chuck. The speed of rotation of the workpiece is variable from 0 to 1,500 rpm in small steps. MAPs through magnetic pressure finish the workpiece. For the present work, the magnetic abrasive powder was prepared through homogeneous mixing of magnetic powder (iron powder of 300 mesh size, 51.4 μm) and abrasive powder (diamond powder of 200 mesh size, 74 μm) which are loosely bonded together by lubricating oil to have composite particles (or conglomerate).

3 Experimental Procedure

Important input process parameters and their working range have been selected on the basis of preliminary experiments conducted using the experimental setup developed for the present work and literature available. The different MAF parameters and their range are tabulated in Table 1.

Table 1 MAF parameters for experimentation

Constant parameters	
Workpiece: the external surface was prepared by conventional turning to get uniform initial roughness	O.D.–37 mm
Pole workpiece gap	1 mm
Grit size of diamond abrasive	74 μm
Grit size of Fe powder	51.4 μm
Machining time	30 Min
Lubricating oil	5 % by wt. of MA
Variable parameters	
Rotational speed of workpiece	400–2,000 rpm
Quantity of magnetic abrasives	10–50 g
Percentage of diamond abrasives in MA	10–50 %
Magnetic flux density	0.4–1.2 T

The finishing characteristics of magnetic abrasives were analyzed by measuring the surface roughness, which was measured at four points before and after finishing using a Mitutoyo surface roughness tester (SJ-210P) having a least count of 0.001 μm (cutoff length = 0.8 mm) and averaged.

4 Results and Discussion

It was observed that mirror-like finish was obtained on steel specimen. Microscopic examination of some test specimen were carried out because it is one of the most important and reliable method for evaluating initial surface integrity through identification of microcracks, microdefects, etc. Surface defects such as microcracks or flaws depend normally on the machining conditions. It is, therefore, necessary to apply special metallographic techniques and instruments like scanning electron microscope (SEM) to examine surface characteristics. The surface roughness alone does not reflect the behavior of abrasive cutting edges. Therefore, the SEM and atomic force microscope images of the magnetic abrasive finished workpieces were taken to provide in-depth comparison of the surfaces generated before and after MAF. All the samples to be subjected to MAF were prefinished by smooth turning. The surface profiles of the selected workpiece surfaces were obtained before and after MAF.

4.1 AFM

Atomic force microscopy (AFM) or scanning force microscopy (SFM) is a very high-resolution type of scanning probe microscopy, with demonstrated resolution of fractions of a nanometer, more than 1,000 times better than the optical

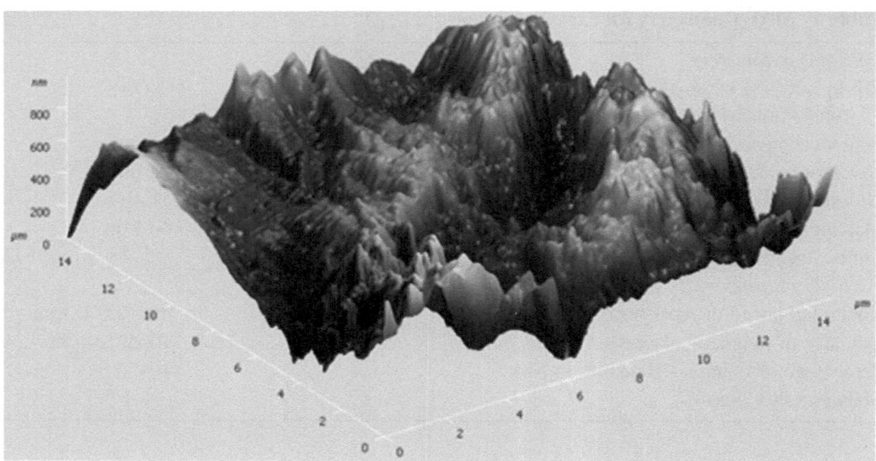

Fig. 2 AFM 3-D micrograph of the surface texture finished by turning (Ra = 800 nm)

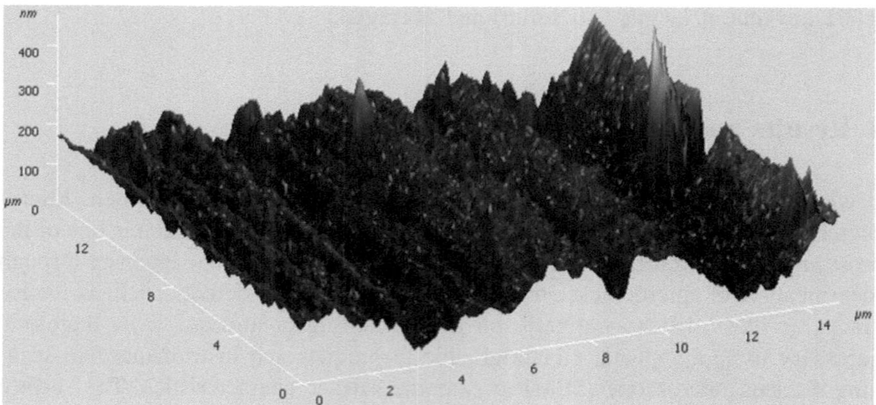

Fig. 3 AFM 3-D micrograph of the surface texture finished by MAF (MFD 1.00 T, quantity 20 gm, speed 800 rpm, and percentage of abrasives 20 %, Ra = 57.69 nm)

diffraction limit. The AFM is one of the foremost tools for imaging, measuring, and manipulating matter at the nanoscale. The information is gathered by "feeling" the surface with a mechanical probe. Piezoelectric elements that facilitate tiny but accurate and precise movements on (electronic) command enable the very precise scanning. The AFM image of the turned surface Fig. 2 shows peaks as high as 800 nm which have been sheared off by the flexible magnetic abrasive brush resulting in heights less than 10 nm (Fig. 4). Also, it is clear from Figs. 3 and 4 that as we increase the quantity of magnetic abrasives and percentage of abrasives, the surface finish improves. Figure 5 shows that at higher speeds, the surface finish is poor.

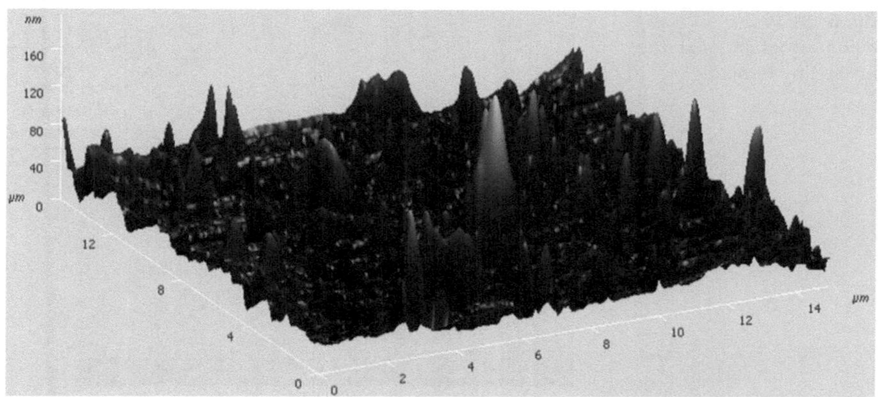

Fig. 4 AFM 3-D micrograph of the surface texture finished by MAF (MFD 0.80 T, quantity 30 gm, speed 1,200 rpm, and percentage of abrasives 30 %, Ra = 10.84 nm)

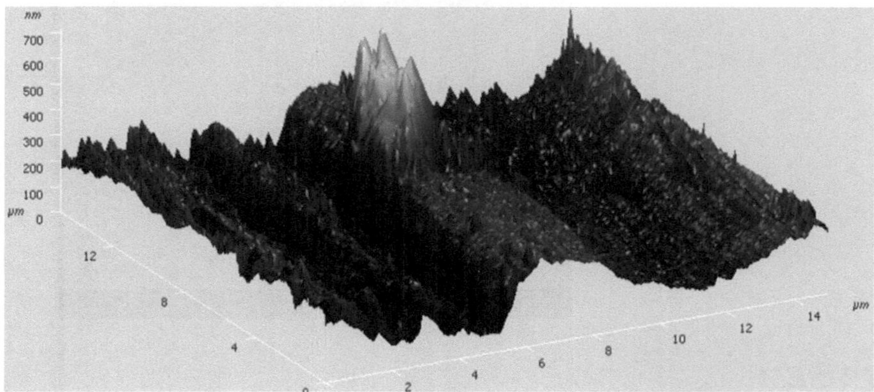

Fig. 5 AFM 3-D micrograph of the surface texture finished by MAF (MFD 0.80 T, quantity 30 gm, speed 2,000 rpm, and percentage of abrasives 30 %, Ra = 68.16 nm)

The heights of the peaks are non-uniform because of magnetic strength as well as abrasive particles protrusion heights are non-uniform. Smaller peaks indicate the area where abrasive particles were held more strongly than other areas and hence could reduce the peak heights more compared to other areas. AFM micrographs show that valley to peak height is reduced considerably by MAF and there is no bad effect of MAF on surface of finished workpiece.

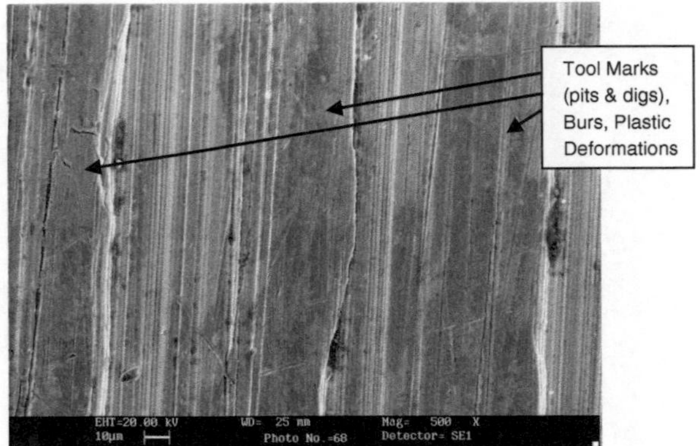

Fig. 6 SEM microphotograph of surface finished by turning

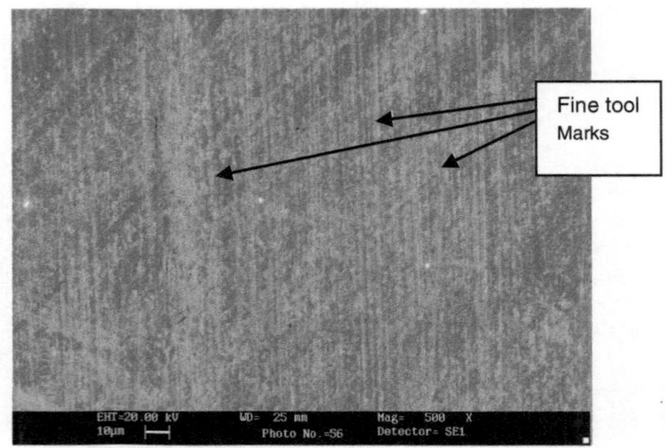

Fig. 7 SEM microphotograph of surface finished by MAF

4.2 SEM

The SEM is a type of electron microscope that images the sample surface by scanning it with a high-energy beam of electrons in a raster scan pattern. The electrons interact with the atoms that make up the sample producing signals that contain information about the sample's surface topography, composition, and other properties such as electrical conductivity.

The figures below show typical SEM micrographs of as received turned surface and magnetic abrasive finished surfaces. The surface generated by turning consists of deep scratches produced by the interaction of abrasive cutting points with the workpiece surface. Huge plastic deformations and burs are also seen. The observations reveal that the finishing of workpiece surface in this process is done by scratching /microcutting. However, marks due to turning, pits and digs, shown in Fig. 6 disappear after MAF as shown in Fig. 7, but fine scratching marks produced by MAF appear on the surface. Tool marks, plastic deformations, and burs decrease by MAF resulting in improved surface finish.

5 Conclusions

In the present work, MAF setup has been designed and fabricated. The performance of loosely bonded MAPs has been studied on non-magnetic stainless steel. The conclusions drawn from this research can be summarized as follows:

1. The SEM micrographs show that tool marks, plastic deformations, and burs are removed by MAF.
2. The best surface finish obtained was 10 nm (Ra) indicated by AFM.
3. AFM micrographs show that valley to peak height is reduced considerably by MAF and there is no bad effect of MAF on surface of finished workpiece.

Acknowledgment All the researchers whose work is referred here are sincerely acknowledged.

References

Feygin S, Kreman G, Igelstyn L (1998) US patent no. 5846270

Fox M, Agrawal K, Shinmura T, Komanduri R (1994) Magnetic abrasive finishing of rollers. Ann CIRP 43(1):181–184

Hanada K, Yamaguchi H, Zhou H (2008) New spherical magnetic abrasives with carried diamond particles for internal finishing of capillary tubes. Diam Relat Mater 17(7–10):1434–1437

Kim J-D (2003) Polishing of ultra-clean inner surfaces using magnetic force. Int J Adv Manuf Technol 21:91–97

Kremen G, Elsayed E, Feygin S, Igelshteyn L (1999) Material removal rate and surface roughness of magnetic abrasive process. Paper presented at the 3rd international machining and grinding conference, Cincinnati, Ohio

Lin C-T, Yang L-D, Chow H-M (2007) Study of magnetic abrasive finishing in free-form surface operations using the Taguchi method. Int J Adv Manuf Technol 34:122–130

Sran LS, Khangura SS (2012) Nano finishing of brass tubes by using mechanically alloyed magnetic abrasives. In: Proceedings of international manufacturing science and engineering conference (MSEC 2012—7264)

Singh DK, Jain VK, Raghuram V, Komanduri R (2005) Analysis of surface texture generated by a flexible magnetic abrasive brush. Wear 259:1254–1261

Singh L, Khangura SS, Mishra PS (2010) Performance of abrasives used in magnetically assisted finishing: a state of the art review. Int J Abras Technol 3(3):215–227

Wang Y, Hu D (2005) Study on inner surface finishing of tubing by magnetic abrasive finishing. Int J Mach Tools Manuf 45:43–49, Shanghai Jiaotong University, Shanghai, China

Yamaguchi H, Handa K (2008) Development of spherical abrasive made by plasma spray. J Manuf Sci Eng 130(3):031107-1–031107-9

Yin S, Shinmura T (2004a) Vertical vibration-assisted magnetic abrasive finishing and deburring for magnesium alloy. Int J Mach Tools Manuf 44:1297–1303

Yin S, Shinmura T (2004b) A comparative study: polishing characteristics and its mechanisms of three vibration modes in vibration-assisted magnetic abrasive polishing. Int J Mach Tools Manuf 44:383–390

Conclusions

In the present work, MAF setup has been designed and fabricated. The performance of this newly fabricated MAF has been studied on non-magnetic stainless steel. The conclusions drawn from the present study are summarized as follows:

1. The MAF [illegible] surface roughness, and improved by MAF.
2. The best surface [illegible] [illegible] [illegible] by MAF.
3. MAF [illegible] [illegible] MAF [illegible] [illegible] [illegible] [illegible].

Acknowledgements [illegible]

References

[illegible], Kremer D, [illegible] (1993) [illegible] [illegible]
Hanada K, Yamaguchi H, [illegible] (2008) [illegible] magnetic abrasive [illegible] finishing [illegible] [illegible] [illegible] 1227–1234
[illegible] (2003) Modeling of [illegible] [illegible] [illegible] finishing [illegible]. Int J Adv Manuf Technol [illegible]
[illegible] [illegible] [illegible] (1996) [illegible] [illegible] [illegible] of magnetic abrasive [illegible] [illegible] [illegible] [illegible] conference [illegible]
[illegible], Yang LD, Chow H M (2005) Study of [illegible] [illegible] [illegible] operations using the [illegible] [illegible]. Int J Adv Manuf Technol [illegible]
[illegible] (2011) [illegible] [illegible] [illegible] magnetic abrasives [illegible] [illegible] [illegible] [illegible] [illegible] [illegible] and engineering congress [illegible]
Singh DK, Jain VK, [illegible] V, [illegible] (2005) [illegible] [illegible] [illegible] a flexible magnetic abrasive brush. Wear 259:[illegible]
[illegible] (2010) Performance of [illegible] [illegible] [illegible] [illegible] finishing: a state of the art review. [illegible]
[illegible] (2003) Study on [illegible] finishing [illegible] Int J Mach Tools Manuf [illegible]
[illegible] (2006) [illegible] [illegible] [illegible] [illegible] Mater Sci Eng [illegible]
[illegible] (2007) [illegible] [illegible] [illegible] of magnetic abrasive [illegible] Int J Mach Tools Manuf [illegible]
[illegible]

Finishing of Aluminum Pipes Using Silica Sand (River Bed)-Based Loosely Bonded Magnetic Abrasives

Arishu Kaushik, Palwinder Singh and Lakhvir Singh

Abstract With the advancements in various fields such as medical research, space research, robotics, and CNC machines, there is dire need of machining processes, which are capable of providing good surface finish in tight tolerances. Besides this, new materials are being developed to realize the need of present industry. The properties of some of the materials are such that they cannot be machined by conventional or traditional methods. A family of new machining techniques called nonconventional machining processes has been developed. These newly developed processes are helping the industry to attain the required degree of accuracy and surface finish. In contrast to these processes, the conventional machining processes have limited performance and productivity. Nonconventional processes are also capable of machining the hard to machine materials and intricate shapes.

Keywords Flexible magnetic abrasive brush · Magnetic abrasive finishing · Material removal rate · Percentage improvement in surface finish

1 Introduction

Abrasive machining is a manufacturing process where material is removed from a workpiece using a multitude of small abrasive particles. Common examples include grinding, honing, and polishing. Abrasive processes are usually expensive,

A. Kaushik (✉) · P. Singh · L. Singh
Department of Mechanical Engineering, B. B. S. B. Engineering College,
Fatehgarh Sahib 140407, Punjab, India
e-mail: arishu_kaushik@yahoo.co.in

P. Singh
e-mail: palwindergill13@gmail.com

L. Singh
e-mail: lakhvir1972@gmail.com

S. S. Khangura et al. (eds.), *Proceedings of the International Conference on Research and Innovations in Mechanical Engineering*, Lecture Notes in Mechanical Engineering,
DOI: 10.1007/978-81-322-1859-3_27,

but capable of tighter tolerances and better surface finish than other manufacturing processes.

The present study is based on one of the nonconventional machining processes called 'Magnetic Abrasive Machining,' which came to the surface in 1938 in a patent by Harry P. Coats. The countries that are involved in the study and development of this process are the USA, CIS, England, France, Bulgaria, Germany, and Japan. There are certain advantages of this process over the conventional ones that make it useful in improving the surface finish of the part. The magnetic field-assisted finishing processes are based on the electromagnetic behavior of the magnetic abrasive particles in the magnetic field.

Magnetic abrasive finishing (MAF) is a micro-finishing process that produces better surface as compared to conventional methods of finishing such as super-finishing, belt grinding, polishing, and honing.

MAF process is suitable for different applications due to small (micro) chip size and self-sharpening of the abrasives. In MAF process, magnetic force plays a dominant role for the formation of flexible magnetic abrasive brush (FMAB) and developing abrasion pressure. One of the merits is that the force of machining can be controlled by controlling the magnetic field strength and percentage of iron in the abrasive. MAF possesses many attractive advantages, and some of them are listed below.

- Self-adaptability.
- Controllability.
- Finishing tool requires neither compensation nor dressing.
- Material surface is free of burns and thermal defects.
- Low energy consumption.
- Nonferrous materials such as aluminum and its alloys, brass and its alloys can also be finished with ease.
- Most efficient from economical point of view.
- Magnetic abrasives have the self-sharpening ability (Fig. 1).

An abrasive is a material, often a mineral, that is used to shape or finish a workpiece through rubbing, which leads to part of the workpiece being worn away. Abrasives generally rely upon a difference in hardness between the abrasive and the material being worked upon, the abrasive being the harder of the two substances. Abrasives are extremely commonplace and are used very extensively in a wide variety of industrial, domestic, and technological applications. This gives rise to a large variation in the physical and chemical composition of abrasives as well as the shape of the abrasive. Common uses for abrasives include grinding, polishing, buffing, honing, cutting, drilling, sharpening, and abrasive machining. MAF is one promising process that is able to remove the material at micro/nano from metallic and nonmetallic surfaces.

A lubricant carries away transport heat (which may affect the physical properties of the workpiece or the abrasive), decreases friction (with the substrate or matrix), suspends worn work material and abrasives allowing for a finer finish, and

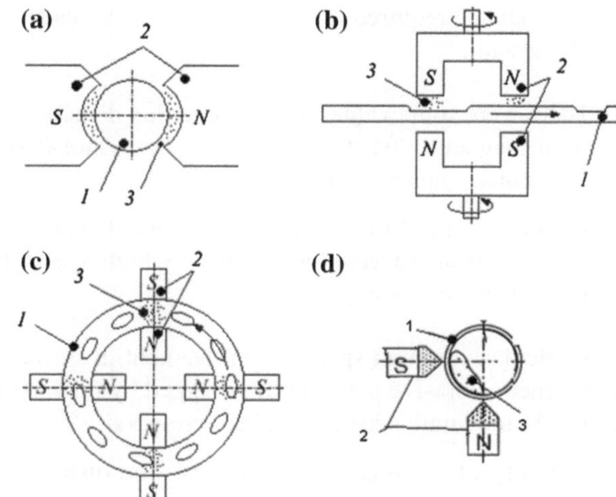

Fig. 1 Schematic diagrams of machine tools for magnetic abrasive finishing (**a**), flat surface components (**b**), and small-sized complex-shaped components (**c**), internal finishing of cylinder-shaped components (**d**). *1* Polished workpiece, *2* pole pieces, and *3* working gap with magnetic abrasive powder

conducts stress to the workpiece. Due to the interaction of the magnetic force, the mechanical action of the abrasive, the effect of individual input parameters controlling the material removal rate, and surface finish in MAF are different from that of other allied processes. The input parameters that significantly affect the material removal rate and surface finish are as follows:

1. Work material
2. Workpiece pole gap distance
3. Size of abrasive
4. Rotational speed of workpiece
5. Quantity of mixture
6. Percentage of abrasives
7. Number of magnetic poles
8. Magnetic flux density
9. Machining time
10. Shape of pole
11. Frequency of supply voltage
12. Axial movement.

2 Control of Abrasion

In MAF, the main factors that control abrasion are as follows:

1. Magnetic field strength: The abrasion of the surface of the tube was greatly affected by magnetic flux density, i.e., magnetic field strength. At low magnetic flux density, the abrasion pressure caused by magnetic abrasive particles was

less than the required for machining, sometimes abrasives were dragged by tube along with.

There were some values of magnetic flux density at which results were the best. At high magnetic flux density, abrasion pressure is so high that it results in less surface finish improvement.

2. Rotational speed of workpiece: Rotational speed of workpiece also affects the surface finish. At very slow and very high speed, the improvement in surface finish was very less.

Particularly at high speeds, improvement in surface roughness was impossible, as magnetic abrasive particles are dragged by tube. At slow speed, time taken was more. At moderate speeds, results were good.

3. Quantity of the magnetic abrasives: The quantity of the abrasives affects the surface finish. Very low and high quantity leads to less surface finish in comparison with the moderate quantity of the abrasives. It happens due to the jumbling of the particles.
4. Composition of the magnetic abrasives: Low composition of the abrasives leads to lesser cutting edges and hence less surface finish. High concentration leads to jumbling of the particles and again decreases the surface finish. High surface finish is obtained at moderate compositions.

In the present work, silica sand (river bed) combined with iron powder was used as magnetic abrasive.

3 Experimentation

Commercially available iron powder of 200 grit size and silica sand (river bed) of grit size 100 were selected. Samples of different compositions and quantity were prepared after weighing. The constituents were thoroughly mixed, and SAE 40 was used as a lubricating oil to make the samples just wet.

The surface finish of each workpiece was measured before and after the experimentation with the help of surface roughness tester.

The following observations and calculations were taken (Table 1).

The effect of speed (rpm) on PISF with current 4 A, percentage of sand 15 %, and weight of abrasives 10 gm is shown in Fig. 2. It has been observed that the PISF increases with the increase in speedup to the speed of 900 rpm and then starts decreasing with the further increase in speed.

The maximum PISF of 63.49 % is observed at a speed of 900 rpm. This happens due to the jumbling of the abrasive particles at higher rpm (Table 2).

Table 1 Observations and calculations

S. no.	Speed (rpm)	Current (A)	Percentage of sand (%)	Weight of abrasives (gm)	PISF (%)
1	600	4	15	10	55.08
2	900	4	15	10	63.49
3	1,200	4	15	10	60.17

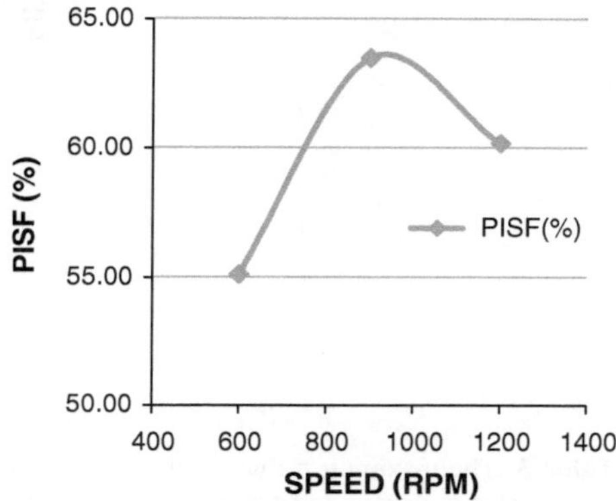

Fig. 2 Effect of speed (rpm) on PISF (%) (current = 4 A, percentage of sand = 15 %, weight of abrasives = 10 gm)

Table 2 Abrasive particles at higher rpm

S. no.	Speed (rpm)	Current (A)	Percentage of sand (%)	Weight of abrasives (gm)	PISF (%)
1	900	3	15	10	45.56
2	900	4	15	10	63.49
3	900	5	15	10	68.75

The effect of current (A) on PISF with speed 900 rpm, percentage of sand 15 %, and weight of abrasives 10 gm being constant is shown in Fig. 3. It has been observed that the PISF increase with the increase in the current.

The maximum value of PISF obtained is 68.75 at 5 A current. With the increase in current, the abrasive pressure also increases and leads to more surface finish (Table 3).

The effect of percentage of sand (%) on PISF (%) with speed 900 rpm, current 4 A, and weight of abrasives 10 gm being constant is shown in Fig. 4. It has been observed that the PISF increases with the increase in the percentage of sand. The maximum value of PISF obtained is 70.01 % at 20 % of sand abrasives. With the increase in percentage of abrasive, we have more cutting edges, hence more finishing.

The effect of weight of abrasives (gm) on PISF (%) with speed 900 rpm, current 4 A, and percentage of sand = 15 % being constant is shown in Fig. 5. It has been observed that the PISF increases with the increase in the weight of magnetic abrasives.

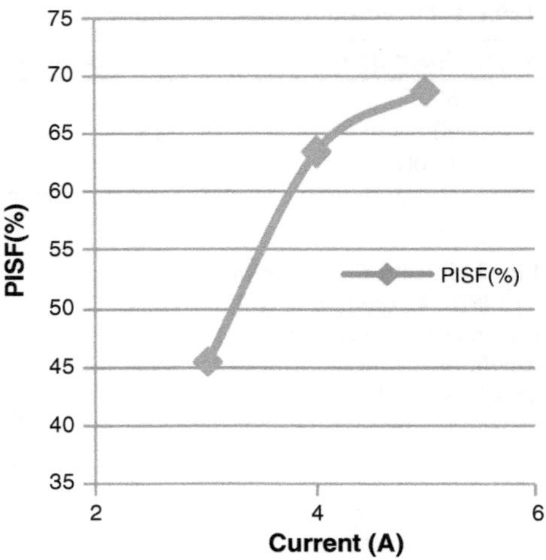

Fig. 3 Effect of current (A) on PISF (%) (speed = 900 rpm, percentage of sand = 15 %, weight of abrasives = 10 gm)

Table 3 The maximum value of PISF

S. no.	Speed (rpm)	Current (A)	Percentage of sand (%)	Weight of abrasives (gm)	PISF (%)
1	900	4	10	10	60.23
2	900	4	15	5	34.08
3	900	4	15	10	63.49
4	900	4	15	15	73.15
5	900	4	20	10	70.01

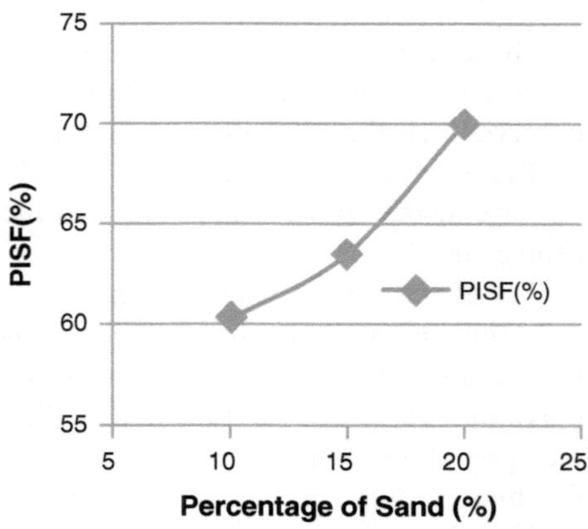

Fig. 4 Effect of percentage of abrasives (%) on PISF (%) (speed = 900 rpm, current = 4 A, weight of abrasives = 10 gm)

The maximum value of PISF obtained is 73.15 % at 15 gm weight of magnetic abrasives. With the increase in weight, we have more cutting edges and hence the surface finish improves.

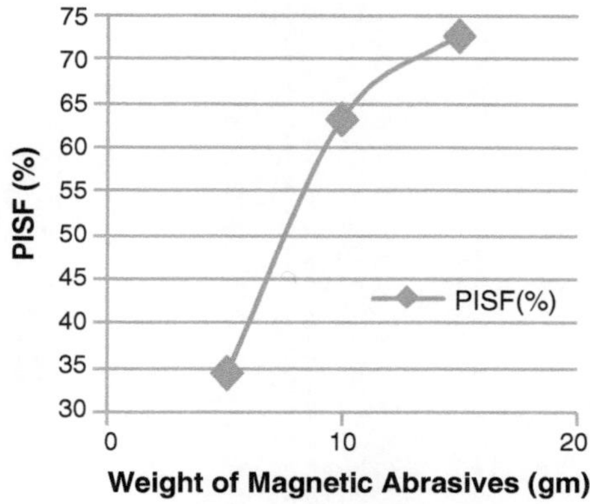

Fig. 5 Effect of weight of abrasives (%) on PISF (%) (speed = 900 rpm, current = 4 A, percentage of sand = 15 %)

4 Conclusion

1. It is concluded from the results that current, rotational speed of workpiece, percentage and weight of the workpiece are affecting the percentage improvement in surface finish.
2. It has been found that the process yields best result of PISF = 73.15 % at rotational speed = 900 rpm, current = 4 A, percentage of sand = 15 %, and weight of the abrasives = 15 gm.
3. The abrasives used are capable of producing a significant improvement in surface finish of the workpiece.

Compression Test Analysis of Cu Super-Hybrid Composite Material

Ahmer Jamal, Rajesh Jain and S. K. Sharma

Abstract A super-hybrid composite material of natural reinforcement and synthetic reinforcement with metal matrix has been developed. Its mechanical behavior has been studied through compression testing. Effect of addition of synthetic reinforcement has been also observed. Addition of natural reinforcement makes this composite as eco-material. Epoxy resin layers are applied in between metal matrix and reinforcements. Copper is used as metal matrix, bamboo strips as natural reinforcement, and glass fabric as synthetic reinforcement.

Keywords Bamboo strips · Compression ratio · Copper sheet · Glass fabric

1 Introduction

The importance of the composite material nowadays is increasing. Along the importance, eco-issues are also a matter of concern. Researchers are trying to make composites materials which must be eco-friendly and are easily available in nature. In this respect, bamboo is having great attention by the researchers. Some work has been done on the bamboo in past recent years.

Nowadays, there is a problem of environmental problems, such as a global climate change and acid rain, due to air, water, and soil pollution. These problems

---

A. Jamal (✉) · R. Jain · S. K. Sharma
Department of Mechanical Engineering, National Institute of Technology,
Kurukshetra 136119, Haryana, India
e-mail: jamal.ahmer@gmail.com

R. Jain
e-mail: jain.rajesh635@gmail.com

S. K. Sharma
e-mail: sksharma49nitk@yahoo.com

S. S. Khangura et al. (eds.), *Proceedings of the International Conference on Research and Innovations in Mechanical Engineering*, Lecture Notes in Mechanical Engineering,
DOI: 10.1007/978-81-322-1859-3_28,

are due to large materials and energy consumption. Although man-made materials have a great impact on the society by bringing advantages and conveniences to human life, they also impose a wide variety of burdens on the environment. Based on such facts, the concept of environmentally conscious materials (eco-materials) is being rapidly accepted by the researchers and countries all over the world. The most important feature of composite materials is that they can be designed and manufactured to meet different requirements. In the past period of composite material development, only mechanical and functional performances were taken into account in the design and processing, but now, concerns also are how material sand related technologies can become less hazardous to the environment (Li et al. 1998).

In this project work, use of wood is replaced by the use of bamboo which can be grown at a faster rate compared to other plant.

Apart from being one of the fastest growing plants, so that harvest time can be short, bamboo has such attractive features as high specific strength and modulus, low density, and, as a natural material, its degradability. To make full use of bamboo, a new technique was developed. According to this technique, bamboo strips are reformed by the help of three steps, i.e., softening, compression, and fixture. For reforming, the bamboo culm was separated into several parts longitudinally, softening is done by steam bath, and was then compressed after softening. According to the different uses, the mechanical properties and compressive ratio (which is defined as $r = (H_o - H_f)/H_o$, where H_o is the original thickness and H_f the thickness after compression) of reformed bamboo (RB) can be designed beforehand and adjusted in the manufacturing process (Li et al. 1994a, b).

In this project work, mechanical testing is performed on the laminates that are prepared by the combination of the copper sheet as metal matrix and bamboo strips with glass fabric as reinforcements. The combination of these three materials makes a hybrid composite. First is the combination of copper and normal bamboo (NB) strips with glass fabric, and the other one is copper and RB with glass fabric. Some specimens with both types of bamboo strips are prepared without glass fabric.

Mechanical behavior and failure modes of aluminum/bamboo sandwich plates under quasi-static loading have been observed already (Sui et al. 2000).

By tensile strength analysis of bamboo and layered laminate bamboo composites, it has been observed that tensile strength and Young's modulus of bamboo increase from inner to outer region across any cross section and from bottom to top of bamboo culms due to increase in volume fraction of fibers. Nodes are the weakest portion of the culm when it comes to tensile loads. Longitudinal cracking is responsible for failure on single lamina (Verma et al. 2012).

A great improvement is found in RB specimens during mechanical testing. Presence or absence of the glass fabric also shows a great influence on the mechanical properties. Orientation of the bamboo strips is also altered during the manufacturing of the composite laminates, and effect of the orientation is also observed.

2 Experimental Details

2.1 Material

2.1.1 Copper Sheet

The copper sheet, which is used in the experiment, is having a thickness of 0.32 mm (28 Gauge). The measured ultimate tensile strength of the used copper sheet is 643.78 MPa. These Cu sheets are marked according to the specimen required as shown in figure.

2.1.2 Bamboo Strips

The bamboo strips are cut by using a flexible hacksaw machine with a range of thickness of 0.3–0.45 mm.

Some mechanical properties (approximately) of this species are as follows:

Tensile strength	123 MPa
Compressive strength	88 MPa
Bending strength	130 MPa

The volume fraction of fibers is dense in the outer region (60–65 %), sparse (15–20 %) in the inner region, and increases linearly with height by about 20–40 % (Verma et al. 2012; Verma and Chariar 2012).

2.1.3 Glass Fabric

The glass fabric sheet, purchased from market, has a tensile strength about 85–90 MPa which is prescribed by the manufacturer.

2.1.4 Adhesive

Araldite consists epoxy resin (grade AW 106) and hardener (grade HV 953 IN). The main properties of this adhesive are great strength, toughness, resilience, excellent resistance to chemical attack, great insulation, nonvolatile, negligible shrinkage, etc.

Tensile strength	30–35 MPa
Young's modulus	3–10 MPa
Density	1.3 gm/cm^3
Curing time	20–24 h (Verma et al. 2012)

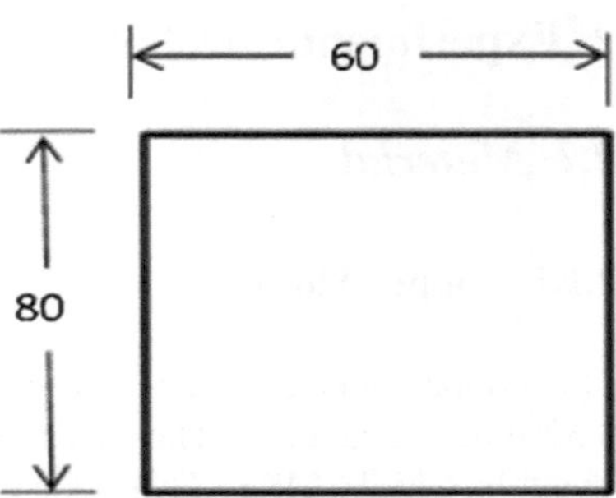

Fig. 1 Dimensions of specimen for compression test

Fig. 2 Finally prepared specimen for compression test

2.1.5 Reformed Bamboo

First of all, the bamboo strips are sliced by a flexible hacksaw (0.5 HP) machine into the above-said thickness. There are three steps for reforming the bamboo strips.

1. *Softening*

Generally, this is done by heating the strips in steam at 120 °C. Before heating the strips, these strips are immersed into water of 80 °C for 24 h.

2. *Compression*

Compression is done under the load of 2,000 tons. Loading and steaming are done simultaneously.

3. *Fixture*

Compression is done until a particular compressive ratio is obtained. Compression ratio achieved is of the range of 2.4–2.9 mm. In this project for 3 h strips are pressed.

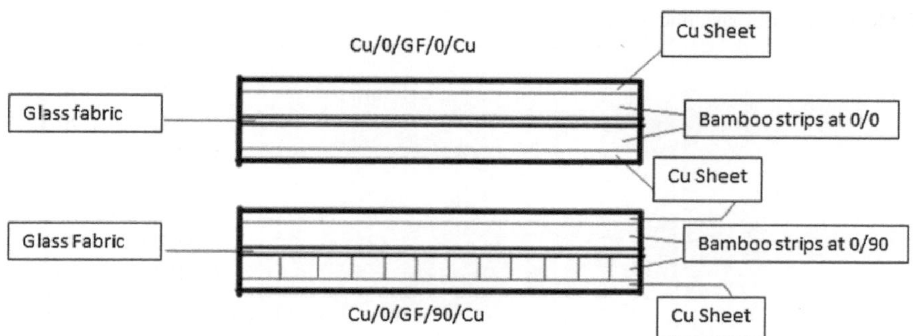

Fig. 3 Assembly Cu sheets, bamboo strips, and glass fabric with different orientation

Fig. 4 Bamboo orientation (0°/0°) (L) and (0°/90°) (R)

2.2 Specimen Fabrication

Hand layup technique is used to make the laminates of composite. After cutting the Cu sheets according to the Fig. 1, glass fabric is also cut as the same shape of Cu sheets. After this, Cu sheets are roughened by the sandpaper (grade 180) and the surface of the Cu sheet is cleaned by petrol or thinner to remove the dirt, oil, or sticky elements. Now, the mixture of epoxy and hardener is prepared by taking epoxy and hardener with the ratio of 2:1. Bamboo strips and glass fabric are also roughened and cleaned. Now, the mixture of the epoxy and hardener is applied in the form of thin layer in between bamboo strips and glass fabric. Now, this composite is placed between two Cu sheets by applying the epoxy resin in the form of layers to make hybrid composite, i.e., copper/bamboo/glass fabric hybrid composite material laminates. Image of the final specimen is given (Fig. 2).

For both types of the bamboo strips specimens are prepared, i.e., for NB strips and RB strips. The orientations, which are used in the present work, are specified by the orientation of the bamboo strips. First orientation is Cu/0°/GF/0°/Cu, i.e., first Cu sheet, then bamboo strips longitudinally, then glass fabric, then again bamboo strips longitudinally, and then again Cu sheet. The second orientation is Cu/0°/GF/90°/Cu, i.e., Cu sheet, then bamboo strips longitudinally, then glass fabric, then again bamboo strips at 90° to the former layer of bamboo strips, and then again Cu sheet, i.e., cross ply. The same orientations are applied for the specimens those are without GF, i.e., Cu/0°/0°/Cu and Cu/0°/90°/Cu. These orientations are used for both types of bamboo strips, i.e., NB and RB strips (Figs. 3, 4).

Table 1 Types of specimens

S. no.	Specimen
1	Cu/NB/GF/NB/(0°/0°)
2	Cu/NB/GF/NB/(0°/90°)
3	Cu/NB/NB/(0°/0°)
4	Cu/NB/NB/(0°/90°)
5	Cu/RB/GF/RB/(0°/0°)
6	Cu/RB/GF/RB/(0°/90°)
7	Cu/RB/RB/(0°/0°)
8	Cu/RB/RB/(0°/90°)

Cu copper; *NB* normal bamboo; *RB* reformed bamboo; *GF* glass fabric; (0°/0°) longitudinal orientation; and (0°/90°) transverse orientation

So the total types of specimens which are produced are as follows in Table 1: For each type of specimen, two samples are prepared.

2.3 Experimental Data

For the compression testing of these specimens, universal testing machine (UTM), having a capacity of 100 tons, is used. Compressive testing is carried out at a pace rate 0.1 kN/sec. There are 2 sets of specimens of each kind.

Data, obtained from UTM, are tabulated below in terms of load (Ton) (Tables 2, 3).

3 Result Analysis and Discussion

The compressive strength can be calculated by

$$\sigma_c = \frac{P_c}{A}$$

where

P_c *Compressive Load* (Newton)
σ_c *Compressive strength* (MPa)
A *Cross-Sectional Area* (mm^2)

Calculated compressive strength is tabulated (Table 4, Fig. 5).

The maximum compressive strength can be observed of RB/GF/(0°/90°). It is due to mainly from the reason that the bamboo strips in case of RB/GF/(0°/90°), which are at 90° to the other layer of bamboo strips, are along to the loading. From the above bar chart, we can observe the compressive strength is highest when the bamboo strips are parallel to the loading and it bears much load before the fracture.

Table 2 For SET 1 of specimens

S. no.	Specimen	Area (mm^2)	Compressive load (Ton)
1	Cu/NB/GF/NB/(0°/0°)	$80 \times 8.1 = 648$	0.96
2	Cu/NB/GF/NB/(0°/90°)	$80 \times 7.3 = 584$	1.01
3	Cu/NB/NB/(0°/0°)	$81 \times 6.8 = 550.8$	0.87
4	Cu/NB/NB/(0°/90°)	$81 \times 7 = 567$	0.93
5	Cu/RB/GF/RB/(0°/0°)	$80 \times 5 = 400$	0.8
6	Cu/RB/GF/RB/(0°/90°)	$82 \times 5.3 = 434.6$	1.1
7	Cu/RB/RB/(0°/0°)	$81 \times 4.8 = 388.8$	0.65
8	Cu/RB/RB/(0°/90°)	$82 \times 4.8 = 393.6$	0.9

Table 3 For SET 2 of specimens

S. no.	Specimen	Area (mm^2)	Compressive load (Tons)
1	Cu/NB/GF/NB/(0°/0°)	$81 \times 7.9 = 639.9$	1.01
2	Cu/NB/GF/NB/(0°/90°)	$81 \times 7.7 = 623.7$	1.24
3	Cu/NB/NB/(0°/0°)	$81 \times 7.2 = 583.2$	0.86
4	Cu/NB/NB/(0°/90°)	$81 \times 7.6 = 615.6$	1.1
5	Cu/RB/GF/RB/(0°/0°)	$81 \times 5.2 = 421.2$	0.92
6	Cu/RB/GF/RB/(0°/90°)	$81 \times 5.3 = 429.3$	1.04
7	Cu/RB/RB/(0°/0°)	$80 \times 4.8 = 392$	0.78
8	Cu/RB/RB/(0°/90°)	$80 \times 5 = 400$	0.87

Table 4 Compressive strength for SET 1 of specimens

S. no.	Specimen	Compressive load (Ton)	Compressive strength (MPa)
1	Cu/NB/GF/NB/(0°/0°)	0.96	14.53
2	Cu/NB/GF/NB/(0°/90°)	1.01	16.97
3	Cu/NB/NB/(0°/0°)	0.87	15.5
4	Cu/NB/NB/(0°/90°)	0.93	16.09
5	Cu/RB/GF/RB/(0°/0°)	0.8	19.62
6	Cu/RB/GF/RB/(0°/90°)	1.1	24.83
7	Cu/RB/RB/(0°/0°)	0.65	16.4
8	Cu/RB/RB/(0°/90°)	0.9	22.43

The other reason for this high strength is reformed bamboo. RB is showing much compressive strength due to increase in fiber fraction volume and hence density. RB/NGF/(0°/90°) is also showing higher strength for the same reasons. But this strength is less than from the former case because of the absence of the glass fabric. The strength of the NB/GF/(0°/0°) specimen is found lowest. The reason for low strength is orientation of the bamboo strips that are at 90° to the loading direction. Strips having (0°/0°) orientation is failed due to cracks in joints along the edges of the strips of butt joints and then de-bonding of Cu sheet, whereas failure in (0°/90°) is mainly due to de-bonding of the Cu sheets and then crack in bamboo

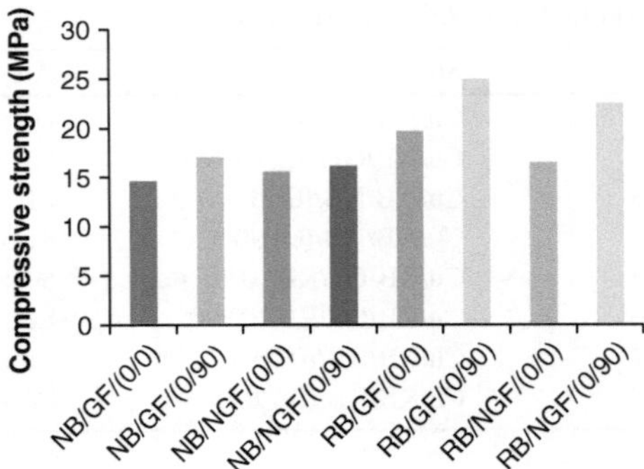

Fig. 5 Comparison chart of compressive strength for SET 1

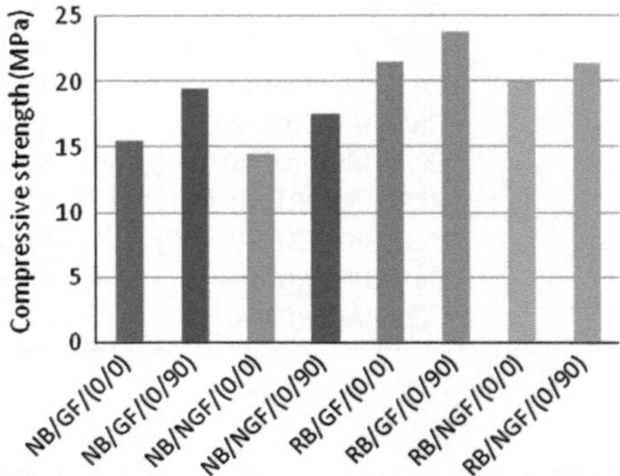

Fig. 6 Comparison chart of compressive strength for SET 2

Table 5 Compressive strength for SET 2 of specimens

S. no.	Specimens	Compressive load (Tons)	Compressive strength (MPa)
1	Cu/NB/GF/(0°/0°)	1.01	15.48
2	Cu/NB/GF/(0°/90°)	1.24	19.5
3	Cu/NB/NGF/(0°/0°)	0.86	14.47
4	Cu/NB/NGF/(0°/90°)	1.1	17.53
5	Cu/RB/GF/(0°/0°)	0.92	21.43
6	Cu/RB/GF/(0°/90°)	1.04	23.77
7	Cu/RB/NGF/(0°/0°)	0.78	19.52
8	Cu/RB/NGF/(0°/90°)	0.87	21.34

strips. At 90°, bamboo strips take all the load and show higher load-bearing capacity hence higher strength. Here, we can see easily that the presence of the glass fiber improving the compressive strength in all the cases of RB, whereas in case of NB/(0°/0°), it is decreasing. This decrement may be because of the uneven thickness of NB strips and misalignment of Cu sheets. But again in case of NB/(0°/90°), presence of glass fiber is increasing the strength (Fig. 6, Table 5).

Table 6 Modes of failure during compression test

Specimen	Mode of failure
Cu/NB/GF/(0°/0°)	De-bonding of Cu sheets and then crack at bamboo strips joints
Cu/NB/GF/(0°/90°)	Crack in transverse bamboo strips and disjoint of longitudinal bamboo strips and then de-bonding of Cu sheet
Cu/NB/NGF/(0°/0°)	De-bonding of the Cu sheets and crack at bamboo strips joints
Cu/NB/NGF/(0°/90°)	Crack in transverse bamboo strips, disjoint of the longitudinal bamboo strips at the butt joint with adhesive and de-bonding of the Cu sheets
Cu/RB/GF/(0°/0°)	De-bonding of Cu sheets and then crack at bamboo strips joints
Cu/RB/GF/(0°/90°)	Crack in transverse bamboo strips and disjoint of longitudinal bamboo strips and then de-bonding of Cu sheet
Cu/RB/NGF/(0°/0°)	De-bonding of the Cu sheets and crack at bamboo strips joints
Cu/RB/NGF/(0°/90°)	Crack in transverse bamboo strips, disjoint of the longitudinal bamboo strips at the butt joint with adhesive and de-bonding of the Cu sheets

The behavior of all these specimens is same as the previous set of specimens. The maximum compressive strength can be observed again of RB/GF/(0°/90°). It is due to mainly from the reason that the bamboo strips in case of RB/GF/(0°/90°), which are at 90° to the other layer of bamboo strips, are in longitudinal position to the loading. From the above bar chart, we can observe the compressive strength is highest when the bamboo strips are parallel to the loading and it bears much load before the fracture. The other reason for this high strength is reformed bamboo. RB is showing much compressive strength due to increase in fiber fraction volume and hence density. RB/GF/(0°/0°) is also showing second highest strength compared to RB/NGF/(0°/90°). But this strength is less than that from the former case because of the orientation of the bamboo strips. The strength of the NB/NGF/(0°/0°) specimen is found lowest. The reason for low strength is orientation of the bamboo strips which are at 90° to the loading direction and absence of the glass fabric. Strips having (0°/0°) orientation is failed due to cracks in joints along the edges of the strips of butt joints and then de-bonding of Cu sheet, whereas failure in (0°/90°) is mainly due to de-bonding of the Cu sheets and then crack in bamboo strips. At 90°, bamboo strips bear much the load and show higher load-bearing capacity hence higher strength. Here, we can see easily that the presence of the glass fabric improves the compressive strength in all the cases.

Modes of failure which are observed are listed (Table 6).

4 Conclusion

RB/GF/(0°/90°) is having maximum strength compared to any other tested in this project work.

Between the NB/GF/(0°/0°) and NB/GF/(0°/90°), compressive strength is more for NB/GF/(0°/90°).

Between the RB/GF/(0°/0°) and RB/GF/(0°/90°), compressive strength is more for RB/GF/(0°/90°).

In case of (0°/90°) for both NB and RB with glass fiber, the compressive strength of (0°/90°) orientation is more compared to (0°/0°) and same result is obtained in both the cases without glass fiber.

Presence of glass fiber is decreasing the compressive strength of NB/(0°/0°), NB/(0°/90°), and RB/(0°/0°) but not in case of RB/(0°/0°).

The maximum value for the minimum strength is shown by RB/NGF/(0°/90°) compared to any other specimen in this project.

The lowest minimum strength is shown by NB/GF/(0°/0°).

The specimen having (0°/90°) orientation has higher strength compared to (0°/0°) orientation of respective specimen except NB/NGF specimen. But this difference is very minor.

Here, presence of glass fiber is decreasing the minimum compressive strength.

References

Li SH, Fu SY, Zhou BL, Zeng QY, Bao XR (1994a) Reformed bamboo and reformed bamboo/aluminium composite, part I manufacturing technique, structure and static properties. J Mater Sci 29:5990–5996

Li S, Fu S, Zhou B, Zeng Q, Bao X (1994b) Enhanced bamboo reinforced aluminium laminates. J Mater Sci Technol 10:335–358

Li SH, De Wijn JR, De Groot K, Zeng QY, Zhou BL (1998) Reformed bamboo/glass fabric/aluminium composite as an ecomaterial. J Mater Sci 33:2147–2152

Sui GX, Yu TX, Kim JK, Zhou BL (2000) Mechanical behavior and failure modes of aluminum/bamboo sandwich plates under quasi-static loading. J Mater Sci 35:1445–1452

Verma CS, Chariar VM (2012) Development of layered laminate bamboo composite and their mechanical properties, composites: part B. Compos B 43:1063–1069

Verma CS, Chariar VM, Purohit R (2012) Tensile strength analysis of bamboo and layered laminate bamboo composites. Int J Eng Res Appl (IJERA) 2(2):1253–1264, ISSN: 2248–9622 www.ijera.com Mar–Apr 2012

Fabrication and Characterization of Jute Fiber-Reinforced–Recycled Polyethylene Composites

Evaran Singh, Dharmpal Deepak, Lakshya Aggarwal and V. K. Gupta

Abstract Due to low cost and high specific mechanical properties, natural fiber represents a good renewable and biodegradable alternative to the most common synthetic reinforcement such as glass fiber. The study involves fabrication of jute fiber-reinforced high-density polyethylene (HDPE) (50 % virgin + 50 % recycled) composites by injection molding technique. The tensile and flexural properties of the composite specimens with 10, 20, and 30 % jute fibers and compared with similar specimens made of 100 % virgin HDPE and 50–50 mixture of virgin and recycled HDPE. It is observed that the reinforcement of jute fibers into HDPE matrix reduces the tensile strength. The tensile strength of the composite decreases from 3.4 to 24.3 % with the increase in fiber content from 10 to 30 % when compared with specimen made of equal proportion of virgin and recycled HDPE. On the other hand, flexural strength increases with the increase in jute fiber content in the composite. The flexural strength of the composite containing 10 and 30 % jute fibers are higher by about 4 and 8.6 %, respectively, when compared with specimen containing 50–50 mixture of virgin and recycled HDPE.

Keywords HDPE · Injection molding · Jute fiber · Recycled HDPE

E. Singh (✉) · D. Deepak · V. K. Gupta
Mechanical Engineering Department, UCoE, Punjabi University, Patiala 147002, Punjab, India
e-mail: evaransingh@gmail.com

D. Deepak
e-mail: deepakbass@yahoo.com

V. K. Gupta
e-mail: guptavk_70@yahoo.co.in

L. Aggarwal
Mahrishi Markandeshwar University, V. P. O. Sadopur, Chandigarh Road, Ambala 133003, Haryana, India
e-mail: lakskiet@gmail.com

S. S. Khangura et al. (eds.), *Proceedings of the International Conference on Research and Innovations in Mechanical Engineering*, Lecture Notes in Mechanical Engineering, DOI: 10.1007/978-81-322-1859-3_29, © Springer India 2014

1 Introduction

Over the past few decades, there is a rapid increase in the demand of fiber-reinforced polymer (FRP) composites due to unique combination of high-performance, great versatility, and processing advantages (Mishra and Biswas 2013). The use of natural fibers, derived from annually renewable resources, as reinforcing fibers in both thermoplastic and thermoset matrix composites provide positive environmental benefits with respect to ultimate disposability and raw material utilization (Rashed et al. 2006). Natural FRP composites combine good mechanical properties with low specific mass and therefore extensively employed in structural materials for housing, railways, and aerospace applications (Nur et al. 2010). The use of natural fiber like jute not only helps in maintaining ecological balance but can also provide employment to the rural populations in countries like India and Bangladesh where jute is abundantly available. Jute fiber can be planted in river flats, depressions, and saline–alkali soil, which are not suitable for planting cotton and most food crops. Further, no pesticides and fertilizers are needed during the growth of jute, thus making jute a kind of "pure green" agro product (Liu et al. 2009). Natural fibers compared to inorganic fibers present some disadvantages like poor hydrophobic polymer matrix compatibility, tendency to form aggregates during processing and low resistance to moisture, which provide a potential reduction as reinforcement for polymers (Mulinari et al. 2009). Several treatments have been used by different researchers to improve fibers/matrix compatibility (Li et al. 2007).

In the present study, natural fiber composites, containing 10, 20, and 30 % jute fibers as reinforcement and mixture of virgin and recycled high-density polyethylene (HDPE) as matrix, have been fabricated. The tensile and flexural properties of the composites are evaluated and compared with those of 100 % virgin HDPE and 50–50 mixture of virgin and recycled HDPE.

2 Experimental Details

2.1 Material and Method

In this study, we have used jute fibers as reinforcement and HDPE as matrix. Both virgin and recycled HDPE were employed for the preparation of composites. Jute fibers were procured from Chandra Prakash & Company from Jaipur (Rajasthan), and HDPE (virgin and recycled) of molding grade was purchased from Goyal Polymers, Chandigarh. The jute fiber/HDPE composite was fabricated by injection molding technique.

Table 1 Description of tensile and flexural specimens

Sample no.	Sample composition (wt.%)			Sample density (g/cc), $[\rho = (\rho_f V_f + \rho_m V_m)/100]$
	Jute fibers (V_f)	HDPE (V_m)		
		Recycled HDPE	Virgin HDPE	
1	0	0	100	0.94
2	0	50	50	0.94
3	10	45	45	0.98
4	20	40	40	1.01
5	30	35	35	1.05

2.2 Chemical Treatment of Jute Fibers

Prior to fabrication of composite, the jute fibers were subjected to chemical treatment with NaOH to improve its adherence with HDPE matrix. Firstly, the jute fibers were manually cut into sizes between 5 and 10 mm followed by washing with distilled water at 80 °C for 1 h and subsequently drying in hot air oven at 100 °C for 5 h (Favaro et al. 2010). Thereafter, mercerization of jute fibers was carried out by soaking them in NaOH solution (5 wt.%) at room temperature for 4 h (Ray et al. 2001). The fibers were washed several times with fresh water to remove traces of NaOH sticking on the surface of jute fibers. Final washing was carried out with distilled water till a pH of 7 was attained. The fibers were then dried at room temperature for 48 h and followed by oven drying at 100 °C for 6 h. A total of 200 gm of jute fibers with initial pH of 10.0 was subjected to chemical treatment. However, due to removal of moisture and fiber loss due to washing, the weight of treated jute fibers decreased to around 150 gm.

2.3 Fabrication of Tensile and Flexural Specimens

In order to investigate the influence of amount of jute fibers on tensile and flexural properties of jute fiber–HDPE composite, the samples for tensile and flexural testing were prepared by varying content of jute fibers as 10, 20, and 30 % (by wt.) in a matrix of HDPE (50 % virgin + 50 % recycled). For the purpose of comparison, the samples, both tensile and flexural, were also prepared by taking 100 % virgin HDPE and 50–50 mixture of virgin and recycled HDPE. The description of samples employed in the study is given in Table 1.

Where $\rho_f = 1.3$ g/cc and $\rho_m = 0.94$ g/cc are the density of jute fibers (Chong et al. 2010) and HDPE (Li et al. 2007), respectively, and V_f and V_m are wt.% of jute fiber and HDPE matrix, respectively.

The tensile and flexural specimens were fabricated by putting mixture of jute fibers and HDPE corresponding to different samples, as described in Table 1, into the injection molding machine (Model-BH100, Make-J.B. Industries Pvt. Ltd.). Five similar specimens, both tensile and flexural, were fabricated corresponding to each composition. The geometry and dimensions of the tensile and flexural specimens were in accordance with ASTM D638 and ASTM D790 standards, respectively.

2.4 Testing of Samples

The tensile and flexural tests were conducted according to ASTM D638 and ASTM D790 standards, respectively, on UTM (Model-SS UTM 1205, Capacity-250KN, Make-P.S.I. Sales Pvt. Ltd.) available at CIPET, Amritsar. Prior to testing, the specimens were conditioned by keeping them in an environment test chamber (Make-KASCO Industries, Pune) for 40 hours at 23 °C and 50 % RH. During the tensile test, dumbbell-shaped specimens were subjected to uniaxial tensile load by gripping their ends in UTM. The gauge length of the specimen was kept 57 mm. The results obtained by testing five similar samples corresponding to each composition were averaged to estimate the tensile strength of the samples.

In order to obtain the flexural strength, short-beam shear (SBS) tests, i.e., 3-point bend test, were performed on the specimens at room temperature. The SBS test was also conducted on the same UTM (maximum loading capacity-50 kN) over a span of 107 mm. The flexural strength (F. S.) of the sample is estimated from the following equation given in ASTM D790:

$$\text{F.S.} = \frac{3PL}{2bt^2}$$

where L is the span (=107 mm), P is the load applied, and b and t are respectively the width and thickness of the specimen.

3 Results and Discussion

3.1 Surface Topography of Jute Fibers

In order to reveal the morphological changes that occurred during the course of chemical treatment of jute fibers, the untreated and treated jute fibers were observed under the optical microscope. The presence of wax, oil, and impurities are clearly revealed on the surface of untreated jute fibers, Fig. 1. The longitudinal view of NaOH-treated jute fibers reveals a cleaned surface. Though, it appears to be quite smooth, but in actual, it is roughened by the NaOH treatment. Further, the surface of treated fiber is free from impurities.

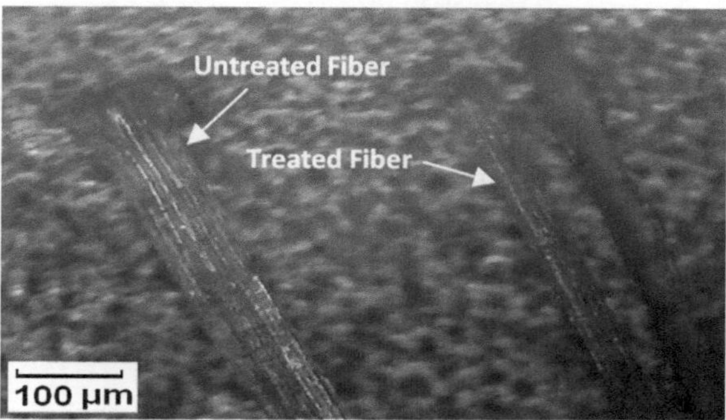

Fig. 1 Optical micrograph of untreated and NaOH (5 %)-treated jute fibers

3.2 Mechanical Properties

The results obtained by conducting tensile and flexural tests on various specimens are summarized in Table 2. Figure 2 shows the tensile strength of different specimens taken in this study. The tensile strength of specimen-B consisting of virgin and recycled HDPE (50 % each) is lower by 5.8 % than the specimen-A consisting of 100 % virgin HDPE. Therefore, the addition of recycled HDPE into fresh HDPE impairs the tensile properties. The tensile strength of the composite decreases with the increase in fiber content from 10 to 30 % when compared to specimens made of HDPE alone (specimens A and B). As compared to specimen-B, the tensile strength of the composite specimen decreases from 3.4 to 24.3 % when the fiber content increases from 10 to 30 %. One of the reasons why the tensile strength of the composites are lower than the specimen made of HDPE alone could be due to poor adhesion between the matrix and the fibers as a result of which the fibers do not share the load. Further, due to the presence of gating system in the center of the mold, the fibers get randomly aligned in the specimen apart from bending and curling and are mainly distributed in the middle of the specimen. As a result, the reinforcement could not play a positive role in improving the tensile strength of the composite.

Unlike tensile strength (Fig. 1), the flexural strength of specimen-B, containing equal proportion of fresh and recycled HDPE, is about 89.8 % higher than the specimen-A having 100 % fresh HDPE (Fig. 2). In general, on reinforcing jute fibers into polymer matrix, the flexural strength increases, except for a slightly lower flexural strength observed for specimen-D with 20 % jute fibers. As compared to specimen-B, the flexural strength of composite containing 10 and 30 % jute fibers are respectively higher by around 4 and 8.6 %, respectively. Aziz and Ansell (2004) also noticed that the incorporation of hemp and kenaf fibers in polymer matrix leads to increase the flexural strength of the composites (Fig. 3).

Table 2 Tensile and flexural strength of specimens

Notation of specimen	Composition	Tensile strength (MPa)	Flexural strength (MPa)
A	100 % fresh	17.39	12.27
B	50 % fresh + 50 % recycled HDPE	16.39	22.56
C	10 % jute + 90 % HDPE (fresh + recycled)	15.84	23.45
D	20 % jute + 80 % HDPE (fresh + recycled)	15.34	23.04
E	30 % jute + 70 % HDPE (fresh + recycled)	12.41	24.51

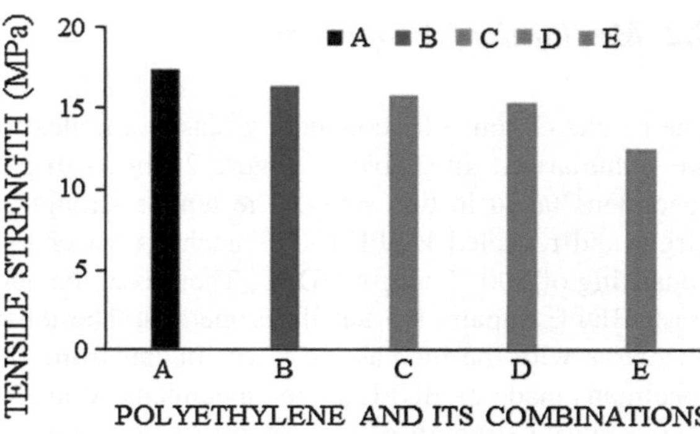

Fig. 2 Tensile strength of various specimens

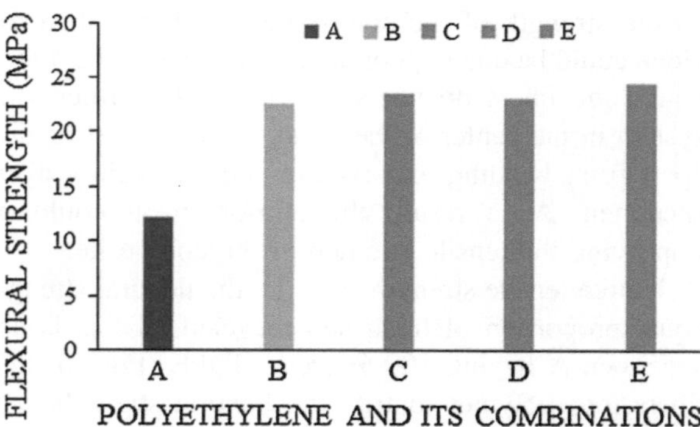

Fig. 3 Flexural strength of various PE specimens

3.3 SEM Analysis of Fractured Surface

SEM characterization of the tensile fractured surface of the jute/HDPE composites has been performed to investigate the interfacial adhesion between the jute fibers and HDPE matrix. The SEM images reveal the fiber pullout phenomenon in bundled form in the fractured composite specimens (Fig. 4a–b). The fiber pullout

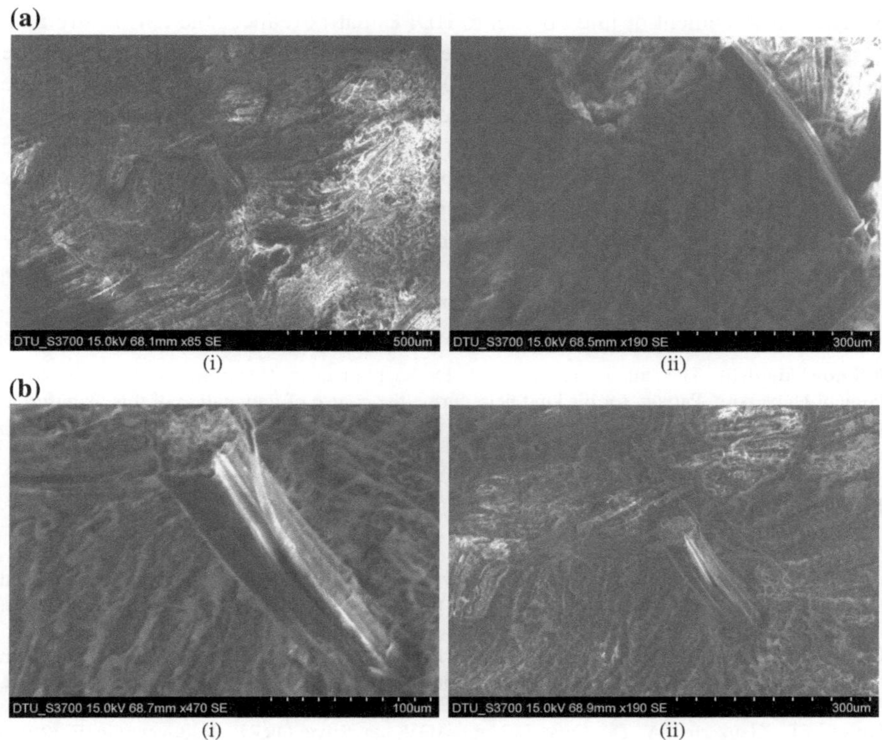

Fig. 4 Scanning electron micrographs of jute fiber/HDPE composite with **a** 10 % jute fibers and **b** 20 % jute fibers (scale is on micrograph)

phenomenon observed in the fractured specimen reveals poor adhesion between the jute fiber and the matrix that results into poor tensile strength of the composites. In addition to fiber pullout, the fiber delamination is also noticed in SEM image shown in Fig. 4b(i).

4 Conclusions

The present study has led to the following conclusions:

1. The study indicates that injection molding can be successfully used to manufacture ecofriendly jute fiber-reinforced HDPE composite containing both recycled and virgin HDPE.
2. The addition of recycled HDPE into virgin HDPE leads to slight reduction in tensile strength but significant increase in flexural strength.

3. The reinforcement of jute fibers into HDPE matrix reduces the tensile strength. The tensile strength of the composite decreases from 3.4 to 24.3 % with the increase in fiber content from 10 to 30 %, when compared with specimen made of 50–50 mixture of fresh and recycled HDPE.
4. The flexural strength increases with the addition of jute fibers into HDPE matrix. The flexural strength of composite containing 10 and 30 % jute fibers are higher by around 4 and 8.6 %, respectively, when compared with specimen made of 50–50 mixture of fresh and recycled HDPE.
5. The primary mechanism responsible for tensile fracture of the composite is fiber pullout with some amount of fiber delamination.

Acknowledgment The authors are thankful to Dr. Sachin Kumar, Department of Biotechnology, Punjabi University, Patiala, for his kind help during the course of preparation of this manuscript.

References

Aziz SH, Ansell MP (2004) The effect of alkalization and fiber alignment on the mechanical and thermal properties of kenaf and hemp bast fiber composites: part 1—polyster resin matrix. Compos Sci Technol 64(9):1219–1230

Chong EL, Ahmad I, Dahlan HM, Abdullah I (2010) Reinforcement of natural rubber/high density polyethylene blends with electron beam irradiated liquid natural rubber-coated rice husk. Radiat Phys Chem 79(8):906–911

Favaro SL, Ganzerli TA, De CarvalhoNeto AGV, da Silva ORRF, Radovanovic E (2010) Chemical, morphological and mechanical analysis of sisal fiber-reinforced recycled high-density polyethylene composites. Express Polym Lett 4(8):465–473

Li X, Tabil LG, Panigrahi S (2007) Chemical treatments of natural fiber for use in natural fiber-reinforced composites: a review. J Polym Environ 15(1):25–33

Liu L, Yu J, Cheng L, Yang X (2009) Biodegradability of poly(butylene succinate) (PBS) composite reinforced with jute fiber. Polym Degrad Stab 94(1):90–94

Mishra V, Biswas S (2013) Physical and mechanical properties of Bi-directional jute fiber epoxy composites. Proc Eng 51:561–566

Mulinari DR, Voorwald HJC, Cioffi MOH, da silva MLCP, Luz SM (2009) Preparation and properties of HDPE/sugarcane bagasse cellulose composites obtained for thermokinetic mixer. Carbohydr Polym 75(2):317–321

Nur HP, Hossain MA, Sultana S, Mollah MM (2010) Preparation of polymer composites using natural fiber and their physico-mechanical properties. Bangladesh J Sci Ind Res 45(2):117–122

Rashed HMMA, Islam MA, Rizvi FB (2006) Effects of process parameters on tensile strength of jute fiber reinforced thermoplastic composites. J Naval Archit Mar Eng 3(1):275–281

Ray D, Sarkar BK, Rana AK, Bose NR (2001) Effect of alkali treated jute fibers on composite properties. Bull Mater Sci 24(2):129–135

Polymer Concrete Composites Made from Industrial Waste Materials: A Review

Gurpreet Singh and Harmesh Kansal

Abstract Polymer concrete composites have been accepted by various manufacturers and widely used as a structural material for machine tools due to their tremendous properties such as compressive stress, flexural strength and damping capacity. In the current scenario, it has been found that many industrial waste materials can be used as aggregates/fillers for polymer concrete (PC) composites. Also by adopting suitable recycling methods, polyester polymer can be obtained from industrial waste and can be suitably used as a binder for the fabrication of the composites.

Keywords Polymer concrete · Polyester polymer concrete · Industrial waste materials · Recycled materials · Fly ash · Polymer concrete bed

1 Introduction

In recent years, considerable progress has been achieved in construction industry and lot of research and development of high-performance and multifunctional constructional materials have been actively persuaded. On the other side as the constructional and production activities are increasing actively in worldwide scenario, these are leading to increase the production of waste materials very rapidly. So, it has been significantly important to develop the technology to treat or recycle the industrial waste. There are several ways to treat the wastes such as

G. Singh (✉)
Department of Mechanical Engineering, B. B. S. B. Engineering College, Fatehgarh Sahib, Punjab, India
e-mail: gurpreet.er@gmail.com

H. Kansal
Department of Mechanical Engineering, UIET, Punjab University, Chandigarh, India

S. S. Khangura et al. (eds.), *Proceedings of the International Conference on Research and Innovations in Mechanical Engineering*, Lecture Notes in Mechanical Engineering, DOI: 10.1007/978-81-322-1859-3_30, © Springer India 2014

landfill, incineration, chemical recycling, material recycling and the utilization of energy from combustion. Most methods excluding material recycling are known to have critical limitations in economic, technical and environmental manners. Material recycling is expected to be more feasible in a way that the simplicity of pretreatment and the reduction in energy consumption and environment pollution can be satisfied (Yang 1996). Polymer concrete (PC) materials, which are having unique applications in construction industry, can be fabricated for the utilization of industrial waste materials such as recycled polyethylene terephthalate (PET), fly ash, red mud and silica fume.

2 Historical Perspectives

The existence of polymer-modified concrete (PMC) and PC came in the 1950s; initially its use was very limited. It was only in the 1970s, after PIC was developed that concrete polymer materials received significant publicity. American Concrete Institute Committee 548, Polymers in Concrete, was formed, and in 1975 the first International Congress on Polymer in Concrete (ICPIC) was held in London (ACI Committee 548 1997). Later RILEM (Réunion Internationale des Laboratoires et Experts des Matériaux) committees were formed to address specific areas in concrete polymer composites. Regular ACI Symposia and ICPIC conferences have helped to make PC materials in the construction industry.

3 Processing of Polymer Concrete

Polymers in concrete have been first overseen by ACI (American Concrete Institute) Committee, since 1971. The committee has sponsored symposium or technical sessions at consecutive convention meetings from 1972 to 1996 on a variety of topics relating to the use of polymers in concrete. The contributions made at these meetings, along with the practical experience gained within the growing industry that applies polymer technology to concrete, form the base of applied concrete technology that is limited only by the ingenuity of the concrete user. "A Guide for the Use of Polymers in Concrete" was published by ACI in 1997, given the direction for producing PC structures worldwide (ACI Committee 548 1997). The guide presented the process for producing the polymer-impregnated concrete (PIC) has been suggested, and PIC is a hydrated Portland cement concrete that has been impregnated with a monomer that is subsequently polymerized in situ. The monomer is introduced into the concrete by soaking at atmospheric pressure or above. After impregnation, the concrete containing the desired amount of monomer then undergoes a treatment to convert the monomer into a polymer. This polymerization reaction causes the molecules of the monomer

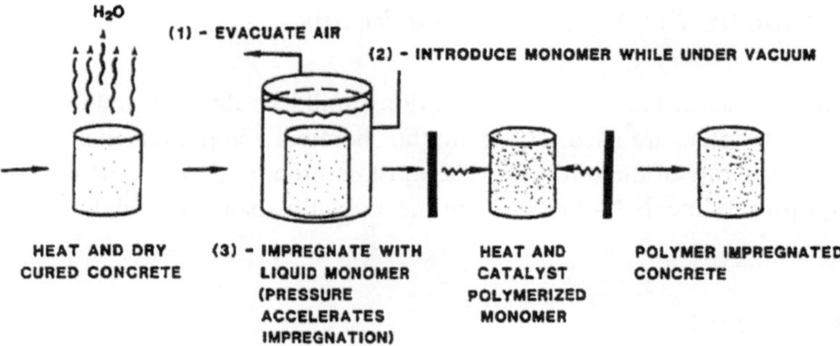

Fig. 1 Polymer concrete processing

to chemically link into a long repeated chain-like structure with higher molecular weight, known as a polymer (Fig. 1).

Polymerization may be done from the three methods available (a) thermal catalytic method, (b) radiation method and (c) promoted catalytic method. Amongst the three methods presented, the thermal catalytic method is more advantageous as compared with the others, as it can be preceded under normal room conditions and process rate is fast. Research scientists have used various components/constituents, whose different combinations have produced unique and different results during their testings. The various components of PC are as follows.

3.1 Monomers

The most common monomers used for producing PCs are methyl methacrylate (MMA), butyl acrylate (BA), styrene and trimethylolpropane trimethacrylate (TMPTMA); these may be used alone or in combined form.

3.2 Initiators for Monomers and Resins

These are also referred as the catalysts whose function is to initiate the growth of polymer chains by decomposing into free radicals. These actually starts the chain's growth, their rate of reaction depends on the concentration and temperature of the initiator. Various types of initiators are used e.g. dibenzoyl peroxide, lauryol and methyl ethyl ketone peroxide.

3.3 Promoters for Monomers and Resins

When it is required that the polymerization should be done at ambient temperatures, the promoters are used. These are the chemical compounds that increase the decomposition rate of the initiators. The most commonly used promoters are N, N-dimethyl-p-toluidine, N,N-dimethyl aniline, cobalt octoate and cobalt naphthenate.

3.4 Aggregates

The aggregates used for the PC's formulation should be of highest quality, dry and free from dust and other organic materials. Moisture may lead to loosen the bond strength. Commonly silica quarts, granite, limestone and other high-quality materials are used as the aggregates.

4 Polymer Resins

A lot of study has been done by the researchers on the various types of resins (a) epoxy, (b) polyester, (c) furan, (d) methyl methacrylate, etc. Vipulanandian and Paul studied the performance of epoxy and polyester PC under various curing conditions, temperatures and strain rates (Vipulanandian and Paul 1990). The influence of aggregate size and distribution on the mechanical properties of PC was investigated in the study. The results showed high durability of PC. O'Connor and Saiidi carried out a comparative study on the basic engineering properties of polyester–styrene PC overlays and Portland cement concrete bridge decks (O'Connor and Saiidi 1993). The results of their study, however, indicated that, in extreme cases with wide temperature ranges and stiff bridge girders, PC overlays may be susceptible to cracking. Solovjov et al. (1994) reported that furan resins were used widely in the Commonwealth of Independent States (some of the former Soviet Republics). They were commonly used in aggressive environments and in construction-related industries. Swamy et al. (1998) studied the performance of concrete slabs coated with acrylic-based coating against chloride and atmospheric carbon dioxide attacks. Their finding was that coated surfaces prevented chloride and carbonation penetration into concrete. They also pointed out that the coating maintained good adhesion with the substrate concrete even after long-term repetitive wetting and drying cycles. Folic and Radonjanin (1998) reported achieving the highest effects of polymer modification on such properties as tensile strength, ductility, bond between reinforcement and concrete, and on the properties contributing to the durability of reinforcements in structures. On the contrary, Silva and Silva pointed out that the effective use and reliability of polymeric mortars require that designers and builders have a better knowledge of the temporal evolution of their mechanical properties (Silva and Silva 2007) (Table 1).

Table 1 Characteristics and applications of PC's

Type of binder	General characteristics	Typical applications
Poly (butadiene styrene)	• High elasticity • Good adhesion • Water proofing • High chemical resistance	• Industrial floor toppings • Repair and levelling of concrete • Fixing bricks/tiles • Water proof mortar
Poly (methyl methacrylate)	•Low tendency to water absorption • High freeze-thaw resistance • High mechanical strength • High outdoor durability	• Used in the manufacture of stair units, facade plates and sanitary products for curbstones.
Polyester	• High mechanical strength	• Panels for public and commercial buildings
	• Good adhesion to other materials	• Floor tiles • Pipes • Stairs
	• Good chemical and freeze-thaw Resistance • High post-setting shrinkage	• Various precast and cast-in applications in construction works
Epoxy	• Strong adhesion to most building materials	• Industrial flooring • Skid-resistant • Overlays in highways
	• Low shrinkage • Superior chemical resistance	• Plastering of exterior walls • Resurfacing of deteriorated structures
	• Good creep resistance • Good fatigue resistance • Low water sorption	• Some machine bases

The study comprised of tests for accelerated aging to quantify property degradation. These authors noted that the degradation was high for epoxy mortar in the cases of temperature cycles and of salt fogging cycles, for both reduced ductility and bending strength. Cycles of dry–wet environment, at a fixed temperature, led to growing reductions in strength with time. The ACI committee published a guide in 1993 for PC overlays providing an overview of thin (less than 25 mm. thick) PC overlay for concrete and steel substrates (ACI Committee 548 1993). The guide emphasizes using PC in the transportation sector, specifically for bridge decks and parking garages. Surface preparation, application, evaluation, maintenance and safety aspects are included. In the same year, the ACI committee presented a similar guide for the selection and use of structural adhesive with concrete. This guide provides the engineer, contractor and architect with a description of the various types of polymer adhesives (epoxy, polyester, acrylic, polyurethane, polysulfide, silicone, vinyl acetate and styrene butadiene) most frequently used for the adhesive bonding of fresh concrete, bonding concrete to other materials, and the adhesive grouting of bolts and other inserts into concrete (ACI Committee 503 1993). It emphasizes the factors that should be considered when selecting a structural adhesive, including its characteristics during installation and in service. The benefits and limitations of adhesive bonding are discussed.

5 Testing of Polymer Concrete Composites

Polymer concrete composites were originally developed as a substitute for decorative stones. They become particularly popular in construction industry during the early 1950s (Miller 2005; Fowler 1999). Superior characteristic of these materials led to their increased utilization during 1980s when they were used to repair damaged Portland cement concrete structures (Aicin 2003; Asthana and Lakhani 2004). PC offers higher strength and ductility and faster curing than the ordinary Portland cement (OPC) concrete. Some of these properties include more rapid curing, higher mechanical strength, better adhesion to substrate, higher resistance to chemicals and lower water impermeability (Miller 2005; EL-Hawary and Abdel-Fattah 2000). PC's are extensively used as structural elements in the machine tools. Many research activities have been done on PC's using industrial waste materials as aggregates, tremendous improvement in the properties has been found using these materials such as fly ash, silica fume, waste glass and rice husk. It was experimentally proved by Mahdi that PC's can be made from the recycled PET with greater strength.

Different types, properties and applications of PCC's have been extensively reported (ACI Committee 548 1997, 1998). Application and performance of PC are dependent upon the specific polymer binder as well as the type of aggregate and its gradation (ACI Committee 548 1996). Abdel found that the distribution of aggregates should be such as to allow for a minimum void volume for dry-packed aggregates that will result in dense packing (Abdel-Jawad and Abdullah 2002). Rao (1993) confirmed that dense packing of aggregates in the PC matrix results in better properties. To achieve this, either the void content of the aggregate mix can be minimized, where the binder requirement for ensuring adequate bonding of all aggregate particles will be less or use loosely packed aggregate mix with higher binder content. Though the aggregate and micro-filler form a major component of the total mass of the PC, there has not been much emphasis on the aggregate and micro-filler mix proportion used in such systems. A review of literature by Fuller and Miller (1907) shows that the aggregates used in PCs are either fine particles or the particle size distribution chosen are based on theoretical basis (Miller 2005) suitable for Portland cement concretes. Since, its applications in structural engineering have increased tremendously because of its good workability, low-temperature curing and early high-strength development. With the expansion of its applications, its chemical resistance has been an important characteristic, shown by Ohama et al. (1986). Gorninski et al. (2007) worked on comparative assessment of isophthalic and orthophthalic polyester PC for different costs, similar mechanical properties and durability. It was seen that orthophthalic polyester PC, when prepared with sand and fly ash, behaves similar to isophthalic polyester polymer concrete, despite the properties of the isophthalic resin being superior to the orthophthalic variety.

PCs possess rapid curing, high mechanical (compressive, flexural and tensile) strength, low water permeability, excellent adhesion to substrate and long-term durability in harsh and corrosive conditions, the conclusions has been stated by EL-Hawary and Abdel-Fattah (2000). It was observed by Ribeiro et al. (2002) that the flexural strength of epoxy PC is slightly affected by the immersion in the solutions of sulphuric acid and chloride sodium, which is an indicator of the good chemical resistance of this kind of concrete to these aggressive agents.

Elalaoui confirmed that these composites also possess good thermal properties, when these composites are exposed to temperature less than 250 °C, epoxy PC are more efficient than cement-based concrete (Oussama and Elhem 2012). Shokrieh (2011) observed that the compressive strength of the optimally designed PC is not affected by heating and cooling cycles. Wang and Wong (2007) found that the elastic modulus remained almost unchanged until 300–400 °C. Moriyoshi (1996) worked on thermal properties of PC using glycerol methacrylate/styrene system at low temperature, in which they described the results of thermal fracture test and flexural test of PC using glycerol methacrylate/styrene system (GM/St) as binder. The flexural strengths of those were found constant regardless of temperatures. On the other hand, cement concrete was fractured by a smaller thermal change from the initial temperature to that of polymer mixtures.

Oyawa (2004) worked on flexural response of PC-filled steel beams. Polymer-based materials were noted to present a wide array of properties that could be tailored to meet specific design requirements, e.g. ductility-based design or strength-based design. The feasibility of the PC in the manufacturing of machine tool beds was investigated by Orak. It was observed that the critical damping ratio of PC was approximately four to seven times higher than that of the cast iron. It has been shown that PC is an appropriate material with respect to damping (Orak 2000).

6 Polyethylene Terephthalate (PET) in Polymer Composites

Utilization of the solid waste materials is the utmost requirement worldwide. Several treatments such as landfill, incineration, chemical recycling, material recycling and the utilization of energy from combustion are available. Most methods excluding material recycling techniques are known to have critical limitations in economic, technical and environmental manners. Material recycling is expected to be more feasible in a way that the simplicity of pre-treatment and the reduction in energy consumption and environment pollution can be satisfied (Yang 1996). Among the plastic wastes, polyethylene forms the largest fraction, followed by polyethylene terephthalate, commonly known as PET. Due to their rapid increase in the use of PET bottles, solid waste problem is raised. In order to find a solution to this problem, in the recent years some works on the re-using of PET wastes have been accelerated. The research on recycling the used PET bottles as construction materials has been conducted worldwide (Choi et al. 2005). One of the methods is to use recycled PET bottles as unsaturated polyester resin to apply to PC. Another method is to use PET as a filling material for lightweight concrete (Yesilata et al. 2009). Other examples include the use of PET fibres as concrete reinforcement materials (Panyakapo and Panyakapo 2008; Markovic et al. 2003. In general, fibre-reinforced concrete or mortar shows excellent performance in tensile and crack resistance; as recycled PET fibres have provided excellent physical characteristics and improved mechanical properties to the concrete (Yesilata et al. 2009; Panyakapo and Panyakapo 2008; Markovic et al. 2003; Ochi et al. 2007). Particularly, in the case of mortar and concrete reinforced with PET fibres, they demonstrate excellent characteristics in suppressing early-age crack generation (Wang et al. 1987). However, because the surface of used PET materials is water-resistant, adhesion performance and dispersion with cement composition materials is poor; thus, water permeation can increase at the interface of the PET fibre–cement matrix. To address this problem of endurance performance, research on the physical characteristics of PET fibre-reinforced concrete exposed to an alkaline environment has been conducted and the results have indicated that the physical characteristics of PET fibre-reinforced concrete exposed to an alkaline environment were good (Silva et al. 2004). Foti (2013) found a possible use of PET for PCC in the form of flat or round bars, or networks for structural reinforcement. Jo (2008) observed that the strength of PC made with a resin based on recycled PET and recycled aggregate increases with increasing resin content.

7 Industrial Waste Materials (Aggregates)

In a study by Wang it was concluded that the fly ash reduces the thermal decomposition of PET, expedites the melting and mixing of PET, reduces the shrinkage of the material during the moulding process and improves the properties of the final product. The compressive strength was far above standards for building materials. The addition of fly ash increased the compressive strength by 31–53 % (Wang et al. 1987). Choi et al. (2005) investigated the effects of waste PET bottles aggregate on the properties of concrete. The waste plastic could reduce the weight by 2–6 % of normal weight concrete. However, the compressive strength was reduced up to 33 % compared with that of normal concrete. Similarly, the results of Batayneh et al. (2007) showed the deterioration of compressive strength with an increase in the proportion of plastic content. For the plastic proportion of 20 % of sand, the compressive strength was reduced up to 70 % compared with that of normal concrete.

Marzouk et al. (2007) studied the effects of PET waste on the density and compressive strength of concrete. The study demonstrates that shredded plastic bottles may be used successfully as sand substitution aggregates in cementitious concrete composites. It was found that the density and compressive strength decreased when the PET aggregates exceeded 50 % by volume of sand. The density and compressive strength of concrete were between 1,000–2,000 kg/m^3 and 5–60 MPa, respectively. Previous studies (Chen 2003; Mahdi et al. 2007; Rebeiz et al. 1994a, b, c; Rebeiz and Fowler 1996; Rebeiz 1996; Vaidya and Nadkarni 1987) have shown that it is possible to use plastic waste in concretes or mortars as a binder thus replacing cement. Most of the studies carried out for the depolymerization of PET have discussed the role of various types of glycols and metal acetates to be used as catalyst in the process (Baliga and Wong 1989; Chen and Chen 1999; Guclu et al. 1998; Kao et al. 1997; Pimpan et al. 2003; Vaidya and Nadkarni 1989). Rebeiz (1996) investigated the strength properties of un-reinforced and reinforced PC using an unsaturated polyester resin based on recycled PET plastic waste. The results showed that the resins based on recycled PET can be used to produce a good quality of precast concrete. Rebeiz and Fowler (1996) also investigated the effect of age and temperature on strength, modulus, shrinkage, thermal expansion and creep.

The investigations were carried out by some researchers to determine the load deflection and moment curvature responses, modes of failure, strength properties etc., of PC using recycled polymer resin obtained from PET (Rebeiz and Fowler 1996; Jo et al. 2007, 2008). Rebeiz and Fowler (1996) found that very good flexural strength can be obtained with reinforced PC using unsaturated polyester resins based on recycled PET. Jo et al. (2007) predicted long-term creep behaviour using short-term creep and to define the characteristics of creep behaviour of PC bound by recycled polyester resin.

Golestaneh (2009) found that utilization of waste silica powder as filler in PC is promising; it may enhance the physical properties and mechanical strength of the PC. The compressive strength of PC with silica powder as filler in comparison with cement concrete was enhanced by four-folds. Also, it has been observed by Kumar that silica fumes and recron fibres enhance the resistance to capillary and porosity problems in concrete, besides enhancing the compressive strength, flexural strength and split tensile strength (Dash 2011). When waste glass was used as aggregate, it was found by Park (2004) that the compressive, tensile and flexural strengths of concretes containing the waste glass aggregates demonstrated a decreasing tendency along with an increase in the mixing ratio of the waste glass aggregates.

The PET plastic waste was depolymerized through glycolysis to produce unsaturated polyester resin (UPER) by Mahdi. It has been found that MEKP when used as catalyst results in giving better strength to the PC (Mahdi 2010). A study has been done using recycled glass fibre-reinforced polymer (GFRP) waste materials obtained by means of milling processes by Ribeio. Obtained results highlight the potential of recycled GFRP waste materials as efficient and sustainable reinforcement and admixture for PC and mortars composites, constituting an emergent waste management solution (Ribeiro and Meixedo 2011). It was found by Tuan (2011) that when rice husk ash (RHA) was used, it can increase the degree of cement hydration. The compressive strength of the sample made with RHA after 7 days was higher than that of the sample made with silica fume.

8 Conclusions

Excellent mechanical properties have been obtained from PC using resin based on recycled PET. Polymer concrete using resin based on recycled PET has high flexural and splitting tensile strengths. Polymer concrete achieves about 70–80 % of its 7-day strength in 1 day. The elastic modulus, flexural strength and splitting tensile strength increase with an increase in compressive strength.

The use of recycled PET in PC helps in reducing the cost of the material, solving some of the solid waste problems posed by plastics and saving energy.

Polymer concrete damping carriage has been found to improve the damping capacity and dynamic characteristics of the CNC table system equivalent of the slide way system. Few other studies report the damping properties of various types of PC, but these studies fail to establish the best mix required for optimum damping.

Recycled polyester derived from PET waste can be used as resin for PCCs; however, very little research work has been reported on the fabrication of PCC entirely made from the waste materials such as fly ash, red mud and silica fume. for machine tool applications.

A lot of research work is still required on vibration damping capacity of PCCs derived from PET waste.

References

Abdel-Jawad YA, Abdullah WS (2002) Design of maximum density aggregate grading. Constr Build Mater 16:495–508

ACI Committee 503 (1993) Guide for selection and use of structural adhesive with concrete. ACI Mater J 89(1):90–105

ACI Committee 548 (1993) Guide for polymer concrete overlays. ACI Mater J 90(5):499–522

ACI Committee 548 (1996) Polymer concrete—structural applications state-of-the-art report—ACI 548.6R. American Concrete Institute, Detroit, pp 1–23

ACI Committee 548 (1997) Guide for the use of polymers in concrete—ACI 548.1R. American Concrete Institute, Detroit, pp 1–29

ACI Committee 548 (1998) Guide for polymer concrete overlays ACI 548.5R. American Concrete Institute, Detroit, pp 1–26

Aicin PC (2003) The durability characteristics of high performance concrete: a review. Cement Concr Compos 25:409–420

Asthana KK, Lakhani R (2004) Development of polymer modified cementitious (polycem) tiles for flooring construction and building materials. Constr Build Mater 18:639–643

Baliga S, Wong WT (1989) Depolymerization of poly(ethylene terephthalate) recycled from post-consumer soft drink bottles. J Polym Sci Part A: Polym Chem 27(6):2071–2082

Batayneh M, Marie I, Asi I (2007) Use of selected waste materials in concrete mixes. Waste Manag 27(12):1870–1876

Chen CH (2003) Study of glycolysis of poly (ethylene terephthalate) recycled from post-consumer soft-drink bottles. III. Further investigations. J Appl Polym Sci 87(12):2004–2010

Chen JW, Chen LW (1999) The glycolysis of poly (ethylene terephthalate). J Appl Polym Sci 73(1):35–40

Choi YW, Moon DJ, Chung JS, Cho SK (2005) Effects of waste PET bottles aggregate on the properties of concrete. Cem Concr Res 35:776–781

Dash AK (2011) Effect of silica fume on engineering properties of fiber reinforced concrete. In: 6th international structural engineering and construction conference on modern methods and advances in structural engineering and construction. Held in Zürich, Switzerland during 21–26 June 2011, pp 1271–1276

EL-Hawary MM, Abdel-Fattah H (2000) Temperature effect on the mechanical behavior of resin concrete. Constr Build Mater 14:317–323

Folic RJ, Radonjanin VS (1998) Experimental research on polymer modified concrete. ACI Mater J 95(4):163–469

Foti D (2013) Use of recycled waste pet bottles fibers for the reinforcement of concrete. Compos Struct 96:396–404

Fowler DW (1999) Polymers in concrete: a vision for the 21st century. Cement Concr Compos 21:449–452

Fuller WB, Thompson SE (1907) The laws of proportioning concrete. J Transp Div Am Soc Civil Eng 59:67–143

Golestaneh M (2009) Evaluation of mechanical strength of epoxy polymer concrete with silica powder as filler. World Appl Sci J 9(2):216–220

Gorninski JP, Dal Molin DC, Kazmierczak CS (2007) Strength degradation of polymer concrete in acidic environments. Cem Concr Compos 29:637–645

Guclu G, Kasgoz A, Ozbudak S, Ozgumus S, Orbay M (1998) Glycolysis of polyethylene terephthalate waste in Xylene. J Appl Polym Sci 69(12):2311–2319

Jo B-W (2008) Mechanical properties of polymer concrete made with recycled PET and recycled concrete aggregates. Constr Build Mater 22:2281–2291

Jo BW, Tae GH, Kim CH (2007) Uniaxial creep behaviour and prediction of recycled-PET polymer concrete. Constr Build Mater 21(7):1552–1559

Jo BW, Park SK, Park JC (2008) Mechanical properties of polymer concrete made with recycled PET and recycled concrete aggregates. Constr Build Mater 22(12):2281–2291

Kao CY, Cheng WH, Wan BZ (1997) Investigation of catalytic glycolysis of polyethylene terephthalate by differential scanning calorimetry. Thermochim Acta 292(1–2):95–104

Mahdi F (2010) Strength characteristics of polymer mortar and concrete using different compositions of resins derived from post-consumer PET bottles. Constr Build Mater 24:25–36

Mahdi F, Khan AA, Abbas H (2007) Physiochemical properties of polymer mortar composites using resins derived from post-consumer PET bottles. Cem Concr Compos 29(3):241–248

Markovic I, Walraven JC, van Mier JGM (2003) Development of high performance hybrid concrete. In: International workshop on high performance fiber reinforced cement composites, pp 277–300

Marzouk OY, Dheilly RM, Queneudec M (2007) Valorization of post-consumer waste plastic in cementitious concrete composites. Waste Manag 27(2):310–318

Miller M (2005) Polymers in cementitious materials. RAPRA Technology, Shrewsbury

Moriyoshi A (1996) Thermal properties of polymer concrete using glycerol methacrylate/styrene system at low temperature. Adv Compos Mater 5(2):161–168

O'Connor DN, Saiidi M (1993) Compatibility of polyester-styrene polymer concrete overlays and Portland cement concrete bridge decks. ACI Mater J 90(1):59–68

Ochi T, Okubo S, Fukui K (2007) Development of recycled. PET fiber and its application as concrete-reinforcing fiber. Cem Concr Compos 29:448–455

Ohama Y, Kobayashi T, Takeuchi K, Nawata K (1986) Chemical resistance of polymethyl methacrylate concrete. Int J Cem Compos Lightweight Concr 8:86–91

Orak S (2000) Investigation of vibration damping on polymer concrete with polyester resin. Cem Concr Res 30(2):171–174

Oussama E, Elhem G (2012) Mechanical and physical properties of epoxy polymer concrete after exposure to temperatures up to 250 °C. Constr Build Mater 27(1):415–424

Oyawa WO (2004) Flexural response of polymer concrete filled steel beams. Constr Build Mater 18(6):367–376

Panyakapo P, Panyakapo M (2008) Reuse of thermosetting plastic waste for lightweight concrete. Waste Manage 28:1581–1588

Park SB (2004) Studies on mechanical properties of concrete containing waste glass aggregate. Cem Concr Res 34:2181–2189

Pimpan V, Sirisook R, Chuayjuljit S (2003) Synthesis of unsaturated polyester resin from post-consumer PET bottles: effect of type of glycol on the characteristics of unsaturated polyester resin. J Appl Polym Sci 88(3):788–792

Rao VVLK (1993) Aggregate mixtures for least void content for use in polymer concrete. Cem Concr Aggregates 15:97–103

Rebeiz KS (1996) Precast use of polymer concrete using unsaturated polyester resin based on recycled PET waste. Constr Build Mater 10(3):215–220

Rebeiz KS, Fowler DW (1996) Flexural strength of reinforced polymer concrete made with recycled plastic waste. ACI Struct J 93(5):524–530

Rebeiz KS, Fowler DW, Paul DR (1994a) Mechanical properties of polymer concrete systems made with recycled plastic. ACI Mater J 91(1):40–45

Rebeiz KS, Serhal SP, Fowler DW (1994b) Structural behavior of polymer concrete beams using recycled plastics. ASCE J Mater Civil Eng 6(1):150–165

Rebeiz KS, Yang S, David WF (1994c) Polymer mortar composites made with recycled plastics. ACI Mater J 91(3):313–319

Ribeiro MCS, Meixedo JP (2011) Mechanical behavior analysis of polyester polymer mortars modified with recycled GFRP waste materials. World Academy of Science Engineering and Technology, p 75

Ribeiro MCS, Tavares CML, Ferreira AJM (2002) Chemical resistance of epoxy and polyester polymer concrete to acids and salts. J Polym Eng 22(1):27–44

Shokrieh MM (2011) Effects of thermal cycles on mechanical properties of an optimized polymer concrete. Constr Build Mater 25:3540–3549

Silva MAG, Silva ZCG (2007) Degradation of mechanical characteristics of some polymeric mortars due to aging. ACI Mater J 104(4):337–343

Silva DA, Betioli AM, Gleize PJP, Roman HR, Gómez LA, Ribeiro JLD (2004) Degradation of recycled PET fibers in Portland cement-based materials. Cem Concr Res 35:1741–1746

Solovjov GK, Trambovetsky V, Kruger D (1994) Furan resin polymer concrete in the commonwealth of independent states (CIS). ACI Mater J 91(2):158–160

Swamy RN, Suryavanshi AK, Tanikawa S (1998) Protective ability of an acrylic-based surface coating system against chloride and carbonation penetration into concrete. ACI Mater J 95(2):101–112

Vaidya UR, Nadkarni VM (1987) Unsaturated polyester resin from poly ethylene terephthalate waste. Ind Eng Chem Res 26(2):194–198

Vaidya UR, Nadkarni VM (1989) Polyester polyols from glycolyzed PET waste: effect of glycol type on the kinetics of polyesterification. J Appl Poly Sci 38(6):1179–1190

Van Tuan N (2011) Hydration and microstructure of ultra-high performance concrete incorporating rice husk ash. Cem Concr Res 41(11):1104–1111

Vipulanandian C, Paul E (1990) Performance of epoxy and polyester polymer concrete. Mat J 87:241–251

Wang YC, Wong PMH (2007) An experimental study of the mechanical properties of fiber reinforced polymer (FRP) and steel reinforcing bars at elevated temperatures. Compos Struct 80:131–140

Wang YJ, Backer S, Li VC (1987) An experimental study of synthetic fiber reinforced cementitious composites. J Mater Sci 22:4260–4281

Yang BS (1996) Industrial processes and waste characterization. Resour Conserv Recycl 16(1–4):93–112

Yesilata B, Isıker Y, Turgut P (2009) Thermal insulation enhancement in concretes by adding waste PET and rubber pieces. Constr Build Mater 23:1878–1882

Effect of Molybdenum Addition in Hardfacing Layer on Erosion Resistance Behavior of Steel

Chamkaur Jindal and Hazoor Singh Sidhu

Abstract In this work, the erosion behavior of hardfaced SS310 steel, used for pulverized coal burner nozzle, is studied using Air Jet Erosion Testing Rig. The hardfacing was done with SMAW by varying the molybdenum (Mo) alloying element percentage from 0 to 5 wt.% in hardfacing electrode powder. Microhardness, microstructure, and SEM/EDS analysis was done to analyze the structural uniformity of hardfaced steel samples. It was found that by hardfacing, the elements' distribution become uniform and the structure of the hardfaced steel become dense and bright. The microhardness of the hardfaced steel was decreased as the Mo content added up to 5 wt.% in electrode powder. The hardfacing shows ductile behavior of erosion at room as well as at 400 °C temperature.

Keywords Shielded metal arc welding · Solid particle erosion · Microhardness · Stainless steel · Molybdenum

1 Introduction

Erosive wear is characterized by successive loss of material from the surface, which is caused by the continuous impact of solid particles (Vicenzi et al. 2008; Camacho et al. 2012). In power plants, the materials of pulverized coal burner nozzle come in contact with continue flow of coal particles at high velocity, pressure, and high temperatures, and the material gets eroded which reduces the

C. Jindal (✉)
Department of Mechanical Engineering, Guru Nanak Dev Engineering College, Ludhiana 141006, Punjab, India
e-mail: mr.chamkaur@gmail.com

H. S. Sidhu
Department of Mechanical Engineering, Yadavindra College of Engineering, Punjabi University, Guru Kashi Campus, Talwandi Sabo, Bathinda 151302, Punjab, India

S. S. Khangura et al. (eds.), *Proceedings of the International Conference on Research and Innovations in Mechanical Engineering*, Lecture Notes in Mechanical Engineering, DOI: 10.1007/978-81-322-1859-3_31, © Springer India 2014

service life of product (Nozzle). This effect plays a major role in equipment reliability and maintenance costs.

The erosion rate depends on several factors, such as size, shape, hardness, angle, and impact velocity of the erodent particles, as well as temperature and properties of the eroded surface. At high temperatures, oxidation can occur simultaneously with erosion (Vicenzi et al. 2008). The material removal increases when temperature elevated, and mostly, the steel material shows maximum value at the shallow angle that is the feature of ductile materials for high-temperature erosion (Shimizu et al. 2011; Chatterjee and Pal 2006). To regain the size and improve the solid particle erosion-resistant properties of material, hardfacing is the best method. Hardfacing is hard surfacing of components which comes in contact with solid particles. Several welding techniques such as oxyacetylene gas welding (OAW), gas metal arc welding (GMAW), shielded metal arc welding (SMAW), and submerged arc welding (SAW) can be used for hardfacing. The most important differences among these techniques lie in the welding efficiency, the weld plate dilution, and materials costs. For example, SMAW is commonly used in industry because of low cost of electrodes and ease in performance in comparison with other welding techniques (Jeshvaghani et al. 2011).

In this work, SS310 steel was used as substrate material, which is generally used for the nozzle material. This has high percentage of chromium and nickel. SMAW has done for hardfacing process. The hardfacing electrodes used for depositing hardfacing layer were chromium carbide tubular cored (ET 075) electrode. Hardfacing was done by the addition of molybdenum (Mo) metal powder in hardfacing tubular electrodes by varying the percentages. The erosion–corrosion resistance was improved by the addition of Mo, and the weight loss or average thickness reduction of the specimens decreased significantly with an increase in Mo (Yu et al. 2002). The material is used under room temperature and high-temperature working conditions. The microhardness and microstructure of the hardfaced layer also affected with the addition of Mo content.

2 Experimentation

2.1 Substrate Material

The material used for the purpose of hardfacing substrate was high-chromium SS310 steel flat. The flat size was 150 $\times$ 50 $\times$ 5 mm. Substrate material chemical composition is given in Table 1.

Table 1 Chemical composition of SS310 steel

Material	C	Mn	Si	P	S	Cr	Al	Ni	Fe
SS310	0.25	2	1.5	0.045	0.030	26	0.1	22	Bal

Table 2 Chemical composition of hardfacing tubular electrodes

Electrode	C	Si	Cu	Ni	Cr	Mo	W	Fe
Alloy 1	3.5	2.2	0.12	0.19	28.5	13.5	7.3	Bal
Alloy 2	3.5	2.2	0.12	0.19	28.5	16.5	7.3	Bal
Alloy 3	3.5	2.2	0.12	0.19	28.5	18.5	7.3	Bal

2.2 Hardfacing Formulation

Hardfacing tubular coated electrode (ET 075) was used for the purpose of hardfacing. ET 075 (Alloy 1) is in the form of tubular cored electrode of chromium carbide type, reinforced with alloying additives. On the flats of substrate material, double layer was deposited.

The two new compositions of alloy powder were prepared by mixing the 3 and 5 wt.% of Mo powder into the commercially available alloy powder of hardfacing tubular coated electrode. The chemical composition of electrode powder is given in Table 2. The electrodes get empty and mixed the alloy powders with 3 and 5 wt.% of Mo (alloy 2 and alloy 3). These compositions again were filled into the tubular electrodes along with potassium silicate as binding agent. After the preparation of new alloy powder and filled in the electrodes, the electrodes were dried at 80 °C for 3–4 days. The binding agent used was volatile in nature and evaporate by continuous heating of the electrodes and prepared alloys of electrode powder, and Mo content remained in the electrodes.

The manual metal arc welding (MMAW) is used for hardfacing purpose onto SS310 steel substrate, after drying the electrodes. The welding parameters given in Table 3 used were kept constant during whole process (Crespa et al. 2008).

2.3 Sample Preparation

Samples were prepared by cutting the hardfaced SS310 steel in 15 × 10 × 6 mm size. The samples were polished using silicon carbide emery papers of grade 220–2,000 and then again polished using 0.3 μm alumina polishing powder suspended in distilled water.

Table 3 Welding parameter

Welding current/A	Welding voltage/V	Electric current mode
120	22	DCEP

Table 4 Parameters for erosion testing

Erodent Material	Alumina, Al_2O_3
Erodent size	50 μm
Particle velocity	50 m/s
Erodent feed rate	4.75 g/min
Nozzle diameter	3 mm
Impact angle	30° and 90°
Test temperature	Room temperature and 400 °C
Stand of distance	10 mm
Shape	Angular

2.4 Erosion Testing

Erosion testing was done with Air Jet Erosion Testing Rig using different parameters as shown in Table 4.

The samples are weighed before and after the erosion test at both the temperature and impact angles. Difference between weight before and after the erosion test is the degradation rate of the hardfaced sample.

2.5 Testing and Analysis

Erosion is a complex wear process requiring outstanding material properties. For parts subjected to abrasive action, the most influential factors are hardness of both hardphases and matrix material, content of hardphases, and nature of precipitations (Zikin et al. 2012).

Microhardness analysis is done on the hardfacing layer using Vicker's microhardness tester. The test is performed under 1 kg load and at various points in hardfacing zone. Microhardness of the hardfaced steel gets changed from substrate material. The percentage change in microhardness is very high.

Microstructure analysis of the hardfaced layer is also done using optical microscopy. The microstructure of the hardfacing layer is bright and dense. The formation of carbide particles changed, which affects the microhardness.

SEM/EDS analysis is done to study the element distribution in the hardfacing layer. The distribution of elements at various points is studied using energy-dispersive spectroscopy. This shows the percentage of elements present at various points in hardfacing layer.

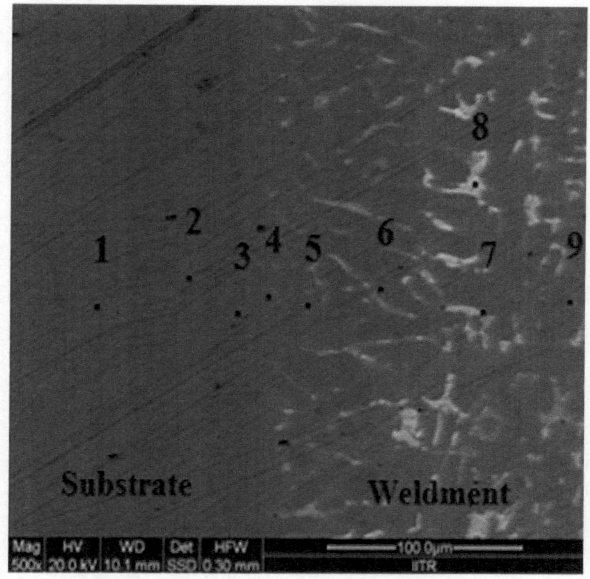

Fig. 1 Cross-sectional EDS point of SS310 steel hardfaced with alloy 3

3 Result and Discussion

3.1 SEM/EDS Analysis

SEM/EDS analysis shows that the elements were uniformly distributed in whole hardfaced layers. Figure 1 shows the characterization of EDS chart of SS310 steel hardfaced with alloy 3 (5 wt.% addition in electrode powder). It reveals that the elements distributed uniformly after hardfacing on substrate metal as nickel from high percentage in substrate dissolve into weldment phase and same in case of chromium. Tungsten (W) also dissolve in substrate metal from the weldment as it has high percentage in electrode powder.

3.2 Microhardness Analysis

Microhardness and erosion test result are shown in Figs. 2 and 3. Figure reveals that microhardness of SS310 steel varies with changing the percentage of Mo in hardfacing electrodes. Hou et al. (2009) also found that the microhardness decreases, but the wear resistance increases with the increase in Mo contents. This decreased microhardness increases the erosion rate at 30° impact angle. As repeated elsewhere, brittle materials are more resistant to low-impact-angle erosion (30°), whereas ductile materials have better performance at high impact angles (90°) (Vicenzi et al. 2008; Sidhu et al. 2007; Lin et al. 2006).

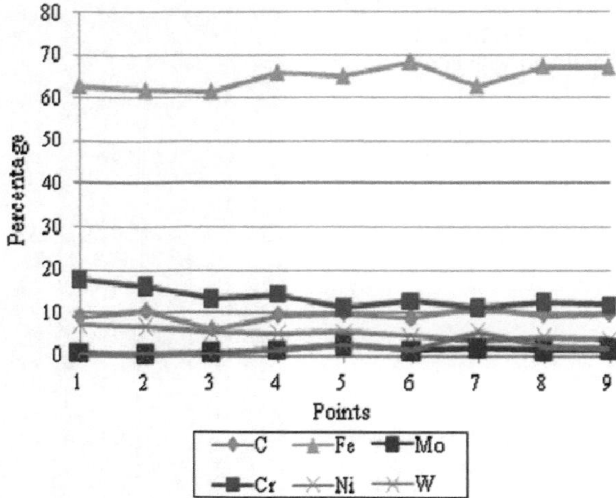

Fig. 2 Cross-sectional EDS line chart of SS310 steel hardfaced with alloy 3

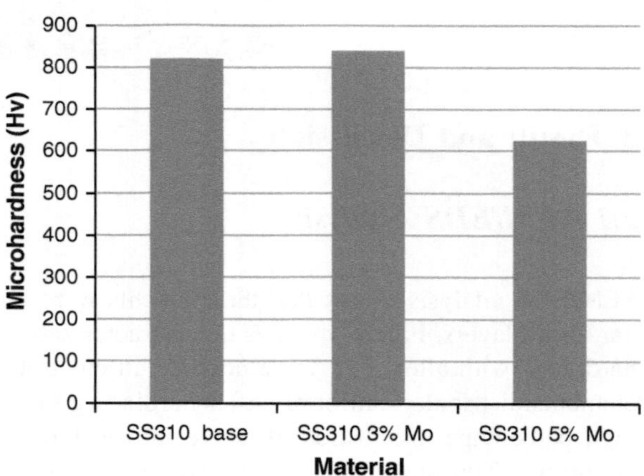

Fig. 3 Microhardness of hardfaced SS310 steel

The microhardness is least when hardfaced with 5 % Mo (alloy 3). It has better performance at 90° impact angle, and Mo-free (alloy 1) hardfacing has high microhardness than alloy 3 so it cannot perform better at same impact angle.

3.3 Erosion Testing Analysis

Solid particle erosion of hardfaced materials depends on many factors including mechanical properties of the target and erodent as well as the impact parameters. The angle of impingement along with other test conditions determines the mechanisms and, therefore, the rates of materials damage (Zikin et al. 2012).

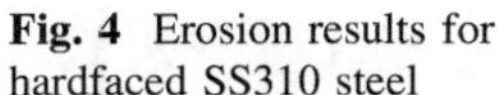

Fig. 4 Erosion results for hardfaced SS310 steel

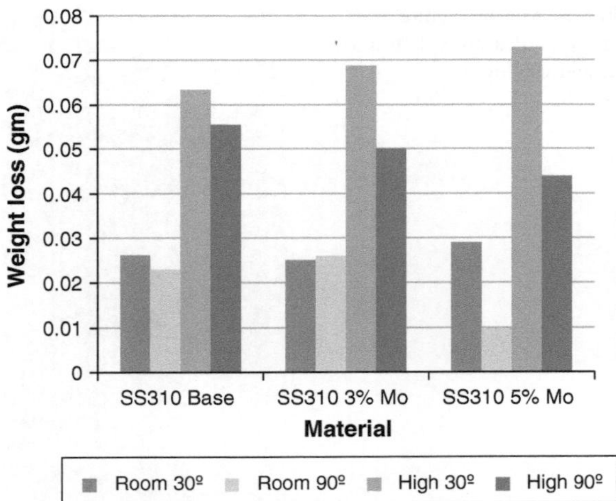

Erosion of ductile materials is essentially a combination of the two processes, with one being dominant over the other depending on impact angle and material properties and particle properties and shape (Wellman and Nicholls 2004). Material properties increase erosion resistance with increasing temperature, e.g., increase in ductility and rate of recovery (Matthews et al. 2009). At high temperature, degradation can also occur due to erosion–oxidation; here, again ductile material has better erosion resistance at high impact angle (90°) than low impact angle (30°).

Figure 4 shows bar chart of erosion rate at different angles at different temperature by varying the Mo content. Degradation rate for room temperature is lower at 90° impact angle than 30° angle, and same degradation is found in case of high-temperature (400 °C) erosion test.

The erosion resistance is increased by addition of Mo content in hardfacing layer. Addition of 5 wt.% Mo in hardfacing layer is the best composition found in this work to improve the maximum erosion resistance. The erosion rate increased when hardfaced by 3 wt.% of Mo addition in electrode powder; this is because the erosion rate is directly proportional to microhardness of material. The microhardness and erosion rate can be compared by Figs. 3 to 4; as the microhardness is increased, the erosion rate decreased at normal angle on both temperature ranges.

3.4 Microstructure Analysis

To reveal the microstructures, the samples were chemically etched in an aqua regia mixture of HNO_3 (30 %), HCl (30 %), and H_2O (40 %).

Microstructure of the hardfacing was uniform and dense. Comparing the hardfaced zone with changing Mo content (Fig. 5), it reveals that the microstructure of hardfacing with alloy 1 has coarse grain structure than hardfaced with

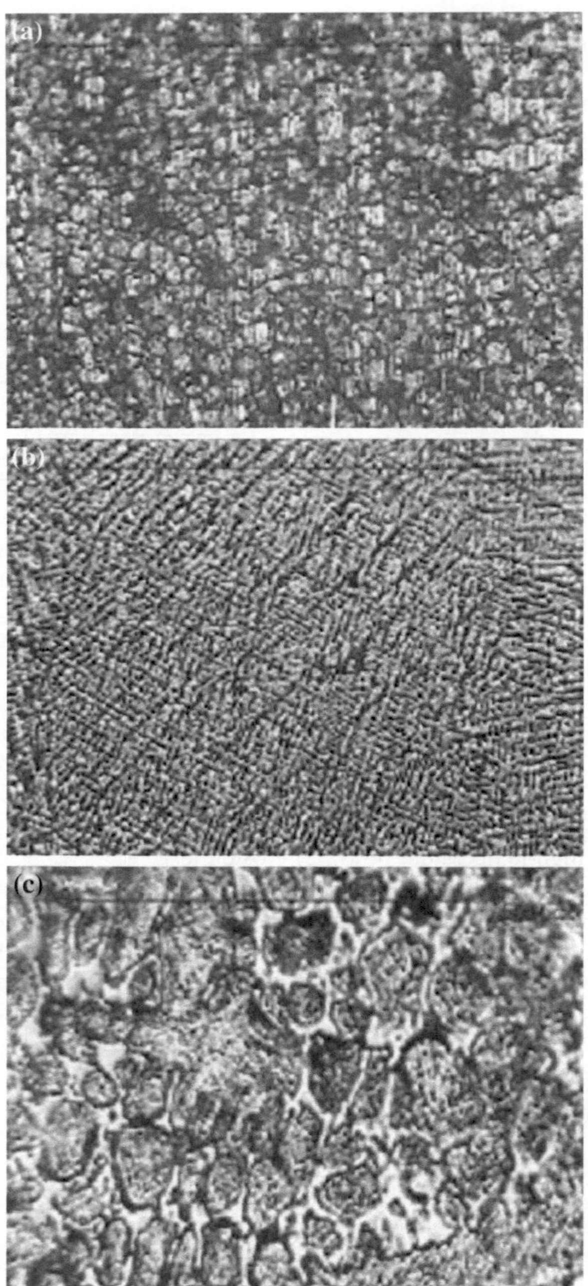

Fig. 5 Microstructure analysis of **a** alloy 1, **b** alloy 2, and **c** alloy 3

alloy 2. Microhardness of hardfacing is influenced by the microstructure of the hardfaced steel. Microhardness with coarse structure is less than fine structure as in Fig. 5b.

Hou et al. (2009) found that the microstructures for Mo-modified hardfacing are brighter than that of the Mo-free coating (alloy 1). At present, although the real reasons for this change are unclear, it may result from the change in the solidification process.

Microstructure of hardfacing with alloy 3 is coarse but dense, and it has low microhardness value than Mo-free hardfacing (alloy 1). The addition of Mo can suppress the formation of M_7C_3-type carbide and promote the formation of $M_{23}C_6$-type carbide, because the microhardness of M_7C_3-type carbide is higher than that of $M_{23}C_6$-type carbide (Hou et al. 2009).

4 Conclusion

The element distribution is uniform when hardfacing with addition of Mo in electrode powder is done. The M_7C_3 carbide formation changes into of $M_{23}C_7$ formation. This decreases the microhardness of the hardfaced steel. The lesser microhardness gives better resistance to solid particle erosion at 90° impact angle. Addition of Mo content improves the uniformity and grain formation of the hardfaced steel. This helps to decrease the degradation (erosion) rate at room temperature as well as at 400° temperature.

References

Borawski B, Todd JA, Singh J, Wolfe DE (2011) The influence of ductile interlayer material on the particle erosion resistance of multilayered TiN based coatings. J Wear 271:2890–2898

Camacho JRL, Chavez AM, Mendez JVM, Torres MV, Hernandez EAG (2012) Solid particle erosion of AISI 304, 316 and 420 stainless steels. J Wear 01–08

Chatterjee S, Pal TK (2006) Solid particle erosion behavior of hardfacing deposits on cast iron—Influence of deposit microstructure and erodent particles. J Wear 261:1069–1079

Crespa AC, Scotti A, Perez MR (2008) Operational behavior assesment of coated tubular electrodes for SMAW hardfacing. J Mater Process Technol 199:265–273

Hidalgo VH, Varela FJB, Menendez AC, Martinez SP (2001) A comparative study of high-temperature erosion wear of plasma sprayed NiCrBSiFe and WC–NiCrBSiFe coatings under simulated coal-fired boiler conditions. Tribol Int 34:161–169

Hou QY, Huang ZY, Shi V, Gao JS (2009) Effects of molybdenum on the microstructure and wear resistance of nickel-based hardfacing alloys investigated using Rietveld method. J Mater Process Technol 209:2767–2772

Jeshvaghani RA, Harati E, Shamanian M (2011) Effects of surface alloying on microstructure and wear behavior of ductile iron surface-modified with a nickel-based alloy using shielded metal arc welding. J Mater Des 32:1531–1536

Lin MC, Chang LS, Lin HC, Yang CH, Lin KM (2006) A study of high-speed slurry erosion of NiCrBSi thermal-sprayed coating. Surf Coat Technol 201:3193–3198

Liu SG, Wu JM, Zhang SC, Rong SJ, Li ZZ (2007) High temperature erosion properties of arc-sprayed coatings using various cored wires containing Ti–Al intermetallics. J Wear 262:555–561

Matthews S, James B, Hyland M (2009) High temperature erosion of Cr3C2-NiCr thermal spray coatings—the role of phase microstructure. J Surf Coat Technol 203:1144–1153

Murthy JKN, Bysakh S, Gopinath K, Venkataraman B (2007) Microstructure dependent erosion in Cr3C2–20(NiCr) coating deposited by a detonation gun. J Surf Coat Technol 202:01–12

Saarivirta EH, Antonov M, Veinthal R, Tuiremo J, Makela A, Siitonen P (2011) Influence of particle impact conditions and temperature on erosion–oxidation of steels at elevated temperatures. J Wear 272:159–175

Shimizu K, Xinba Y, Ishida M, Kato T (2011) High temperature erosion characteristics of surface treated SUS410 stainless steel. J Wear 271:1349–1356

Shin JC, Doh JM, Yoon JK, Lee DY, Kim JS (2003) Effect of molybdenum on the microstructure and wear resistance of cobalt-base Stellite hardfacing alloys. J Surf Coat Technol 166:117–126

Sidhu HS, Sidhu BS, Prakash S (2007) Solid particle erosion of HVOF sprayed NiCr and Stellite-6 coatings. Surf Coat Technol 202:232–238

Tylczak JH (2013) Erosion–corrosion of iron and nickel alloys at elevated temperature in a combustion gas environment. J Wear 01–09

Vicenzi J, Marques CM, Bergmann CP (2008) Hot and cold erosive wear of thermal sprayed NiCr-based coatings: influence of porosity and oxidation. Surf Coat Technol 202:3688–3697

Wellman RG, Nicholls JR (2004) High temperature erosion–oxidation mechanisms, maps and models. J Wear 256:207–217

Yu XQ, Fan M, Sun YS (2002) The erosion–corrosion behavior of some Fe_3Al-based alloys at high temperatures. J Wear 253:604–609

Zikin A, Antonov M, Hussainova I, Katona L, Gavrilovic A (2012) High temperature wear of cermets particle reinforced NiCrBSi hardfacings. Tribol Int 1–11

Tribological Properties of Different Micro-abrasives in Friction Materials

Ram Kumar and Deepak Jain

Abstract The tribological properties of abrasive-based novel friction materials, viz., Alumina, SiC, and SiO_2, were characterized for physical and chemical properties. The wear and frictional performance were evaluated under various operating conditions like speed and pressure. Reduced scale prototype was set up to evaluate performance parameters. Worn surface of friction materials was analyzed over SEM to understand the surface wear. Result showed that under the different operating conditions, SiO_2 has higher wear resistance followed by SiC and alumina that had least wear resistance. Apart from this, better friction properties were observed for SiO_2 that was followed by alumina and SiC.

Keywords Tribological properties · Friction materials · Wear analysis

1 Introduction

Brake system plays a major role in today's fast-moving transportation industry as the liberty to speed depends on its efficient control. When the brake is applied, the friction material lining on the brake pad presses against the rotating component (disc or drum) fixed on the wheel, thereby converting kinetic energy into heat energy due to friction process. The heat thus generated at the sliding interface of the rotor and stator is dissipated primarily by conduction through various components of the brake, by convection and radiation to the atmosphere and adjacent

R. Kumar (✉) · D. Jain
Mechanical Engineering Department, Lovely Professional University, Phagwara 144402 Punjab, India
e-mail: rkiitd12@gmail.com

D. Jain
e-mail: deepak.jain79.in@gmail.com

S. S. Khangura et al. (eds.), *Proceedings of the International Conference on Research and Innovations in Mechanical Engineering*, Lecture Notes in Mechanical Engineering, DOI: 10.1007/978-81-322-1859-3_32, © Springer India 2014

components. Parts of this heat are absorbed by chemical, metallurgical, and wear processes occurring at the interface. The most important function of friction material is to provide adequate friction with minimal damage to the pad surface. Abrasive particles of these materials enhance the friction property even at elevated temperature and help in achieving an optimum frictional coefficient (μ_{avg}). The frictional material is thus the heart of a braking system.

Friction materials are multi-ingredient composites containing binder, fibers, functional fillers, and space fillers. The composition of friction materials is complex and normally consists of more than 10 ingredients among which abrasives are used for its good performance in friction (Blau 2001; Bijwe 1997; EL-Tayeb and Liew 2009). Each ingredient is incorporated to perform desired function without affecting the behavior of other ingredients.

Influence of various abrasives such as SiO_2 and SiC in brake friction material was studied for their friction and wear performance (Crosa et al. 1991). Friction coefficient of four different abrasives was evaluated in the order of silicon carbide > zircon > quartz > magnesia (Kim et al. 2011) and was independent of the load unless the abrasives were fractured due to low fracture roughness. Also, the friction coefficient decreased with time and was stabilized due to blunting of the abrasives and clogging by wear debris. It was noticed that surface damage against a counter disc was severe in the case of using silicon carbide, whereas magnesia polished the surface. The friction material with silicon carbide highlighted the considerable effect of pressure on the friction coefficient. All friction material specimens produced stick–slip during the slow-speed drag tests, which was more significant when silicon carbide and zircon were used as abrasives. Composites containing the finest SiC particles showed the highest values of friction coefficient and reduced specific wear rate and lowest fade (Vlastimil et al. 2010). Friction materials with 2 % abrasives (by weight) offered adequate friction (Satapathy and Bijwe 2000).

The present study aims to analyze the tribological performance of abrasives, viz., alumina, SiC, and SiO_2 of micron size in brake friction material for its wear performance and friction coefficient.

2 Experiment Details

2.1 Composite Fabrication

Friction composites prepared on the basis of fixed material batch contained 10 different master ingredients, i.e., phenolic resin binder, lapinus Rb220, steel wool, PAN, fiber, vermiculite, graphite, cashew dust and brass swarfs, and space fillers. These master ingredients amount to 98 % by weight composition. Rest 2 % of composition constituted abrasives that were mixed with master ingredients. The mixture was prepared using plough shear mixer at the chopper speed of 2,800 rpm

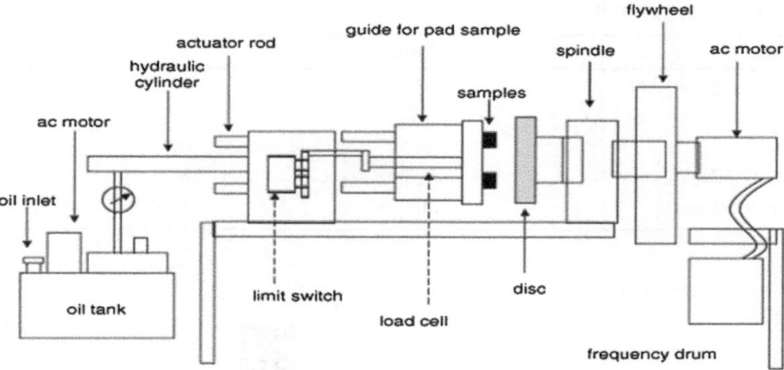

Fig. 1 Schematic diagram of tribo test rig

and feeder speed of 1,450 rpm that ensured the physical isotropy of the composites. Later, mixture was cured in a mold supported by the adhesive-coated back plate. Each composite specimen was prepared with 85 g of mixture and heated at a temperature of 140 °C for 8–10 min at 10 MPa. The surface of brake pad was then grinded in a grinding machine after the post-curing of composites in an oven for 6 h.

2.2 *Design of Experiment*

The friction and wear test were conducted on a reduced scale prototype disc of original vehicle (0.12 m radius), and specimens of size 25 × 25 cm^2 were cut from brake pads. Three speed (800, 1,000, and 1,200 rpm) and three pressure (2, 3, and 4 MPa) combinations were chosen for design of experiments. For a given specimen, test consists of 25 brakings of duration 2 s each. For each experiment, new disk was used from the same lot of disks with new pad specimens. During experiment, temperature varies from room temp to 80 °C and it was ensured to be maintained below 80 °C at start of each brake (Fig. 1).

3 Result and Discussion

Physical (density and porosity) characterization and chemical (acetone extraction) characterization were done as per standard practice (Replacement Brake Lining Assemblies 1993). Hardness testing was done on a Rockwell hardness tester using S scale.

Following conclusions are made from the Table 1:

1. acetone extraction (uncured resin)—alumina > SiO_2 > SiC
2. density—alumina = SiO_2 > SiC

Table 1 Characteristics of the developed composites

Properties	Alumina	SiC	SiO_2
Acetone extraction (%)	0.86	0.23	0.52
Density (g/cc)	2.14	2.14	2.09
Porosity (%)	1.22	0.96	1.91
Hardness (BHN)	84	88	89
Water absorption (%)	1.38	1.09	1.76

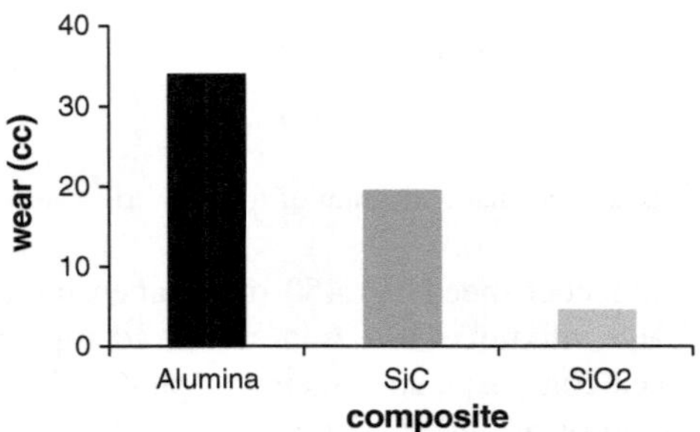

Fig. 2 Wear performance of composites

3. porosity—SiO_2 > alumina > SiC
4. hardness—SiO_2 > SiC > alumina
5. water absorption—SiO_2 > alumina > SiC.

During acetone extraction, the uncured region gets washed out. The highest value of this parameter (0.86) for alumina indicates that it had the maximum uncured region that leads to its least hardness when compared with SiO_2 and SiC. Similarly it was noticed that if porosity is more, density decreases and water absorption increases. Highest water absorption was noticed for SiO_2 that had much higher porosity as compared to the other two abrasive-based composites.

3.1 Wear Analysis

Wear performance of composites was evaluated using standard weight reduction method. Figure 2 shows the wear performance of composites. Wear of the composites followed the order alumina > SiC > SiO_2. SiO_2 shows the minimum wear under different speed and pressure followed by SiC and then alumina. Alumina is the poorest choice from wear point of view, and SiO_2 is the better choice.

Figure 3 shows the SEM of worn surface of worn-out composites. Proper distribution of abrasives and good adhesion of the ingredients can be noticed. Alumina surface shows cracks along with a peeled-off fiber, which support higher wear of these composites. SiC showed a lot of primary ingredients that were exposed during wearing. SiO_2 was the best performer in the series of composites and showed a least number of exposed ingredients.

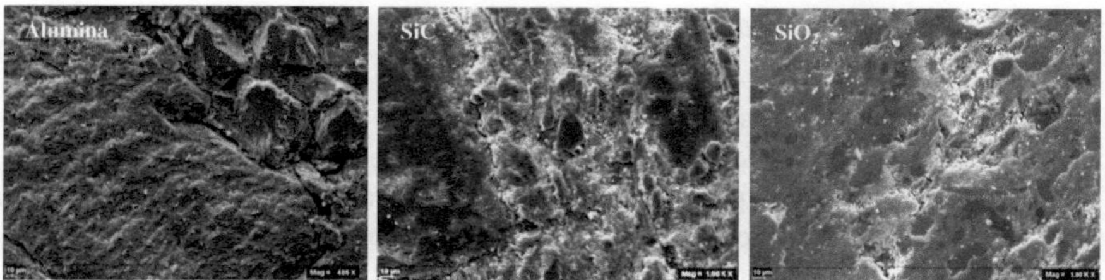

Fig. 3 SEM of worn surfaces of composites

Fig. 4 Friction performance of composites under **a** 2 MPa, **b** 3 MPa, and **c** 4 MPa

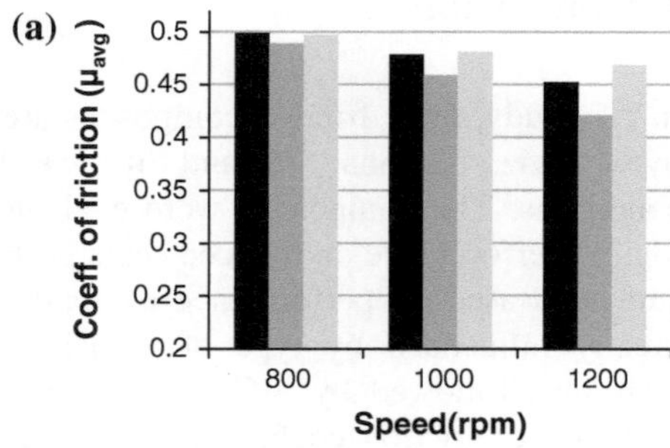

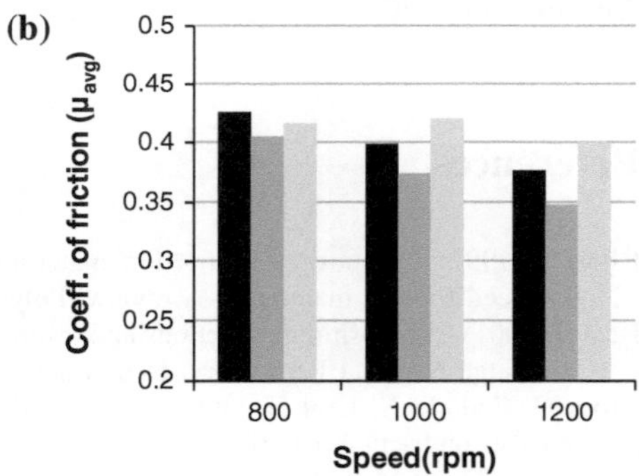

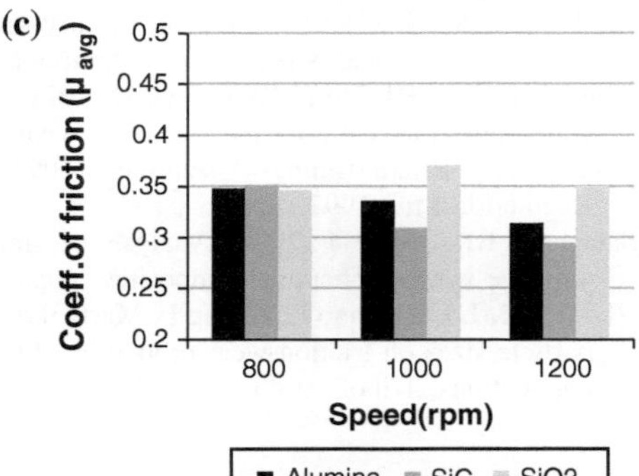

3.2 Friction Performance of Composite

Figure 4 shows the decreased coefficient of friction (μ_{avg}) with the increase in pressure, but the trend is not continuous with speed. At higher speed, μ_{avg} decreased rapidly due to high temperature within a small period of time. At higher pressure, μ_{avg} decreased because it is inversely proportional to load. Composites containing SiO_2 show least variation in μ_{avg} followed by alumina and SiC.

4 Conclusion

In this study, three friction composites are developed using micro-abrasives (2 % by wt.), viz., alumina, SiC, and SiO_2, as abrasive parts under identical processing conditions. The composites were evaluated for physical, mechanical, and tribological performance on reduced scale prototype. Frictional and wear behavior was studied. Based on performance evaluation, it was observed that performance is greatly influenced by type of abrasive. SiO_2 proved to impart highest wear resistance followed by SiC and then alumina apart from this SiO_2 shows less variation of friction coefficient and hence offer superior properties followed by Alumina and SiC.

References

Bijwe J (1997) Composites as friction materials: recent developments in non-asbestos fiber reinforced friction materials—A review. Polym Compos 183:378–396

Blau P (2001) Compositions, functions and testing of friction brake materials and their additives. Oak Ridge National Laboratory, Department of Energy, Tennessee, USA

Crosa G, Enderle N, Leal H, Oliveira A, Stedile FC, Baumvol IJR (1991) Study on effect of abrasives on friction material composites. In: Proceedings of ICAM 91, pp 87–93. Strasbourg, France, 27–29 May 1991

EL-Tayeb NS, Liew KW (2009) On the dry and wet sliding performance of potentially new frictional brake pad materials for automotive industry. Wear 266(1–2):275–287

Kim SS, Hwang HJ, Shin MWK, Jang H (2011) Friction and vibration of automotive brake pads containing different abrasive particles. Wear 271:1194–1202

Replacement Brake Lining Assemblies (1993) E.C.E. Regulation No-90, INTEREUROPE Regulation Ltd., 1997, UN

Satapathy BK, Bijwe J (2000) Analysis of simultaneous influence of operating variables on abrasive wear of phenolic composites. Wear 253:787–794

Vlastimil MJ, Lu Y, Jiao L, Huang L, Martynkova GS, Toma V (2010) Effects of silicon carbide particle sizes on friction-wear properties of friction composites designed for car brake lining applications. Tribol Int 43:144–151

Investigation on Process Parameter Optimization for Coating Density and Microhardness of Cold-Sprayed Coatings

Tarun Goyal, R. S. Walia and T. S. Sidhu

Abstract Most of the existing multi-response optimization approaches focus on subjective and practical know-how of the process. As a result, some confusions and uncertainties are introduced in overall decision-making process. In this investigation, an approach based on a utility theory and Taguchi quality loss function (TQLF) has been applied to low-pressure cold spray (LPCS) process to deposit copper coatings, for simultaneous optimization of more than one response characteristics. In the present paper, two potential response parameters, i.e., coating density (CD) and microhardness (MH), have been selected. Utility values based on these response parameters have been analyzed for optimization by using Taguchi approach.

Keywords Cold spray (CS) · Optimization · Taguchi method · Utility concept

1 Introduction

The cold gas dynamic spray method (CGSM), hereafter referred to simply as cold spray (CS), is a relatively new process by which coatings of ductile materials (or composite materials with significant ductile phase content) can be produced

T. Goyal (✉)
Mechanical Engineering Department, Shaheed Udham Singh College of Engineering and Technology, Tangori, Mohali, Punjab, India
e-mail: goyaltarun1@gmail.com

R. S. Walia
Delhi Technological University, Delhi 110042, India
e-mail: waliaravinder@yahoo.com

T. S. Sidhu
Shaheed Bhagat Singh State Technical Campus, Ferozepur 152004, Punjab, India
e-mail: tssidhu@rediffmail.com

S. S. Khangura et al. (eds.), *Proceedings of the International Conference on Research and Innovations in Mechanical Engineering*, Lecture Notes in Mechanical Engineering,
DOI: 10.1007/978-81-322-1859-3_33, © Springer India 2014

without significant heating of the sprayed powder. The kinetic energy of the particles is sufficient to produce large deformations and high interfacial pressures and temperatures, which appear to produce a solid-state bond (Papyrin 2006). Cold spray processing was developed in the former Soviet Union more than a decade ago as an offshoot of supersonic wind tunnel testing (Davis et al. 2004; Papyrin et al. 2007).

CS is a process of applying coatings by exposing a metallic or dielectric substrate to a high velocity (300–1,200 m/s) jet of small (1–50 μm) particles accelerated by a supersonic jet of compressed gas. The two main clear-cut distinctions of the low-pressure cold gas dynamic spray (LPCGDS) system from the high-pressure cold gas dynamic spray (HPCGDS) system are as follows: the utilization of low-pressure gas (5–10 bars instead of 25–30 bars) and the radial injection of powder instead of axial injection. The accelerating gas (usually air or N_2) is injected at low pressure (5–10 bars) and preheated within the gas heater to temperatures up to about 400 °C to optimize its aerodynamic properties. Solid powder particles are radially introduced downstream of the throat section of the supersonic nozzle, thus eliminating the need for a high-pressure delivery system, which increases system portability and operational safety and significantly reduces spraying costs. Within the nozzle, static pressure is maintained below the atmospheric pressure, ensuring that feedstock particles are effectively drawn in from the powder feeder by Venturi effect (Maev and Leshchynsky 2008). Figure 1 shows a schematic of the LPCGDS system.

To obtain good quality coatings, these spray parameters should be selected carefully and then optimized. In the process of optimization of process parameters, it is seen that one particular setting of input parameters for a response characteristics may not be suitable for other characteristics of the process/product. In most of the manufacturing processes, more than one quality characteristics have to be considered for the optimization of process parameters making it necessary that several response characteristics have to be simultaneously optimized. Based on the foregoing discussions, in this paper, Taguchi method is briefly reviewed for the multi-response optimization. The multi-response optimization of the response parameters of low-pressure cold spray (LPCS) process is presented by using the experimental data. Optimization models have been developed by the combination of the Taguchi method and the utility concept. The multi-response optimization of quality characteristics, i.e., coating density (CD) and microhardness (MH), of LPCS has been carried out.

2 Experimental Procedure

Table 1 shows the process parameters that were identified as potential important in affecting the quality characteristics of the LPCS process under consideration (Goyal et al. 2012). The process parameters, their designated symbols, and ranges are also given in Table 1.

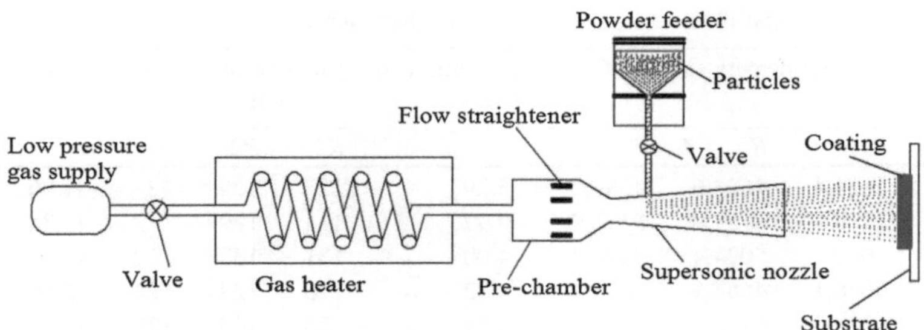

Fig. 1 A typical LPCGDS device (Grujicic et al. 2004)

Table 1 Process parameters and their range

Symbol	Process parameters	Range	Level 1	Level 2	Level 3
A	Feed type	Gravity, argon	Gravity	Argon	–
B	Substrate material	Al alloy, brass, Ni alloy	Al alloy	Brass	Ni alloy
C	Stagnation pressure	104–120 psi	104	112	120
D	Stagnation temperature	350–400 °C	350	375	400
E	Standoff distance	2.5–7.5 mm	2.5	5.0	7.5

Nozzle type: converging–diverging, carrier gas: air, powder size <45 μm

The Taguchi's mixed-level design was selected as it was decided to keep two levels of powder feeding arrangement. The rest four parameters were studied at three levels. The effect of selected process parameters was studied on the following response characteristics of LPCS process:

a. CD
b. MH.

CD and MH are "higher the better" type of quality characteristics. A simplified multi-criterion methodology based on Taguchi's approach and utility concept (given below) is used to achieve the objective of this study. The observed values of response parameters are given in Table 2.

2.1 Utility Concept

Utility can be defined as the usefulness of a product or a process in reference to the expectations of the users. The overall usefulness of a process/product can be represented by a unified index termed as *utility* which is the sum of the individual utilities of various quality characteristics of the process/product. The methodological basis for utility approach is to transform the estimated response of each quality characteristic into a common index.

Table 2 Experimental results of various response characteristics

Exp No.	Coating density (kg/m^3) CD			S/N ratio (dB)	Microhardness (H $_{v0.3}$) MH			S/N ratio (dB)
	*R*1	*R*2	*R*3		*R*1	*R*2	*R*3	
1	4654.4	4623.6	4593.2	73.29	122	126	124	41.86
2	3128.5	3157.9	3116.9	69.92	134	129	132	42.38
3	5012.1	5034.4	4997.4	74.00	131	132	122	42.15
4	4638.1	4654.1	4614.4	73.32	126	128	124	42.00
5	6229.5	6169.3	6125.0	75.81	124	125	128	41.98
6	9016.9	8992.2	9021.0	79.09	126	127	128	42.07
7	27354.5	27106.4	27500.7	88.72	129	131	132	42.32
8	8948.0	8887.9	8877.1	78.99	137	133	134	42.58
9	9440.4	9442.0	9474.6	79.51	134	135	139	42.66
10	3075.1	3111.0	3084.0	69.79	123	118	111	41.36
11	3259.0	3211.9	3245.9	70.20	119	115	117	41.36
12	3422.3	3457.9	3437.5	70.72	126	131	129	42.18
13	5273.6	5261.5	5225.3	74.40	125	121	122	41.77
14	5206.4	5299.1	5216.8	74.38	110	119	113	41.12
15	2741.4	2758.3	2749.8	68.78	122	120	124	41.72
16	8615.5	8628.4	8671.6	78.72	129	125	127	42.07
17	5796.5	5874.8	5818.7	75.31	125	127	128	42.05
18	30161.0	30513.7	30244.6	89.63	123	128	132	42.11
Total	145973.1	146184.4	146014.3	1364.7	2265	2270	2266	755.80
	$\overline{T}_{CD}$ = overall mean of CD = 8114.29				$\overline{T}_{MH}$ = overall mean of MH = 125.94			

*R*1, *R*2, and *R*3 represent repetitions

If X_i is the measure of effectiveness of an attribute (or quality characteristic) i and there are n attributes evaluating the outcome space, then the joint utility function can be expressed (Bunn Derek 1982) as

$$U\,(X_1,\,X_2,\,\ldots X_n) = f(U_1(X_1),\,U_2(X_2)\ldots U_n\,(X_n)) \tag{1}$$

where $U_i(X_i)$ is the utility of the ith attribute.

The overall utility function is the sum of individual utilities if the attributes are independent and is given as follows:

$$U(X_1, X_2, \ldots X_n) = \sum_{i=1}^{n} U_i(X_i). \tag{2}$$

The attributes may be assigned weights depending upon the relative importance or priorities of the characteristics. The overall utility function after assigning weights to the attributes can be expressed as

$$U(X_1,\,X_2,\,\ldots X_n) = \sum_{i=1}^{n} W_i U_i(X_i) \tag{3}$$

where W_i is the weight assigned to the attribute i, the sum of the weights for all the attributes must be equal to 1.

2.2 Determination of Utility Value

A preference scale for each quality characteristic is constructed for determining its utility value. Two arbitrary numerical values (preference number) 0 and 9 are assigned to the just acceptable and the best value of the quality characteristic, respectively. The preference number (P_i) can be expressed on a logarithmic scale as follows (Gupta and Murthy 1980; Kumar et al. 2000):

$$P_i = A \times \log\left(\frac{X_i}{X_i'}\right) \tag{4}$$

where

X_i Value of any quality characteristic or attribute i

X_i' Just acceptable value of quality characteristic or attribute i

A Constant

The value of A can be found by the condition that if $X_i = X^*$(where X^* is the optimal or best value), then $P_i = 9$

Therefore,

$$A = \frac{9}{\log \frac{X^*}{X_i'}}$$

The overall utility can be calculated as follows:

$$U = \sum_{i=1}^{n} W_i P_i \tag{5}$$

subject to the condition: $\sum_{i=1}^{n} W_i = 1$

Among various quality characteristics type, viz. smaller the better, higher the better, and nominal the better, suggested by Taguchi, the utility function would be higher the better type. Therefore, if the utility function is maximized, the quality characteristics considered for its evaluation will automatically be optimized (maximized or minimized as the case may be).

3 Analysis and Discussions

Based on the methodology developed in the previous section, following case has been considered to obtain the optimal settings of the process parameters of LPCS for predicting the optimal values of combined responses. Two quality characteristics, i.e., CD and MH, have been included in utility response.

Taguchi L_{18} orthogonal array (OA) (Roy 1990) has been adopted for conducting the experiments. Powder feeding arrangement (*A*), substrate material (*B*), air stagnation pressure (*C*), air stagnation temperature (*D*), and standoff distance (*E*) were selected as input parameters. Response parameters (quality characteristics) were CD and MH, when they are optimized individually; the summary of results is produced in Table 3.

Following is the stepwise procedure for transforming experimental data into utility data.

3.1 Construction of Preference Scales

1. Preference scale for CD (P_{CD}):

X^* Optimal value of CD = 27584.59 (refer Table 3)
X_i' Just acceptable value of CD = 2,700 (All the observed values of CD are greater than 2,700)

Following equation is obtained from Eq. (4):

$$P_{CD} = 8.91 \times \log\left(\frac{X_{CD}}{2700}\right) \quad (6)$$

2. Preference scale for MH (P_{MH}):

X^* Optimal value of MH = 138.36 (refer Table 3)
X_i' Just acceptable value of MH = 110 (All the observed values of MH are greater than 110).

Following equation is obtained from Eq. (4):

$$P_{MH} = 90.34 \times \log\left(\frac{X_{MH}}{110}\right). \quad (7)$$

3.2 Calculation of Utility Value

Equal weights (1/2 each) have been assigned to the selected quality characteristics, assuming both the quality characteristics are equally important. However, these weights can be varied depending upon the case or user requirements, if any.

Table 3 Optimal setting and values of process parameters

Response characteristics	Optimal level of process parameters	Significant process parameters	Predicted optimal value of quality characteristics
CD	*A*1, *B*3, *C*3, *D*2, *E*1	*A*, *B*, *C*, *D*, *E*	27584.59 kg/m^3
MH	*A*1, *B*3, *C*3, *D*2, *E*2	*A*, *B*, *C*, *D*, *E*	138.36 $H_{v0.3}$

The following relation was used to calculate the utility function based on the experimental trials:

$$U(n,r) = P_{\mathrm{CD}}(n,r) \times W_{\mathrm{CD}} + P_{\mathrm{MH}}(n,r) \times W_{\mathrm{MH}} \tag{8}$$

where $W_{CD} = \frac{1}{2}$; $W_{MH} = \frac{1}{2}$. *n* is the trial number (*n* = 1, 2, 3, ···,18), and *r* is the repetition number (*r* = 1, 2, 3). The calculated utility values are shown in Table 4.

3.3 Analysis of Utility Data for Optimal Settings of Process Parameters

The average and main response in terms of utility values and S/N ratio (Tables 5 and 6) are plotted in Fig. 2. It can be observed from Fig. 2i–v that the first level of powder feed arrangement (*A*1), third level of substrate material (*B*3), third level of air stagnation pressure (*C*3), second level of air stagnation temperature (*D*2), and first level standoff distance (*E*1) are expected to yield a maximum values of the utility and S/N ratio within the experimental space.

The pooled version of ANOVA for utility data and S/N ratio are given in Tables 7 and 8, respectively. It can be noticed from Table 7 that all the input parameters have significant effect (at 95 % confidence level) on the utility function. Similarly, it had been found from Table 8 that all the chosen parameters in study have significant effect on the S/N ratio of utility function.

3.4 Optimal Values of Quality Characteristics (Predicted Means)

The optimal values of utility and thus the optimal values of response characteristics in consideration are predicted at the above levels of significant parameters.

The average values of all the response characteristics at the optimum levels of significant parameters with respect to utility function are recorded in Table 9.

The optimal values of the predicted means (μ) of different response characteristics can be obtained from the following equation:

$$\mu = A1 + B3 + C3 + D2 + E1 - 4T \tag{9}$$

Table 4 Calculated utility data based on responses CD and MH

Trial number	Utility values			S/N ratio (dB)
	*R*1	*R*2	*R*3	
1	3.09	3.71	3.38	10.53
2	4.16	3.43	3.85	11.54
3	3.63	3.78	3.22	10.93
4	3.71	4.03	3.39	11.32
5	3.97	4.11	4.56	12.45
6	5.00	5.15	5.31	14.23
7	7.61	7.89	8.07	17.90
8	6.63	6.03	6.18	15.94
9	6.30	6.44	7.02	16.34
10	2.44	1.65	2.44	6.30
11	2.91	3.21	2.57	9.12
12	3.12	3.91	3.59	10.87
13	3.80	3.16	3.31	10.61
14	2.27	2.85	3.80	8.90
15	2.61	2.75	2.39	8.19
16	4.3	4.76	4.08	12.82
17	3.99	4.32	4.46	12.55
18	6.86	7.67	8.26	17.54

*R*1, *R*2, *R*3 = repetitions of experiments against each of the trial conditions

Table 5 Average and main effects (raw data: CD and MH)

Process parameter designation	Average utility values			Main effects		Difference
	*L*1	*L*2	*L*3	*L*2 − *L*1	*L*3 − *L*2	(*L*3 − *L*2) − (*L*2 − *L*1)
A	4.95	3.76	–	−1.19	–	−1.1
B	3.23	3.67	6.16	0.45	2.49	2.04
C	4.16	4.07	4.83	−0.09	0.76	0.85
D	3.90	5.07	4.09	1.17	−0.99	−2.16
E	4.98	4.05	4.03	−0.92	−0.02	0.90

*L*1, *L*2, and *L*3 represent average values of raw data of corresponding parameters at levels 1, 2, and 3, respectively. *L*2 − *L*1 is the average main effect when the corresponding parameter changes from level 1 to level 2. *L*3 − *L*2 is average main effect when the corresponding parameter changes from level 2 to level 3

A—powder feed arrangement, *B*—substrate material, *C*—air stagnation pressure, *D*—air stagnation temperature, *E*—standoff distance

where *A*1—first level of powder feed arrangement, *B*3—third level of substrate material, *C*3—third level of air stagnation pressure, *D*2—second level of air stagnation temperature, and *E*1—first level of standoff distance.

Table 6 Average S/N values and main effects (raw data: CD and MH)

Process parameter designation	S/N average values			Main effects (dB)		Difference
	*L*1	*L*2	*L*3	*L*2 − *L*1	*L*3 − *L*2	(*L*3 − *L*2) − (*L*2 − *L*1)
A	13.46	10.77	–	−2.71	–	−2.71
B	9.88	10.95	15.51	1.07	4.56	3.49
C	11.58	11.75	13.01	0.17	1.26	1.09
D	11.34	13.48	11.52	2.14	−1.96	−4.10
E	13.04	11.78	11.53	−1.26	−0.25	1.01

*L*1, *L*2, and *L*3 represent average values of S/N data of corresponding parameters at levels 1, 2, and 3, respectively. *L*2 − *L*1 is the average main effect when the corresponding parameter changes from level 1 to level 2. *L*3 − *L*2 is average main effect when the corresponding parameter changes from level 2 to level 3

A—powder feed arrangement, *B*—substrate material, *C*—air stagnation pressure, *D*—air stagnation temperature, *E*—standoff distance

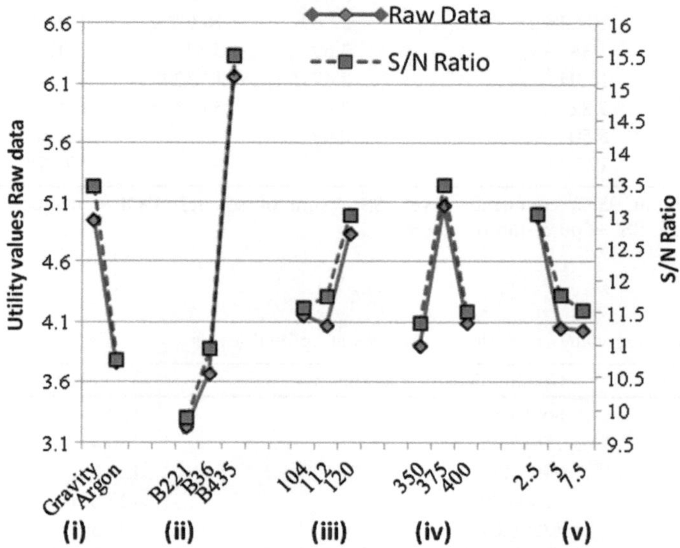

Fig. 2 Average and main response in terms of utility values and S/N ratio

The 95 % confidence interval of confirmation experiments (CI_{CE}) can be computed (Roy 1990) by using the following equation:

$$CI_{CE} = \sqrt{F_{\alpha}(1,f_e)V_e\left[\frac{1}{n_{eff}} + \frac{1}{R}\right]} \tag{10}$$

where $F_{\alpha}(1, f_e)$ = the *F*-ratio at the confidence level of $(1-\alpha)$ against DOF 1 and error degree of freedom f_e, R = sample size for conformation experiments,

Table 7 Pooled ANOVA (raw data: CD and MH)

Source	SS	DOF	V	F-ratio	SS′	P %
A	19.07	1	19.07	155.58*	18.94	12.71
B	90.07	2	45.04	367.50*	89.83	60.27
C	6.25	2	3.12	155.58*	6.00	4.03
D	14.28	2	7.14	58.25*	14.03	9.41
E	14.00	2	7.00	57.11*	13.75	9.22
E (Pooled)	5.39	44	0.12	–	6.50	4.36
Total (*T*)	149.06	53	–	–	149.06	100

* Significant at 95 % confidence level. SS = sum of squares, DOF = degree of freedom, V = variance, SS′ = pure sum of squares

Table 8 S/N pooled ANOVA (raw data: CD and MH)

Source	SS	DOF	V	F-ratio	SS′	P %
A	32.72	1	32.72	47.58*	32.03	2.96
B	107.38	2	53.69	78.07*	106.00	58.81
C	7.38	2	3.69	5.37*	6.01	17.16
D	16.94	2	8.47	12.32*	15.57	12.94
E	7.82	2	3.91	5.68*	6.44	4.60
E (Pooled)	5.50	8	0.69	–	11.69	3.51
Total (T)	177.74	17	–	–	177.74	100

* Significant at 95 % confidence level. SS = sum of squares, DOF = degree of freedom, V = variance, SS′ = pure sum of squares

Table 9 Average values of various responses at optimal levels

Levels	Coating density, CD (kg/m^3)	Microhardness MH ($H_{v0.3}$)
*A*1	8696.68	129.33
*B*3	15075.35	130.44
*C*3	9995.41	128.28
*D*2	12604.70	127.83
*E*1	13290.07	123.39

Note The above average values are taken from experimental data

V_e = error variance, $n_{\text{eff}} = \frac{N}{1+\text{DOF}}$ N = total number of trials, and DOF = total degrees of freedom associated in the estimate of mean response.

1. *For CD*

$$\mu_{\text{CD}} = A1 + B3 + C3 + D2 + E1 - 4T_{\text{CD}} = 27205.06$$

where $A1 = 8696.68$, $B3 = 15075.35$, $C3 = 9995.41$, $D2 = 12604.70$, $E1 = 13290.07$ (Table 9):

Table 10 Observed values of quality characteristics (confirmation experiment)

Exp. No.	CD (kg/m³)			MH ($H_{v0.3}$)		
	r_1	r_2	r_3	r_1	r_2	r_3
1	28375.7	27818.6	29763.5	134	136	139
2	26583.8	25983.2	29432.4	140	138	137
3	29985.3	24643.7	27798.5	137	140	133
Overall average	27820.5			137.1		

$T_{CD} = 8114.29$ (Table 2)

The following values have been obtained by the ANOVA:

$N = 54, f_e = 44, v_e = 4349657.9, n_{eff} = 5.4, R = 3, F_{0.05}\ (1, 44) = 4.064$

From Eq. (10), $CI_{CE} = \pm 3027.51$

The predicted optimal range (for conformation runs of three experiments) for CD is given by

CI_{CE}: $24177.55 < \mu_{CD} < 30232.57$

2. *For MH*

$$\mu_{MH} = A1 + B3 + C3 + D2 + E1 - 4\,T_{MH} = 135.51$$

where $A1 = 129.33$, $B3 = 130.44$, $C3 = 128.28$, $D1 = 127.83$, $E2 = 123.39$ (Table 9):

$T_{MH} = 125.94$ (Table 2)

The following values have been obtained by the ANOVA:

$N = 54, f_e = 44; v_e = 19.77, n_{eff} = 5.4, R = 3, F_{0.05}\ (1, 44) = 4.064$

From Eq. (10), $CI_{CE} = \pm 6.45$

The predicted optimal range (for conformation runs of three experiments) for MH is given by

CI_{CE}: $129.06 < \mu_{MH} < 141.96$.

3.5 Confirmation Experiment

For confirmation of experimental results, three experiments were performed at optimal settings as suggested by Taguchi analysis of utility data. The observed values of various response characteristics have been given in Table 10. It can be noticed that overall average of the observed values of the response characteristics fall well within the 95 % CI_{CE} of the optimal range of the respective response characteristics.

4 Conclusions

The important conclusions of this research work are enlisted below:

1. The optimal setting of the process parameters were predicted for optimization of CD and MH using the model. The optimal settings were *A*1—first level of powder feed arrangement, *B*3—third level of substrate material, *C*3—third level of air stagnation pressure, *D*2—second level of air stagnation temperature, and *E*1—first level of standoff distance.
2. The decreasing order of percentage contribution of the parameters to achieve a higher value of utility function is as follows: substrate material (60.27 %), powder feeding arrangement (12.71 %), air stagnation temperature (9.41 %), standoff distance (9.22 %), and air stagnation pressure (4.03 %).
3. The overall average of quality characteristics was found to be 27820.5 kg/m^3 for CD and 137.1 $H_{v0.3}$ for MH, which falls well within the 95 % CI_{CE} of the optimal range of the respective response characteristics.

References

Bunn Derek W (1982) Analysis for optimal decisions. Wiley, New York, p 275

Davis JR et al. (eds) (2004) Handbook of thermal spray technology, 1st edn. ASM International® Materials Park, OH

Goyal T, Walia RS, Sidhu TS (2012) Effect of parameters on coating density for cold spray process. J Mater Manuf Processes 27(2):193–200

Grujicic M, Zhao CL, Tong C, DeRosset WS, Helfritch D (2004) Analysis of the impact velocity of powder particles in the cold-gas dynamic-spray process. Mater Sci Eng A 368:222–230

Gupta V, Murthy PN (1980) An introduction to engineering design methods. Tata McGraw Hill, New Delhi, pp 78–98

Kumar P, Barua PB, Gaindhar JL (2000) Quality optimization (multi-characteristics) through Taguchi technique and utility concept. J Qual Reliab Eng Int 16:475–485

Maev RG, Leshchynsky V (2008). Introduction to low pressure gas dynamic spray physics and technology. Wiley2VCH, Weinheim

Papyrin AN (2006) Cold spray: state of the art and applications. In: Cold spray technology. Elsevier, Amsterdam, pp 1–2

Papyrin A, Kosarev V, Klinkov S, Alkhimov A, Fomin V (2007) Cold spray technology. Elsevier, London

Roy RK (1990) A primer on Taguchi method. Van Nostrand Reinhold, New York

Effect of Heat Treatment on Wear Behavior of Hardfaced Steel

Satnam Singh Sekhon and Hazoor Singh

Abstract This research work was done with an aim for the modification of wear resistance of hardfaced steel by heat treatment. The proposed material was alloy steel, because of its various applications in manufacturing the rotating parts where large amount of abrasive wear takes place. The manual metal arc welding (MMAW) process was used to deposit the hardfacing layers with two different hardfacing electrodes. Heat treatment was done to further increase the wear resistance of the material. The chemical composition of the samples was determined with the help of the spectrometer. The pin-on-disk apparatus was used for the wear analysis of all the samples. It was seen that hardfacing of the samples improved its hardness and wear resistance, but the heat treatment of the hardfaced samples has further improved the hardness and wear resistance properties.

Keywords Manual metal arc welding · Pin-on-disk wear · Hardfacing · Heat treatment

Abbreviations

SUT Substrate metal (untreated)
SHT Substrate metal heat treated
AIUT Sample hardfaced by alloy 1 and untreated
A1HT Sample hardfaced by alloy 1 and heat treated
A2UT Sample hardfaced by alloy 2 and untreated
A2HT Sample hardfaced by alloy 2 and heat treated

S. S. Sekhon · H. Singh (✉)
Yadavindra College of Engineering, Punjabi University, Guru Kashi Campus, Talwandi Sabo, Bathinda 151302, Punjab, India
e-mail: hazoors@yahoo.com

S. S. Khangura et al. (eds.), *Proceedings of the International Conference on Research and Innovations in Mechanical Engineering*, Lecture Notes in Mechanical Engineering, DOI: 10.1007/978-81-322-1859-3_34, © Springer India 2014

1 Introduction

Wear is a process of progressive and undesirable loss or degradation of material from the surface of mechanical components, and therefore, it is considered as one of the most significant surface failures of engineering parts (Sabet et al. 2011). The abrasive wear is probably the most significant cause of mechanical damage of equipment components coming in contact with abrasive/erosive bodies. Hard abrasive particles penetrate the components and cause damage in the form of material loss (Kumar et al. 1999). Hardfacing is generally used to deposit a wear resistant alloy on either a worn component or new item which is subjected to wear in service. The combination of wear resistant surface with an inexpensive structural material, which can often be resurfaced many times, can offer considerable economic advantages compared with component manufactured from where resistant material alone (Gregory 1978). Welding is a key technology to apply hardfacing alloys. The common methods for improving the wear resistance of a softer material include ion implantation, laser cladding, thermal oxidation, heat treatment, chemical vapor deposition, and physical vapor deposition (Lin and Chen 2013). The welding hardfacing not only has a high wear and impact resistance, anticorrosive behavior of the deposited metal, but also can restore the dimensions of worn out components. In addition, the welding hardfacing may produce a thick-deposited layer with a high deposition rate, and the resulting hardened layer has a high bonding strength (Bing-quan et al. 2003). Normalizing is a process of heating the steel above the critical temperature, holding for a period of time long enough for transformation to occur. It establishes a more uniform carbide size and distribution. A fast cooling rate depresses the transformation temperature and refines precipitate size, and hence increases the hardness of microalloyed steels. The microstructure is changed from ferrite/pearlite to bainitic ferrite with an increase in cooling rate, and such microstructural change is responsible for the high strength–toughness combination of microalloyed steels at high cooling rate (Zhao et al. 2013). Considering the matrix microstructure and the amounts, morphology, and distribution of various second phases can be adjusted through heat treatment technology, so it is possible to improve the steel properties if we are able to select the heat treatment parameters reasonably (Wang et al. 2013).

2 Experimentation

2.1 Substrate Material

The alloy steel material of the tractor spindle was used to deposit the alloying element layer over its surface to improve its wear resistance properties. The surface of the spindle was made flat 1 inch wide throughout the length with the help

Table 1 Chemical composition of base materials (wt.%)

C	0.37
Si	0.25
Mn	0.66
Ni	0.8
Cr	0.15
Mo	0.01
Cu	0.15
W	0.02
S	0.03
Al	0.01
P	0.03
Fe	Bal.

of surface grinder, to use it as a substrate material. The chemical composition of substrate material according to weight's percentage is shown in Table 1.

2.2 *Hardfacing Process*

The manual metal arc welding (MMAW) was used for hardfacing over the flat surface of the substrate material. Double layer of hardfacing material was deposited over the substrate material. This hardfacing was done with two different hardfacing electrodes over different sections of substrate material.

2.3 *Hardfacing Electrodes*

Two different types of tubular coated hardfacing electrodes were used for the hardfacing purpose. One of the electrodes used for hardfacing was CPET 071 (alloy 1), and it was a tubular cored electrode of chromium carbide type, reinforced with alloying additives. The other electrode used for hardfacing was CPET 075 (alloy 2). The chemical compositions of two hardfacing electrodes determined by spectrometer are shown in the Tables 2 and 3.

2.4 *Preparation of Samples for Testing*

The Samples were divided into three groups

- The substrate material (group 1)
- The hardfaced material with alloy 1 (group 2)
- The hardfaced material with alloy 2 (group 3).

Table 2 Chemical composition of hardfacing electrode (alloy 1)

C	3.5
Si	2.22
Ni	0.192
Cr	28.5
Mo	13.5
W	7.24
Cu	0.174
Fe	Bal.

Table 3 Chemical composition of hardfacing electrode (alloy 2)

C	0.12
Mn	0.18
Si	7.79
P	0.48
S	5.61
Cr	27.73
Ni	0.21
W	8.5
Fe	Bal.

Eight samples from each group were cut with the help of wire cut machining with similar shape and dimensions. The dimension of the samples was $8 \times 8 \times 36\ mm^3$.

2.5 Heat Treatment Process

Normalizing of four samples from each group was done to study the changes taking place in the samples after the heat treatment process. In this process, samples were heated above the critical temperature, that is, about 900 °C for a time period long enough for transformation to occur, and then it was cooled slowly in the air or the room temperature. The cooling time of the samples was 30 min for 1 mm^2.

2.6 Microstructural Analysis

The samples were polished across the cross-section of the weldment. The polishing was done with the help of emery papers of silicon carbide of different grades ranging between 100 and 2,000. Then, it was again polished with 0.3 μm alumina polishing powder suspended in distilled water for better finishing, and then, etching of the samples was done using the solution of HCl, HNO_3, and distilled water. The microstructure of different samples was analyzed with the help of optical microscopy to analyze the changes in the microstructure of different samples after hardfacing and heat treatment process.

2.7 Microhardness Analysis

The microhardness of different samples was compared to analyze the changes in microhardness due to different hardfacing materials and heat treatment. Microhardness was measured across the cross-section by taking different points on weldment, HAZ, and substrate.

2.8 Chemical Analysis

It is very important to know the presence and percentage composition of various elements which are the part of the metal, to analyze its various properties and to know the effect of addition of some specific element on its properties. The chemical composition of the material was done with help of spectrometer at CTR, Ludhiana.

2.9 Wear Analysis

The samples were rubbed on the emery paper to remove the upper layer from 1 to 2 mm. This was done to get the highest microhardness point of the sample on top as the top surface may be annealed during cutting of the samples. The distance of the highest microhardness point from the top surface was measured by Vickers microhardness tester. The test was carried out at two different velocities. Load was kept constant at 3 kg, and distance travelled by each sample was 2,000 m. The samples were weighed before and after the process, and the change in weight due to wear is measured.

3 Results and Discussion

3.1 Microhardness Analysis

It was observed that the microhardness of the substrate material throughout lies between 220 and 240 HV. The microhardness of heat-treated substrate material (386 HV) was more as compared with the untreated substrate material.

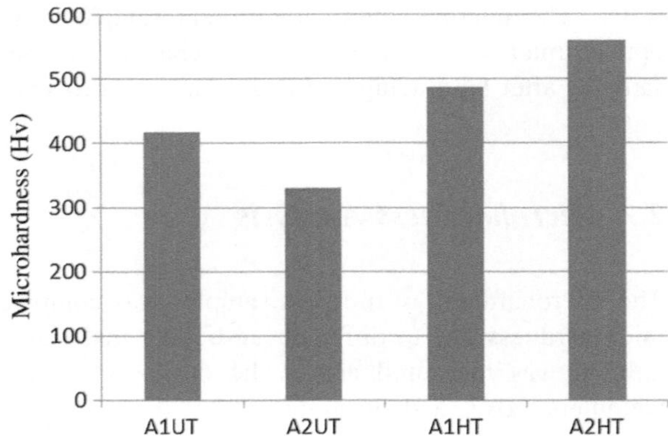

Fig. 1 Average microhardness value at the top of hardfaced surface

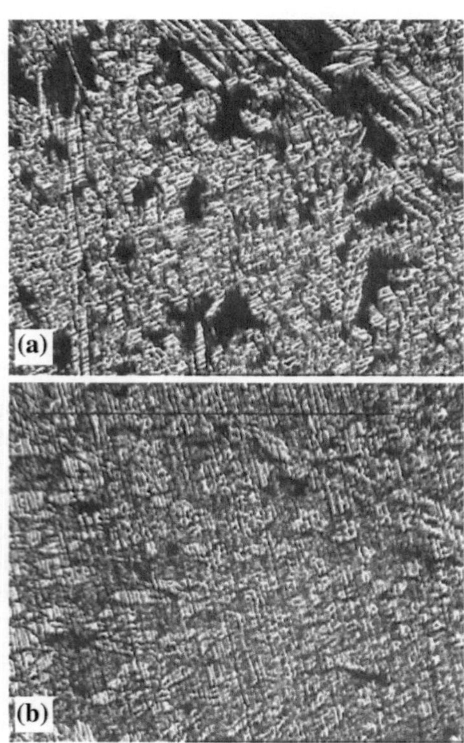

Fig. 2 Microstructure of weldment of steel hardfaced with alloy 1 **a** untreated and **b** heat-treated

Figure 1 shows that average microhardness of the hardfaced surfaced with alloy 1 was 417 HV at the top surface of the weldment, whereas after heat treatment, it was increased to 488 HV. In the same way, the average microhardness of the hardfaced surface with alloy 2 was 330 HV at surface of the weldment, and it was increased to 560 HV after heat treatment.

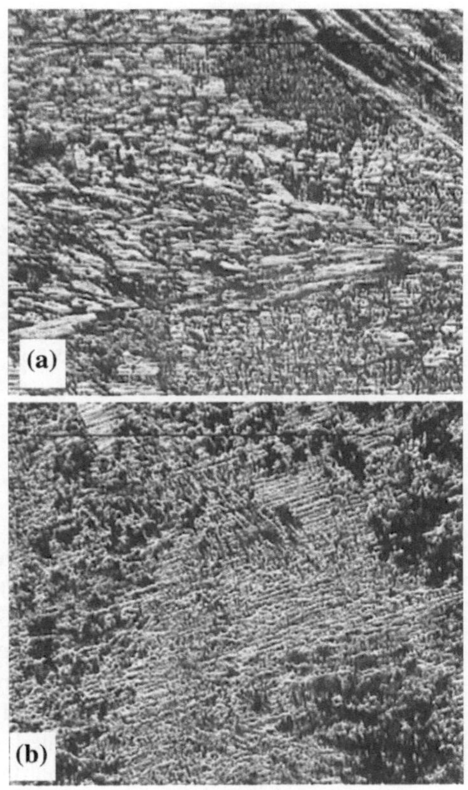

Fig. 3 Microstructure of weldment of steel hardfaced with alloy 2 **a** untreated and **b** heat-treated

3.2 Microstructural Analysis

The structure of heat-treated metal was slightly finer than the other, this refinement of the elements leads to the greater microhardness of heat-treated sample.

The structures of samples hardfaced by alloy 1 are shown in Fig. 2a, b. It was evident that the microstructures showed a dendritic pattern. The sample with heat treatment here showed a fine and dense structure than the untreated sample, which increased the microhardness of this sample.

The weldments of samples hardfaced by alloy 2 are shown in the Fig. 3a, b. The structures were dendritic structures. The sample without heat treatment showed the coarse structure. After the heat treatment, the structure became fine by the proper distribution of the elements through the heat treatment, and thus, the microhardness of the sample was increased.

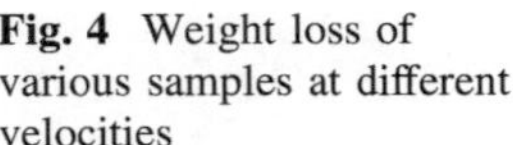

Fig. 4 Weight loss of various samples at different velocities

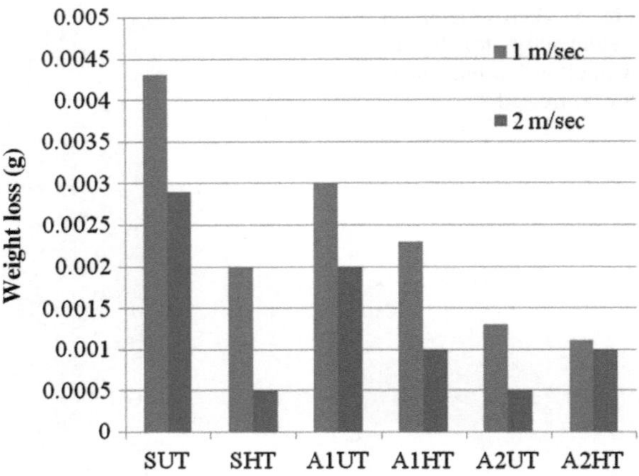

3.3 Wear Test

It showed that the wear of both the untreated and heat-treated metals was reduced when the velocity was increased. It also showed that the weight loss in the heat-treated sample was less than the weight loss of untreated sample, and hence, the wear resistance of the heat-treated sample was increased, it was due to the structure refinement during the heat treatment of the samples.

It is observed that the sample hardfaced by alloy 1 and heat-treated sample has more wear resistance than the untreated sample. This may be due to structure refinement which increased its microhardness and wear resistance.

The weight loss of the sample heat-treated and hardfaced by alloy 2 was reduced because when the sample undergoes heat treatment, the structure was refined, and thus, the microhardness and wear resistance of the sample were increased. It is also observed that the wear of this sample was least from all the samples. So, it proved to be good wear resistance material (Fig. 4).

4 Conclusion

1. It was observed that heat treatment increased the hardness value and the wear resistance of the samples.
2. The hardfacing with the alloy 1 increased the hardness and the wear resistance of the samples.
3. The values of hardness and the wear resistance were increased to a more extent in the samples hardfaced with alloy 2 than the samples hardfaced with alloy 1.
4. Refined structure and formation of carbides are responsible for the improvement in wear resistance.

References

Bing-quan C, Cang-xiu C, Jun-bo P (2003) A high-Fe Aluminum matrix welding filler metal for hardfacing Aluminum–Silicon alloys. J Wuhan Univ Technol Mater Sci 18(1):25–28

Gregory EN (1978) Hardfacing. J Tribol Int 11:129–134

Kumar S, Mondal DP, Khaira HK, Jha AK (1999) Improvement in high stress abrasive wear property of steel by hardfacing. J Mater Eng Perform 8(6):711–715

Lin YC, Chen YC (2013) Reinforcements affect mechanical properties and wear behaviour of WC clad layer by gas tungsten arc welding. J Mater Des 45:6–14

Sabet H, Khierandish S, Mirdamadi S, Goodarzi M (2011) The microstructure and abrasive wear resistance of Fe–Cr–C hardfacing alloys with the composition of hypoeutectic, eutectic, and hypereutectic at Cr C 1/4 6. J Tribol Lett 44:237–245

Wang SS, Ping DL, Chang L, Hui XD (2013) Enhanced mechanical properties induced by refined heat treatment for 9Cr–0.5Mo–1.8W martensitic heat resistant steel. J Mater Des 50:174–180

Zhao J, Lee JH, Kim YW, Jiang Z, Lee CS (2013) Enhancing mechanical properties of a low-carbon microalloyed cast steel by controlled heat treatment. J Mater Sci Eng A 559:427–435

Experimental and Characterization of Aluminium Over Mild Steel by Friction Surface Processing

Sardar Jaspal Singh, M. Prakash and P. Laxminarayana

Abstract Friction surfacing is an advanced manufacturing process, which has been successfully developed over the past decade. The process is used for corrosion- and wear-resistant coatings and for reclamation of worn engineering components. The major requirement is for flexibility to enable rapid changes in process parameters in order to develop new applications, with variations in materials and geometries in terms of cost and reliable manner. So, the present work deals with the solid-state coating by friction surfacing process where aluminium is coated over mild steel substrate. The effects of traverse speed on the geometry, interfacial bond characteristics and mechanical properties of coatings are studied. Metallurgical studies were made using optical microscopy; mechanical tests included bend tests and microhardness tests.

Keywords Friction surfacing · Steel · Al · Tool profile · Bend test · Microstructure · Microhardness · Shear test

1 Introduction

The material surfaces are engineered to impart specific properties, which could be different from those of the core material. These surface modifications are generally carried out to impart wear and corrosion resistance of the substrate materials. Various surface engineering process techniques are adopted for surface

S. J. Singh (✉) · M. Prakash · P. Laxminarayana
Department of Mechanical Engineering, University College of Technology, Osmania University, Hyderabad 500007, Andhra Pradesh, India
e-mail: jaspal85@gmail.com

M. Prakash
e-mail: malothprakash@gmail.com

S. S. Khangura et al. (eds.), *Proceedings of the International Conference on Research and Innovations in Mechanical Engineering*, Lecture Notes in Mechanical Engineering, DOI: 10.1007/978-81-322-1859-3_35, © Springer India 2014

Table 1 Chemical composition of aluminium 6082

Element	Present (%)
Si	0.7–1.3
Fe	0.0–0.5
Cu	0.0–0.1
Mn	0.4–1.0
Mg	0.6–1.2
Zn	0.0–0.2
Ti	0.0–0.1
Cr	0.0–0.25
Al	Balance

modifications that include chemical deposition process, high-temperature diffusion-related process involving chemical reactions, deposition by fusion routes such as weld overlay and flame spray techniques, which are widely employed. Friction surfacing, which is related to friction welding, utilizes the frictional energy dissipated during operation and generates a layer of plasticized metal without the need for external heat source.

The friction surfacing of aluminium presents an inherent problem due to the high thermal conductivity of the metal. However, with aluminium, heat quickly escapes upwards through the rod material.

Effectively the cooling rate of the substrate is increased, thus achieving thermal balance between the rod and the substrate to retain more heat in the metal transfer zone.

2 Experimental Procedure

2.1 Material Specification

A mild steel plate of 250-mm-length, 120-mm-width, and 10-mm-thickness plate was taken and for obtaining good friction between two metals; i.e., this process is done on conventional vertical milling machine, and it can also be done on CNC machines also, but it should be of spring-loaded control machine, because in this CNC machines, the *Z*-axis will remain fixed according to the programmed made by the operator.

2.2 Chemical Composition of Al 6082

See Table 1.

Table 2 Chemical composition of mild/low carbon steel

Element	Content (%)
Carbon, C	0.14–0.20
Iron, Fe	98.81–99.26 (as remainder)
Manganese, Mn	0.60–0.90
Phosphorous, P	≤0.040
Sulphur, S	≤0.050

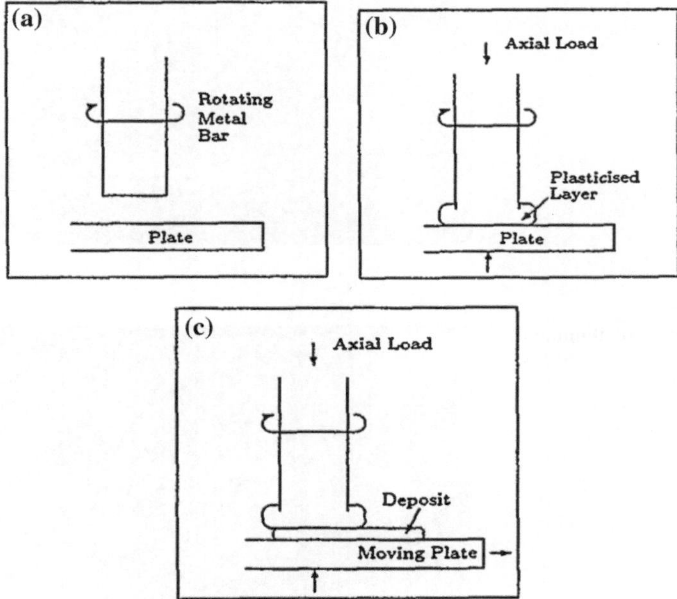

Fig. 1 Principle mechanism of friction surfacing process. **a** First step, **b** second step, **c** final step

2.3 Chemical Composition of Mild/Low Carbon Steel

See Table 2.

3 Principle Mechanism of Friction Surfacing Process

Friction surfacing is a solid-phase cladding technique; it uses a combination of heat and deformation to clean surfaces and metallurgically bond metals together. In its simplest arrangement, a rotating consumable bar is brought into contact, under low load, with stationary substrate. At initial contact, the rotating bar is preferentially heated to form plasticized layer by the frictional motion (Figs. 1, 2, 3, 4).

Fig. 2 Frictional surfacing set-up on conventional milling machining

Fig. 3 Coating of aluminium over mild steel

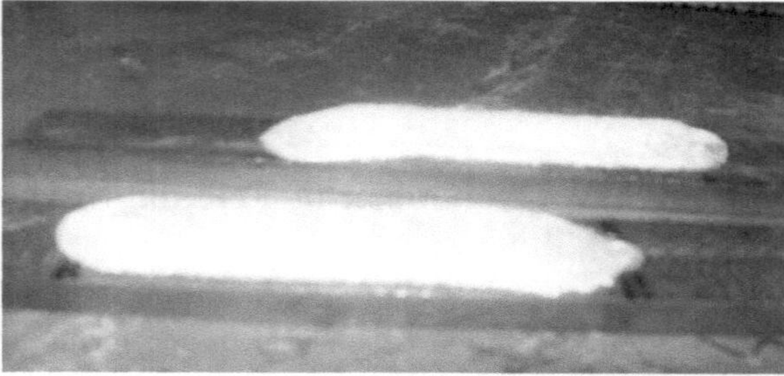

Fig. 4 The formation of layers along the plate after coating of aluminium over mild steel

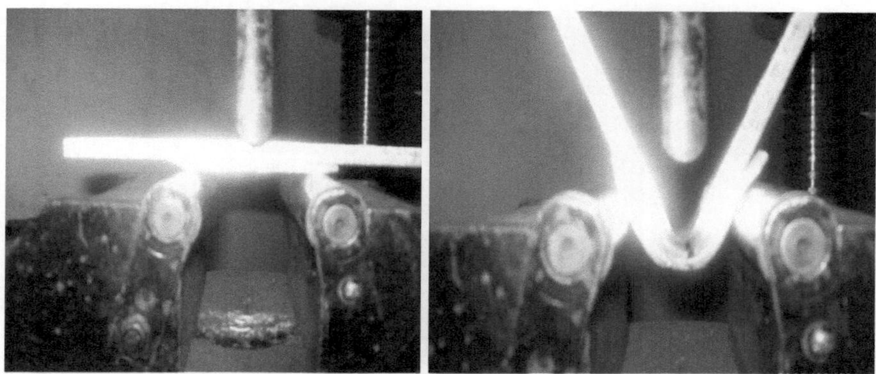

Fig. 5 Before and after testing the specimen

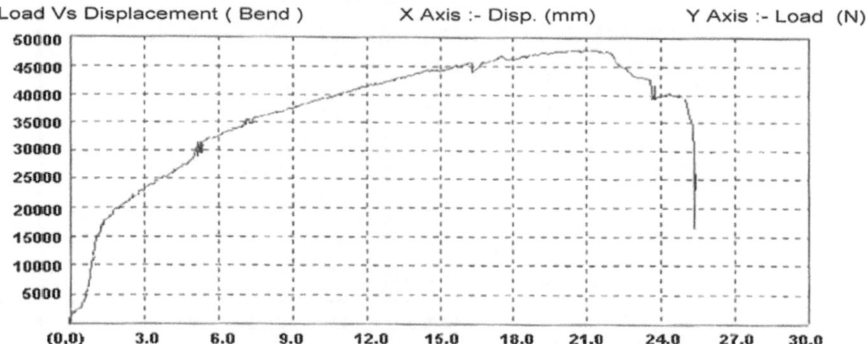

Fig. 6 Graph of load versus displacement

Table 3 Standard Vickers scales

Microhardness scales	Test force F (N)	Low-force hardness scales	Test force F (N)	Macrohardness scales	Test force F (N)
HV 0.01	0.09807	HV 0.2	1.961	HV 5	49.03
HV 0.015	0.1471	HV 0.3	2.942	HV 10	98.07
HV 0.02	0.1961	HV 0.5	4.903	HV 20	196.1
HV 0.025	0.2452	HV 1	9.807	HV 30	294.2
HV 0.05	0.4903	HV 2	19.61	HV 50	490.3
HV 0.1	0.9807	HV 3	29.42	HV 100	980.7

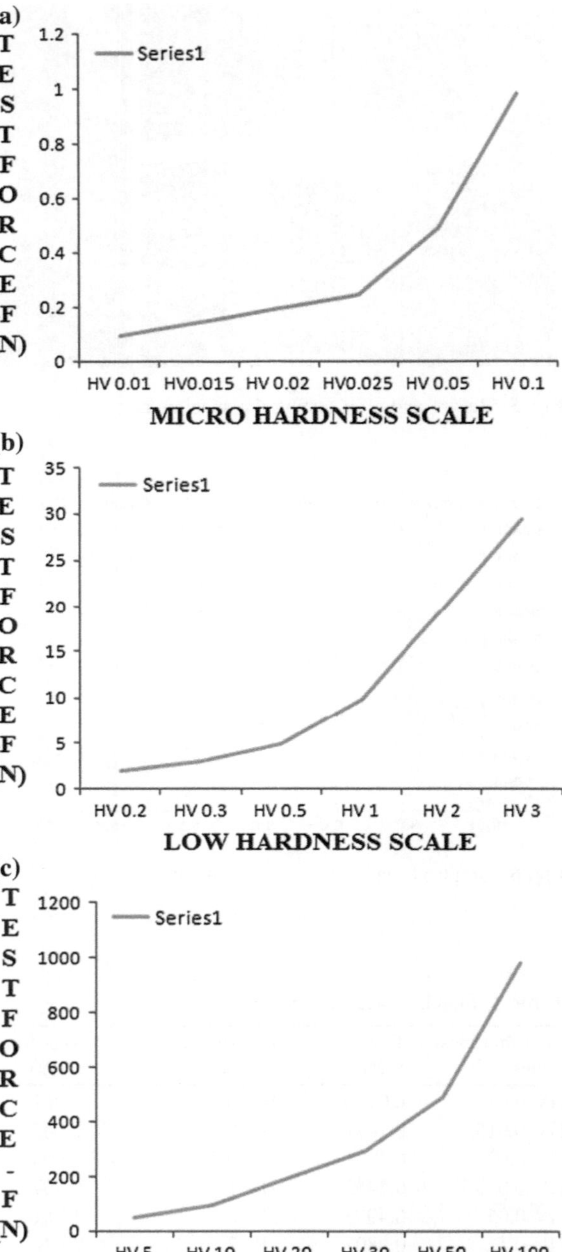

Fig. 7 Graphs of micro hardness scales versus test forces (F), low-force hardness scales versus test forces (F), and macro hardness scales versus test forces

4 Results and Discussion

4.1 Bending Test

This test was conducted on the universal testing machine as the specimen standard is fixed between two supporting blocks and the load is applied on the specimen by gradually increasing the load over the work piece until the breakage occurs in the specimen (Fig. 5).

Graph between load (N) versus displacement (mm) (Fig. 6).

4.2 Vickers Hardness Test

The test is conducted on machines with an indentor of a square-based diamond pyramid having an angle of 136 degrees between opposite faces, and the specimen is carefully prepared and placed on the stage. It is raised upwards to make contact with indentor. The applied load is (5–120 kg in increment of 5 kg) selected according to the thickness and hardness of the specimen (Table 3, Fig. 7).

5 Conclusion

It is observed from friction surfacing, excellent bonding is obtained with no defects such as porosity and oxidation. The material properties such as bond integrity, bending strength and hardness distribution will not change, and there is an increase as compared with before the friction surfacing process. No cracking occurred in the heat-affected zone. Melting of material will not take place as the coat forms at recrystallization temperature only. More advisable process among the existing process used in wide range of applications. From the tests, it is concluding that these friction surfacing products can be used efficiently.

Analysis of Roughness in Turning of EN-8D Alloy Steel Using Cryogenically Treated HSS Tool

Harchand Singh, Jagtar Singh and Anil Kumar Singla

Abstract The purpose of this research was to examine the surface roughness in turning EN-8d alloy steel using HSS tool of grade M-42. Cryogenic treatment of tool was done, and comparison was made for surface roughness between untreated and cryogenic treated tool (CT). In this research work, central composite face-centered design approach of response surface methodology was applied to study the impact of machining parameters on surface roughness for CT. Mathematical model for surface roughness was developed to achieve optimum selection of machining parameters.

Keywords HSS tool · Cryogenic treatment · EN-8d steel · Turning · Surface roughness · Mathematical model

1 Introduction

The increasing importance of turning operations is gaining new dimensions in the present industrial age, in which the growing competition calls for all the efforts to be directed toward the economical manufacture of machined parts. EN-8d is a popular grade of through-hardening medium carbon steel, which is suitable for the

H. Singh (✉)
Sant Longowal Institute of Engineering and Technology, Longowal, Sangrur 148106 Punjab, India
e-mail: harchand987@gmail.com

J. Singh · A. K. Singla
Department of Mechanical Engineering, Sant Longowal Institute of Engineering and Technology, Longowal, Sangrur 148106 Punjab, India
e-mail: jagtarsliet@gmail.com

A. K. Singla
e-mail: anil_singla@yahoo.com

S. S. Khangura et al. (eds.), *Proceedings of the International Conference on Research and Innovations in Mechanical Engineering*, Lecture Notes in Mechanical Engineering, DOI: 10.1007/978-81-322-1859-3_36, © Springer India 2014

manufacture of parts such as general-purpose axles and shafts, spindles, gears, bolts, and studs. So the machining of this kind of material required a cutting tool which is having good strength, toughness, wear resistance, cutting ability, and ability to retain its hardness at high temperatures. High-speed steel is a special alloy steel which contains the alloying elements like tungsten, chromium, vanadium, cobalt, and molybdenum up to 25 %, which is capable to operate safely at higher cutting speeds (Ramji et al. 2010). M-42 super-high-speed steel is a premium cobalt high-speed steel with a chemical composition designed for high hardness and superior hot hardness. It exhibits excellent wear resistance by virtue of high heat-treated hardness (68–70 HRC), and the high cobalt content (8 %) imparts the hot hardness. As such, the cutting edges on tools made from M-42 super-high-speed steel stay sharp as well as hard in heavy-duty and high-production cutting applications.

Quality and productivity play significant role in today's manufacturing market. Surface roughness is one of the most crucial quality measures because the extent of quality of finished product influences the degree of satisfaction of the consumers during product usage. Surface roughness of workpiece is directly related to the tool wear. So the latter can be improved by reducing the former. The tool wear rate or hence the surface roughness can be improved with the help of cryogenic treatment of tool (−125 to −196 °C) which can also improve the metallurgical properties of material (Huang et al. 2003). The main reason for this is the complete transformation of retained austenite into hard martensite plus the formation of very fine carbides uniformly dispersed in the tempered martensitic structure (Molinari et al. 2001). Cryogenic treatment is a supplementary subzero treatment to conventional heat treatment process which is being used for many years but is truly in its infancy when compared to heat treating. Cryogenic treatment is the process in which a material is submitted to subzero temperatures (below 0 °C) in order to enhance the service life through metallurgical changes that occurs during treatment. It can gives surprisingly good results that depend upon the application as some reports showing 92–817 % increases in tool lives after they have being cryogenically treated at −196 °C (Sidhu et al. 2010; Kalsi et al. 2010; da Silva et al. 2006).

After going through the past research, it is concluded that the cryogenic treatment can reduce tool wear, thereby improving the surface roughness. The cryogenic treatment influenced surface roughness to the extent of 34 %. Cutting velocity and feed rate influenced surface roughness by 25.88 and 37.73 %, respectively (Dhar et al. 2009). The comparison of the performance of cryogenically treated and untreated inserts revealed that for the depth of cut 0.8 mm, the surface roughness of the work specimen using the cryogenically treated insert at 200, 250, and 300 m/min was 22.35, 7.58, and 5.94 %, respectively, less than that of the untreated insert, whereas for the depth of cut 1.6 mm, the surface roughness of the work specimen using the cryogenically treated insert at 200, 250, and 300 m/min was 23.62, 20.96, and 21.62 %, respectively, less than that of the untreated insert. Also for the feed rate of 0.15 mm/rev and the depth of cut 0.8 mm, the surface roughness of the work specimen using the cryogenically treated insert at 200, 250, and 300 m/min was 22.17, 13.97, and 31.87 %

respectively, less than that of the untreated insert. When the depth of cut was further increased to 1.6, the flank wear of the cryogenically treated insert was found to be 15.95, 8.54, and 5.04 less than that of the untreated insert. The above comparisons clearly show that the cryogenically treated inserts provide better surface finish than that of the untreated inserts. As the cryogenic treatment imparts higher wear resistance on cutting tool inserts, the keenness of the cutting edge was maintained in the case of cryogenically treated inserts, and this was the reason for better surface finish (Vadivel and Rudramoorthy 2009).

Improved machinability and surface roughness with cryogenic cooling is due to reduction in diffusion wear mechanism and less degradation of tool hardness or a reduction in yielding strength at lower temperatures (Hong 1991). Turning of AISI 1060 steel was performed by carbide inserts using cryogenic cooling with liquid nitrogen jet which results in the reduction in tool wear and improvement in surface finish as compared to wet and dry machining. The use of liquid nitrogen in cryogenic cooling is helpful in the retention of tool hardness and favorable interactions of chip and tool as well as workpiece and tool (Paul et al. 2001).

The metallurgical investigation of cryogenic treated HSS tool reveals that the retained austenite structure was completely converted into martensite structure, after cryogenic treatment and hardness was improved by 17 %. The process also promotes the precipitation of small carbide particles in tool steels with proper alloying metals. The fine carbides act as hard areas with a low coefficient of friction in the metal that increases the wear resistance of the metals, thereby improving surface roughness (Sendooran and Raja 2011). It can be observed that there is an increase in grain size of P-32 tungsten carbide inserts due to cryogenic treatment, thereby increasing toughness of tool (Sreerama Reddy et al. 2007) which may result in improving the surface roughness of workpiece. After the turning of EN-31 steel (at cutting speeds of 1,000, 1,500, and 2,000 rpm, feed rate of 0.10, 0.15, and 0.20 mm/rev and depth of cut of 0.1, 0.2, and 0.3 mm), it was investigated that the surface roughness decreased with the decrease in feed and depth of cut but constant with increase in speed (Barik and Mandel 2010).

2 Methodology

2.1 Preparation of Cutting Tool and Workpiece

The tool bit of M-42 HSS was selected for preparing single-point cutting tool. The chemical composition of tool material was (weight %) 1.1 C, 3.9 Cr, 9.2 Mo, 1.4 W, 7.8 Co, and 1.2 V. The tool material was purchased in the form of square bits having length 75 mm and width and thickness 12.5 mm each, with both the ends beveled. The geometry of tool was prepared (using standard angles) on tool and cutter grinder.

Work material selected for the research work was EN-8D, which is a very popular grade of through-hardening medium carbon steel. The long rod of EN-8d steel was cut into 36 numbers of pieces. Each piece was having length and diameter of 75 mm and 24.5 mm, respectively. The chemical composition of workpiece was (weight %) 0.430 C, 0.220 Si, 0.720 Mn, 0.30 P, 0.119 Cr, 0.118 Ni, 0.125 Cu, and 98.06 Fe.

2.2 Cryogenic Treatment

One of the tools was cryogenically treated (using standard cycle) at Institute of Auto Parts, Ludhiana. The recommended cycle for this material consists of cooling the material to −196 °C followed by the three cycles of heating to +196 °C for tempering.

Following steps were taken for the cryogenic treatment of tool

- Step 1: Cooling to −196 °C (6 hours at the rate of 0.5 °C/min).
- Step 2: Cold soaking at −196 °C for 24 hours.
- Step 3: Heating to +196 °C (12 hours at the rate of 0.5 °C/min).
- Step 4: Hot stabilization at +196 °C for 2 hours.
- Step 5: Cooling to room temperature (1 hour average).
- Step 6: Stabilization at room temperature for 2 hours.
- Step 7: Heating to +196 °C (I hour average).
- Step 5–7 were repeated three times.

2.3 Experimental Procedure

The turning operation was performed on CNC lathe machine at Central Tool Room, Ludhiana. Initially, the trail experiments were performed in order to specify the lower and upper limit of machining parameters. Cutting speed, feed, and depth of cut were taken as variable parameters whose effect on the surface roughness had to be studied.

The lower and upper limits of variable parameters along with middle level are shown in Table 1. Central composite face-centered design approach of response surface methodology (Montgomery 2012) was used to develop the design matrix for experimentation. Numbers of experiments to be conducted were calculated by adding the number of factorial, axial, and center point runs (Montgomery 2012). So the total number of experiments to be conducted was calculated as 18. The design matrix used for the experimentation is shown in Table 2. The complete set of eighteen experiments was repeated two times, one for untreated tool (UT) and other for cryogenic treated tool (CT) in order to compare the “surface roughness” of workpieces. The experiments were performed in a random order so as to avoid any systematic error.

Table 1 Machining parameters

Parameters	Units	Symbols	Lower limit	Middle value	Upper limit
Speed	r.p.m	S	400	500	600
Feed	mm/rev	F	0.06	0.08	0.1
Depth of cut	mm	D	0.2	0.4	0.6

Table 2 Design matrix used for experimentations

Std. order	Run order	Speed (rpm)	Feed (mm/rev.)	Depth of cut (mm)
1	14	400	0.06	0.2
2	6	600	0.06	0.2
3	9	400	0.1	0.2
4	5	600	0.1	0.2
5	11	400	0.06	0.6
6	3	600	0.06	0.6
7	18	400	0.1	0.6
8	15	600	0.1	0.6
9	13	400	0.08	0.4
10	12	600	0.08	0.4
11	8	500	0.06	0.4
12	4	500	0.1	0.4
13	10	500	0.08	0.2
14	1	500	0.08	0.6
15	2	500	0.08	0.4
16	16	500	0.08	0.4
17	7	500	0.08	0.4
18	17	500	0.08	0.4

Surface roughness for all the set of specimens was measured with the help of Surfcorder SE-1200. Variation in the surface roughness of the workpieces was observed due to different combinations of parameters, such as speed, feed, and depth of cut, used for turning. The center line average value of surface roughness, i.e., R_a was obtained from the surface roughness readings on the digital screen of Surfcorder SE-1200 as shown in Table 3. Comparison between surface roughness with CT and UT (at same combination of parameters) was made as well as percentage improvement in response (surface roughness) was also calculated. Mathematical model for surface roughness was developed, and the influence of machining parameters on response was investigated.

2.4 Mathematical Modeling

The mathematical model was constructed using Design-Expert software for the surface roughness of workpiece (related to cryogenically treated tool), after which

Table 3 Observed surface roughness for untreated tool (UT) and cryogenic treated tool (CT)

Std. Order	SR (UT) in μm	SR (CT) in μm
1	2.404	2.019
2	4.859	3.167
3	5.263	4.697
4	7.682	4.742
5	4.572	4.309
6	7.569	4.611
7	6.216	5.679
8	4.415	4.206
9	5.829	4.381
10	4.258	4.111
11	3.841	3.185
12	6.53	4.146
13	6.884	3.797
14	7.182	4.596
15	6.042	3.146
16	7.419	3.719
17	5.89	3.775
18	5.237	3.06

Table 4 Regression coefficients for model

Coefficient of regression	Factor	Value
b_0	Combined effect of all factors	−4.19236
b_1	Speed (S)	−0.020934
b_2	Feed (F)	230.18964
b_3	Depth of cut (D)	10.88986
b_{12}	Interaction of S and F	−0.17987
b_{13}	Interaction of S and D	−0.014775
b_{23}	Interaction of F and D	−102.75

the effect of machining parameters on the response was determined. The model development can be accomplished after the evaluation of regression coefficients as shown in Table 4, by using least square method (Montgomery 2012).

Adequacy of the model can be tested by analysis of variances, and the significance test was carried out to check the significance of regression coefficients (Montgomery 2012) as shown in Table 5. The F values obtained from model were compared with F values of standard table at 95 % level of confidence. The null hypothesis H_0 was to be accepted when the F values obtained by model were less than the tabulated F values, whereas H_0 was to be rejected if the F values of model exceed the tabulated F values. The F statistic that was derived from the mean squares was converted into its corresponding P values which gave us an indication that, whether the corresponding coefficients of regression were significant or not. Some important guidelines about P values statistic, which were followed while developing the model [one f], are as follows:

Table 5 Analysis of variance

Source	SS	*DF*	MS	*F* value	*P* value	Status
S	0.0062	1	0.0062	0.048	0.8318	Insignificant
F	3.82	1	3.82	29.89	0.0006	Significant
D	2.48	1	2.48	19.41	0.0023	Significant
S^2	0.46	1	0.46	3.56	0.0958	Insignificant
F^2	0.079	1	0.079	0.62	0.4546	Insignificant
D^2	0.35	1	0.35	2.75	0.1356	Insignificant
SF	1.04	1	1.04	8.11	0.0216	Significant
SD	0.7	1	0.7	5.47	0.0475	Significant
FD	1.35	1	1.35	10.58	0.0117	Significant
Model	10.92	9	1.21	9.5	0.0021	Significant
Lack of fit	0.6	5	0.12	0.86	0.5895	Insignificant

- If $p < 0.05$, then the model is statistically significant.
- If $p < 0.05 < 0.10$, then the model might be significant.
- If $p < 0.10$, then the model is not significant.

Final model was obtained by neglecting all the insignificant terms from the model. So an adequate and precise model equation for surface roughness can be written as follows:

$$Y_{SR} = -4.19236 + 230.18964 \times F + 10.88986 \times D - 0.17987 \times SF - 0.014775 \times SD - 102.75 \times FD$$

3 Results and Discussion

The surface roughness with UT and CT was measured and compared with each other, for same combinations of machining parameters. It was concluded that the surface roughness with CT was improved by 27.56 % with respect to UT. The influence of different significant parameters on surface roughness was studied.

3.1 Influence of Feed Rate on Surface Roughness

The relation between feed and surface roughness is shown in Fig. 1. It was observed that the surface roughness increases with increase in feed rate. This was due to the fact that as the feed rate increases, the cutting forces required for machining would also increase and hence more heat would be generated between the tool–chip interfaces which in turn increase surface roughness. The relation between surface roughness and feed rate can be written as follows:

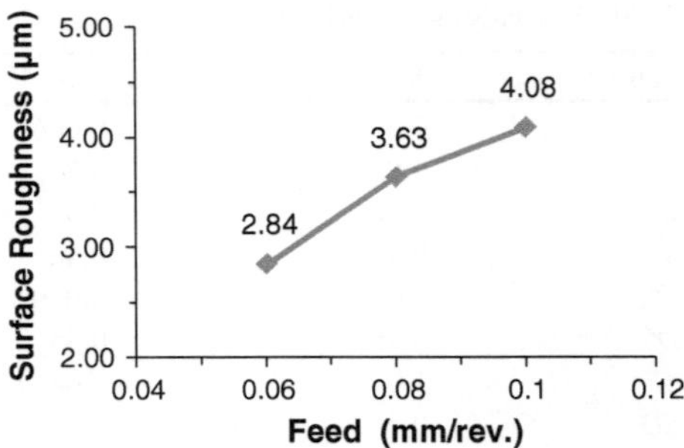

Fig. 1 Surface roughness versus feed

$$R_a = \frac{F^2}{32\,r}$$

where R_a is average surface roughness, F is feed rate, and r is nose radius. So we can say that at fixed nose radius, the surface roughness of workpiece increases with the square of feed rate during operation. The minimum value of surface roughness obtained was 2.84 µm at the feed rate of 0.06 mm/rev, whereas the maximum value of surface roughness obtained is 4.08 µm at the feed rate of 0.1 mm/rev, which implies that the surface roughness increases with the increase in feed rate.

3.2 Influence of Depth of Cut on Surface Roughness

The relation between surface roughness and depth of cut is shown in Fig. 2. It can be concluded that with increase in depth of cut, surface roughness also increases. As the depth of cut increases, the area of contact between the tool and workpiece increases, which results in increased friction. The high value of friction is responsible for the increase in temperature and thus generates a deformation zone which results in increasing surface roughness. The minimum value of surface roughness obtained was 3.49 µm at the depth of cut of 0.2 mm, whereas the maximum value of surface roughness obtained was 4.49 µm at the depth of cut of 0.6, which implies that the surface roughness increases with the increase in depth of cut.

3.3 Influence of Interaction of Feed and Cutting Speed on Surface Roughness

The combined effect of surface roughness and feed rate is shown in Fig. 3. It can be observed that at constant cutting speed, the surface roughness increases with increase in feed rate. It was also concluded that the surface roughness decreases

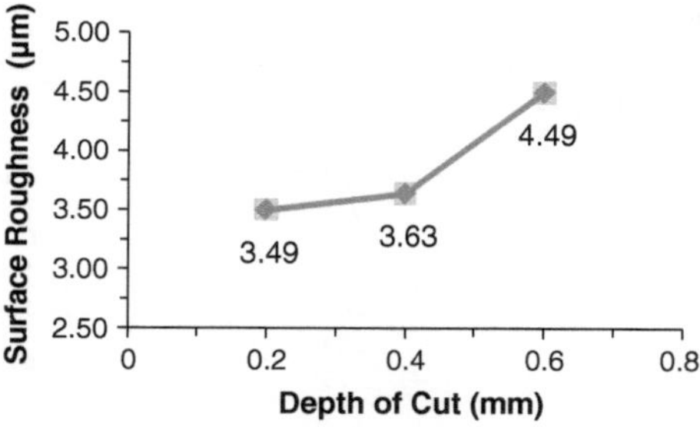

Fig. 2 Surface roughness versus depth of cut

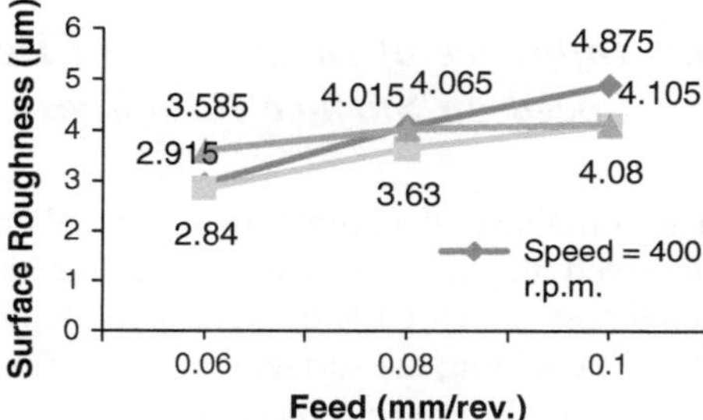

Fig. 3 Surface roughness versus feed and cutting speed

with increase in cutting speed up to the certain limit called critical speed, but once the critical speed is reached, the surface roughness starts increasing. This was due to the reason that low cutting speed contributes to the formation of built-up edges, which results in the welding of chip material with tool tip and hence rough turning operation. Also at low cutting speed, the chip thickness was large and curl radii were smaller, which increase machining force that results in higher surface roughness.

At higher cutting speeds up to critical limit, the temperature between workpiece and tool increases, which facilitates reconstruction of machined surface and thereby wiping of the flaws and defect. The chip thickness was small, and curl radii were higher, which reduce machining force, thereby decreasing surface roughness. But for the cutting speed higher than critical limit, high heat was generated due to which plastic deformation of workpiece occurred, which results in degradation of workpiece and causes vibrations or tool chattering, thereby increasing surface roughness.

It was also investigated that the minimum value of surface roughness obtained was 2.84 μm at cutting speed of 500 rpm and feed rate of 0.06 mm/rev, whereas the maximum value of surface roughness obtained was 4.875 μm at cutting speed of 400 rpm and feed rate of 0.1 mm/rev, which gave the good indication of decreasing surface roughness with decrease in feed rate and increase in cutting speed. Here, 500 rpm is the critical speed above which surface roughness starts to decrease.

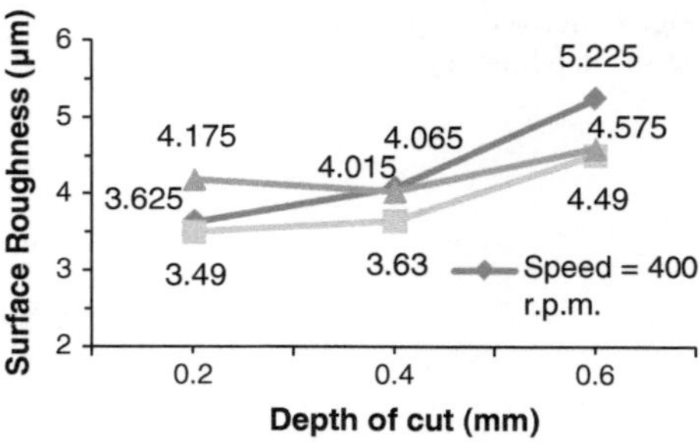

Fig. 4 Surface roughness versus depth of cut and cutting speed

3.4 Influence of Interaction of Depth of Cut and Cutting Speed on Surface Roughness

The combined effect of depth of cut and cutting speed is shown in Fig. 4. It can be observed that at constant cutting speeds up to the critical limit, the surface roughness increased with increase in depth of cut. At higher cutting speeds above the critical limit, the surface roughness first decreased with increase in depth of cut because higher cutting speed dominated over the depth of cut up to a certain limit and then increased with further increase in depth of cut where the effect of higher depth of cuts dominated over the cutting speed. It was also concluded that at constant depth of cuts, the surface roughness decreased with increase in cutting speed up to critical limit.

It was also investigated from the Fig. 4 that the minimum value of surface roughness obtained was 3.49 µm at cutting speed of 500 rpm and depth of cut of 0.2 mm, whereas the maximum value of surface roughness obtained was 5.225 µm at cutting speed of 400 rpm and depth of cut of 0.6 mm, which reveals that the surface roughness decreases with decrease in depth of cut and increase in cutting speed up to critical limit above which it starts increasing.

3.5 Influence of Interaction of Feed Rate and Depth of Cut on Surface Roughness

The combined effect of feed rate and depth of cut is shown in Fig. 5. It was observed that at a constant depth of cuts, the surface roughness increased with increase in feed rates and at constant feed rates up to a certain limit, the surface roughness increased with increase in depth of cut, whereas for higher feed rates, the surface roughness first decreased with increase in depth of cut up to a certain limit called critical limit and then increased with increase in depth of cut beyond that critical limit. That type of variation was due to the fact that large cutting forces required for higher feed rates were overcome by lower depth of cuts and also the effect of cutting speed may dominate the effect of feed rate and depth of cut. At

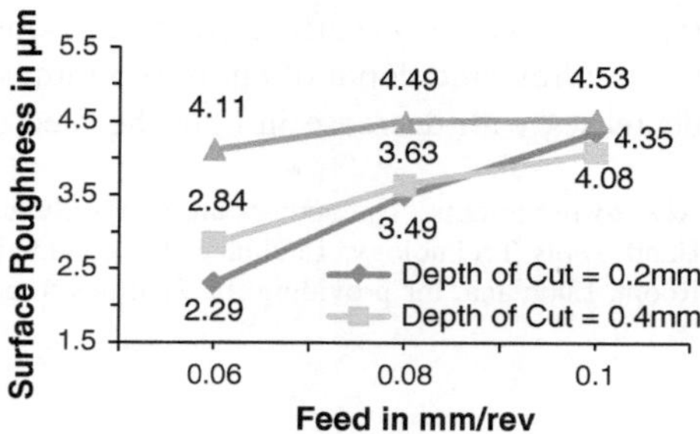

Fig. 5 Surface roughness versus feed and depth of cut

higher feed rates and depth of cuts, the surface roughness increased due to the positive effect of both the parameters.

It is also observed in the Fig. 5 that the minimum value of surface roughness obtained was 2.29 μm at feed rate of 0.06 mm/rev and the depth of cut of 0.2 mm, whereas the maximum value of surface roughness obtained was 4.53 μm at feed rate of 0.1 mm/rev and the depth of cut of 0.6 mm, which reveals that the surface roughness increases with increase in both the feed rate and depth of cut.

4 Conclusion

The surface roughness was improved by 27.56 % when CT was used as compared to UT. The surface roughness increases with increase in feed rate, as the minimum and maximum values of surface roughness obtained were 2.84 and 4.08 μm at feed rate of 0.06 and 0.1 mm/rev., respectively.

Surface roughness also increases with increase in depth of cut as the minimum and maximum values of surface roughness obtained were 3.49 and 4.49 μm at the depth of cut 0.2 and 0.6 mm, respectively.

It was also concluded that the surface roughness increased with the combined effect of feed and cutting speed as the minimum value of surface roughness obtained was 2.84 μm at feed rate of 0.06 mm/rev. and cutting speed of 500 rpm, whereas maximum value of surface roughness obtained was 4.875 μm at feed rate of 0.1 mm/rev. and cutting speed of 400 rpm. So we can say that surface roughness decreases with decrease in feed rate and increase in cutting speed.

The increase in surface roughness was also examined with the combined effect of depth of cut and cutting speed as the minimum value of surface roughness obtained was 3.49 μm at depth of cut 0.2 mm and cutting speed of 500 rpm, whereas the maximum value of surface roughness obtained was 5.225 μm at depth of cut 0.6 mm and cutting speed of 400 rpm, which indicates that the surface roughness decreases with decrease in depth of cut and increase in cutting speed.

Increase in surface roughness was also observed due to the combined effect of feed rate and depth of cut as the minimum value of surface roughness obtained was 2.29 μm at feed rate of 0.06 mm/rev. and depth of cut of 0.2 mm, whereas maximum value of surface roughness obtained was 4.53 μm at the feed rate of 0.1 mm/rev. and depth of cut of 0.6 mm, which reveals that the surface roughness decreases with decrease in both the feed rate and depth of cut.

Acknowledgments The authors are gratefully acknowledged to the institute of Auto Parts and Hand Tools Technology, Ludhiana, for conducting the cryogenic treatment and Central Tool Room, Ludhiana, for providing the facilities to conduct the experimentation.

References

Barik CR, Mandel NK (2010) Parametric effect and optimization of surface roughness of EN 31 in CNC dry turning. Int J Lean Thinking 3(2):1–13

da Silva FJ, Franco SD, Machado AR, Ezugwu EO, Souza A Jr (2006) Performance of cryogenically treated HSS tools. Int J Wear 261:674–685

Dhar NR, Paul S, Chattopadhyay AB (2009) Influence of cryogenic cooling on tool wear, dimensional accuracy and surface finish in turning AISI, 1040 and E4340C steels. Wear 249:932–942

Hong SY (1991) Economical cryogenic machining high speed cutting of difficult- to-machine materials. In: Proceedings of first international conference on manufacturing technology, Hong Kong, 27–29 Dec 1991

Huang JY, Zhu YT, Liao XZ, Beyerlein IJ, Bourke MA, Mitchell TE (2003) Microstructure of cryogenic treated M2 tool steel. Mater Sci Eng A339:241–244

Kalsi NS, Sehgal R, Sharma VS (2010) Cryogenic treatment of tool materials: a review. Mater Manuf Process 25:1077–1100

Molinari A, Pellizzari M, Gialanella S, Straffelini G, Stiasny kH (2001) Effect of deep cryogenic treatment on the mechanical properties of tool steels. J Mater Process Technol 118:350355

Montgomery DC (2012) The text book of design and analysis of experiments. Wiley, Singapore, pp 394–456

Paul S, Dhar NR, Chattopadhyay AB (2001) Beneficial effects of cryogenic cooling over dry and wet machining on tool wear and surface finish in turning AISI, 1060 steel. J Mater Process Technol 116:44–48

Ramji BR, Narasimha Murthy HN, Krishna M (2010) Analysis of roughness and flank wear in turning grey cast iron using cryogenically treated cutting tool. Res J Appl Sci Eng Technol 2(5):414–417

Sendooran S, Raja P (2011) Metallurgical investigation on cryogenic treated HSS tool. Int J Eng Sci Technol (IJEST) 5(5):3992–3996

Sidhu HS, Gaurav K, Bhatia R (2010) Life enhancement of single point cutting tool by hard facing and cryogenic treatment. Int J Eng Sci Technol 4(1):19–23

Sreerama Reddy TV, Ajaykumar BS, Venkatarama Reddy M, Venkataram R (2007) Improvement of tool life of cryogenic treated P 30 tools. In: International conference on advanced materials and composites (ICAMC-2007), 24–26 Oct 2007

Vadivel K, Rudramoorthy R (2009) Performance analysis of cryogenically treated coated carbide inserts. Int J Adv Manuf Technol 42:222–232

Rice Husk-Reinforced Composites: A Review

Rajendra Kumar and Tejeet Singh

Abstract Due to the environmental considerations, cost reduction, and high performance of engineering application, the demand of natural fibers is increasing day by day. Between these natural fibers, the rice husks have been extensively exploited. Therefore, this article overviews the used area of rice husk composites in terms of mechanical, electrical, thermal properties, and coupling agents. For good bonding, coupling agent such as titanate, silane, and zirconate can be used for enhanced mechanical properties. Application of RHPP composite is improving biodegradability, for automobile (interior parts, cabin lining, exterior parts, etc.), electrical (electrical connector, microwave containers, etc.), and food packaging.

Keywords Rice husk · Polypropylene · Coupling agent mechanical · Electrical · Thermal properties

1 Introduction

A composite is a structural material that of two or more combined constituents that is combined at macroscopic level and is not soluble in each other. Thus, the recent research and development efforts have led to new products based on natural resources. Some of these are biodegradable polymers like polylactic acid (PLA), cellulose esters, polyhydroxyalkanoates, and starch polymers. Furthermore, natural fiber-reinforced polymers made with natural fibers like flax, hemp, kenaf, jute, or

R. Kumar (✉)
Government Polytechnic, Firozabad, Uttar Pradesh, India
e-mail: radaur_rudra@yahoo.co.in

T. Singh
Saheed Bhagat Singh Technical Campus, Firozepur, Punjab, India
e-mail: tejeetsingh@rediffmail.com

S. S. Khangura et al. (eds.), *Proceedings of the International Conference on Research and Innovations in Mechanical Engineering*, Lecture Notes in Mechanical Engineering,
DOI: 10.1007/978-81-322-1859-3_37,

cotton fibers are important research and development (R&D) achievements. Natural fibers as reinforcement for petrochemical polymers like PP (polypropylene) have already been established in the automotive (Bax and Mussig 2008). Wood plastic composite is very suitable to achieve durability without using toxic chemicals. These are light in weight, non-toxic, low cost, and safe (Ashori 2008). Polymer composites consist of resin and a reinforcement, two main constituents chosen according to the desired mechanical properties, and the application for which they are to be employed (Chauhan et al. 2011).

In this scientific review article, the overall characteristics of rice husk (RH)-reinforced composites, in terms of mechanical properties, thermal properties, electrical properties, and water absorption properties, will be reviewed. We conclude that a specific review article on the overall characteristics of rice husk-reinforced composites has not yet been published; it is believed that such an article should be of significant value to the composites' research area.

1.1 Natural Fiber Composites

Over the last few decades, the natural fiber area has been growing interest in composite applications. These types of composite present many advantages compare to synthetic fibers, such as low tool wear, low density, cheaper cost availability, and biodegradability (Wambua et al. 2003). Some other important advantages of natural fibers are low specific weight, environmental friendly, carbon dioxide sequestration, producible low investment, good thermal and acoustic insulating properties, no skin irritation, thermal recycling is possible (Kumar et al.), and high specific mechanical performance (Lu et al. 2003). The most common natural plants used in applications are blast fibers, such as hemp, jute, flax, kenaf, and sisal (Wambua et al. 2003). Some also other important natural plants from agricultural sources are banana, pineapple, rubber wood, palm oil fruit bunch, rice husk (Kumar et al. 2010). Rice husk is one of such major agro-waste products, which contains cellulose 50 %, lignin 25–30 %, and ash 15–20 % silica (Ismail and Waliuddint 1996). Natural fibers found in all life cycles of all walks of life. These fibers are subdivided based on their origins, that is, whether they are derived from plants, animals, or minerals or consumer waste as shown in Fig. 1. The chemical composition, moisture content as well as the structure of plant fibers, is fairly complicated (Akil et al. 2011; Bismarck et al. 2005).

The main constituent of any plant fiber is cellulose (Chawla 1998). Cellulose is hydroscopic. Most polymeric fibers swell due to moisture absorption. This absorption leads to alterations in weights and dimensions, as well as in strengths, and stiffness (Akil et al. 2011). High moisture absorption could also cause deterioration in mechanical properties and loss of dimensional stability (Alvarez et al. 2004; Baiardo et al. 2004). The advantages of natural fibers as reinforcement for polymers include low density, wide availability, and biodegradability (Karnani et al. 1997).

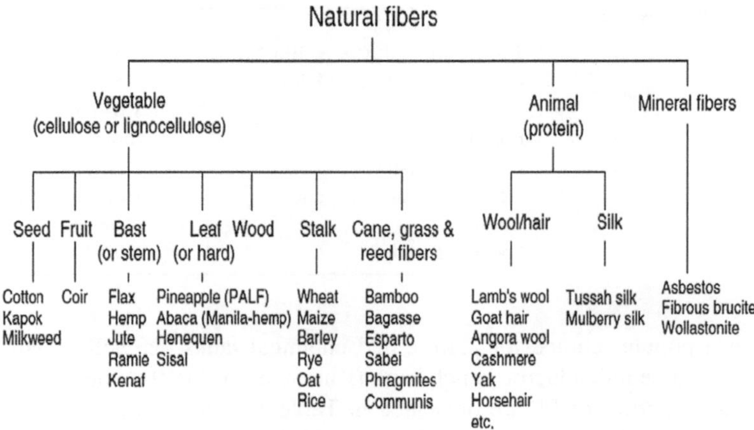

Fig. 1 Classification of natural fibers

Rice husk is available in abundance in nature with several important applications: as pet food fiber, pillow stuffing, fertilizer, etc. Rice hulls can be utilized for brewing beer to increase the lautering ability of a mash. (Kumar et al. 2010). Rice husk and other agricultural residues are generated as a byproduct during the different industries process (Rosa et al. 2012), (Satyanarayana et al. 2005). Plastics such as polyethylene, PP, and polystyrene can be used as matrices having reinforcements of glass, carbon, aramid fibers, or fillers such as talc, $CaCO_3$, and mica. Every filler has a specific property that becomes incorporated into the polymer matrix. Some of the common beneficial properties are lower cost, lower coefficient of expansion, and reduced cycle time due to high conductivity of filler (Ashori 2008).

1.2 Polypropylene (PP)

Polypropylene is an economical material that offers a combination of outstanding physical, chemical, mechanical, thermal, and electrical properties. Some of the important properties of PP are as non-toxic, low weight, easily fabricated, low moisture absorption, excellent dielectric properties, melting temperature, molding temperature, specific heat, heat required to melt, and heat required to cooling (Kumar et al. 2010; Whelan and Goff 1986). The PP is higher temperature resistance material, due to which it suitable for many items such as funnels, bottle, trays, pails, and instrument jars that have to be frequently used for medical environment.

It has an excellent resistance to acids and alkalis, but poor resistance to aromatic, aliphatic, and chlorinated solvents. The superior chemical resistance of PP

Table 1 Polypropylene property

Elongation (%)	200–700
Tensile strength (68 × 106 N/m^2)	4.3–5.5
Impact strength (53.14 J/m)	0.5–2.0
Tensile modulus (105)	1.6–2.3
Burning rate	Slow
Effect of alkalis	Resistant
Effect of organic solvents	Resistant below 80 °C
Clarity	Opaque
Specific gravity	0.90–0.91

makes it a popular choice for plating and chemical tanks, as well as laboratory cabinetry and semiconductor bench tops (Kumar et al. 2010). Some more other important properties of PP are presented in Table 1 (Kumar et al. 2011).

Reason of excellent properties is the use of PP increasing day by day. PP can be reinforced with filler such as $CaCO_3$, talc, mica, zirconia, wollastonite, etc., to improve the mechanical and thermal properties (Dearmitt and Breese 2001; Jilken et al. 1991).

2 Properties of Rice Husk Polypropylene (Rhpp)-Reinforced Composites

2.1 Mechanical Properties

The performance of materials is always presented in terms of their mechanical characteristics, such as tensile properties, flexural properties, compression properties, impact properties, and wear behavior. These characteristics are important to determine material ability, especially under extreme and critical conditions, which are directly connected with engineering performance.

Rice husk is one of the natural plants' fiber used as reinforcement in polymer matrix composites (PMC). Rice husk when burnt in the open air outside the rice mill yields two types of fillers, that is, white rice husk ash (WRHA) and black rice husk ash (BRHA). These ashes filled into epoxidized natural rubber (ENR) (50 mol % epoxidation grade). The effect of coupling agent and tensile, tear, and hardness properties was studied and found that WRHA exhibits the best overall properties as compared with BRHA (Ishak and Bakar 1994). The flexural modulus increased with filler content, while tensile strength, elongation at break, and Izod impact strength showed a decrease (Fuad et al. 1995). Residues from plants are increasing alternatives as cellulose sources for several applications. Cellulose content expected for rice husk resulted in a yield (Sud et al. 2008). Tensile strengths of the composites decreased with filler loading, but these properties were significantly improved by addition of compatibilizing agent (Yang et al. 2005).

WRHA-filled PP and ethylene-diene monomer (EPDM) blends were prepared using five different mixing sequences in two types of internal mixers. The mixing sequences were studied in order of the addition of (i) WRHA followed by EPDM to molten PP, (ii) EPDM followed by WRHA to molten PP, (iii) master batch of EPDM and WRHA, (iv) PP followed by WRHA to molten EPDM, and (v) WRHA followed by PP to molten EPDM. All prepared blends had similar ultimate tensile strength, tear strength, and hardness except for the blend prepared by the fourth mixing sequence; it had the highest tensile strength but the lowest elongation at break. The blend formed in the fifth sequence showed the highest elongation at break (%) and comparable tensile strength (Siriwardena et al. 2000). Four levels of filler loading (10, 20, 30, and 40 wt%) sample were prepared. In tensile testing, six levels of test temperature (−30, 0, 20, 50, 80, and 1,108 °C), and five levels of crosshead speed (2, 10, 100, 500, and 1,500 mm/min) were performed and found that the tensile strengths of composites slightly decreased with increases in filler loading, whereas tensile modulus improved with the higher filler loading. Notched and unnotched Izod impact strengths were lowered by addition of rice husk flour. The composite became brittle at higher crosshead speed and showed plastic deformation with increasing test temperature (Yang et al. 2004). Improvement in the hardness and impact strength of the composites are by the use of these additives. Possible applications of the rice husk PP composites are in the building and automotive sectors (Kumar et al. 2012). Mechanical properties such as Young's modulus and flexural modulus increased, whereas yield strength and elongation at break decreased with the increase in filler loading for unmodified and ground talc and rice husk composite. The RHP composites exhibited lower yield strength, Young's modulus, flexural modulus, and higher elongation at break talc composites (Premalal et al. 2002). Mechanical properties decreased as filler loading increased, but the composites had an acceptable level. Rice husk polyolefin composites are suitable for bathroom and food packaging (Yang et al. 2005). The flexural, tensile, and impact properties increased with increases in RH percentage in most of the tests. However, after exceeding a threshold value, properties tended to decrease. RH size also played a significant role; smaller RH size produced composites of higher strength (Rozman et al. 2002). Tensile strength, elongation at break, and mass swell decreased, but tensile modulus and hardness increased by increasing WHRA loading in NR/LLDPE blends (Ismail et al. 2000).

Mechanical properties showed that tensile strength of PLA/ MRSF composite was lower than PLA/RSF composites (Qin et al. 2011).

In summary, the mechanical properties of rice husk-reinforced composite and other natural fibers are collected together and shown Fig. 2. Data for the other natural fillers were taken from existing literature, where these composites were prepared using PP, with natural fiber randomly spread between them (Akil et al. 2011; Rozman et al. 2004; Nachtigall et al. 2007).

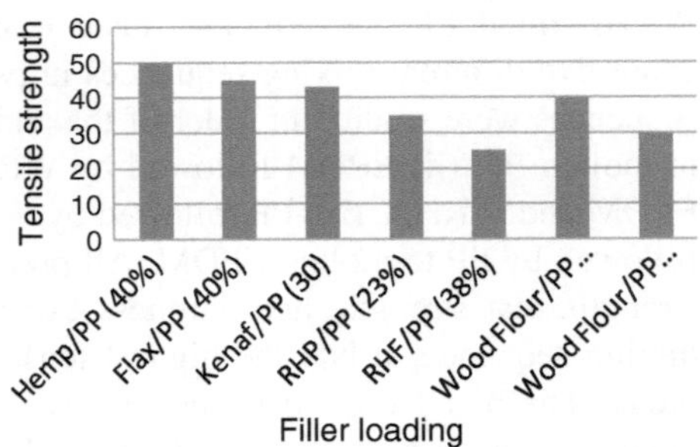

Fig. 2 Comparison on tensile strength (MPa) of rice husk/PP composite to the other natural fiber composites

2.2 Thermal Properties

Thermal property studies are another essential characterization, which need to be considered to fully distinguish the overall behavior of rice husk (i.e., ash, powder, straw) fiber-reinforced composite. For this reason, three types of regular characterization methods were presented, such as different scanning calorimetric (DSC), thermogravimetric analysis (TGA), and dynamic mechanical analysis (DMA).

Thermal stability of rice husk cellulose helps to improve the knowledge on the complex behavior of cellulose degradation (Sud et al. 2008).

Microscopic techniques, such as light microscopy, scanning electron microscopy, and field-emission SEM (FE-SEM), were used to observe the surface and internal structure of rice husk. Microscopic examination showed that two main components of husk, lemma, and palea consisted of outer epidermis, layers of fibers, vascular bundles, parenchyma cells, and inner epidermis, in sequence from the outer to the inner surface. Histochemical staining showed that epidermal and fiber cell walls were lignified, and the walls of parenchyma and lower epidermal cells were not lignified. The outer epidermal walls were extremely thick, highly convoluted, and lignified (Park et al. 2003). Experiments carried out on Brunei rice husk (BRH) to obtain bio-oil using a fixed-bed pyrolysis rig. ZSM-5, Al-MCM-41, Al-MSU-F, and Brunei rice husk ash (BRHA) were used as the catalysts for the catalytic pyrolysis experiments. The catalytic runs increased the calorific value and water content in the bio-oil, while viscosity, density, and acid number were decreased (Bakar and Titiloye 2012). Uniformly sized ultrafine silica powder can be obtained by non-isothermal decomposition of rice husk in an oxidizing atmosphere. The properties of reactant and product including morphology, particle size, surface area, pore volume, and pore distribution, have been investigated by TEM, SEM, XRD, FTIR, ICP-MS, and EA. By using a TGA technique, a mechanism including two reaction stages was observed for the thermal decomposition of rice husk in air. The activation energy was found to be 166 ± 10 kJ/mol (Liou 2004). Rice husk was used as a raw material to prepare nanosilica without adding an extra surfactant. Experimental results showed that silica produced by hydrochloric acid possesses higher surface area than that of sulfuric, oxalic, and citric acids. Surface characteristics of the sample depend mainly upon gelation pH. The highest surface area and pore volume of silica samples were 634 m^2/g and 0.811 cm^3/g, respectively. Pore diameters were controllable from 3 to 9 nm by adjusting the solution pH value. Particles had a uniform size of 5–30 nm. The objective of this study was to develop a method of nanosilica preparation that enhances the economic benefits of reusing rice husk waste (Liou and Yang 2011). Effect of temperature on the mineralogical compositions of rice husk ash subjected to different grinding time studied by TGA/DTA technique. It was found that the greater the weight loss, the less the crystallinity of the RHA (Ramadhansyah et al. 2012). The unidirectional thermal insulation composite can be designed by choosing the natural fiber (Liu et al. 2012).

2.3 Electrical Properties

Using alkali hydroxide from rice husk, three types of carbon samples were prepared and analyzed. The first three samples were prepared by NaOH activation, the other three by KOH activation and remaining two of commercial carbons. The double layer capacitance strongly depends on the pore structure and functional groups. A specific capacitance larger than 200 F g-1 was achieved by using porous carbon prepared with NaOH (activation temperature: 7,508 °C, activation time: 30 min). All carbons prepared with rice husk have larger double layer capacitance (125–210 F g-1) than commercial grade carbons (78–100 F g-1) (Guo et al. 2003). Direct current conductivity measurements by the two-probe method were carried out for RHAC. Samples were prepared at 700, 800, and 9,008 °C. Samples were compressed in the pressure range of 75.9–578.2 KPa in a hollow glass cylinder using two metal plungers connected to a Keithley ammeter. The conductivity at room temperature was found to increase from $3.28 = 10^{-4}$ to $2.02 = 10^{-3}$ S cm^{-1} with increases in compression pressure. Dense packing of material, collapse of pores, and decrease in air gap between carbon particles could be factors for increases in conductivity with compression pressure. The temperature-dependent conductivity study revealed that the energy gap values of samples were 0.1023, 0.0745, and 0.052 eV, respectively, determined from the linear plot of I versus $1/T(K - 1)$. The unit cost of electricity using rice husk gasifier-based power generation systems has been calculated and its financial feasibility assessed in comparison with utility-supplied and diesel-generated electricity. With the cost and efficiency data assumed here, the unit cost of electricity produced by rice husk gasifier dual fuel engine-generator system varies between Rs 2/kWh and Rs 7/kWh (Kalderis et al. 2008). Semiconductor grade carbon from rice husk can be prepared, rice husk on pyrolysis in a fixed-bed reactor produces solid char, when mixed with metallic magnesium powder at a certain ratio and being heated in a closed steel tube reactor at a high temperature results in formation of silicon-doped carbon. The resistivity versus temperature characteristics showed that material was semiconducting in nature. Suitable doping with phosphorous and boron resulted in "n" and "p" type characteristics in silicon-doped carbon. The *I–V* characteristic of a p–n junction prepared by suitable doping of the silicon-doped carbon showed a current ramp in forward direction and a sharp rise in reverse current like a diode (Kapur et al. 1996).

2.4 Effect of Coupling Agent

The mechanical properties of composites, such as tensile strength, tear strength, hardness, and tensile modulus, improved with the addition of silane coupling agents (Maiti et al. 2008). The tensile strengths of the composites decreased as the filler loading increased, but the tensile properties were significantly improved with the addition of the compatibilizing agent (Yang et al. 2005). PROSIL 2020 coupling agent improved the tensile strength of the composites most, while the impact properties were enhanced by the LICA 38 and the PROSIL 9,234 coupling agents. None of the coupling agents increased the stiffness of the composites except for NZ 44 at maximum filler loading (Fuad et al. 1995). Coupling agent increased in the Si/O ratio for drying ground husk surface, while this ratio deceased for steam-exploded husk particles, and result indicated that coupling agent might the linked to the rice husk surface through chemical reactions (Park et al. 2003).

2.5 Effect of Water Absorption

White ash was then mixed with kaolin and starch to make pellet adsorbents with reasonable strength to be utilized in a packed column. Both ash and pellet samples showed good adsorption capacities toward organic substances in wastewater. Corresponding changes in silanol concentrations were successfully correlated to changes in adsorption capacity toward either Congo red or vacuum pump oil molecules. Rice husk was used to reduce the concentration of ferric ions present as inorganic impurities in crude. Silica was extracted by alkaline treatment of rice husk to carry out the exchange of ions. Simultaneous determination of Fe^{3+} ions in the presence of uranium ions before and after treatment revealed that the concentration of uranium was reduced by 18 % during the removal of Fe^{3+} (Chou et al. 2001). The capacity of rice husk activated carbon (RHAC) to adsorb refractory sulfur compounds of dibenzothiophenes (DBTs) from commercial kerosene was evaluated in terms of their textural and chemical characteristics (Kumagai et al. 2009). Experiments have been carried out on BRH to obtain bio-oil using a fixed-bed pyrolysis rig. ZSM-5, Al-MCM-41, Al-MSU-F, and BRHA were used as the catalysts for the catalytic pyrolysis experiments, and comparison was done to analyze the changes in the bio-oil properties and yield. Properties of the liquid catalytic and non-catalytic bio-oil were analyzed in terms of water content, pH, acid number, viscosity, density, and calorific value. The bio-oil chemical composition shows that ZSM-5 increases the production of aromatic hydrocarbons and light phenols, while Al-MCM-41 reduces the acetic acid production. The catalytic runs increased the calorific value, and water content in the bio-oil, while viscosity, density, and acid number decreased (Bakar and Titiloye 2012). The adsorption of pyridine (Py) from synthetic aqueous solution by RHA and commercial grade granular activated carbon (GAC) is possible. The maximum removal of Py was found to be 96 and 97 % at lower concentrations (-50 mg dm^{-3}) and 79.5 and 84 % at higher concentrations (600 mg dm^{-3}) using 50 and 30 kg m^{-3} of RHA and GAC dosage, respectively, at 30 ± 1 °C. Adsorption of Py is endothermic in nature, and equilibrium data can be adequately represented by Toth and Redlich-Peterson isotherm equations. Py can be recovered from the spent adsorbents by using acidic water and 0.1 N $H2SO_4$. The overall adsorption of Py on RHA and GAC was found to be in the order of GAC > RHA (Lataye et al. 2008).

3 Conclusion

Rice husk is annually renewable agricultural raw material, and their availability is more. Rice husk is good source of silica (SiO_2). Mechanical strength decreases by addition of rice husk because of poor bonding between rice husk and polymers. For good bonding, coupling agent such as titanate, silane, and zirconate can be used for enhanced mechanical properties. The extensive use of rice husks is because of their properties such as low density, relative high strength, and modulus, high level of filler loadings, biodegradability, and safe working environment. Composites enhanced mechanical strength and acoustic performance, light in weight, fuel saving, lower cost of production, shatterproof performance. Composites thus prepared have applications in the construction of building (laminated, door, roofing, beam, sheet, etc.), improving biodegradability, for automobile (interior parts, cabin lining, exterior parts, etc.), electrical (electrical connector, microwave containers, etc.), and food packaging.

References

Akil HM, Omar MF, Mazuki AAM, Safiee S, Ishak ZAM, Abu Bakar A (2011) Kenaf fiber reinforced composites: a review. Mater Des 32:4107–4121

Alvarez V, Fragaand A, Vazquez A (2004) Effects of the moisture and fiber content on the mechanical properties of biodegradable polymer–sisal fiber biocomposites. J Appl Polym Sci 91:4007–4016

Ashori A (2008) Wood-plastic composites as premising green composites for automotive industries. Bioresour Technol 99:4661–4667

Baiardo M, Ziniand E, Scandola M (2004) Flax fiber–polyester composites. Compos Part A: Appl Sci Manuf 35:703–710

Bakar SMA, Titiloye JO (2012) Catalytic pyrolysis of rice husk for bio-oil production. J Anal Appl Pyrol 362–368

Bax B, Mussig J (2008) Impact and tensile properties of PLA/Cordenka and PLA/flax composites. Compos Sci Technol 68:1601–1607

Bismarck A, Mishra S, Lampke T (2005) Plant fibers as reinforcement for green composites. CRC Press, Boca Raton

Chauhan SR, Gaur B, Dass K (2011) Effect of fiber loading on mechanical properties, friction and wear behavior of vinyl ester composites under dry and water lubricated conditions. IJMS 1:1–8

Chawla KK (1998) Fibrous materials. Cambridge University Press, United Kingdom

Chou KS, Tsai JC, Lo CT (2001) The adsorption of Congo red and vacuum pump oil by rice hull ash. Bioresour Technol 78:217–219

Dearmitt DC, Breese KD (2001) Filled polypropylene: a cost-performance comparison of common fillers. Plast. Addit. Compound 3:28–33

Fuad MYA, Ishakand ZAM, Omar AKM (1995) Application of rice husk ash as fillers in PP: effect of titanate, zirconate and silane coupling agents. Eur Polym J 31:885–893

Guo Y, Qi J, Jiang Y, Yang S, Wang Z, Hongding X (2003) Performance of electrical double layer capacitors with porous carbons derived from rice husk. Mater Chem Phys 80:704–709

Ishak ZAM, Bakar AA (1994) An investigation on the potential of rice husk ash as fillers for epoxidized natural rubber (ENR). Eur Polym J 31:259–269

Ismail MS, Waliuddint AM (1996) Effect of rice husk ash on high strength concrete. Constr Build Mater 10:521–526

Ismail H, Nizam JM, Abdul Khalil HPS (2000) The effect of a compatibilizer on the mechanical properties and mass swell of white rice husk ash filled natural rubber/linear low density polyethylene blends. Polym Test 20:125–133

Jilken L, Malhammar G, Selden R (1991) The effect of mineral fillers on impact and tensile properties of polypropylene. Polym Test 10:329–344

Kalderis D, Bethanis S, Paraskeva P, Diamadopoulos E (2008) Production of activated carbon from bagasse and rice husk by a single-stage chemical activation method at low retention times. Bioresour Technol 99:6809–6816

Kapur T, Kandpal TC, Garg HP (1996) Electricity generation from rice husk in Indian rice mills: potential and financial viability. Biomass Bioenergy 10:393–403

Karnani R, Krishnan M, Narayan R (1997) Biofiber-reinforced polypropylene composites. Polym Eng Sci 37:476–483

Kumagai S, Shimizu Y, Enda YTY (2009) Removal of dibenzothiophenes in kerosene by adsorption on rice husk activated carbon. Fuel 88:1975–1982

Kumar V, Sinha S, Saini MS, Kanungo BK, Biswas P (2010) Rice husk as reinforcing filler in polypropylene composites. Rev Chem Eng 26:41–53

Kumar V, Tyagi L, Sinha S (2011) Wood flour reinforced plastic composites: a review. Rev Chem Eng 27:253–264

Kumar V, Saini MS, Kanungo BK, Sinha S (2012) Effect of various additives on mechanical properties of rice husk polypropylene (RHPP) composites. J Polym Eng 32:163–166

Lataye DH, Mishra IM, Mall ID (2008) Pyridine sorption from aqueous solution by rice husk ash (RHA) and granular activated carbon (GAC). Parametric, kinetic, equilibrium and thermodynamic aspects. J Hazard Mater 154:858–870

Liou TH (2004) Preparation and characterization of nano-structured silica from rice husk. Mater Sci Eng 364:313–323

Liou TH, Yang CC (2011) Synthesis and surface characteristics of nanosilica produced from alkali-extracted rice husk ash. Mater Sci Eng 176:521–529

Liu K, Takagi H, Osugi R, Yang Z (2012) Effect of lumen size on the effective transverse thermal conductivity of unidirectional natural fiber composites. Compos Sci Technol 72:633–639

Lu X, Zhang MQ, Rong MZ, Shi G, Yang GC (2003) Self-reinforced melt processable composites of sisal. Compos Sci Technol 63:177–186

Maiti S, Banerjee P, Purakayastha S, Ghosh B (2008) Silicon-doped carbon semiconductor from rice husk char. Mater Chem Phys 109:169–173

Nachtigall MS, Cerveira GS, Rosa SML (2007) New polymeric coupling agent for polypropylene/ wood flour composites. Polym Test 26:619–628

Park BD, Wi SG, Lee KH, Singh AP, Yoon T, Kim YS (2003) Characterization of anatomical features and silica distribution in rice husk using microscopic and micro-analytical techniques. Biomass Bioenergy 25:319–327

Premalal HGB, Ismail H, Baharin A (2002) Comparison of the mechanical properties of rice husk powder filled polypropylene composites with talc filled polypropylene composites. Polym Test 21:833–839

Qin L, Qiu J, Liu M, Ding S, Shao L, Lu S, Zhang G, Zhao Y, Fu X (2011) Mechanical and thermal properties of poly(lactic acid)Composites with rice straw fiber modified by poly(butyl acrylate). Chem Eng J 166:772–778

Ramadhansyah PJ, Mahyun AW, Salwa MZM, Abu Bakar BH, MegatJohari MA, Wan Ibrahim MH (2012) Thermal analysis and pozzolanic index of rice husk ash at different grinding time. Procedia Eng 59:101–109

Rosa SML, Rehman N, Miranda MIG, Nachtigal SB, Bica CID (2012) Chlorine-free extraction of cellulose from rice husk and whisker isolation. Carbohydr Polym 37:1131–1138

Rozman HD, Yeo YS, Tay GS, Abubakar A (2002) Mechanical and physical properties of polyurethane composites based on rice husk and polyethylene glycol. Polym Test 22:617–623

Rozman HD, Ahmadhilmi KR, Abubakar A (2004) Polyurethane (PU)-oil palm empty bunch (EFB) composites the effect of EFBG reinforcement in mat form and isocyanate: treatment on the mechanical properties. Polym Test 23:559–565

Satyanarayana KG, Ramos LP, Wypych F (2005) Development of new materials based on agro and industrial wastes towards ecofriendly society. In: Ghosh TN, Chakrabarti T, Tripathi G (eds) Biotechnology in energy management. APH Publishing Corporation, New Delhi, pp 583–624

Siriwardena S, Ismail H, Ishiaku US (2000) Effect of mixing sequence in the preparation of white rice husk ash filled polypropylene/ethylene-propylene—diene monomer blend. Polym Test 20:105–113

Sud D, Mahajan G, Kaur MP (2008) Agricultural waste material as potential adsorbent for sequestering heavy metal ions from aqueous solutions—a review. Bioresour Technol 99:6017–6026

Wambua P, Ivens J, Verpoest I (2003) Natural fibers: can they replace glass in fiber reinforced plastics. Compos Sci Technol 63:1259–1264

Whelan A, Goff JP (1986) Paper presented to the PRI mould making. In: 86 Conference at Solihull, England, Jan 1986

Yang HS, Kim HJ, Son J, Park HJ, Lee BJ, Hwang TS (2004) Rice-husk flour filled polypropylene composites; mechanical and morphological study. Compos Struct 63:305–312

Yang HS, Kim HJ, Park HJ, Lee BJ, Hwang TS (2005a) Effect of compatibilizing agents on rice husk flour reinforced polypropylene composites. Compos Struct 77:45–55

Yang HS, Kim HJ, Park HJ, Lee BJ, Hwang TS (2005b) Water absorption behavior and mechanical properties of lignocellulosic filler-polyolefin bio-composite. Compos Struct 72:429–437

Effect of Glass Fiber and Filler Volume Fraction Variation on Mechanical Properties of GFRP Composite

Anurag Gupta, Hari Singh and R. S. Walia

Abstract Unsaturated polyesters are important matrix resins used for glass fiber-reinforced composites/plastics. The strength of glass fiber-reinforced polyester composite is mainly related to the glass content of the material and the arrangement of glass fibers, but the mechanical properties can be altered by introducing some filler materials in glass fiber-reinforced polymer (GFRP) composites. In this paper, an effort is made to study the effect of different % of filler material on mechanical properties of the GFRP. Hand layup process is used for manufacturing the GFRP composite strips for testing. E-type glass fiber chopped strand mat (CSM) is used as reinforced material, unsaturated polyester resin is used for matrix, and calcium carbonate ($CaCO_3$) is used as filler material. During manufacturing the specimens, the polyester resin % is kept fixed and effect of filler and glass fiber percentage variation is evaluated in terms of tensile strength, impact strength and toughness variation. In the experiments, the increase in the volume fraction of E-glass fiber CSM increased the material flexural, tensile, and impact strength. The consequences show that appropriate procedures are applied with the proper choices. When the $CaCO_3$ ratio increases, particles could cluster, and this causes deviation in strength variation curves, but it has also been observed that the finishing of specimen is better in case of higher filler volume fraction.

Keywords GFRP · Composites · Mechanical properties

A. Gupta (✉) · H. Singh
Department of Mechanical Engineering, National Institute of Technology, Kurukshetra, Haryana, India
e-mail: anurag_k2@yahoo.com

H. Singh
e-mail: hsingh_nitk@rediffmail.com

R. S. Walia
Department of Mechanical Engineering, Delhi Technological University, New Delhi, India
e-mail: waliaravinder@yahoo.com

S. S. Khangura et al. (eds.), *Proceedings of the International Conference on Research and Innovations in Mechanical Engineering*, Lecture Notes in Mechanical Engineering,
DOI: 10.1007/978-81-322-1859-3_38, © Springer India 2014

1 Introduction

Polymers and their composites are emerging as viable alternative products to metal-based ones in many common and advanced engineering applications (ASM Handbook 1992). Composite materials are made by combining two materials where one of the materials is reinforcement (fiber) and the other material is a matrix (resin). The combination of the fiber and matrix provides characteristics superior to either of the materials alone. Some examples of composite materials are plywood, reinforced concrete, glass fiber and polyester resin, and graphite and epoxy resin. The principal advantage of these materials is the very high strength-to-weight ratio, which makes them attractive in aircrafts, spacecrafts, cars, boats, and sport equipment.

The ease of fabrication, the availability of a good choice of materials from both thermoplastic and thermoset varieties and economic viability have made the advent of these newer materials for industries ranging from automobile to sports goods (Derrien et al. 2000). The many uses of these materials in space and aeronautical industry-related applications are well known (Merhi et al. 2006).

Just as any chef knows that "secret herbs and spices" can turn an ordinary recipe into a culinary masterpiece, proper selection of resin fillers and additives used to formulate a resin system can result in the fabrication of polymeric composites with optimal properties. In commercial applications, and often in research studies, little consideration is given to specific ingredients in resin formulation as composites are specified based exclusively on reinforcement and resin constituent materials. However, general specification of only the reinforcement and resin is not sufficient because seemingly insignificant variations can have profound effects on the properties of composite materials.

Orientation and distribution of the reinforcement in the components have significant consequences on the mechanical properties and the final dimensions of the components (Nageotte et al. 2000; Rosato and Rosato 2005). Many do not realize what significant effects seemingly minor formulation changes can have. Just as careful consideration is given to the selection of appropriate fiber reinforcement and matrix materials for a given application, similar consideration should be given to the selection of fillers and additives used to formulate the mixed resin systems. Significant economic implications can result if these materials are not fully considered.

The purpose of use of fillers can be divided into two basic categories (1) to improve the properties of the material and (2) to reduce the cost of component. Common fillers used for thermoplastics include mineral fillers, such as calcium carbonate, talc, and wollastonite. Filler materials are also used extensively with polyester resins for a variety of reasons, such as cost reduction of the molding, facilitation of the molding process, and imparting specific properties to the molding. Fillers are often added in quantities up to 50 % of the resin weight although such an amount affects the flexural and tensile strengths of the laminates (Cheon et al. 1999). The use of fillers can be beneficial in laminating or casting of thick components where otherwise considerable exothermic heating can occur. Addition of certain fillers can also contribute to increase the heat resistance of the laminate (Hart-Smith 1996). Fillers also affect processing by increasing the viscosity of resin mixes. The viscosity rise is related to the particle size distribution. Increased viscosity may be beneficial or detrimental, depending on the composite manufacturing applications (Gupta et al. 1999). In the present work, the effect of silica and calcium carbonate fillers in the presence of glass fiber reinforcement on tensile strength, impact strength, and flexural strength of woven glass fiber composites has been studied. The glass fiber volume has been kept minimal so that the effect of fillers is not subdued.

The study on such economical filler is necessary to ensure that mechanical properties of the product are not affected adversely by their addition. A large number of materials have been studied for their use as filler in polymer, but only few of them are found to deal with the material system containing fibers simultaneously (Gupta et al. 2001). The most commonly used categories of fillers include calcium carbonate, kaolin clay, and alumina trihydrate (ATH). In the composite industry, the selection of filler is typically based on general properties associated with the categories of fillers. Here in this paper, the effect of calcium carbonate is under consideration. Fillers are necessary for the reduction in cost as well as the improvement of specific properties such as chemical resistance, heat resistance, dimensional stability, hardness, surface smoothness, and shrinkage (Dalenberg 1984).

2 Manufacturing Technique

Hand layup method was used to fabricate the composite. A transparent polyester film is used as liner and unsaturated polyester resin manufactured by Morex Petrochem Pvt. Ltd premixed with some amount of accelerator, i.e., Cobalt naphthalate; initiator, i.e., methyl ethyl keton peroxide (MEKP); and filler $CaCO_3$ is applied on the surface of the polyester liner. Then, sheet of glass fiber CSM manufactured by Goa Glass Ltd is laid down on it and again CSM is wetted by pre mixed resin. Different number of layers was added depending upon the glass content required by placing layers one by one. Fibers were assured for complete impregnation in the resin using washer rollers. Samples were then cured at room temperature for 2 days.

2.1 Composite Specimen Composition

Batch size Unsaturated Polyester resin—1 kg.
Cobalt naphthalate (accelerator)—2 % of resin mass.
Initiator (MEKP)—5 % of resin mass.
$CaCO_3$ (Filler)—20 to 60 % as per composition of specimen.

Table 1 Composition of composite

	Polyester (%)	Filler (%)	E-glass (%)
1	30	60	10
2	30	55	15
3	30	50	20
4	30	45	25
5	30	40	30
6	30	35	35
7	30	30	40
8	30	25	45
9	30	20	50

CSM (300 gm/m^2)—10 to 60% as per composition of specimen.
The compositions of different test performed are given in Table 1.

2.2 Mechanical Testing

The tensile, flexural, and impact strengths of these composite materials were evaluated at 25 °C according to standard procedures (Colclough and Daniel 1998; Wypych 2000; Gardner 2000; ASTM International 2010). Tensile testing of specimens was carried out on a universal tensile testing machine (ENKAY Enterprises, New Delhi) having capacity of 600 KN. While the tensile strength of samples was calculated using Eq. (1)

$$S = P/BD \tag{1}$$

where S = tensile strength (N/mm^2 = MPa), P = load KN or Newton, B = width of sample (15 mm), D = thickness of sample (6 mm) (Fig. 1).

Flexural strength of the composites was tested on a universal testing machine (ENKAY Enterprises, New Delhi) having capacity of 600 tons. The flexural strength of samples was calculated using Eq. (2).

$$S = 3PL/2BD^2 \tag{2}$$

where S = flexural strength (N/mm^2 = MPa), P = load (KN or Newton), B = width of sample (15 mm), D = thickness of sample (10 mm), L = span length of flexural fixture (100 mm) (Fig. 2).

Impact strength of composites was tested on universal impact testing machine (Paras Enterprises, New Delhi) having capacity of 29.4 J. Equation (3) was used to calculate impact strength of specimens:

$$S = \text{Impact energy}/\text{Area of Sample} \tag{3}$$

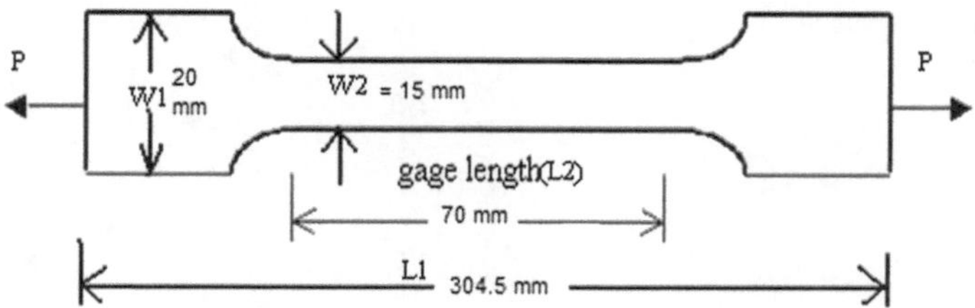

Fig. 1 Specimen for tensile test

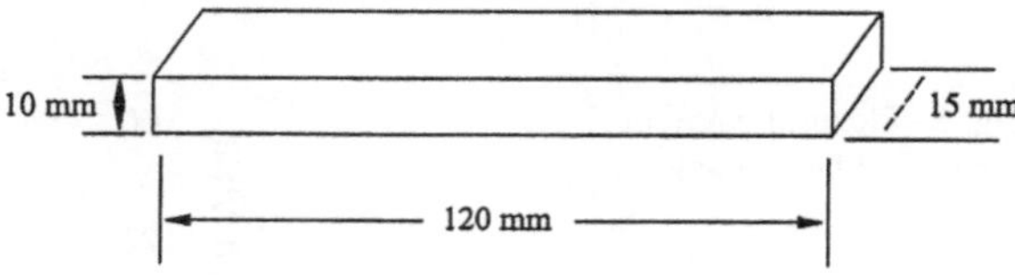

Fig. 2 Specimen for flexural test

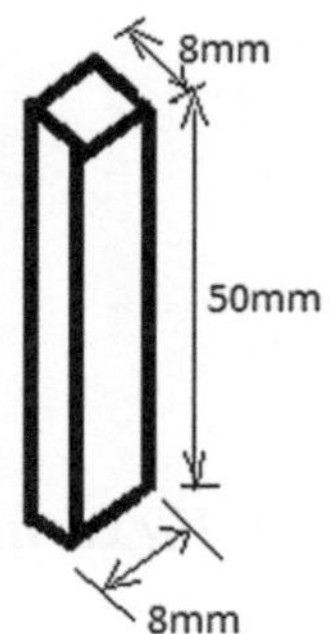

Fig. 3 Specimen for impact test

where $S =$ impact strength (MPa·mm), impact energy is in joule and the area (width $\times$ thickness) of the sample is in mm^2 (Fig. 3).

Four specimens of each composition were manufactured for each testing and the average results are shown in next section.

3 Results and Discussion

3.1 Density

Calculated specific gravity values of the specimens are given in Fig. 4. As shown in the figure, densities of the composites decreased with the increase in E-glass content. The change in density showed a linear fall except some exceptions.

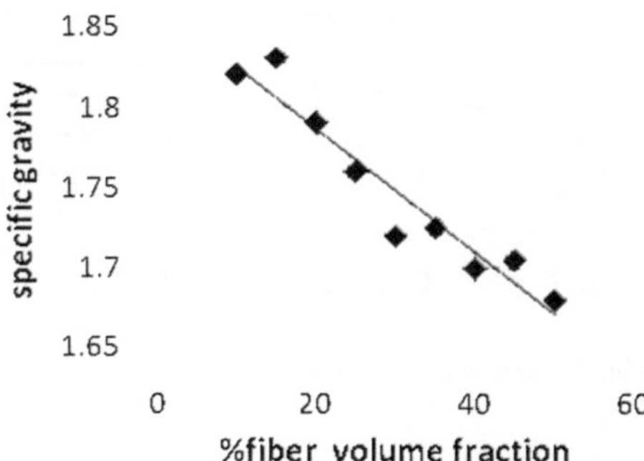

Fig. 4 Density versus % fiber volume

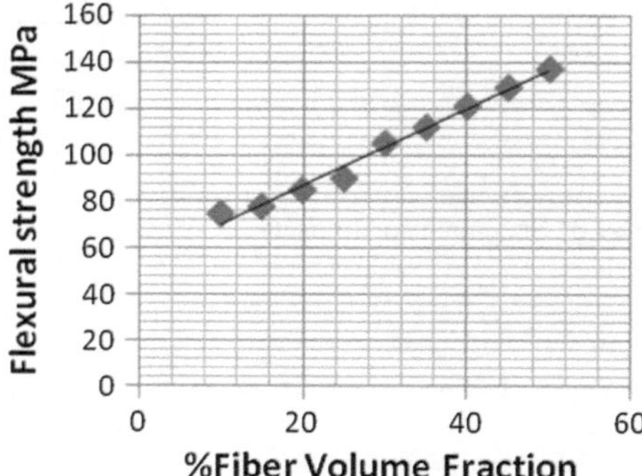

Fig. 5 Flexural strength versus % fiber volume

3.2 Flexural Strength

Flexural strengths were calculated, and results are given in Fig. 5. As the fiber volume fraction increases, flexural strength increases too. In this study, resin was used 30 % by weight, which seems adequate to gain higher strength to composite in higher fiber concentrations. Lower resin concentrations could not provide enough wetting properties, which would cause a dramatic drop in strength values.

3.3 Tensile Strength

As expected, fiber incorporation in matrix increased the tensile strength of the material. Tensile strengths of the composites were found to increase with fiber as shown in Fig. 6, resembling to flexural strengths. Clustering problem can also be seen in tensile test results. In these results, it was also seen that particular shaped filler has no distinctive effect on strength values of the composite. Here, strength was under control of the fiber content.

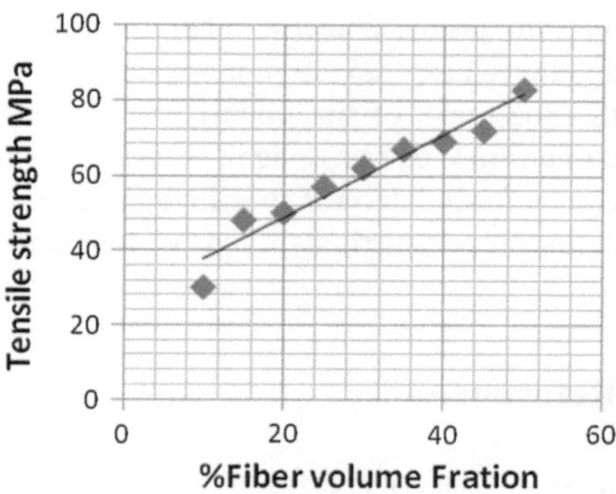

Fig. 6 Tensile strength versus % fiber volume

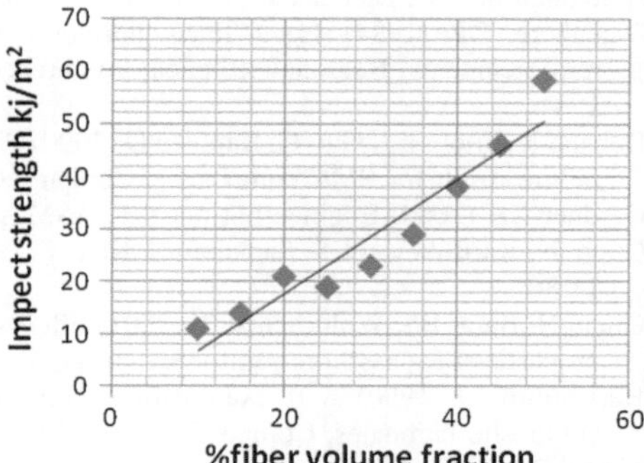

Fig. 7 Impact strength versus % fiber volume

3.4 Impact Strength

It is known that particular reinforcements cause fall in toughness of the composite. So increase in filler concentration was expected to decrease the impact strength of the composite, whereas fiber provided an improved impact resistance. Same as the flexural and tensile strength, dimensional advantages of fibers could improve the impact strength. The increase in impact strength with increasing fiber concentration is observed in Fig. 7.

4 Conclusion

It is has been seen that as the glass fiber concentration increases, the mechanical properties of composites also improved. In the experiments, the increase in the volume fraction of E-glass fiber CSM increased the material flexural, tensile, and impact strength. The consequences show that appropriate procedures are applied with the proper choices. When the $CaCO_3$ ratio increases, particles could cluster, and this causes deviation in strength variation curves, but it has also been observed that the finish of specimen is better in case of higher filler volume fraction. So a compromise can be done in between finish and strength, and optimum composition can be formulated for specific application. To avoid cluster formation of $CaCO_3$ in resin, proper stirring should be done and particle size should be small.

References

ASM International (1992) ASM handbook, vol 18. ASM International, Materials Park, USA

ASTM International (2010) D638: 2010: Standard test method for tensile properties of plastics

Cheon SS, Lim TS, Lee DG (1999) Impact energy absorption characteristics of glass fiber hybrid composites. Compos Struct 46:267–278

Colclough GW Jr, Daniel DP (1998) Engineered materials. Compos ASM 1:161–163

Dalenberg DP (1984) Long fiber thermoset molding compounds. In: Plastic fillers and reinforcements, Regional technical conference, Atlanta, Society of Plastics Engineers, Oct 1984

Derrien K, Fitoussi J, Guo G, Baptiste D (2000) Prediction of the effective damage properties and failure result for SMC composites. Comput Methods Appl Mech Eng 185:93–107

Gardner CH (2000) Engineering plastics. ASM, p 222

Gupta N, Karthikeyan CS, Sankaran S, Kishor (1999) Material characterization. J Mater Sci Lett 11:86

Gupta N, Barar BS, Woldesenbet E (2001) Effect of filler addition on the compressive and impact properties of glass fibre reinforced epoxy. Bull Mater Sci 24(2):219–233

Hart-Smith LJ (1996) A re-examination of the analysis of in-plane matrix failures in fibrous composite laminates. Compos Sci Tech 56:107–121

Merhi D, Michaud V, Comte E, Manson J-AE (2006) Predicting sizing dependent bending rigidity of glass fibre bundles in SMC. Compos A 37:1773–1786

Nageotte VM, Maazouz A, Peix G, Bres S (2000) Methodologies for the characterization of glass fibre orientation and distribution in large components molded from SMC. Polym Testing 22:867–873

Rosato D, Rosato D (2005) Reinforced plastics handbook, 3rd edn. Elsevier, Oxford, pp 127–137

Wypych G (2000) Handbook of fillers: A definitive user's guide and databook, 2nd edn. ChemTec Publishing, Toronto, pp 824–829

Slurry Erosion Testing of Detonation Gun-Sprayed Cr_3C_2–25NiCr and Cr_2O_3 Coatings on CF8M Steel

Sanjeev Bhandari, Harmesh Kumar Kansal, Harpreet Singh, Sanjiv Kumar and Maninder Kaur

Abstract In the current investigation, cermet coating (Cr_3C_2–25NiCr) and ceramic coating (Cr_2O_3) were deposited on CF8M steel by detonation gun (D-gun) thermal-spraying process. Subsequently, the slurry erosion behavior of these coated and uncoated specimens was investigated using a high-speed erosion test rig. Slurry collected from an actual hydropower plant was used as the abrasive media; the analysis of eroded specimens was done using SEM and stylus profilometry. Signatures of microcutting in binder phase, fracture of carbide grains on the eroded surface of Cr_3C_2–25NiCr coating, grain removals, and fatigue-induced lateral crack chipping were observed on the eroded surface of Cr_2O_3 coating, while signatures of formation of ploughing, lips, shearing of platelet, formation of crater, and microcutting were observed on the eroded surface of CF8M steel.

Keywords D-gun · Microhardness · Slurry erosion · Surface roughness

S. Bhandari (✉) · S. Kumar · M. Kaur
Mechanical Engineering Department, Baba Banda Singh Bahadur Engineering College, Fatehgarh Sahib, Punjab, India
e-mail: sanjeevbhandari2007@gmail.com

H. K. Kansal
Department of Mechanical Engineering, University Institute of Engineering and Technology, Panjab University, Chandigarh, India
e-mail: shaarut@yahoo.com

H. Singh
School of Mechanical, Materials and Energy Engineering, Indian Institute of Technology Ropar, Rupnagar, Punjab, India
e-mail: harpreetsingh@iitrpr.ac.in

S. S. Khangura et al. (eds.), *Proceedings of the International Conference on Research and Innovations in Mechanical Engineering*, Lecture Notes in Mechanical Engineering, DOI: 10.1007/978-81-322-1859-3_39,

1 Introduction

Erosion due to solid particles (carried by water) is the most common and unavoidable problem in hydroturbine power plants. Major components of hydroturbine power plants that get eroded are runner, impeller, guide vanes, needle, valves, etc. Due to increase in silt content (10,000 ppm compared to 100 ppm in normal season), this problem becomes more severe during monsoon season for hydropower plants situated in the Himalayan region in India and often leads to shutdown of these hydropower plants (Mann 2000). If unattended, the erosion of hydroturbines can lead to loss of turbine efficiency as high as 5–10 %, resulting in a heavy revenue loss per year for existing hydropower stations. Besides heavy loss of material, the erosive silt causes number of "operation and maintenance" problems.[1] Conventional steels used in hydropower plants are not able to sustain the slurry erosion problems that persist in hydroturbines (Xu and Luo 1998; Hattori 2002). To overcome this problem of slurry erosion wear, wear-resistant alloys or suitable wear-resistant coatings (Kanjlia and Gupta 2008) deposited by various advanced techniques are generally used. From the literature, it has been found that the erosion of components working in water loaded with sediments cannot be prevented; however, it can be minimized by cermet and ceramic coatings on the surfaces of underwater parts. Detonation gun (D-gun) spray coatings are gaining popularity due to exceptional hardness, wear resistance, cost effectiveness, marginal thermal deformation, and chemical changes in the substrate materials. Moreover, very dense (99 % of theoretical density) and adherent coatings can be produced by this technique (Davis 2004). The D-gun spray process involves the entrainment of powdered materials with the high-velocity combustion products of a detonation wave as it propagates through a water-cooled barrel. The two-phase mixture of molten particles and detonation products exits the barrel and impinges against a target substrate, where the hot particles bond in overlaying platelets (Knapp 1997). The evaluation of slurry erosion behavior of materials under actual service conditions is often a difficult task, due to interactive effects of different parameters such as slurry concentration, velocity, and properties of abrasive medium on wear rate. Therefore, accelerated erosion testing of materials can be performed by increasing load, velocity, and other operational parameters in a laboratory test rig, where real contact conditions can be simulated. Moreover, to study the failures or material removal mechanisms, a lot of research has been done in this field by various researchers. In most of the studies, simple rectangular shape specimens fitted at a particular angle at a time have been simulated in the laboratory using jet or pot-type tester. Little information is available in the open literature on the cylindrical shape specimen of hydroturbine materials when exposed to slurry erosion conditions in a high-speed slurry erosion tester. The condition of erosion in this test rig is expected to be very similar to those prevailing in actual hydroturbines. Therefore, the aim of the current work is to

[1] Retrieved from http://www.vncold.vn, accessed on 20 May 2009.

Table 1 Chemical composition of CF8M (ASTM A743-grade) steel

Grade	C max	Mn max	Si max	P max	S max	Cr	Ni	Mo	Fe
CF8M	0.08	1.50	1.50	0.04	0.04	17.0–21.0	9.0–13.0	2.0–3.0	Balance

Table 2 Characteristics of Cr_3C_2–25NiCr and Cr_2O_3 powders

Powder	Make and commercial code	Chemical composition, wt %	Morphology	Particle shape	Powder size (μm)
Cr_3C_2–25NiCr	H.C.Starck Amperit 584.054	C 9–11 % Cr 65.9–73.0 % Ni 18–22 % Fe max. 0.5 % O max. 0.6 %	Agglomerated sintered	Almost spherical	−88 + 10
Cr_2O_3	H.C.Starck Amperit 707.001	SiO_2 max. 0.25 % TiO_2 max. 0.15 % Fe_2O_3 max. 0.1 % Cr_2O_3 balance	Fused	Regular and angular block	−88 + 15

investigate the slurry erosion behavior of D-gun-sprayed Cr_3C_2–25NiCr and Cr_2O_3 coatings on CF8M steel using a high-speed erosion test rig, in order to explore the possibility of use of this coating system in actual hydraulic turbines.

2 Experimentation

2.1 Substrate Material

CF8M stainless steel (ASTM A743-grade), which is commonly used in hydropower plants, was selected as a substrate material for the research work. The chemical composition of CF8M steel is given in Table 1. Hollow cylindrical specimens of external diameter 12 mm, internal diameter 6 mm, and length 10 mm were prepared from the steel.

2.2 Coating Deposition

The feedstock powders used in this study were commercially available Cr_3C_2–25NiCr and Cr_2O_3 powders, details of which are given in Table 2. To improve the adherence of coating to the substrate, samples were grit-blasted using alumina grit

Table 3 Process parameters for D-gun spray process

Parameters (units)	Cr_3C_2–25NiCr coating	Cr_2O_3 coating
O_2 rate (SLPH)	2,720	5,120
Pressure (MPa)	0.2	0.2
C_2H_2 rate (SLPH)	2,400	2,140
Pressure (MPa)	0.14	0.14
N_2 flow rate (SLPH)	960	720
Pressure (MPa)	0.4	0.4
Spray angle (°)	90	90
Spray distance (mm)	165	180
Power (VA)	450	450

of 60 mesh before the deposition of the coating. The blasting was performed at an air pressure of approximately 0.6 MPa. Samples were then coated by detonation gun (D-gun) spray process available with SVX powder M Surface Engineering Pvt. Ltd., Noida, India. The coatings were deposited using the standard process parameters reported in Table 3.

The surface morphology of as-sprayed coatings was studied by scanning electron microscopy (JEOL JSM-6610LV).

2.3 Slurry Erosion Testing

A high-speed erosion test rig (DUCOM TR401) was used to study the slurry erosion behavior of coated and bare steel specimens. The test rig consists of various components such as slurry abrasion chamber, slurry tank, rotor, and controller as shown in Fig. 1. The slurry abrasion chamber encloses the test specimen and slurry. Slurry tank is a cylindrical stainless steel vessel, in which the slurry is prepared to the required concentration. Three inlet and three outlet pipes are provided between slurry abrasion chamber and slurry tank for re-circulation of slurry. The rotor attached with testing specimens was rotated in the abrasion chamber. The rotation of the rotor attached with testing specimens inside the chamber leads to centrifugal action, which forces the slurry to flow out of abrasion chamber through the outlet pipes connected to lower part of abrasion chamber. It results in partial vacuum in the erosion chamber due to which, slurry is sucked in through the inlet pipes. Thus, samples are rotated in a chamber full of moving slurry. The rotation of rotor is such that slurry from slurry tank flows from top chamber toward downward direction and used slurry leaves from bottom chamber to enter the slurry tank, thus ensuring continuous re-circulation of slurry. This high-speed erosion test rig is capable of creating accelerated erosion of standard test specimens with water containing abrasive particles of controlled size and composition (slurry). One of the advantages of this rig is that twelve specimens may be tested at a time. This ensures zero tolerance to the change of experimental

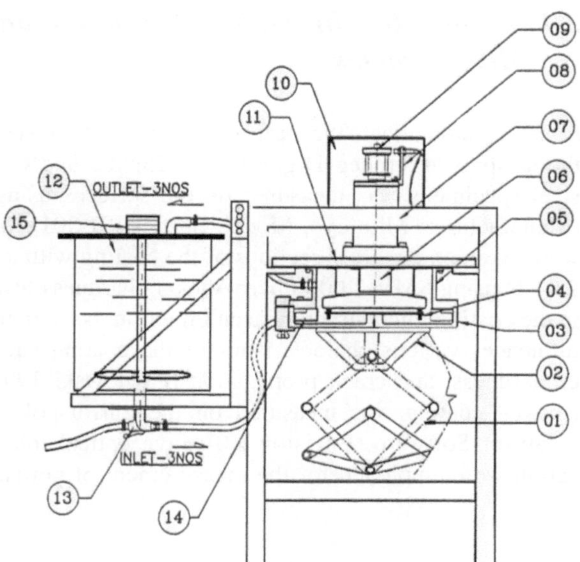

Fig. 1 Schematic diagram of high-speed erosion test rig used for current investigation. *01* Jack, *02* Jack top frame, *03* Slurry chamber, *04* Specimen, *05* Supporting frame, *06* Rotor, *07* Spindle assembly, *08* Sensor hold, *09* Sensor disc, *10* Belt guard, *11* Baffle ring, *12* Slurry tank, *13* Valve for drain out, *14* Baffle, *15* Stirrer motor

conditions during the comparison of slurry erosion testing of different specimens under similar experimental conditions. More information regarding the slurry preparation, average particle size of the erodent, and different phases present in the slurry can be found elsewhere (Bhandari et al. 2012a, b). Test was carried out at 2,250 rpm with slurry of 10,000 ppm concentration having erodent particles of average particle size 100 μm to study the erosion performance of D-gun-sprayed coatings and substrate steel.

A typical erosion test cycle began with mounting of machined uncoated and coated specimens to the rotor assembly in slurry chamber. Care was taken to prevent over-tightening of clamping screws, which could crack coating or induce compressive stresses in the specimens. After completing the slurry erosion testing cycle of 1 h, specimens were removed from the rotor assembly, brushed gently, and cleaned with acetone in order to remove attached sand particles if any. The specimens were weighed before and after each slurry erosion cycle. The loss in mass of each specimen was recorded with the help of precision microbalance having an accuracy of 0.1 mg. As erosion is a surface phenomenon, so the peripheral surface area of each specimen was calculated by measuring the external diameter and length with the help of digital micrometer having an accuracy of 0.01 mm. Specific mass loss was calculated by using the following relation:

$$\text{Specific mass loss}\left(\frac{\text{g}}{\text{m}^2}\right) = \frac{\text{Mass loss (g)}}{\text{Peripheral surface area (m}^2\text{)}}$$

The slurry erosion process was repeated for six cycles for each of the samples.

2.4 Surface Roughness, Microhardness, and Porosity Measurements

Slurry erosion affects the surface roughness of the specimen. Surface roughness of the as-sprayed, eroded D-gun Cr_3C_2–25NiCr and Cr_2O_3 coated, and bare CF8M steel specimens was measured on the surfaces using a surface roughness tester (Manufacturer: Mitutoyo, Model: Surftest SJ 301). Microhardness measurements were taken on the cross sections of the coating with a Vickers indenter at a load of 2.94 N using SMV-1,000 micro-Vickers hardness tester. The presence of porosity in the coating microstructure is an important issue in thermal-sprayed coatings as it influences various characteristics of the coating such as intersplat bonding, microhardness, and crack propagation (Pawlowski 1995). Apparent porosity of as-sprayed coating was measured on the mirror-polished cross section by Image Analyzer Software (Envision 3.0). Five optical micrographs at 1,000X magnification were employed for the measurement of porosity.

3 Results and Discussion

3.1 SEM Analysis of the As-sprayed Coatings

Surface morphology of as-sprayed Cr_3C_2–25NiCr coating is shown in Fig. 2. Solidified molten splats and splats of deformed particles of the powder as a consequence of heat due to combustion of acetylene and high energy impacts can be seen in the microstructure of as-sprayed Cr_3C_2–25NiCr coating. SEM micrograph also exhibits some spherical particles. Some micropores are also evident on the microstructure of Cr_3C_2–25NiCr coating. Figure 3 shows the surface morphology of as-sprayed Cr_2O_3 coating. The characteristic microstructure of the as-sprayed Cr_2O_3 coating shows interconnecting splats due to cohesion between Cr_2O_3 particles and chaotic build-up of single-phase splats of material. Some spherical particles and pores can also be seen in the Cr_2O_3 coating microstructure.

3.2 Microhardness, Porosity, and Surface Roughness Measurements

An observed range of microhardness values for a given material has been reported in Table 4. The observed range of apparent porosity of as-sprayed coatings is also reported in Table 4. Similar results of porosity for Cr_3C_2–NiCr coating on structural steel DIN 12CrMo44 by detonation spray system (Wang et al. 2002), Cr_3C_2–NiCr coating on mild steel substrate by hypersonic velocity oxygen fuel process

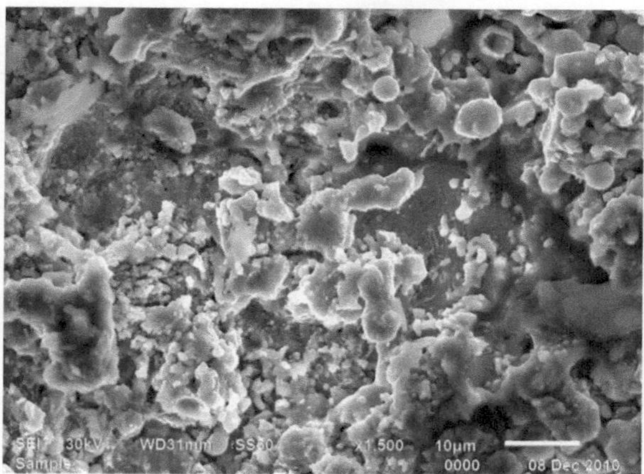

Fig. 2 Surface SEM analysis of the D-gun-sprayed Cr_3C_2–25NiCr coating on CF8M steel in as-sprayed condition

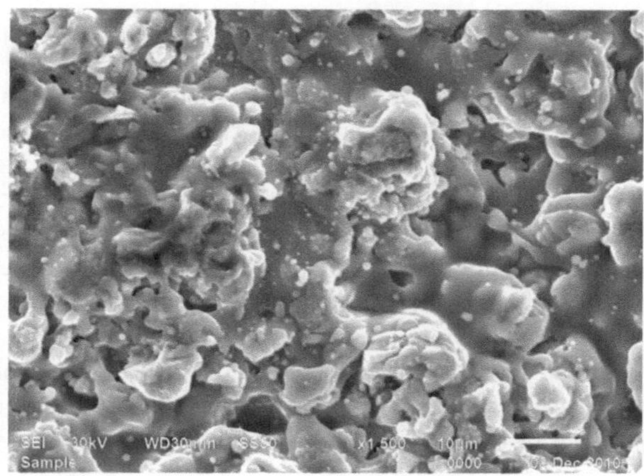

Fig. 3 Surface SEM analysis of the D-gun-sprayed Cr_2O_3 coating on CF8M steel in as-sprayed condition

(Wang 1995, 1996), Cr_3C_2–NiCr coating on mild steel (0.25 % C, 0.7 % Mn, 0.25 % Si, and 0.05 % S) substrates by detonation gun (D-gun) spraying (Sundararajan et al. 1998), Cr_2O_3 coating deposited by air plasma spray thermal process (Barbezat and Nicoll 1993) in their studies. The surface roughness profiles for the as-sprayed coatings, as prepared steel and eroded coatings, eroded steel have been shown in Figs. 3 and 4, respectively, whereas the data have been compiled in Table 4. From the compiled data, it has been observed that the surfaces have become smoother as a consequence of slurry erosion. If the extent of decrease in average surface roughness is measured for the bare (15.3 %) and Cr_2O_3 coated steel (2 %), it showed that the bare steel has undergone higher percentage (%) reduction in Ra values in comparison with its coated counterpart.

Table 4 Various characteristics of D-gun-sprayed Cr_3C_2–25NiCr and Cr_2O_3 coatings on CF8M steel in as-sprayed and eroded condition

Property/material	Microhardness (HV 2.94 N)	Porosity (%)	Surface roughness (Ra) [μm]
CF8M steel	190 ± 20 HV	–	1.30 ± 0.2
Eroded CF8M steel	260 ± 20	–	1.10 ± 0.1
As-sprayed Cr_3C_2–25NiCr coating	860–980 HV	1–2	5.50 ± 0.4
Eroded Cr_3C_2–25NiCr coating	–	–	4.60 ± 0.2
As-sprayed Cr_2O_3 coating	1220–1440 HV	<3	3.40 ± 0.1
Eroded Cr_2O_3 coating	–	–	3.31 ± 0.1

3.3 Slurry Erosion Testing

The cumulative specific mass loss curves for the test materials are shown in Fig. 5. It has been observed that specific mass loss was higher for CF8M steel in comparison to Cr_3C_2–25NiCr and Cr_2O_3 coatings. This may due to its lower hardness in comparison with the coatings. This indicated that coatings are useful to develop erosion resistance in the base steel. Moreover, the rate of specific mass loss (g/m^2 h) [calculated by division of specific mass loss (g/m^2) and time (h)] was initially higher, which generally decreased during later hours of testing. Enhanced wear rate during initial hours of exposure could also have been partially due to rapid erosion of the micro-hills (surface roughness) present on the surfaces. Initial higher roughness may provide number of sites for slurry particles to shear. This may be regarded as the transient state of erosion. However, once the surfaces are smooth, the erosion may achieve what we call steady state of erosion. During steady state, the rate of erosion is found to be constant.

3.4 Material Removal Mechanism

In the current investigation, cylindrical specimens were rotated in the chamber of high-speed slurry erosion test rig having slurry prepared with distilled water. This may cause erosion (effect of slurry flow) and abrasion damage (effect of slurry particles) to the specimen surfaces simultaneously or in quick succession. It may be assumed that during the rotation of the specimen, the interaction angle with slurry flow ranges from −90° to +90°. As reported in literature, the angle of incidence of the particles influences the wear of materials substantially and can change the wear mechanism Fig. 6.

In high-speed erosion test rig, some of the slurry particles hit against the curved surface of specimen with certain angle along the flow. Hitting force can be divided into two components:

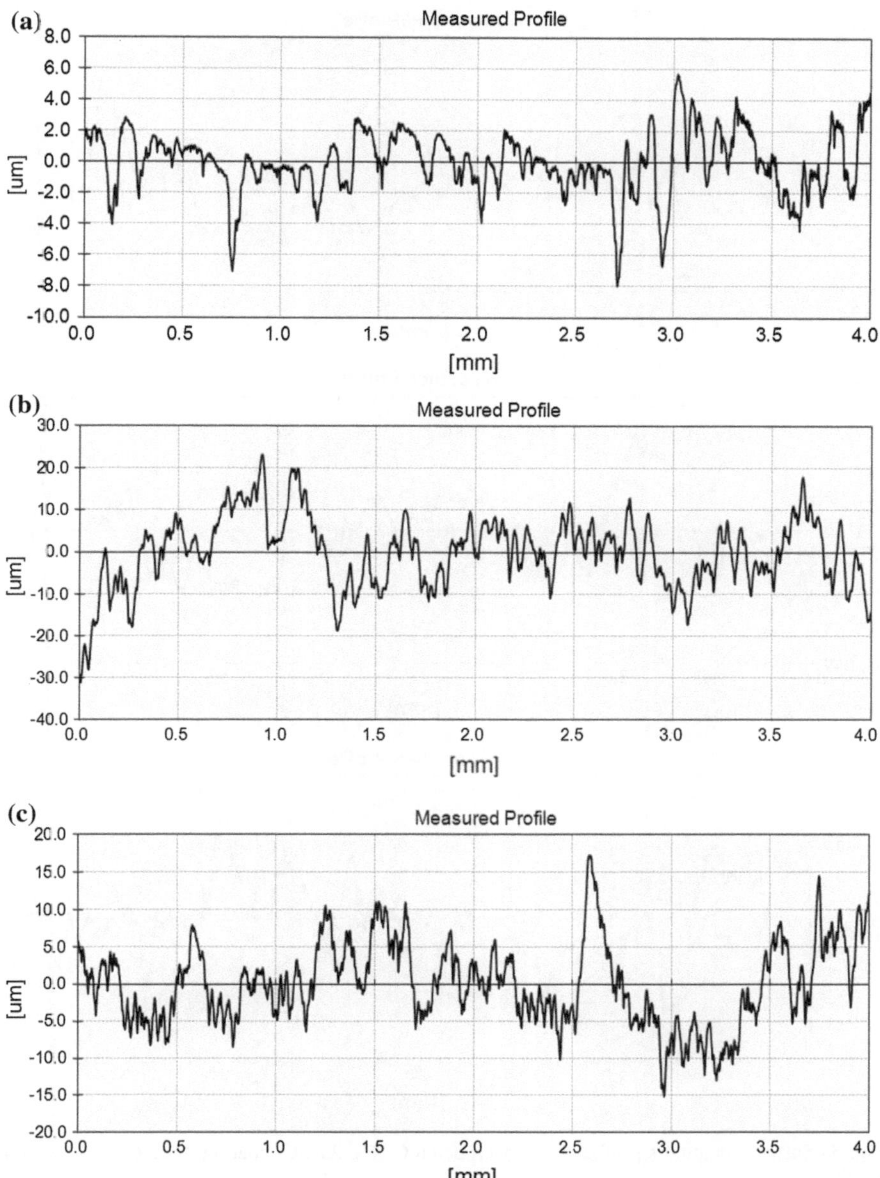

Fig. 4 Surface roughness profile of **a** as-sprayed D-gun Cr_3C_2–25NiCr coating, **b** as-sprayed D-gun Cr_2O_3 coating, and **c** as prepared CF8M steel

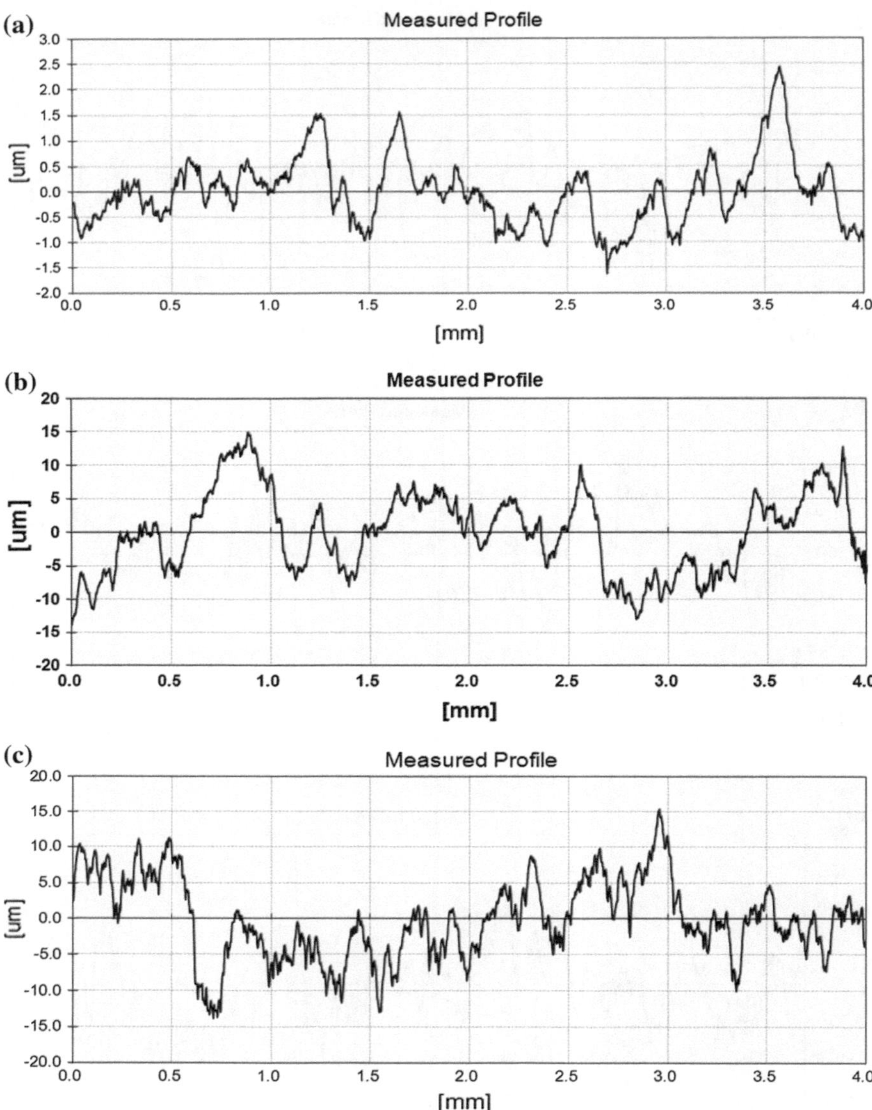

Fig. 5 Surface roughness profile of slurry eroded **a** Cr_3C_2–25NiCr coating, **b** Cr_2O_3 coating, and **c** CF8M steel

- Tangential component which is parallel to the surface of specimen has a cutting effect at the surface.
- Whereas the perpendicular component can make the material to fail by fatigue or under plastic deformation, the normal component of the impact velocity may also cause brittle cracking, particularly in brittle materials.

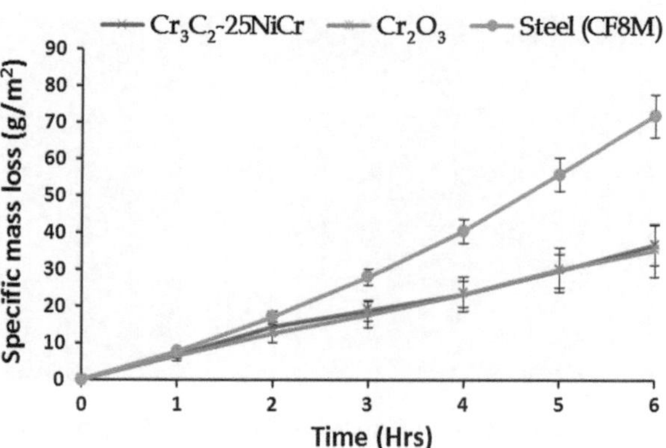

Fig. 6 Cumulative mass loss versus time plots for bare, D-gun Cr_3C_2–25NiCr and Cr_2O_3 coated CF8M steel at 2,250 rpm with slurry of 10,000 ppm concentration and average particle size 100 μm

Therefore, it can be thought that slurry erosion of the specimens is occurring by the simultaneous action of abrasive wear (cutting or ploughing) and impact wear.

However, the predominance of either of the mentioned modes of material loss would depend on material-related and slurry flow factors. The material-related factors are chemical composition (constituents, grain size, etc.), surface condition, ductility, brittleness, hardness, toughness, etc., and slurry flow factors are concentration, average size, velocity, etc. (Prasad et al. 2001).

To investigate the material removal mechanism, the eroded specimens were examined under scanning electron microscope. The SEM features of eroded surfaces of D-gun spray Cr_3C_2–25NiCr, Cr_2O_3 coated CF8M steel, and bare CF8M steel are shown in Fig. 7a, b, respectively. The texture of the surface of coatings (Fig. 7) which are exposed to liquid–solid particle eroding media and as-sprayed coating (Figs. 2, 3) are quite different. It is observed that the surface of each eroded sample is relatively smooth. These observations are also in agreement with the profilometry data (Table 4). This may be attributed to the removal of protuberances present on the surfaces of as-sprayed specimens because of their low adherence to the substrate. Moreover, repeated impacts might have also removed these protuberances. This initial observation is in good agreement with Lathabai et al. (1998) who studied the slurry erosion behavior of Cr_2O_3 coating deposited on mild steel samples using flame spray process using SiC erodents and observed that protuberances had been knocked off during early stages of erosion testing, resulting in a smoother surface. It can be seen from Fig. 7 and that the dominating mechanism of material removal for the coating sprayed with Cr_2O_3 powder has a combination of (1) grain removals and (2) fatigue-induced lateral crack chipping. The SEM micrographs for this coating show several evidences of microcutting (feature observed at low angle impact), which indicate that the material is removed by the cutting action. The microcutting may predominantly occur within the softer binder region. It may further cause weakening of bonding of the carbide grain in the binder phase, which may get loosened and eventually pulled out. There are also evidences of indentation impressions (feature observed at high angle impact) due

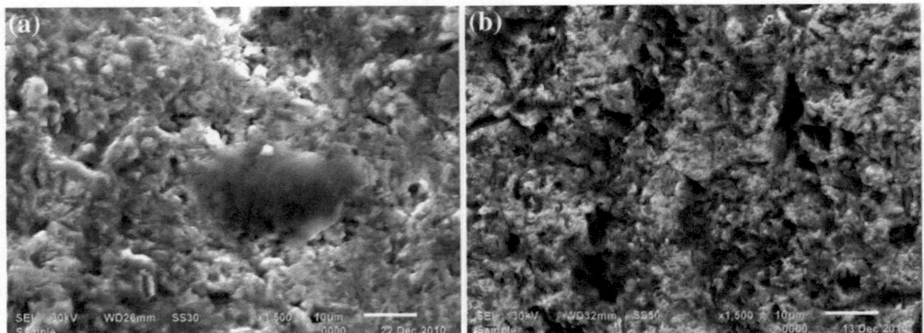

Fig. 7 SEM features of eroded surfaces at 2,250 rpm with slurry of 10,000 ppm concentration and average particle size 100 μm

to the repeated impact of erodent on the surface. These observations show that surface of the coating behaves neither as purely ductile nor purely brittle and has a composite behavior as suggested by Murthy and Venkataraman (2006). The SEM features (Fig. 7b) of CF8M steel show microcutting, narrow craters, deep surface cavities, and ploughing along with lip formation that are created by irregular-sharp-edged slurry particles.

4 Conclusions

- For identical test conditions, D-gun-sprayed Cr_3C_2–25NiCr and Cr_2O_3 can reduce the erosion rates of the CF8M steel. This may be due to their higher hardness in comparison with substrate steel.
- While uncoated CF8M steel showed the ductile behavior during slurry erosion, the D-gun-sprayed Cr_3C_2–25NiCr coating exhibited a mixed behavior. On the other hand, Cr_2O_3 coating mainly exhibited brittle behavior during the slurry erosion testing.

References

Barbezat G, Nicoll AR (1993) Abrasion, erosion and scuffing resistance of carbide and oxide ceramic thermal sprayed coatings for different applications. Wear 64:529–537

Bhandari S, Singh H, Kumar H, Rastogi V (2012a) Slurry erosion behaviour of detonation gun spray Al_2O_3 and Al_2O_3–$13TiO_2$-coated CF8M steel under hydro accelerated conditions. Tribol Lett 45:319–331

Bhandari S, Singh H, Kumar H, Rastogi V (2012b) Slurry erosion performance study of detonation gun-sprayed WC-10Co-4Cr coatings on CF8M steel under hydro-accelerated conditions. J Therm Spray Technol 21(5):1054–1064

Davis JR (2004) Handbook of thermal spray technology. ASM International, Materials Park, OH (Chap. 1)

Hattori S (2002) Mechanism of erosion, fundamentals of cavitation and slurry erosion. Turbomachinery 30(11):644–651

Kanjlia VK, Gupta AC (2008) 3rd international conference, in proceedings of silting problems in hydropower projects. Central Board of Irrigation and Power New Delhi, India, pp 27–28

Knapp JK (1997) Fine-particle slurry wear resistance of selected tungsten carbide thermal spray coatings. Tribol Int 30(3):225–234

Lathabai S, Ottmüller M, Fernandez I (1998) Solid particle erosion behaviour of thermal sprayed ceramic, metallic and polymer coatings. Wear 221:93–108

Mann BS (2000) High-energy particle impact wear resistance of hard coatings and their application in hydroturbines. Wear 237(1):140–146

Murthy JKN, Venkataraman B (2006) Abrasive wear behaviour of WC–CoCr and Cr_3C_2–20(NiCr) deposited by HVOF and detonation spray processes. Surf Coat Technol 200:2642–2652

Pawlowski L (1995) The science and engineering of thermal spray coatings, Wiley, New Jersey (West Susse ex P109 IUD)

Prasad BK, Modi OP, Jha AK, Patwardhan AK (2001) Effects of some material and experimental variables on the slurry wear characteristics of Zinc–Aluminum alloys. J Mater Eng Perform 10(1):75–80

Sundararajan G, Prasad KUM, Rao DS, Joshi SV (1998) A comparative study of tribological behavior of plasma and D-gun sprayed coatings under different wear modes. J Mater Eng Perform 7(3):343–351

Wang B (1995) Erosion-corrosion of coatings by biomass-fired boiler fly ash. Wear 188:40–48

Wang B (1996) Erosion-corrosion of thermal sprayed coatings in FBC boilers. Wear 199:24–32

Wang J, Sun B, Guo Q, Nishio M, Ogawa H (2002) Wear resistance of a Cr_3C_2–NiCr detonation spray coating. J Therm Spray Technol 11:261–265

Xu H, Luo X (1998) In: Proceedings of the 19th IAHR symposium on hydraul. Mach. Cavitation, pp 544–555

Sliding Wear Performance of Plasma-Sprayed Ni–20Cr Coating on AISI 309 SS Steel

Maninder Kaur and Sanjeev Bhandari

Abstract Ni–20Cr (metallic powder) was coated on AISI 309 SS steel by shrouded plasma-spray process. Subsequently, the coating was characterized by XRD and SEM analyses. The XRD analysis indicated the formation of γ-Ni phase for the Ni–20Cr coating. The wear behavior of the bare, Ni–20Cr-coated AISI 309 SS steel was investigated according to ASTM standard G99-03 on a pin-on-disk wear test rig. The wear tests were carried out at normal loads of 30 and 50 N with a sliding velocity of 1 m/s. Cumulative wear rate (CWR) and coefficient of friction (μ) were calculated for these cases. The worn-out surfaces were then examined by SEM analysis. The as-sprayed coating exhibited typical splat morphology. It has been concluded that the plasma-sprayed Ni–20Cr coating can be useful to reduce the wear rate of AISI 309 SS steel. Moreover, coating was found to be adherent to the substrate steel during the wear tests.

Keywords Plasma spray · Sliding distance · Coefficient of friction · SEM

1 Introduction

Wear is a process of removal of material from one or both of two solid surfaces in solid-state contact (Bhushan and Gupta 1991). These wear-related problems can be minimized mostly by using high-cost wear resistant alloys/metals better than the existing low-cost alloys or by improving the wear and corrosion resistance of the existing metals and alloys by surface modifications. As the wear is a surface phenomenon and occurs mostly at outer surfaces, therefore, it is more appropriate

---

M. Kaur (✉) · S. Bhandari
Mechanical Engineering Department, Baba Banda Singh Bahadur Engineering College, Fatehgarh Sahib, Punjab, India
e-mail: maninder.kaur@bbsbec.ac.in

S. S. Khangura et al. (eds.), *Proceedings of the International Conference on Research and Innovations in Mechanical Engineering*, Lecture Notes in Mechanical Engineering, DOI: 10.1007/978-81-322-1859-3_40, © Springer India 2014

and economical to use latter method of making surface modification than using the former one.

Among the various commercially viable surface coating techniques, plasma spraying fosters progress in both development of materials and modern coating technology because of advances in powder and wire productions. The advanced plasma technique has many advantages such as high productivity for thick coating films of more than 100 μm and good applicability for a wide range of coating materials including ceramic powder, further the process does not cause degradation of the mechanical properties of the alloy substrate (Gill and Tucker 1986). Plasma spraying is a well-established means of forming thick coatings used, for instance, for their resistance to wear, corrosion friction, and ionic conduction properties. They are used in many industrial applications, to improve the abrasive, erosive, and sliding wear of machine components. Developments in plasma-spraying techniques as well as advances in powder and wire production have resulted in surface coatings with excellent properties under service conditions, thus enlarging the field of its application (Steffens and Nassenstein 1993). In this presented work on wear study, the use of plasma spraying has been made to improve the wear properties of AISI 309 SS steel material.

In the current study, AISI 309 SS steel has been studied with reference to its wear behavior. AISI 309 SS steel is extensively used in thermal power plant components, especially in the boiler parts and the hoppers for handling pulverized coal from the coal crushers. AISI 309 SS steel plates are fixed on the inner sides of the hoppers, where they come in contact with bulk of free-flowing pulverized coal mass. Therefore, suffer heavy erosive wear and need replacement at regular intervals. This eventually causes economic loss and downtime also. To reduce this wear problem, the plasma-spray coating has been deposited on AISI 309 SS steel and was investigated with regard to their wear characteristics. Ni–20Cr coating was selected to be deposited on AISI 309 SS steel substrate after a comprehensive literature survey.

2 Experimentation

The substrate material for the present study has been selected after discussion with Guru Gobind Singh Super Thermal Plant, Ropar (Punjab), India. The chosen material is designated as AISI 309 SS. Tables 1 and 2 show some chemical and mechanical properties of the AISI 309 SS. Small pins having circular cross-section of 8 mm, and length 50 mm were prepared from AISI 309 SS. These pins were required to perform pin-on-disk experiment at room temperature. The faces of the pins were grinded, followed by polishing with emery papers down to 1,000 grit. Nickel–20Chromium (Ni–20Cr) powder was chosen for plasma-spray deposition on the substrate specimens. The particle size for the powder was 45 ± 5 μm.

The coating work was carried out by a commercial firm, namely Anod Plasma Ltd. Kanpur (India).The specimens were grit blasted with Al_2O_3 powder before

Table 1 Chemical properties of AISI 309 SS

Element	Wt%
C	0.20
Mn	2.00
P	0.045
S	0.03
Si	0.75
Cr	22 (min) and 24 (max)
Ni	12 (min) and 15 (max)
Fe	Balance

Table 2 Mechanical properties of AISI 309 SS

Density ($\times$1,000 kg/m^3)	8
Poisson's ratio	0.27–0.30
Elastic modulus (GPa)	200
Tensile strength (Mpa)	515
Yield strength (Mpa)	205
Elongation (%)	40 (%)
Reduction in area (%)	50 (%)
Hardness (HRB)	95 (HRB)

being plasma sprayed. A total of 40 kW Miller Thermal (USA) plasma-spray apparatus was used to apply the coatings. Argon was used as powder carrying, as well as, shielding gas. The process parameters for the shrouded plasma-spray process employed for applying the coatings are summarized in Table 3. The thickness of coating was monitored during the process of plasma spraying with a thickness gauge; Minitest-2000 made in Germany. Efforts were made to obtain coatings of uniform thickness of 100 μm.

Surface morphology of the as-sprayed coatings was studied with the help of scanning electron microscope (LEO 435VP). The XRD analysis of the coating was also carried out to identify various phases formed on their surfaces. Dry sliding wear tests for the uncoated and plasma-sprayed AISI 309 SS were conducted using a pin-on-disk machine according to ASTM standard G99-03 (model: wear and friction monitor tester TR-20), supplied by M/S DUCOM, Bangalore (India). A complete description of the wear testing has been reported elsewhere (Singh et al. 2008). The wear tests have been carried out for a total sliding distance of 2,400 m (10 cycles of 4 min duration each), so that only top coated surface was exposed for each plasma-sprayed sample.

Tangential force was monitored continuously during the wear tests. Weight losses for pins have been measured at different intervals of time to determine wear loss. The wear rate data for the coated as well as uncoated specimens were plotted with respect to sliding distance to establish the wear kinetics. The specific wear rates for the materials were obtained by

Table 3 Parameters of the argon shrouded plasma-spray process

Parameter	Value
Arc current (A)	700
Arc voltage (V)	35
Powder flow rate (rev/min)	3.2
Spraying distance (mm)	90–110
Plasma arc gas (Argon) (psi)	59
Carrier gas (psi)	40
Spray gun nozzle diameter (mm)	6

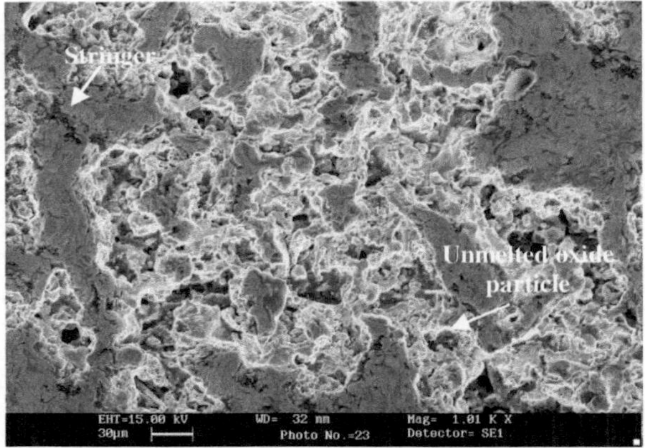

Fig. 1 SEM micrographs showing surface morphology of plasma-sprayed Ni–20Cr coating on substrate AISI 309 SS

$$W = \Delta w / L\rho F \tag{1}$$

where W denotes specific wear rates in, mm^3/N m, Δw is the weight loss measured in g, L the sliding distance in, m, ρ the density of the worn material in g/mm^3 and F the applied load in N. The coefficient of friction (μ) has been plotted against the sliding time to give the friction behavior of the materials, which was calculated as below:

$$m = \text{Frictional Force (N)}/\text{Applied Normal Load (N)} \tag{2}$$

Some of the worn-out surfaces were analyzed by SEM analysis.

3 Results

3.1 Characteristics of As-sprayed Ni–20Cr Coating

SEM analysis: SEM morphologies for the plasma-sprayed Ni–20Cr coating on AISI 309 SS steel substrates are shown in Fig. 1. Microstructures revealed are

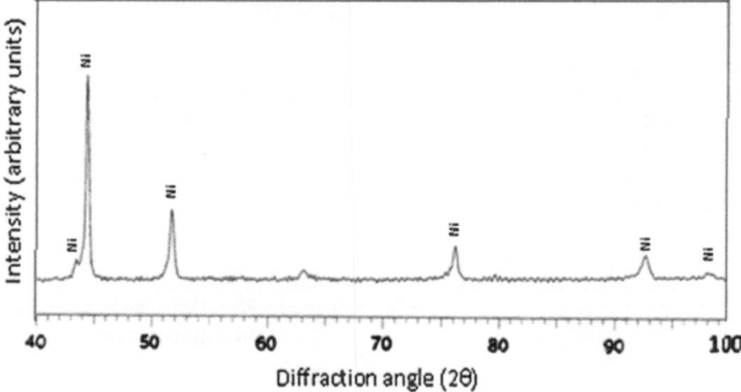

Fig. 2 X-ray diffraction patterns for the plasma-sprayed Ni–20Cr coating on substrate AISI 309 SS

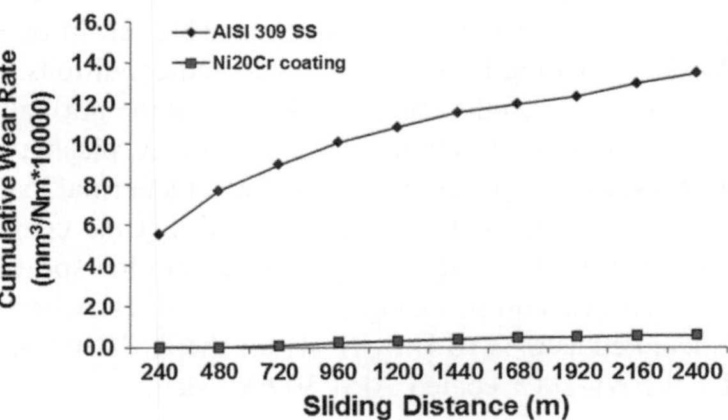

Fig. 3 Variation in CWR with sliding distance for the uncoated and plasma-sprayed Ni–20Cr-coated AISI 309 SS steel subjected to wear at normal load of 30 N and sliding velocity of 1 m/s

typical for a plasma-spray process consisting of splats which are irregular shaped with distinct boundaries. Most of the splats are well formed without any sign of disintegration. Presence of some oxide stringers as well as open pores has also been noticed in general in the coating. The Ni–20Cr coating has shown fine size splats in its structures, Fig. 1.

XRD analysis: As shown in Fig. 2, the XRD of the Ni–20Cr coating has revealed γ-Ni as the main phase without formation of any intermetallic phase.

3.2 Wear Behavior at a Normal Load of 30 N

The variation in the cumulative wear rate (CWR) with the sliding distance for plasma-sprayed Ni–20Cr-coated and uncoated AISI 309 SS steel has been plotted in Fig. 3. It is evident from the plots that the uncoated steel has shown much higher wear rates as compared to its coated counterparts. Further, the wear rate for

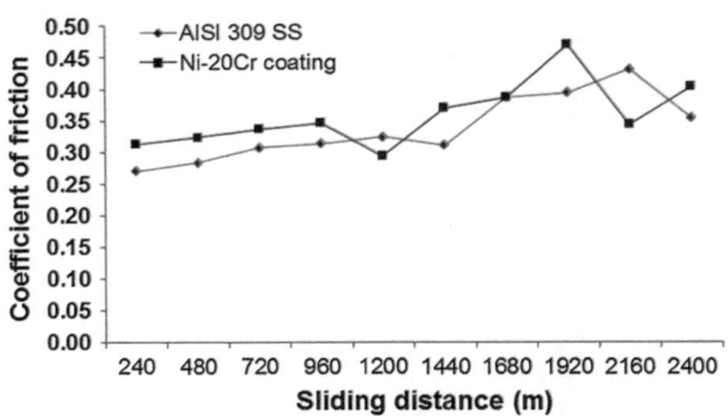

Fig. 4 Variation in coefficient of friction with sliding distance for the uncoated and plasma-sprayed Ni–20Cr-coated AISI 309 SS steel subjected to wear at normal load of 30 N and sliding velocity of 1 m/s

the former case shows an approximately linearly increasing trend with increase in sliding distance, whereas in the latter cases, the wear rate has shown the tendency to become uniform. A uniform wear is a favorable trend from the point of view of wear resistance. Therefore, it can be concluded that the coatings are useful to enhance the wear resistance of the base steel. In other words, the plasma-sprayed Ni–20Cr coating is better than uncoated conditions. The coefficient of friction (μ) determined from the frictional force, and the normal load has been plotted against the sliding time for the uncoated, as well as, plasma-sprayed Ni–20Cr-coated AISI 309 SS steel subjected to wear testing at a normal load of 30 N and sliding velocity of 1 m/s in Fig. 4. It can be observed that the coefficients of friction (μ) for the uncoated AISI 309 SS steel, as well as, plasma-sprayed Ni–20Cr AISI 309 SS steel have shown abrupt changes in their values, whereas. The mean value of μ is calculated as 0.34 (S.D.: −0.04) for the AISI 309 SS steel and 0.44 (S.D.: −0.04) for the Ni–20Cr-coated AISI 309 SS steel.

3.3 Wear Behavior at a Normal Load of 50 N

The variation in CWR with sliding distance for the plasma-sprayed Ni–20Cr-coated, as well as, uncoated AISI 309 SS steel has been shown in Fig. 5. It is evident from the graph that the wear rates for the Ni–20Cr plasma-spray-coated samples do not show significant change with wear distance as compared to that for the uncoated AISI 309 SS steel; rather, for the uncoated AISI 309 SS steel, the CWR goes on increasing with the sliding distance. This shows that coatings are successful in enhancing the wear resistance of the base steel. It is clear that CWR for the plasma-sprayed Ni–20Cr coating do not differ significantly; however, as compared to that for the AISI 309 SS steel case, the values are insignificant for the coated samples. The plots between the coefficient of friction (μ) and the sliding time for the uncoated, as well as, plasma-sprayed Ni–20Cr-coated AISI 309 SS steel have been shown in Fig. 6. The mean value of μ for the uncoated, Ni–20Cr-coated case is found to be 0.38 (S.D.: 0.06), 0.40 (S.D.: 0.05).

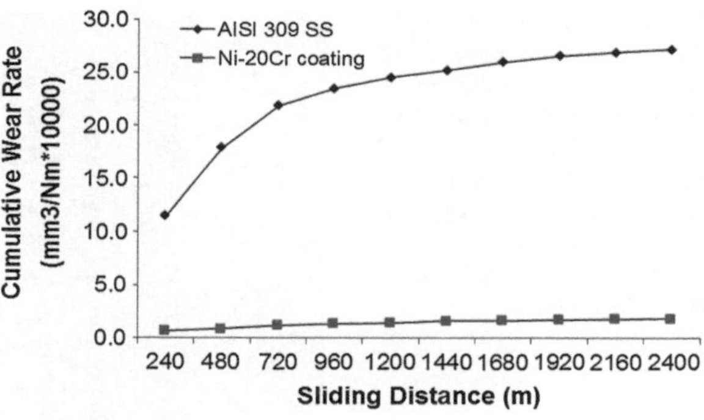

Fig. 5 Variation in coefficient of friction with sliding time of uncoated, the plasma-sprayed Ni–20Cr-coated AISI 309 SS steel subjected to wear at normal load of 50 N and sliding velocity of 1 m/s

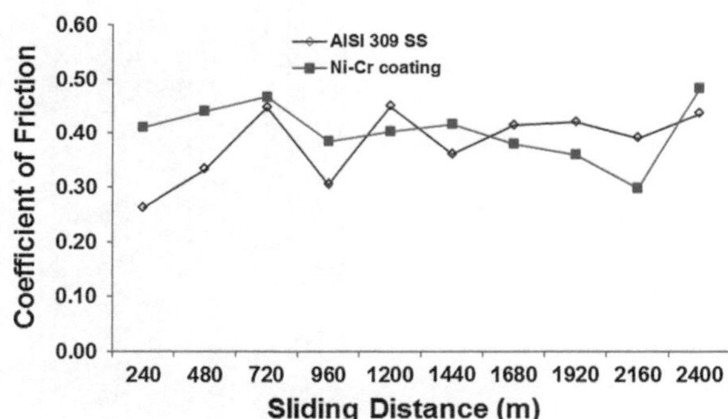

Fig. 6 Variation in cumulative wear volume with sliding distance for the uncoated and plasma-sprayed Ni–20Cr-coated AISI 309 steel subjected to wear at normal load of 50 N and sliding velocity of 1 m/s

3.4 SEM Analysis

The SEM micrographs for the worn-out surfaces for the uncoated AISI 309 SS steel subjected to wear at normal load of 50 N with a sliding velocity of 1 m/s after a sliding distance of 2,400 m have been reported in Fig. 7. The SEM micrographs clearly show the presence of wear tracks on the surfaces. The surfaces have become rougher with unidirectional growth of the structure, probably along the direction of rotation. Further, it looks as a surface has lost the material in the form of microchips, probably due to plowing of the surface by the wear debris between the contact surface of the pin and the disk, Fig. 7 shows the surface morphology for the plasma-sprayed Ni–20Cr-coated steel samples subjected to wear at a normal load of 50 N with a sliding velocity of 1 m/s, respectively, after a total sliding distance of 2,400 m. It can be perceived from comparison of these micrographs with those of corresponding as-sprayed specimens Fig. 1 that the coating has retained its original microstructure, by and large, even after testing. There are cases, where very marginal deformation is indicated in the case of the plasma-sprayed Ni–20Cr-coated case. This shows that the coated samples have shown significantly higher wear resistance in comparison with their uncoated counterparts.

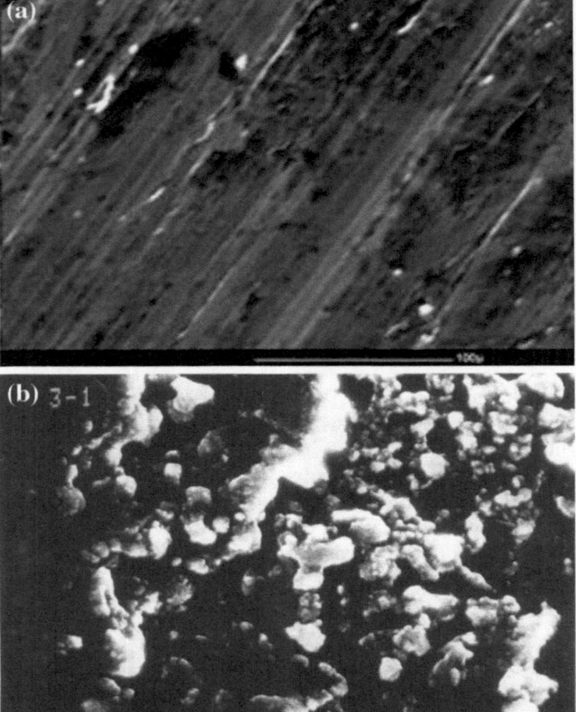

Fig. 7 SEM micrograph of the uncoated and plasma-spray-coated AISI 309 SS steel subjected to wear at a normal load and sliding velocity of 1 m/s for a total sliding distance of 2,400 m at normal load 50 N. **a** Uncoated AISI 309 SS. **b** Ni–20Cr coated

4 Discussion

The Ni–20Cr powder was successfully deposited on AISI 309 SS substrate steel by the plasma-spray process. The thickness of the coating was 100 μm. The XRD analysis of the Ni–20Cr coating revealed γ-Ni as the main phase without formation of any intermetallic phase. This is exactly what is expected from the Ni–20Cr phase diagram. The XRD analysis results are in good agreement with those reported by Singh (2003, 2005) whereas Choi et al. (2002) have identified only γ-Ni phase by XRD in air plasma-sprayed Amdry 962. Similar observation has been made by Sundararajan et al. (2003) for air plasma-sprayed Ni–20Cr coating, where they have indexed XRD pattern for the coating to mainly Ni solid solution γ (fcc) phase. The similar XRD peaks have also been reported by Singh (2003), Sidhu and Prakash (2003), La et al. (1999), Liu and Gao (2001). The SEM analysis for the as-sprayed Ni–20Cr has revealed typical splat-like morphology for the plasma-spray coating with distinct splat boundaries. Singh (2003) has reported similar splat-like morphologies for Ni–20Cr coating which were deposited on some boiler steels, while Singh (2005) observed similar microstructure features for the coating on some Fe- and Ni-based super alloys. Sampath et al. (2004) have reported that the deposit microstructures are a direct result of splat morphologies, while Wigren and

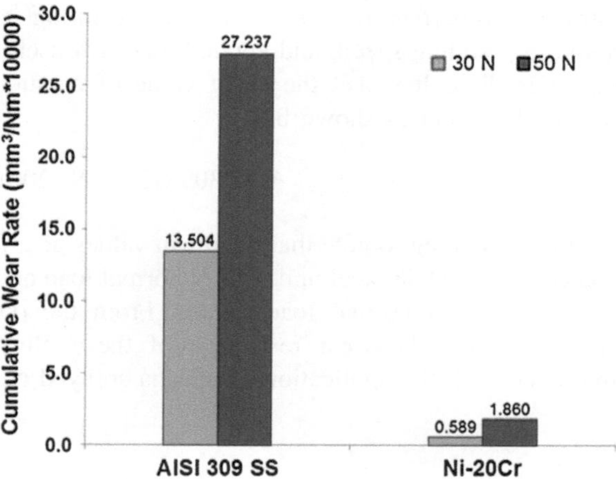

Fig. 8 Bar chart showing variation in CWR for the uncoated and Ni–20Cr-coated steel under normal load conditions of 30 and 50 N at a constant sliding velocity of 1 m/s after a total sliding distance of 2,400 m

Tang (2001) have also reported that the microstructure of a thermally sprayed coating typically consists of a multiphase matrix (often a mix between hard, soft and amorphous), pores, oxides, delaminations, crack, grit residues, and unmelted particles. The usefulness of the coating with regard to wear was investigated with the help of pin-on-disk wear test rig. According to ASTM standard, G99-03 has already been stated. The uncoated AISI 309 SS steel showed higher CWR, as well as, cumulative wear volume loss in comparison with the plasma-sprayed Ni–20Cr and coated AISI 309 SS steel under all the investigated load variants of 30 and 50 N. Moreover, this CWR, as well as, cumulative wear volume went on increasing with increase in sliding distance for the uncoated steel, whereas the same become nearly uniform in the plasma-spray-coated cases. This indicates that the wear resistance of the AISI 309 SS steel got increased significantly after the application of the coating. Although the coefficient of friction has increased for the coated case, yet this increase is marginal.

It has been observed from the overall results of investigation Fig. 8 that the Ni–20Cr coating has shown comparatively lower CWR in comparison with the uncoated steel under the normal load testing of magnitudes of 30 N as well as on 50 N.

Therefore, it may be concluded that the Ni–20Cr coating is more wear resistant than the uncoated. The surface of the uncoated steel has suffered damage of its contact surface in the form of micropits, which most likely may have occurred due to microplowing effect of the wear debris between the contact surface of steel and rotating disk. Similar observation has also been reported by Mishra (2006), Singh et al. (2008) coatings, whereas no such damage of the contact surfaces of the coatings has been observed. Moreover, the coating was found to be successful in keeping their surface contact with the substrate steel when subjected to wear tests.

Further, the coefficient of friction (μ) for the uncoated AISI 309 SS steel is less in general when compared to plasma-sprayed coating under all normal load conditions. The variation in coefficient of friction is due to several factors, such as intrinsic properties of the materials, operating conditions (temperature and humidity), sliding speed, and applied load. Khruschovhave (1957) also suggested that it is difficult to get the exact value of coefficient of friction. The μ value follows the order as shown below:

$$\text{AISI 309 SS} < \text{Ni-20Cr}$$

It is interesting to note that the mean values of μ for the uncoated and Ni–20Cr coated AISI 309 SS steel under 50 N normal load are found to be higher than their respective 30 N normal load values. From the ongoing discussion, it can be concluded that the wear resistance of the AISI 309 SS can significantly be improved with the application of plasma-sprayed Ni–20Cr coatings.

5 Conclusion

- The plasma process provides the possibility of deposition of Ni–20Cr powders on AISI 309 SS steel. A uniform coating thickness of 100 mm was achieved.
- The XRD analysis of the as-sprayed coating indicated the presence of desired phases in the coatings. γ-Ni was identified for Ni–20Cr coating.
- The SEM analysis revealed splat-like morphology with distinct boundaries for the as-sprayed coatings which is a typical characteristic factors of plasma-sprayed coatings.
- The wear resistances for AISI 309 SS steel plasma-sprayed Ni–20Cr followed a general trend irrespective of the value of normal load as given below:

$$\text{AISI 309 SS} < \text{Ni-20Cr.}$$

References

Bhushan B, Gupta BK (1991) Handbook of tribology: material coating and surface treatments. McGraw-Hill, New York

Choi H, Yoon B, Kim H, Lee C (2002) Isothermal oxidation of air plasma spray NiCrAlY bond coatings. Surf Coat Technol 150(2–3):297–308

Gill BJ, Tucker RC (1986) Plasma spray coating processes. J Mater Sci Technol 2(3):207–213

Khruschovhave MM (1957) Resistance of metals to wear by abrasion, as related to hardness. In: Proceedings conference on lubrication and wear, Institute of Mechanical Engineering, London

La P, Bai M, Xue Q, Liu W (1999) A study of Ni_3Al coating on carbon steel surface via the SHS casting route. Surf Coat Technol 113:44–51

Liu Z, Gao W (2001) Oxidation behaviour of cast Ni_3Al alloys and microcrystalline Ni_3Al + 5 % Cr coatings with and without Y doping. Oxid Met 55(5–6):481–504

Mishra SB (2006) Development of erosion–corrosion wear resistant on super alloys. PhD thesis, Department of Metallurgical and Materials Engineering, Indian Institute of Technology Roorkee, Roorkee

Sampath S, Jiang XY, Matejicek J, Prchlik L, Kulkarni A, Vaidya A (2004) Role of thermal spray processing method on the microstructure, residual stress and properties of coatings: an integrated study of Ni-5 Wt% Al bond coats. Mater Sci Eng A-Struct 364:216–231

Sidhu BS, Prakash S (2003) Evaluation of the corrosion behaviour of plasma-sprayed Ni_3Al coatings on steel in oxidation and molten salt environment at 900 °C. Surf Coat Technol 166(1):89–100

Singh B (2003) Studies on the role of coatings in improving resistance to hot corrosion and degradation. PhD thesis, Department of Metallurgical and Materials Engineering, Indian Institute of Technology Roorkee, Roorkee

Singh H (2005) Hot corrosion studies on plasma spray coatings over some Ni-and Fe-based super alloys. PhD thesis, Department of Metallurgical and Materials Engineering, Indian Institute of Technology Roorkee, Roorkee

Singh H, Grewal MS, Sekhon HS, Rao RG (2008) Sliding wear performance of high-velocity oxy-fuel spray Al_2O_3/TiO_2 and Cr_2O_3. J Eng Tribol 222:601–610

Steffens HD, Nassenstein K (1993) Thermal spraying: a review of 1993. J Therm Spray Technol 25(6):280–284

Sundararajan T, Kuroda S, Itagaki T, Abe F (2003) Steam oxidation resistance of Ni-Cr thermal spray coatings on 9Cr-1Mo steel: part 2: 50Ni-50Cr. ISIJ Int 43(1):104–111 (38)

Wigren J, Tang K (2001) Some considerations for the routine testing of thermal sprayed coatings. In: Proceedings of international thermal spray conference, Singapore, pp 1221–1227, 28–30 May 2001 [illegible] W, [illegible] Surf Coat [illegible]

Samadi S, [illegible] N, [illegible] (2011) [illegible] processing method on the microstructure, residual stress and properties of coatings [illegible] NiCrAlY bond coat. Surf Coat [illegible]

Sidhu BS, Prakash S (20[illegible]) [illegible] coatings [illegible] and molten salt environments [illegible]

Singh D (2005) Studies on the role of coatings in improving [illegible] PhD thesis, Department of [illegible] Engineering, Indian Institute of Technology Roorkee, Roorkee, India

Singh H (2005) [illegible] sprayed coatings [illegible] superalloys. PhD thesis, Department of Metallurgical and Materials Engineering, Indian Institute of Technology Roorkee, Roorkee, India

Singh H, Grewal MS, Sekhon HS, Rao RG (2008) Sliding wear performance of high velocity [illegible] Al_2O_3/TiO_2 and Cr_2O_3 coatings [illegible]

Stokes J, [illegible] (20[illegible]) Thermal spray [illegible] Surf Coat Technol [illegible]

Sundararajan T, Kuroda S, Itagaki T, Abe F (2003) Steam oxidation resistance of [illegible] Ni-Cr [illegible] coatings on [illegible] ISIJ Int [illegible]

Tiwari A, [illegible] (20[illegible]) [illegible] In: Proceedings of [illegible] thermal spray [illegible]

Effect of Natural Fillers on Wear Behavior of Glass-Fiber-Reinforced Epoxy Composites

Kishore Debnath, Vikas Dhawan, Inderdeep Singh and Akshay Dvivedi

Abstract Fiber-reinforced plastics have entered into the engineering marketplace few decades ago owing to their excellent mechanical properties and cost-effective high-quality manufacturing. The use of polymer matrix composites for the production of mechanical components such as gears, brakes, cams, bearings and bushes has grown tremendously in the recent years. Therefore, from the theoretical and practical engineering point of view, the study of the wear behavior of polymer matrix composites becomes highly decisive. The present experimental work endeavors at the wear behavior of glass-fiber-reinforced epoxy composites filled with three different natural fillers, such as, rice husk, wheat husk and coconut coir under various sliding conditions. The wear performance analysis of the composites was carried out on pin-on-disk wear test machine under ambient conditions (27 °C and 60 % humidity). The weight loss was measured by applying normal loads of 10, 20 and 30 N under varying sliding speeds of 1, 2 and 3 m/s. The tests were conducted for a constant sliding distance of 1,000 m. The specific wear rate for glass-fiber-reinforced epoxy laminates under dry sliding condition was of the order of 10^{-8} mm^3/N mm. Further, the morphology of the worn surfaces was examined by using scanning electron microscope (SEM) to analyze the wear mechanism of the developed composites.

Keywords Glass–epoxy laminates · Natural fillers · Specific wear rate · SEM

K. Debnath (✉) · V. Dhawan · I. Singh · A. Dvivedi
Department of Mechanical and Industrial Engineering, Indian Institute of Technology Roorkee, Roorkee 247667 Uttarakhand, India
e-mail: debnath.iitr@gmail.com

V. Dhawan
e-mail: vikas251999@gmail.com

I. Singh
e-mail: dr.inderdeep@gmail.com

A. Dvivedi
e-mail: akshaydvivedi@gmail.com

S. S. Khangura et al. (eds.), *Proceedings of the International Conference on Research and Innovations in Mechanical Engineering*, Lecture Notes in Mechanical Engineering, DOI: 10.1007/978-81-322-1859-3_41,

1 Introduction

The consumption of polymer matrix composites is increasing continuously in various engineering applications due to their excellent combination of mechanical properties. Low density and high strength to weight ratio are the two main characteristics which makes these materials attractive candidate particularly for the automotive and aircraft industries (Pihtili and Tosun 2002; Bajpai et al. 2012a, b, 2013; Debnath et al. 2012a, b). These emerging materials are widely used in industries for production of number of mechanical components, such as, gears, cams, wheels, brakes, clutches, bearings and bushes, most of which are subjected to tribological loading conditions (Chauhan et al. 2010). The load-carrying capacities of thermoset polymer composites are more stimulating compared to thermoplastic polymer composites. Hence, the sliding components made up of glass-fiber-reinforced epoxy composites are tribologically more adequate. To make these fibrous systems tribologically even more viable, the possibilities of developing filled fiber-reinforced plastics need to be thoroughly examined (Sampathkumaran et al. 2005). It has been cited in various research initiatives that specific wear rate (SWR) and the coefficient of friction of polymer composites sliding against steel counter faces can be reduced by incorporating synthetic fillers (Suresha and Chandramohan 2006). Ahmed et al. (2012) explored that jute-fiber-reinforced epoxy composites filled with ceramic fillers, such as, Al_2O_3 and SiC, significantly enhance the tribological properties of jute–epoxy composites. Srivastava and Wahne (2007) stated that the filler materials in particulate form improve the mechanical properties and as well as the wear resistance of glass-fiber-reinforced polymer (GFRP) composites. This can be explained by the fact that the fillers in particle form enrich the bonding strength between the fiber and the epoxy resin. Furthermore, incorporation of fillers with different amounts has a notably significant effect on reducing wear and frictional performance of random orientation GFRP composites.

Nowadays, engineers and scientists are more concerned with sustainability; therefore, for developing polymeric composites, raw lingo-cellulosic materials, such as wood fillers, jute, sisal, pine apple, wheat straw, almond husk or rice husk, are added to the polymers (Pande 1998; Pan and Sano 1999). Incorporation of natural fillers in enhancing the tribological performance of polymeric composites is a continuing upsurge of research. Addition of such materials introduces several advantages compared to traditional inorganic fillers, such as biodegradability, low density, reasonable strength and stiffness and low cost (Pande 1998). However, despite the numerous advantages, the use of natural fillers in polymer composites for improving the tribological properties has not been extensively investigated. Therefore, the present research endeavor is an attempt to experimentally investigate the effect of addition of natural fillers on wear characteristics of glass–epoxy laminated composites.

Table 1 Composition and designation of the samples

S. No.	Composition	Sample
1.	Glass fabric + epoxy + rice husk	A
2.	Glass fabric + epoxy + wheat husk	B
3.	Glass fabric + epoxy + coconut coir	C

2 Experimental Procedure

2.1 Laminates Fabrication

The laminates of 4 mm thickness were prepared by conventional hand lay-up technique. The epoxy resin LY556 and hardener HY 951 were mixed and stirred mechanically in a ratio of 10:1 by weight, and then, this mixture was used for coating on ECR-glass fiber fabric. A predefined amount (5 wt.%) of natural filler was evenly distributed over the fabric, and the second layer of fabric was placed on the previous layer. The process was continued till the desired thickness was achieved. Finally, the complete mold was placed in a press and a compression load of 15 ton was applied. The mold was left for 24 h at room temperature to complete the curing process. The material used to develop the three different composites with their corresponding designation is summarized in Table 1. Specimens of suitable dimension were cut using a diamond cutter. The prepared sample used for the test is shown in Fig. 1.

2.2 Wear Test Details

The dry sliding wear tests were conducted on pin-on-disk wear tribometer as per ASTM G 99. The top view of the pin-on-disk wear tribometer (Ducom India TR20LE) is shown in Fig. 2. The test was carried out by applying normal loads of 10, 20 and 30 N under varying sliding velocities 1, 2 and 3 m/s. The sliding distance kept constant to 1,000 m for all the experiments. The counter face is a metallic disk (140 mm $\times$ 8 mm) made up of ground hardened steel (EN-31, 64 HRC, Ra $=$ 0.7 μm). The contact surfaces of all the samples were polished with an emery paper of 800 grit size to ensure proper intimate contact between the specimens and counter face. Before the test, both the rotating disk and specimens were cleaned with acetone. The weight loss of the samples was measured by taking the weight difference of the sample before and after each test. The weight of the samples was measured by precision electronic balance Shimdzu-AUW220D with an accuracy of 0.0001 g. The wear performance of composites was expressed in terms of specific wear rate. The SWR of the specimens was calculated by using the Eq. 1.

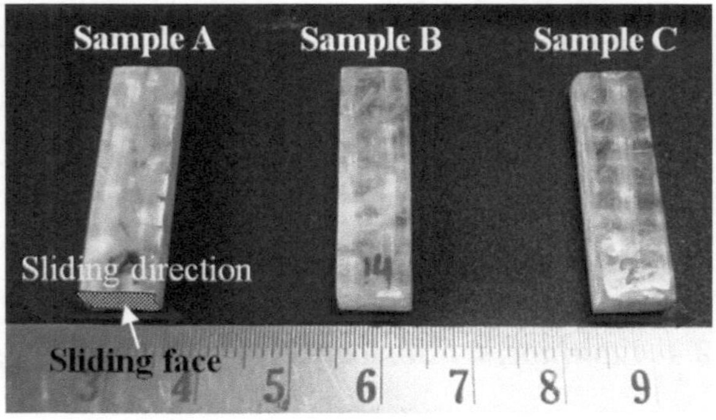

Fig. 1 Samples prepared for sliding wear test

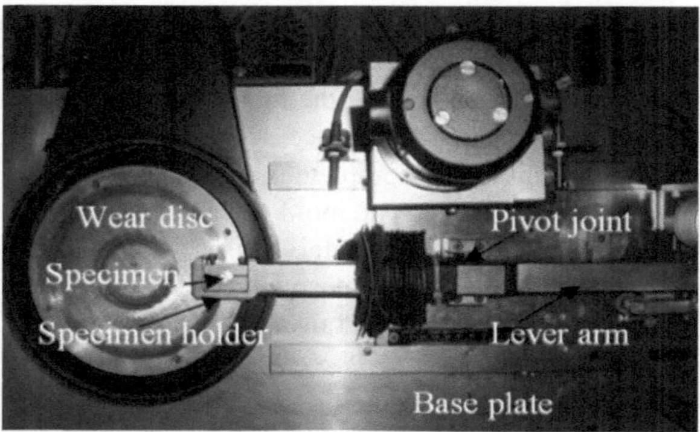

Fig. 2 Top view of pin-on-disk tribo test machine

$$\mathrm{SWR} = \Delta m / (\rho D F_{\mathrm{N}}) \quad (1)$$

where

Δm mass loss during test duration (gm),
ρ density of the specimen (gm/mm^3),
D sliding distance (m) and
F_{N} applied normal load (N).

2.3 Morphology Study

The photomicrographs of the worn surfaces were taken by scanning electron microscope (LEO 435VP). The samples were cleaned with acetone and air-dried, and a thin film of gold was coated using sputter coater (BALTEC SCD 005).

3 Results and Discussion

3.1 Effect of Normal Load on Specific Wear Rate

Figures 3 and 4 show the variation of SWR with applied normal load for the sample A, B and C. Figures also demonstrate that the SWR decreased with increase in applied normal load for all the samples. The highest value of SWR was observed for sample C with a value of 21.85×10^{-8} mm^3/N mm, which was obtained at sliding velocity of 2 m/s when subjected to applied normal load of 10 N (presented graphically in Fig. 3), whereas the maximum SWR observed for sample A was 11.48×10^{-8} mm^3/N mm under the same parameter settings, and for sample B the value was 19.59×10^{-8} mm^3/N mm. The SWR for the sample A was much lower than the sample C. Therefore, from the experimental data, it can be inferred that rice-husk-filled laminated composites have the superior wear resistance ability than the other developed composites. The decreasing trend of SWR with increase in applied normal load can be attributed to the fact that as the load increases, the temperature at the interface becomes high which results in formation of thin film on the counter surface and again this film is transferred back to the sample mating surface which is also known as back film transfer. The accumulation of such thin film on the samples rubbing surface in turns results in decrease in SWR as it acts as a shield and protects the surface from severe wear. Yousif (Yousif 2009) investigated the wear behavior of coir-fiber-reinforced polyester composite (CFRP) under dry sliding condition and concluded that the SWR was low at higher applied normal loads and longer sliding distances due to the formation of back film transfer. The minimum SWR obtained during the tests for all the samples is shown in Fig. 4. The lowest value of SWR was observed in sample A out of all the samples with a value of 2.95×10^{-8} mm^3/N mm obtained at sliding velocity of 3 m/s and an applied normal load of 30 N.

3.2 Effect of Sliding Velocity on Specific Wear Rate

The variation of SWR with respect to sliding velocity is shown in Figs. 5 and 6. The experimental data show that the SWR increased initially with sliding velocity (up to 2 m/s) and decreased with further increase in sliding velocity. The increase in SWR with sliding velocity may be due to the rise in temperature at the interface of the specimen and disk, and when this temperature reaches to the softening point of the polymer, fiber–resin debonding starts which in turn results in easy shearing of fibers or fillers due to the repeated axial thrust force. Sometimes fiber alignment in the direction of sliding motion is also observed. But the SWR again reduced due to the formation of thin film which sticks to the specimen mating surface. The maximum SWR was observed at sliding velocity of 2 m/s when samples were

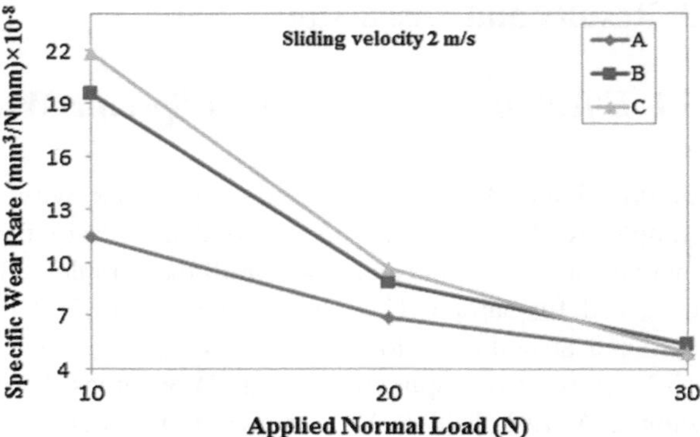

Fig. 3 Variation of SWR with normal load at sliding velocity of 2 m/s

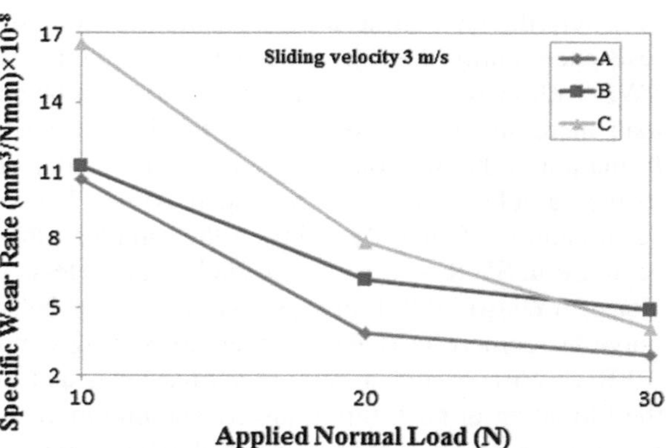

Fig. 4 Variation of SWR with normal load at sliding velocity of 3 m/s

subjected to a normal load of 10 N. It can be envisaged from the figures that applied normal load is a more significant factor than the sliding velocity as the slope of the lines are more steeper in Figs. 3, 4, 5, and 6.

3.3 Worn Surface Analysis

Figure 7, 8 and 9 represent the SEM images of the worn surfaces for the samples A, B and C under the applied normal load of 10 N and sliding velocity of 2 m/s. The morphology of the sample A presented in Fig. 7 indicates a predominant mild wear in terms of micro-cracks and patches of thin polymer film formed over the fibers due to plastic deformation which shields the contact surface and contributes to higher wear resistance. The surface pattern of the sample B is shown Fig. 7b

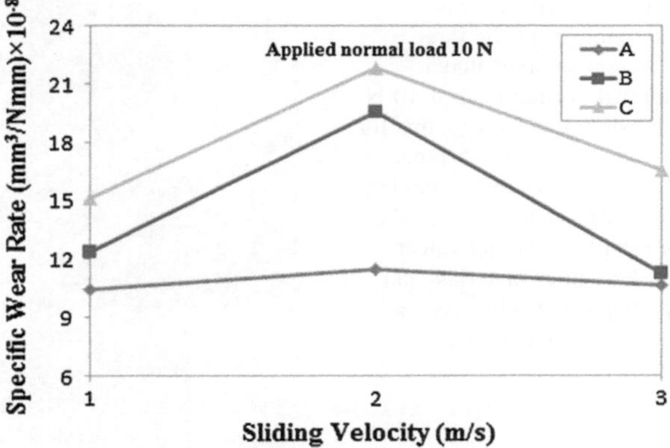

Fig. 5 Variation of SWR with sliding velocity at normal load of 10 N

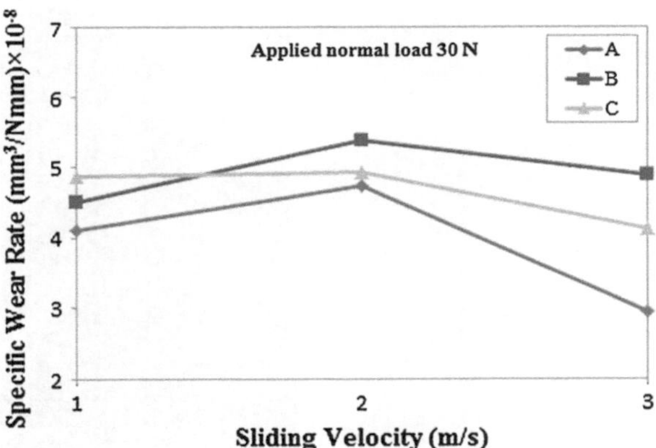

Fig. 6 Variation of SWR with sliding velocity at normal load of 30 N

where fiber breakage, debris formation, matrix cracking and debonding are the main reasons of surface deterioration. In case of sample C, the surface damage was more severe than the other two samples as seen in Fig. 7c. A severe wear of sample C includes high amount of fiber breakage, fiber pull-out and debris that can be visibly confirmed in Fig. 7c.

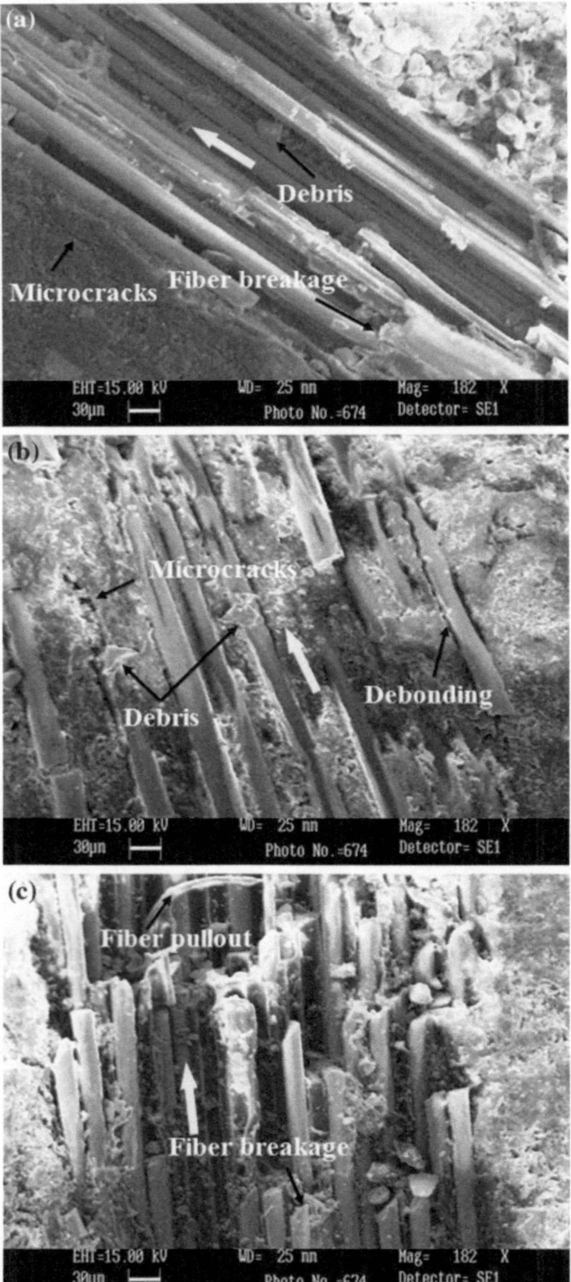

Fig. 7 Worn surfaces morphology of the three different samples under applied normal load of 10 N and sliding velocity 2 m/s for a constant sliding distance of 1,000 m. **a** sample A having rice husk as filler materials; **b** sample B having wheat husk as filler materials and **c** sample C having coconut coir as filler materials

4 Conclusions

The following conclusions can be drawn from the present study:

1. The dry sliding wear behavior of the glass–epoxy laminated composites can be significantly improved by incorporating natural fillers.
2. The SWR of the developed composites decreased with the increase in applied normal load for all the samples, whereas both increasing and decreasing trends were observed in SWR with increase in sliding velocity.
3. The maximum SWR was observed for sample A (11.48×10^{-8} mm^3/N mm) that was much lower than the sample B (19.59×10^{-8} mm^3/N mm) and C (21.85×10^{-8} mm^3/N mm). The maximum wear resistance was observed in sample A followed by the sample B and C (SWR followed the order of $C > B > A$).
4. The worn surface images of the specimens displayed that back film transfer, fiber fracture, micro-cracks and formation of debris were the main reasons of surface damage. Debonding of fibers from matrix, matrix cracking and fiber breakage was also observed on the worn surfaces.

References

Ahmed KS, Khalid SS, Mallinatha V, Amith Kumar SJ (2012) Dry sliding wear behavior of SiC/Al_2O_3 filled jute/epoxy composites. Mater Des 36:306–315

Bajpai PK, Singh I, Madaan J (2012a) Comparative studies of mechanical and morphological properties of PLA and PP-based natural fiber composites. J Reinf Plast Compos 31(24):1712–1724

Bajpai PK, Singh I, Madaan J (2012b) Finite element model for microwave heating of thermoplastic composites. Int J Mater Eng Innov 3(3/4):247–258

Bajpai PK, Singh I, Madaan J (2013) Frictional and adhesive wear performance of natural fibre reinforced polypropylene composites. Proc Inst Mech Eng Part J: J Eng Tribol 227(4):385–392

Chauhan SR, Kumar A, Singh I (2010) Sliding friction and wear behaviour of vinylester and its composites under dry and water lubricated sliding conditions. Mater Des 31(6):2745–2751

Debnath K, Singh I, Dvivedi A (2012) Development and tribological characterization of GFRP laminates with natural fillers. In: 4th international and 25th all India manufacturing technology, design and research conference (AIMTDR 2012), vol II, pp 771–775, 14–16 Dec 2012

Debnath K, Dvivedi A, Singh I (2012) Wear behavior of glass/epoxy composites filled with rice husk. In: 3rd international multicomponent polymer conference (IMPC-2012) 23–25 Mar 2012

Pan HJ, Sano YI (1999) Atmospheric acetic acid pulping of rice straw II. Behavior of ash and silica in rice straw during atmospheric acetic acid pulping and bleaching. J Reinf Plast Compos 34:34–40

Pande H (1998) Non-wood fibre and global fibre supply. UNASYLVA-FAO, vol 44, pp 44–50

Pihtili H, Tosun N (2002) Investigation of the wear behaviour of a glass-fibre-reinforced composite and plain polyester resin. Compos Sci Technol 62(3):367–370

Sampathkumaran P, Seetharamu S, Thomas P, Janardhana M (2005) A study on the effect of the type and content of filler in epoxy–glass composite system on the friction and slide wear characteristics. Wear 259(1):634–641

Srivastava VK, Wahne S (2007) Wear and friction behaviour of soft particles filled random direction short GFRP composites. Mater Sci Eng A 458(1):25–33

Suresha B, Chandramohan G (2006) Friction and wear characteristics of carbon epoxy and glass-epoxy oven roving fiber composites. J Reinf Plast Compos 25(7):771–782

Yousif BF (2009) Frictional and wear performance of polyester composites based on coir fibres. Proc Inst Mech Eng Part J: J Eng Tribol 223(1):51–59

SiC_p-Reinforced Al6063 MMCs: Mechanical Behavior and Microstructural Analysis

Sarbjit Singh, Inderdeep Singh, Akshay Dvivedi and J. Paulo Davim

Abstract The growing steel prices have significantly affected the manufacturing expenditure in automobile and household industries, thus making a strong case for substituting steels with materials having lightweight and high-strength to weight ratio. The aluminum and its alloys have outstanding properties such as lightweight, wear, and corrosion resistance that make them suitable in numerous industrial applications. The present research investigation explores the microstructural and mechanical characterization of Al6063 SiC-reinforced metal matrix composites (MMCs). The MMCs were developed in-house using modified stir casting route. The effect of smaller size SiC reinforcement of 5, 10, and 15 μm has been investigated.

Keywords MMCs · Stir casting · Microstructural behavior · Mechanical behavior

S. Singh (✉)
Department of Mechanical Engineering, PEC University of Technology, Chandigarh, India
e-mail: sarb1234@gmail.com

I. Singh · A. Dvivedi
Department of Mechanical and Industrial Engineering, Indian Institute of Technology Roorkee, Roorkee 247667, Uttarakhand, India
e-mail: dr.inderdeep@gmail.com

A. Dvivedi
e-mail: akshaydvivedi@gmail.com

J. Paulo Davim
Department of Mechanical Engineering, University of Aveiro, Campus Santiago, 3810-193 Aveiro, Portugal
e-mail: pdavim@ua.pt

S. S. Khangura et al. (eds.), *Proceedings of the International Conference on Research and Innovations in Mechanical Engineering*, Lecture Notes in Mechanical Engineering, DOI: 10.1007/978-81-322-1859-3_42, © Springer India 2014

1 Introduction

The MMCs will possess the desirable attributes of metal and ceramics if they are designed and developed correctly. The high strength, ductility, and high-temperature resistance are some of the major attributes of metals, and on the other hand, the ceramic reinforcements have high stiffness and hardness with highly brittle behavior. Combining the ceramic reinforcement with the ductile metal matrix produces a material that has the mechanical properties of both the metallic matrix and the ceramic reinforcement (Singh 2013).

The synthesis of MMCs by stir casting (liquid route) comprises producing a melt of desired metal matrix followed by the addition of ceramic reinforcement into the melt (Hashim et al. 1999). The next step involves the solidification of the melt to obtain the desired distribution of the ceramic reinforcement. There are numerous factors that require considerable attention for the production of MMCs using stir casting route. These factors include (Hashim et al. 2001, 2002a, b)

- Uniform distribution of ceramic reinforcement
- Wettability of the ceramic reinforcement with the matrix
- Porosity in the developed metal matrix composites (MMCs)
- Reaction between the ceramic reinforcement and metal matrix at higher temperature.

For obtaining the desired or the optimum properties of the MMCs, the above-said issues must be addressed during the production of MMCs using stir casting route.

2 Development of Metal Matrix Composites

Metal matrix composite materials are presently undergoing active development in the whole world. The benefits of using MMC materials lie in the advantage of attaining a combination that can result in a number of service benefits. The development of MMC may encompass primary issues such as selection and distribution of constituent phases, the characteristics of the interface, and the possible tailoring of constituents (Dvivedi 2008; Singh et al. 2013). There are numerous production methods available for the production of the MMC materials; there is no unique route in this respect. Owing to the choice of matrix material and reinforcement and of the type of reinforcement, the production methods can vary substantially. Processing of MMCs can be classified according to whether the matrix is in solid, liquid, or vapor phase, while it is being combined with the reinforcement. The comprehensive grouping for processing of MMCs is as follows:

1. Solid state: diffusion bonding, powder blending, and consolidation.
2. Liquid state: liquid metal infiltration squeezes casting, stir casting, spray deposition, etc.
3. Vapor state: physical vapor deposition.

Normally, the liquid-state production technique is more effective than the solid-state and vapor-state production techniques because solid-state processing necessitates a longer time. The matrix metal is used in various forms in different production methods. Generally, the powder is used in powder metallurgy technique, and a liquid matrix is used in liquid metal infiltration, plasma spray, spray casting, squeeze casting, pressure casting, and compocasting. A molecular form of matrix is used in physical vapor deposition and chemical vapor deposition.

2.1 Liquid-State Production Processes

Liquid-state production processes are more efficient and economical as compared to the solid-state production processes, since they require more time for fabrication. Majority of the MMCs are fabricated by liquid metallurgy route to take the cost advantage.

2.2 Liquid Metal Infiltration

Liquid metal infiltration comprises infiltration of a fibrous or particulate reinforcement preform by a liquid metal. Liquid-phase infiltration of MMCs is not a straight forward, essentially because of complications with wetting the ceramic reinforcement by the molten metal. When the infiltration of a fiber preform happens freely, reactions among the fiber and the molten metal may take place, which considerably degrade of the properties of the fiber. Fiber coatings applied before infiltration increase the wetting and allow better control over the interfacial reactions. Such coatings have been developed and are producing inspiring results. However, the drawback is that the fiber coatings need not to be exposed to air before infiltration since surface oxidation of the coating takes place (Singh et al. 2013).

2.3 Squeeze Casting

Most of the reinforcement materials such as graphite, silicon carbide, and aluminum oxide do not wet optimally in the molten matrix, and hence, it is problematic to produce the MMC by liquid infiltration process. In contrast, in the

Table 1 Comparative evaluation of the different fabrication techniques used for MMCs (Dvivedi 2008)

Route	Cost	Range of shape/size	Damage to reinforcement	Range of volume fraction
Powder metallurgy	High	Wide range/restricted size	Reinforcement fracture	–
Squeeze casting	Medium	Limited by preform shape	Severe damage	up to 45 %
Spray casting	High	Limited shape/large size	–	30–70 %
Liquid metallurgy	Low	Wide range of shapes/larger size up to 500 Kg	No damage	up to 30 %

squeeze casting process, the molten metal will be forced-infiltrated into the fiber bundles, ejecting all absorbed and entrapped gases.

Squeeze casting is defined as the capability to forcibly charge liquid metal into a preheated ceramic fiber or any reinforcement preform that is set in a metal die and then to allow the liquid metal to solidify whist applying a high pressure, thereby squeezing the molten metal (Hashim et al. 2001).

Composites produced through this process have the benefit of negligible reaction between the reinforcement and the molten metal because of the shorter processing time involved. Such composites are also naturally free from common casting flaws such as porosity and shrinkage cavities.

2.4 Stir Casting

Most stir casting comprises incorporation of ceramic particles into the liquid aluminum melt and allowing the mixture to solidify. Here, the critical thing is to generate good wetting among the particulate reinforcement and the liquid aluminum alloy melt. The simplest and most commercially used process is known as vortex method or stir casting method. The vortex method involves the addition of pretreated ceramic particles into the vortex of molten alloy formed by the rotating impeller. Microstructural inhomogeneity can cause particle accumulation and sedimentation in the melt and during solidification. Inhomogeneity in reinforcement dispersal in the cast composites might also be a problem as a result of interaction between the suspended ceramic particles and moving solid–liquid interface during solidification. Generally, it is possible to incorporate up to 30 % ceramic particles in the size range 5–100 μm in a variety of molten aluminum alloys. The method is not appropriate for the incorporation of submicron-size ceramic particles or whiskers. Another variant of stir casting process is compocasting. Here, ceramic particles are incorporated into the alloy in the semisolid state (Kalaiselvan et al. 2011) (Table 1).

2.5 Development and Fabrication of Stir Casting Setup for Production of MMCs

The present section includes the development and fabrication of a stir casting setup using melt–stir–squeeze–bottom pouring route to yield the MMCs. The main design consideration for the development of stir casting setup is to obtain the objectives that are given as follows (Singh 2013; Dvivedi 2008):

- Stirring mechanism design with variable speed
- Type of stirrer, blade angles, and position of blade in the melt
- Temperature measurement and control
- Squeezing unit design and placement for bottom pouring
- Design of crucible for bottom pouring
- Inert mixing atmosphere to prevent oxidation of the melt
- Particle addition arrangement in the melt
- Pretreatment of reinforcement to enhance wettability
- Use of allowing elements to enhance wettability
- Self-alignment of all mating parts for better control on the process
- Design of mold for near-net manufacturing
- Selection of material for stirrer, stirrer blade, crucible and squeezer
- Reduction in process time to reduce settlement of particles.

The valuation of the developed setup for the production of MMC was judged by the characteristics of the product (MMC) produced. Trials and errors led to the overall development in the design and manufacturing of various components in the setup which enhanced the performance of the setup. The porosity and wettability are the most challenging issues faced by the research fraternity (Lin et al. 2010). The optimal control of the process parameters and conditions, such as operating temperature, stirring speed, stirrer blade angles, pretreatment of SiC particles, addition of alloying elements, reduction in time gap between phases of stirring and squeezing, led to the successful stir casting procedure (Su et al. 2010; Balasivanandha et al. 2008). A schematic diagram of the developed stir casting setup is shown in Fig. 1.

3 Microstructural Characterization of the Developed MMCS

Microstructural characterization intended to find out the distribution of SiC particles in the aluminum matrix was carried out using optical microscopy. The samples for microstructure characterization were prepared using emery papers of different mesh size (100–1,000). Kerosene was applied on the emery paper for removing the abrasive particles and thus avoiding their deposition on the surface of the sample. Final polishing of the sample was done on velvet cloth by means of

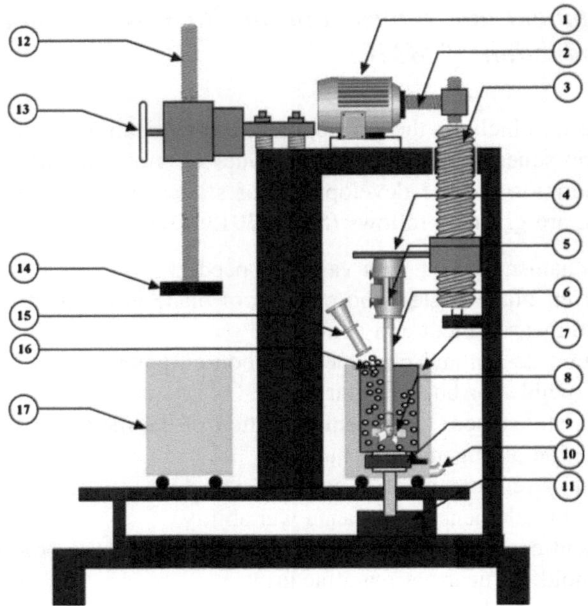

Fig. 1 Schematic of stir casting setup. *1* Motor1 (M1), *2* rotor, *3* lead screw, *4* support and gear for M2, *5* Motor2 (M2), *6* stirring rod, *7* primary furnace, *8* impeller, *9* flow control valve, *10* argon supply, *11* metallic mold, *12* lead screw, *13* handle, *14* squeezer, *15* nozzle, *16* abrasive particles, *17* secondary furnace

alumina paste. The resulting microstructure of the polished samples has been examined under optical microscope (Leitz MM6 large field metallographic microscope) at different magnifications.

The properties of MMCs depend comprehensively on the distribution of the reinforcing particles and the interfacial bonding between the particle and the matrix (Sahin and Acilar 2003; Aigbodion and Hassan 2007). In practice, to attain uniform distribution of the reinforcing particles in the MMC is a challenging task. The optical micrographs, SEM, and FE-SEM images of the Al6063 alloy reinforced with SiC particles of sizes 5, 10, and 15 μm are shown in Figs. 2, 3, and 4. Micrographs are taken for the samples both in etched and in un-etched conditions in order to distinguish the matrix and reinforcement phase. Dark areas visible in the micrographs indicate the porosity present in the developed composites.

Figure 2a–c shows the optical micrographs, SEM, and FE-SEM images of Al6063/10 % SiC composites reinforced with 5-μm-size SiC particles. A uniform distribution of SiC particles is observed from Fig. 2a, with certain level of porosity. The SEM image shown in Fig. 2b also reveals the uniform distribution of SiC particles. A fairly good interfacial bonding of SiC particles with Al6063 matrix is evident from FE-SEM images as shown in Fig. 2c. The angular and irregular shape of SiC particles is apparent from the micrographs and FE-SEM photograph.

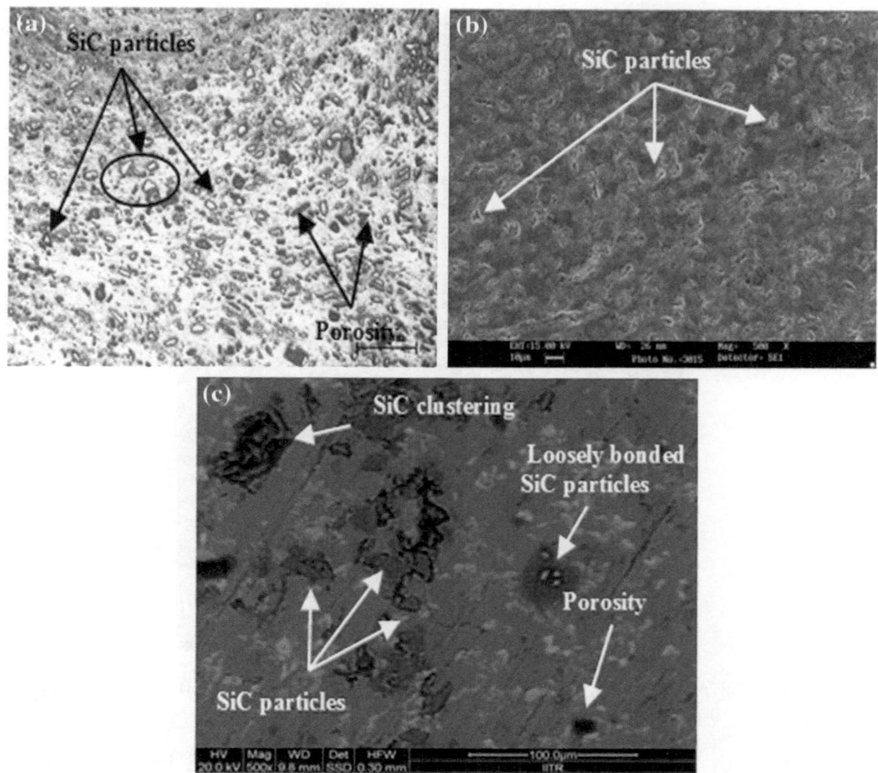

Fig. 2 **a** Optical micrographs, **b** SEM, **c** FE-SEM images of Al6063/10 % SiC composites reinforced with 5-μm-size SiC particles

Figure 3a–c shows the optical micrographs, SEM, and FE-SEM images of Al6063-/10 % SiC-reinforced composites with SiC particles of size 10 μm. Reasonably uniform distribution is identified from the micrographs and SEM images. Good interfacial bonding between the SiC particles and aluminum matrix is observed from the FE-SEM analysis as shown in Fig. 3c. Microstructure observed from SEM micrographs, Fig. 3b, points out that the distribution of 10-μm-size SiC was more uniform as compared to 5- and 15-μm-size SiC particles. This is due to the proper wetting of SiC particles with the matrix. Further, the 5-μm-size SiC particles have a tendency to agglomerate more and 15-μm-size particles have a tendency to settle down due to density difference.

Figure 4a–c shows the micrographs, SEM, and FE-SEM images of Al6063/10 % SiC reinforced with SiC particles of size 15 μm. Figure 4a establishes the better distribution of SiC particles with reasonably good number of SiC particles. Uniform distribution of particles is also seen from the SEM image as shown in Fig. 4b. Porosity content is found to be more with Al6063/10 % SiC composites having 15-μm-size SiC particles as compared to Al6063/10 % SiC composites

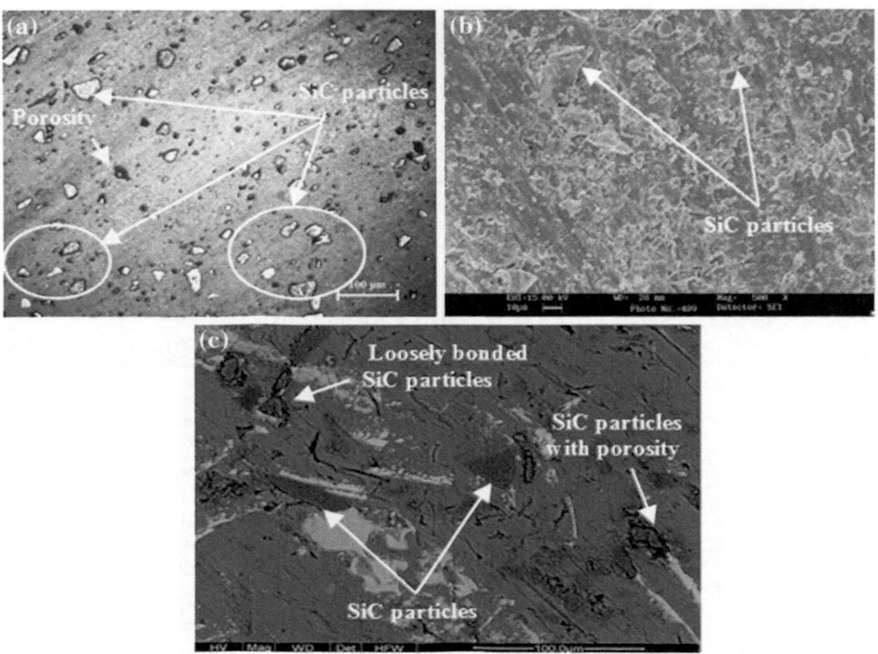

Fig. 3 **a** Optical micrographs, **b** SEM, **c** FE-SEM images of Al6063/10 % SiC composites reinforced with 10-μm-size SiC

reinforced with 5- and 10-μm-size SiC particles, as observed from the micrographs. The formation of porosity in the cast composites prepared by stir casting may be attributed to the higher dissolution of gases.

The FE-SEM EDAX analysis of the developed MMCs reveals the different elemental composition and their quantity present in Al6063-/SiC-reinforced composites. Peaks of silicon (Si) and carbon (C) as shown in Fig. 5 indicate the presence of SiC reinforcing phase within the matrix.

4 Tensile Behavior of Developed MMCS

Tensile strength tests were performed on developed Al6063 MMCs reinforced with different SiC particles of different mesh sizes. Tensile specimens from each Al6063 casting reinforced with SiC particles of different size were prepared from the top, middle, and bottom portions of the billet. The tensile specimens are prepared as per ASTM standard of B557M-10, having gauge length of 25 mm and width 6 mm as shown in Fig. 6. The specimens were subjected to uniaxial tension on a Hounsfield Monsanto (H25KS/05) universal testing machine (UTM). The average value of the ultimate tensile strength (UTS) and percentage elongation or ductility of the specimens was recorded.

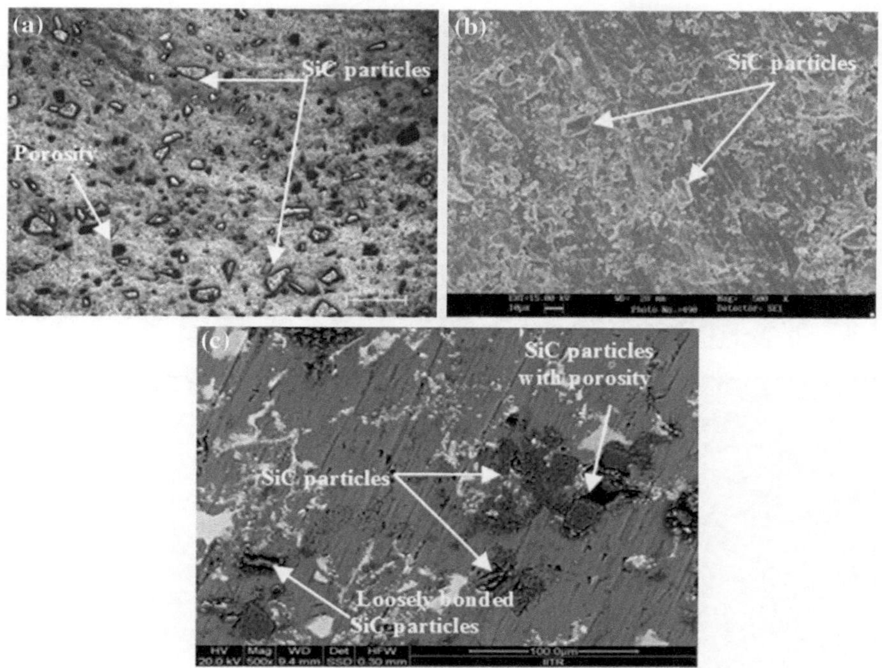

Fig. 4 **a** Optical micrographs, **b** SEM, **c** FE-SEM images of Al6063/10 % SiC composites reinforced with 15-μm-size SiC particles

The variation in tensile strength of Al6063/SiC composites reinforced with 10 % SiC particles of sizes 5, 10, and 15 μm is shown in Fig. 7. It can be shown from Fig. 7 that the composites reinforced with 10 μm size of SiC particles show increased tensile strength as compared to 5 and 15 μm size of SiC-reinforced composites. The average tensile strength of 5-, 10-, and 15-μm-particle-size reinforced composites was found 124.4, 143.4, and 129.6 MPa, respectively, with minimum and maximum values of 107–134 MPa for 5 μm, 136–149 MPa for 10 μm, and 119–139 MPa for 15-μm-size SiC particles. The increase in the tensile strength for 10-μm-size SiC particles reinforced composites may be attributed to the proper interfacial bonding of the SiC particles with the matrix, which causes better load transfer between the two phases, causing improved resistance to the tensile load. The micrograph and FE-SEM images of the 10-μm-size SiC-reinforced composites as shown in Fig. 3 also establishs the proper interfacial bonding of SiC particles with the Al6063 matrix. The lower tensile strength with the 5-μm-size reinforced composites may be because of the agglomeration of SiC particles and increased porosity.

The percentage elongation or ductility of the Al6063/SiC particles reinforced composites depends upon the reinforcement size and is found in the range of 7–13 %. The 10-μm-size SiC particles reinforced composites shows highest

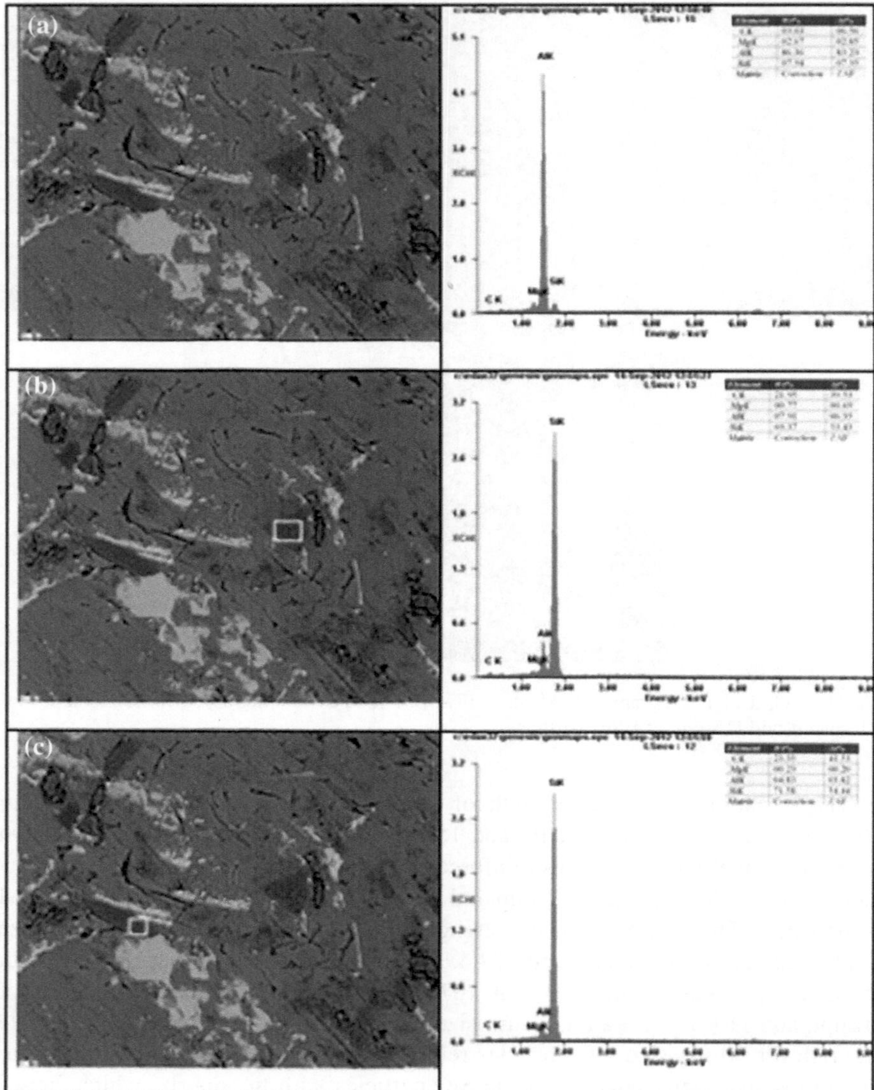

Fig. 5 Chemical composition of the developed MMCs (**a**) MMC sample (**b**) and (**c**) at SiC particles at different location

elongation (13 %) followed by the 15-μm (11 %)- and 5-μm (7 %)-size SiC particles reinforced composites. The fractography analysis of the surface is shown in Fig. 8, which reveals dimples, porosity, and SiC particles on the fractured surface. The semiductile behavior of MMCs was because of the de-bonding between the matrix and SiC reinforcement and porosity present in the MMC (Emamy et al. 2008).

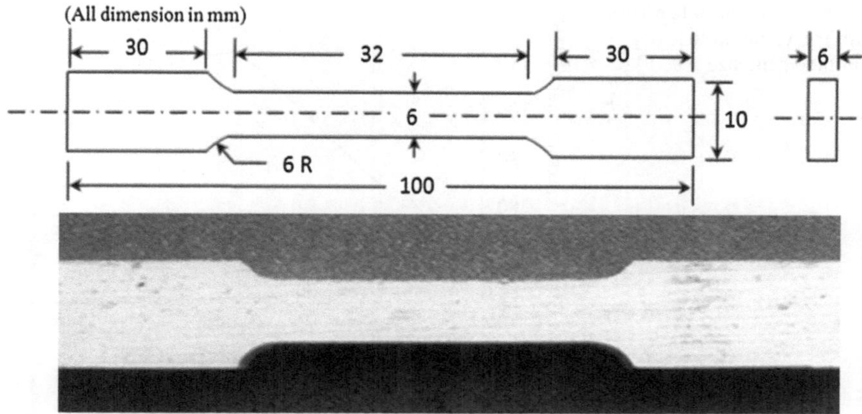

Fig. 6 Tensile specimen as per ASTM standard

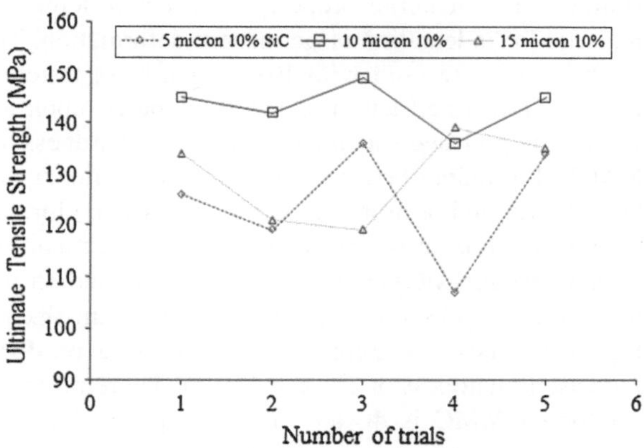

Fig. 7 Variation of tensile strength with variation in size of reinforcement

Fig. 8 Fractography analysis of the tensile specimen

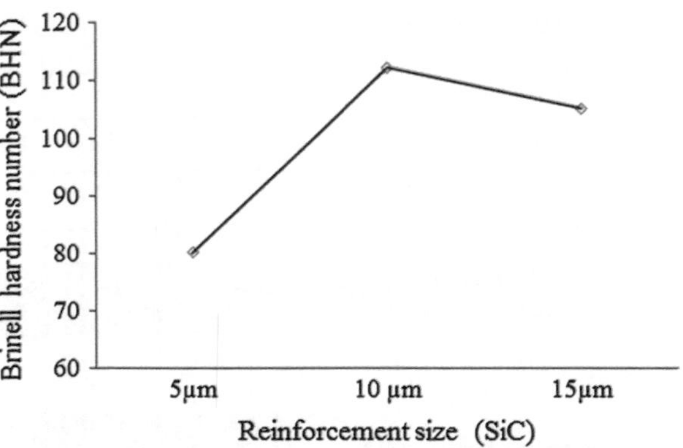

Fig. 9 Variation in hardness with the variation in reinforcement size

5 Hardness Measurement of the Developed MMCs

Hardness is a material property, and it is a defined as resistance offered by the material for indentation or permanent deformation. The hardness of the developed MMCs was measured with Brinell hardness tester (Democratic Republic, Germany). The Brinell hardness testing is based upon a simple indentation test performed on polished specimen to find their hardness, and it is extensively used for MMCs. It provides better average hardness value over a large area, which contains both matrix and reinforcing phase. In order to eliminate the agglomeration effect, the mean of four tests was taken for each specimen. Figure 9 presents the variation in hardness for Al6063/10 % SiC for 5, 10, and 15 μm reinforcement sizes. From the figure, it is observed that the 10- and 15-μm-size reinforced composites exhibit higher hardness as compared to 5-μm-size reinforced MMC. The percentage increase in hardness of 10- and 15-μm-size reinforced MMC with respect to 5 μm reinforced MMC is 40 and 33 %, respectively. The increase in hardness with increase in reinforcement size indicates the better interfacial bonding with the matrix and reinforcement. However, a slight decrease in hardness for 15 μm reinforce MMCs is attributed to larger metal reinforcement interface area and increased gravitational effect on the SiC particles because of density difference. Porosity present in the developed MMCs also affects the hardness, because increase in hardness is attributed by less amount of porosity present in the composites (Kalaiselvan et al. 2011).

6 Conclusions

The present experimental research investigation deals with the design and development of a stir casting setup for the production of near-net shape or net shape of the SiC-reinforced Al6063 MMC parts. A range of SiC reinforcement sizes have been used for the production of MMCs. The developed MMCs have been characterized for microstructure, tensile strength, and hardness. The following specific conclusions can be drawn from the experimental investigation:

1. A stir and squeeze casting setup has been designed and developed successfully for the production of near-net shape or net shape of MMC parts. The developed setup has additional features of bottom pouring and squeezing. The casting route used for the production of MMCs is melt–stir–squeeze–bottom pouring–quenching.
2. Uniform distribution of SiC particles has been found in the developed MMCs. Optical microscopy, SEM, and FE-SEM have been used to substantiate the uniform distribution of SiC particles in the Al6063 matrix. The EDAX analysis has been used to measure the chemical composition of the developed MMCs. From the EDAX analysis, it was found that the different constituents are in appropriate proportion in the developed MMCs.
3. Universal testing machine has been employed to find out the tensile behavior of the developed MMCs. It was found that the tensile strength of the developed MMCs was in the range of 107–148 MPa with 10 % SiC reinforcement of sizes 5, 10, and 15 μm. The percentage elongation or ductility of the developed MMCs has been found in the range of 7–13 %. The semiductile behavior of the developed MMCs has been revealed by the fractography analysis.

References

Aigbodion VS, Hassan SB (2007) Effects of silicon carbide reinforcement on microstructure and properties of cast Al-Si-Fe/SiC particulate composites. Mater Sci Eng, A 447:355–360

Balasivanandha PS, Karunamoorthy L, Kathiresan S, Mohan B (2008) Influence of stirring speed and stirring time on distribution of particles in cast metal matrix composite. J Mater Process Technol 171:268–273

Dvivedi A (2008) Electric discharge machining of Al6063-SiC MMC produced by stir casting process. Ph.D. Thesis, Indian Institute of Technology Roorkee, India

Emamy M, Razaghian A, Lashgari HR, Abbasi R (2008) The effect of Al-5Ti-1B on the microstructure, hardness and tensile properties of Al2O3 and SiC-containing metal-matrix composites. Mater Sci Eng, A 485:210–217

Hashim J, Looney L, Hashmi MSJ (1999) Metal matrix composites: production by the stir casting method. J Mater Process Technol 92–93:1–7

Hashim J, Looney L, Hashmi MSJ (2001) The enhancement of wettability of SiC particles in cast aluminium matrix composites. J Mater Process Technol 119(1–3):329–335

Hashim J, Looney L, Hashmi MSJ (2002a) Particle distribution in cast metal matrix composites-Part II. J Mater Process Technol 123(2):258–263

Hashim J, Looney L, Hashmi MSJ (2002b) Particle distribution in cast metal matrix composites-Part I. J Mater Process Technol 123(2):251–257

Kalaiselvan K, Murugan N, Siva P (2011) Production and characterization of AA6061–B4C stir cast composite. Mater Des 32:4004–4009

Lin G, Zhang H, Li H, Guan L, Huang L (2010) Effects of Mg content on microstructure and mechanical properties of SiCp/Al-Mg composites fabricated by semi-solid stirring technique. Trans Nonferrous Soc China, 1851–1855

Sahin Y, Acilar M (2003) Production and properties of SiC reinforced aluminium alloy composites. Compos A 34:709–718

Singh S (2013) Primary and secondary processing of metal matrix composites. PhD. Thesis, Indian Institute of technology Roorkee, Roorkee

Singh I, Singh S, Singh A (2013) Conventional and unconventional hole making in metal matrix composites. In: Machining and machine tools: research and development, Woodhead Publishing House, USA. ISBN: 978-0-85709-154-3 (print), ISBN: 978-0-85709-219-9 (online)

Su H, Gao W, Zhang H, Liu H, Lu J, Lu Z (2010) Optimization of stirring parameters through numerical simulation for the preparation of aluminum matrix composite by stir casting process. J Manuf Sci Eng 132:061007-1-7

Effects of Humidification on Proton and Electron Conductivity of an Activated Carbon–Nafion Composite Electrode

Amandeep Singh Oberoi, John Andrews and Baljit Singh Bathal Singh

Abstract An activated carbon–Nafion composite electrode with dual proton and electron conductivity has been developed, which can be employed for applications for storage of hydrogen electrochemically. Electrochemical impedance spectroscopy was used to measure proton conductivity of the composite electrode. The measured dc resistance was used to calculate electron conductivity of the composite electrode. The effect of humidification on proton and electron conductivity of the composite electrode was studied by varying the level of hydration. The activated carbon used was a common form of activated charcoal powder. As a source of proton-conducting medium within the composite, Nafion-117 solution was employed. Electron conductivities in the range of 0.000177–0.019597 S/cm were recorded, while the proton conductivities were in the range of 0.014775–0.042273 S/cm. It was found that the electron conductivity drops with the increased level of hydration, while proton conductivity goes up.

Keywords Electrochemical impedance spectroscopy · Composite electrode · Activated carbon · Nafion · Electron conductivity · Proton conductivity

1 Introduction

The development of a safe, simple, and economical energy storage system are some of the major challenges to use hydrogen as a future fuel. Conventional hydrogen vessels including high-pressure storage systems or cryogenic systems have various restrictions in terms of cost-effectiveness and net-energy losses (Andrews and Seif Mohammadi 2013; Andrews and Doddathimmaiah 2008;

A. S. Oberoi (✉) · J. Andrews · B. S. B. Singh
School of Aerospace, Mechanical and Manufacturing Engineering, RMIT University, Melbourne, Australia
e-mail: s3364207@student.rmit.edu.au

S. S. Khangura et al. (eds.), *Proceedings of the International Conference on Research and Innovations in Mechanical Engineering*, Lecture Notes in Mechanical Engineering, DOI: 10.1007/978-81-322-1859-3_43, © Springer India 2014

Ma et al. 2006; Sakintuna et al. 2007; Shinnar 2003; Yu et al. 2012; Sone et al. 1996). One of the promising methods of hydrogen storage is its absorption in materials like metal hydrides, but gravimetric densities of less than 2 wt% only have been obtained so far.

Recently, a novel concept of a 'proton flow battery' has been proposed by Andrews and Seif Mohammadi (2013) based on a reversible proton exchange membrane (PEM) fuel cell with an integrated solid-state electrode capable of storing hydrogen in atomic form. Some preliminary results have been reported by these authors on the performance of an experimental proton flow battery using a composite metal hydride–Nafion electrode for storing hydrogen.

For efficient performance, it is important for the composite electrode to have dual proton conductivity and electron conductivity on higher side for various applications like in PEM fuel cell. The present paper therefore describes a special method to measure the proton conductivity of a composite aC–Nafion electrode using electrochemical impedance spectroscopy. The electron conductivity of the same electrode is also measured. The effects on these conductivities of the level of hydration of the aC–Nafion composite are investigated.

2 Experimental

The equipment and materials used in the experimental measurement of the proton and electron conductivities are presented in Table 1.

An activated charcoal powder mixed with Nafion-117 solution (perfluorosulfonic acid) was chosen as the material to fabricate the composite aC–Nafion composite electrode. The aC–Nafion solution was stirred well before pouring it into a mold to attain the desired shape after evaporation of the solvent in which the Nafion ionomer was dissolved. The mold containing this mixture was kept at room temperature for 12 h and then heated in an electric oven at 110 °C for 2 h, to evaporate the solvent and form the solid electrode.

The aC–Nafion composite electrodes obtained from the mold were exposed to atmospheres with different relative humidity to hydrate the electrodes to various levels. Five samples named as aCN-RC, aCN-RH35, aCN-RH75, aCN-RH100, and aCN-WC with specifications shown in Table 2 are prepared. Sample aCN-RC was stored in room conditions (relative humidity around 60 %) for 24 h. Samples aCN-RH35, aCN-75, and aCN-100 were stored in a humidity chamber maintained at 35, 75, and 100 % relative humidity, respectively. Sample aCN-WC was stored in deionized water for 24 h before testing. All samples were then subjected to electrochemical impedance spectroscopy to measure their proton conductivities.

To measure its proton conductivity, each sample was sandwiched between two Nafion-117 membranes backed by two gold-plated metal electrodes in, as shown in Fig. 1. The usage of the two Nafion membranes on either side of the composite electrode ensured that there was no direct electrical (that is, electron) current

Table 1 List of equipment and materials used in this research

S. No.	Equipment and materials		
	Equipment/materials	Manufacturer	Model/registered product name
1	aC powder	Sigma-Aldrich	aC-K-6010
2	Ion exchange membrane	Sigma-Aldrich	Nafion®117
3	Nafion solution	Alfa Aesar	Perfluorosulfonic acid
4	Electrochemical interface	CH instruments	EIS workstation
5	Humidity chamber	Thermoline Scientific	TCH-150-SD
6	Oven	Thermoline Scientific	TEO-24G

Table 2 Specifications of the prepared samples

Sample	Condition	Width of sample in cm W	Length of sample in cm L	Area in cm^2 'A' $A = W \times L$	Thickness in cm 'Th'
aCN-RC	Room condition	0.119	0.512	0.060928	0.074
aCN-RH35	RH—35 %	0.55	0.42	0.231	0.074
aCN-RH75	RH—75 %	0.588	0.239	0.140532	0.074
aCN-RH100	RH—100 %	0.42	0.475	0.1995	0.074
aCN-WC	Wet condition	0.54	0.445	0.2403	0.074

between the two electrodes of the EIS instrument, given that the composite electrode is itself a good electron conductor.

The two Nafion membranes used in the sandwich were made from a single piece of Nafion-117 membrane treated with H_2SO_4 and boiled in deionized water to remove dirt, grease, and stains and hence to obtain a clear membrane for testing. The cleaned membrane was cut in two pieces to match the size of the gold-plated metal electrodes, which are to be used in the EIS technique. Two gold-plated metal electrodes were cleaned with acetone and ethanol in order to remove any accumulated dirt or grease.

The positive and negative leads of the EIS workstation were applied to the overall test assembly (Fig. 1). Nyquist plots, i.e., Z' versus $-Z''$, were obtained for frequencies varying from 0.1 to 10^6 Hz.

Proton resistances obtained from the EIS technique, as the high-frequency intercepts on the real impedance (Z') axis, were used to calculate proton conductivity of the sample along with their cross-sectional areas and thicknesses. The tests were repeated with all samples, i.e., aCN-RC, aCN-RH35, aCN-RH75, aCN-RH100, and aCN-WC, and hence, their respective proton conductivities were calculated.

A standard ohm meter was employed to measure electron resistance using outer electrodes that had the same cross-sectional area as the sample electrodes. Hence, electron conductivity for all five different samples was calculated using the known dimensions of the electrodes.

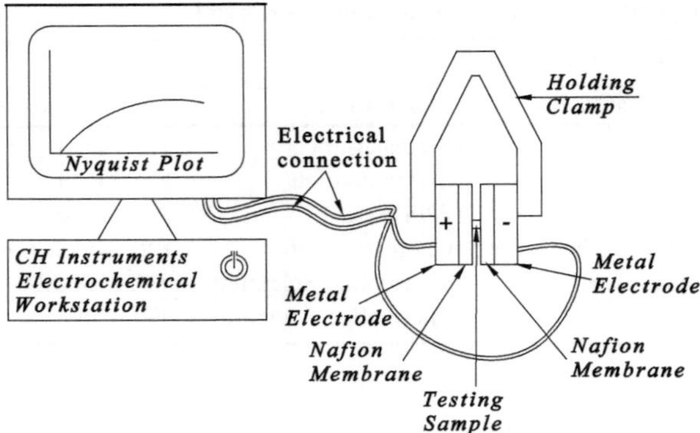

Fig. 1 Measurement of proton resistance of the samples using electrochemical impedance spectroscopy technique

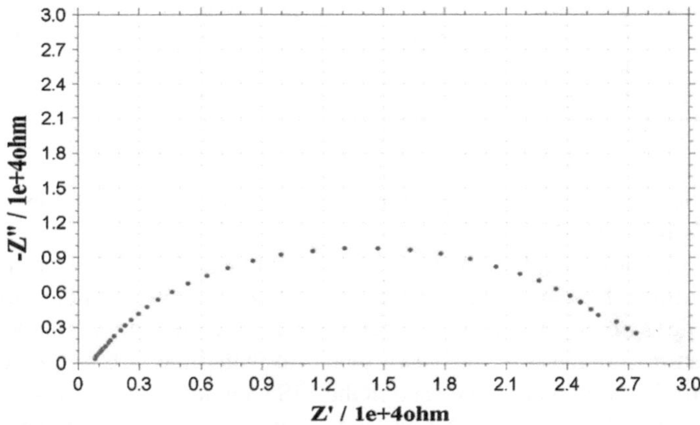

Fig. 2 Nyquist plot for the composite electrode sample aCN-RC

3 Results and Discussion

When the AC voltage is applied through the electrochemical workstation in the frequency range of 0.1–10^6 Hz, a Nyquist plot is obtained as shown in Fig. 2 for the aCN-RC electrode. The intercept of the Nyquist plot on the higher-frequency end of the curve is a measure of the total series resistance (real component of the total impedance). This intercept can best be estimated from the zoomed view of the high-frequency end of the Nyquist, as shown in Fig. 3. The proton resistances of the Nafion membranes were measured separately using EIS. The measured

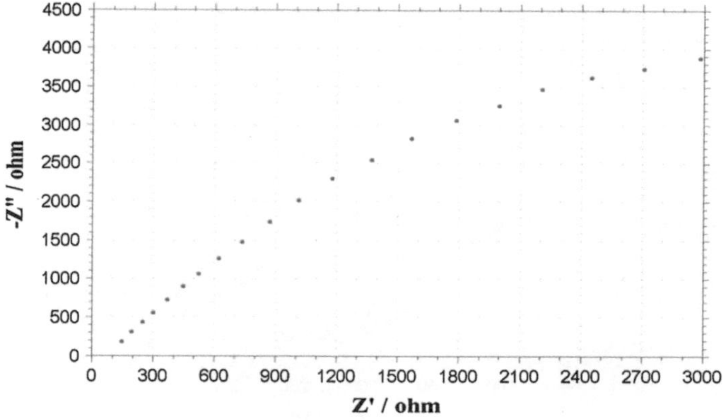

Fig. 3 Zoomed view of the Nyquist plot for composite electrode sample aCN-RC

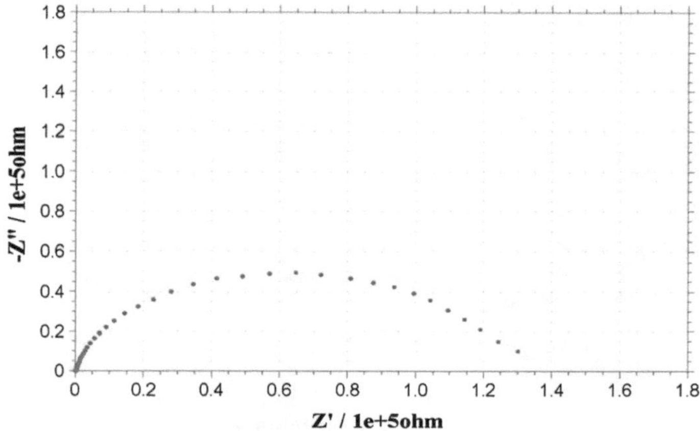

Fig. 4 Nyquist plot for the composite electrode sample aCN-RH35

proton resistances of Nafion membranes and resistances of the gold-plated metal electrodes employed in the setup were subtracted from the total proton resistance of the whole setup to obtain the proton resistance of the sample.

The Nyquist plot and zoom view for the aCN-RH35, aCN-RH75, and aCN-RH100 samples are shown, respectively, in Figs. 4, 5, 6, 7, 8, 9, and the corresponding plots for the aCN-WC electrode are given in Figs. 10 and 11.

The proton conductivities for all samples at differing levels of hydration obtained by this method are presented in Table 3. The proton conductivities of the

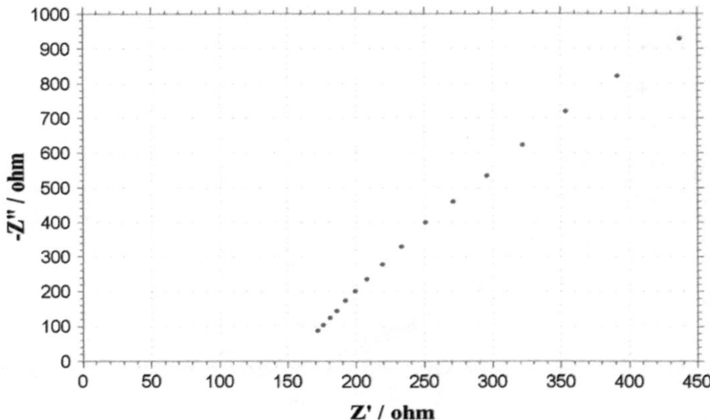

Fig. 5 Zoomed view of the Nyquist plot for composite electrode sample aCN-RH35

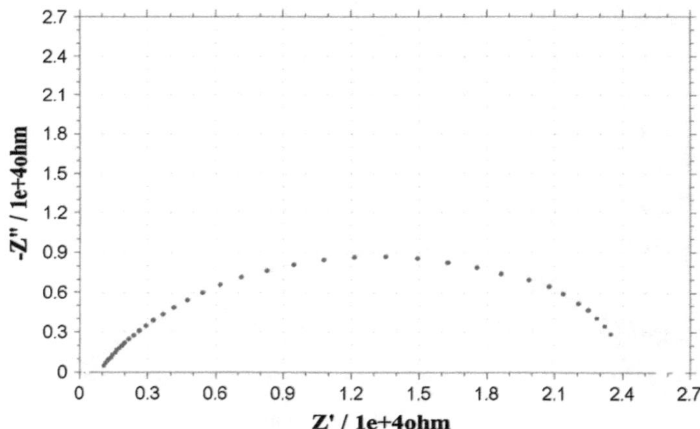

Fig. 6 Nyquist plot for the composite electrode sample aCN-RH75

electrodes were calculated from their known cross-sectional areas, thicknesses, and measured proton resistances across the electrodes using the following equation:

$$\sigma_{\mathrm{p}} = t/A \times R$$

where

- σ_{p} proton conductivity of the sample
- t thickness of the sample
- A area of the sample
- R proton resistance of the sample.

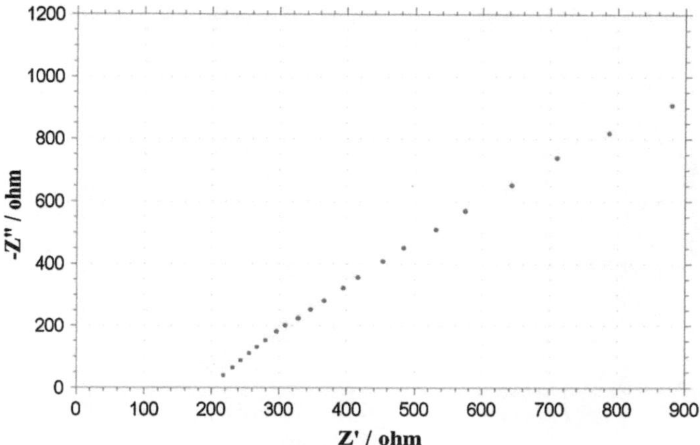

Fig. 7 Zoomed view of the Nyquist plot for composite electrode sample aCN-RH75

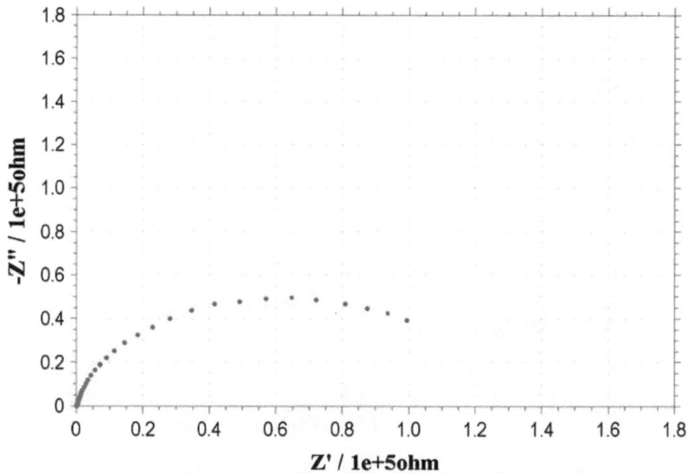

Fig. 8 Nyquist plot for the composite electrode sample aCN-RH100

At room condition, the proton conductivity of the electrode was found to be less (equal to 0.014775 S/cm) as compared to the electron conductivity (equal to 0.019597 S/cm). As the level of hydration of the electrode goes up at relative humidity of 35 %, the proton conductivity rises (equal to 0.020465 S/cm) and the electron conductivity gets lower (equal to 0.006449 S/cm) as compared to the

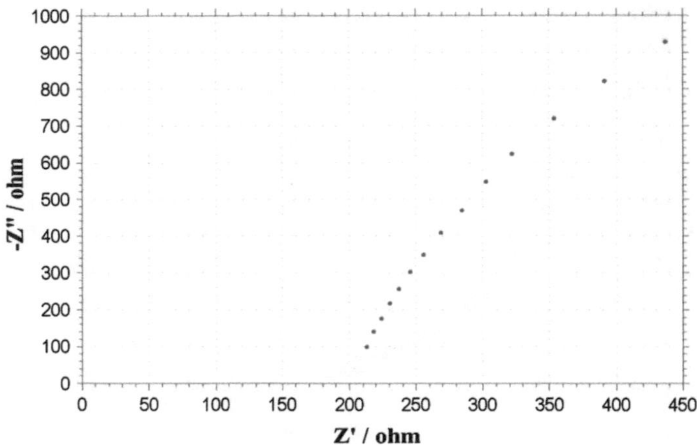

Fig. 9 Zoomed view of the Nyquist plot for composite electrode sample aCN-RH100

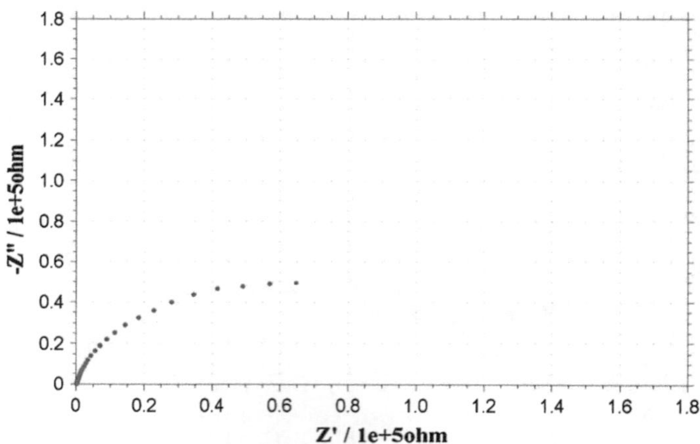

Fig. 10 Nyquist plot for the composite electrode sample aCN-WC

respective conductivities at room conditions. Under wet conditions, the proton conductivity further rises up (equal to 0.042273 S/cm) and electron conductivity gets further down (equal to 0.000177 S/cm).

The electron conductivity directly across each electrode was calculated by the measured electron resistance using a standard ohm meter and is given in Table 3, along with the corresponding calculated proton conductivities.

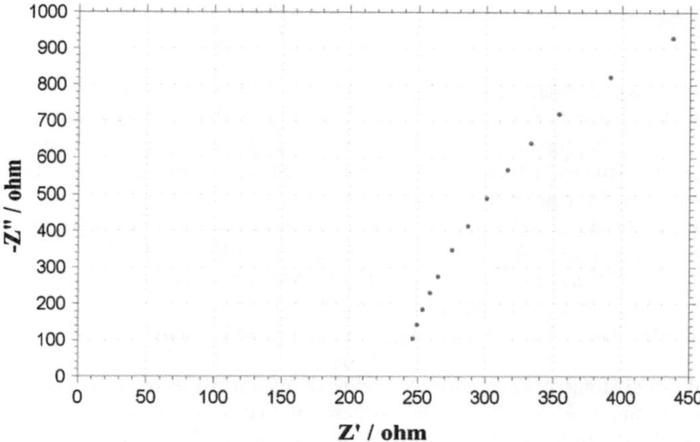

Fig. 11 Zoomed view of the Nyquist plot for composite electrode sample aCN-RH100

Table 3 Proton and electron conductivities of the samples

Sample	Proton conductivity of the sample in S/cm 'σp'	Electrical conductivity of the sample in S/cm 'σe'
aCN-RC	0.014775	0.019597
aCN-RH35	0.020465	0.006449
aCN-RH75	0.030718	0.001645
aCN-RH100	0.037269	0.000219
aCN-WC	0.042273	0.000177

4 Conclusion

The proton conductivity of the composite aC–Nafion electrode rose 0.014775–0.042273 S/cm with increasing hydration. The measured electron conductivities recorded were in the range 0.000177–0.019597 S/cm and were found to fall with increasing hydration. The increase in proton conductivity with more water content is consistent with the usual behavior of Nafion. A possible explanation of the fall in electron conductivity with increasing hydration is that Nafion swells as its water content increases, so that the area of contact between the aC particles embedded in the Nafion is likely to be diminished.

Importantly for their potential application in a proton flow battery and in other areas, the composite aC–Nafion electrode made and tested in this study has shown acceptable levels of both proton and electron conductivities.

References

Andrews J, Doddathimmaiah A (2008) Regenerative fuel cells. In: Gasik M (ed) Fuel cell materials. Woodhead Publishing, Helsinki University, Finland, pp 344–385

Andrews J, Seif Mohammadi S (2013) Towards a 'proton flow battery': investigation of a reversible PEM fuel cell with integrated metal-hydride hydrogen storage. Int J Hydrogen Energy (submitted for publication)

Ma S, Kuse A, Siroma Z, Yasuda K (2006) Measuring conductivity of proton conductive membranes in the direction of thickness. Espec Technol Rep 20:12–20

Sakintuna B, Lamari-Darkrim F, Hirscher M (2007) Metal hydride materials for solid hydrogen storage: a review. Int J Hydrogen Energy 32:1121–1140

Shinnar R (2003) The hydrogen economy, fuel cells, and electric cars. Int J Technol Soc 25:45–476

Sone Y et al (1996) Ionic conductivity of an extrude nafion. J Electrochem Soc 143:1254

Yu Y, Zhao N, Shi C et al (2012) Electrochemical hydrogen storage of expanded graphite decorated with TiO_2 nanoparticles. Int J Hydrogen Energy, XXX:1–7

Combating Wear by Surface Modification Techniques

Vineet Shibe and Vikas Chawla

Abstract Surface modification techniques are used to enhance the service life of several engineering components by combating wear. Material loss due to wear in various industries is significantly high. Surfacing is a cost-effective and proven method of depositing protective coating. Research is going on over years to reduce the corrosion, erosion, and wear either in the form of using a new corrosion-, erosion-, and wear-resistant material or by improving these properties in the existing material by using surface modification techniques such as hardfacing and surface coating. The economic success of the hardfacing process depends on selective application of hardfacing material and its chemical composition for a particular application. In this paper, an attempt has been made to discuss the various types of surface modification techniques such as hardfacing and surface coating used for combating wear. Surface protection by different hardfacing techniques and surface coatings employed on the substrate surface of material by different methods and their advantages have been discussed.

Keywords Wear · Hardfacing · Surface coatings · Thermal spraying

1 Introduction

In well-designed tribological systems, the removal of material is usually a very slow process, but it is very steady and continuous (Archard 1980). The modes or different types of wear are abrasion, erosion, corrosion, adhesion, impact, and

V. Shibe (✉)
Department of Mechanical Engineering, Punjab Technical University, Jalandhar, Punjab, India
e-mail: shibevineet@gmail.com

V. Chawla
D.A.V. College of Engineering and Technology, Kanina, Mohindergarh, Haryana, India
e-mail: vikkydmt@gmail.com

S. S. Khangura et al. (eds.), *Proceedings of the International Conference on Research and Innovations in Mechanical Engineering*, Lecture Notes in Mechanical Engineering,
DOI: 10.1007/978-81-322-1859-3_44, © Springer India 2014

surface fatigue. The surface characteristics of engineering materials have a significant effect on the serviceability and life of a component thus cannot be neglected in design. Surface engineering can be defined as the branch of science that deals with methods for achieving the desired surface requirements and their behavior in service for engineering components. The surface of any component may be selected on the basis of texture and color, but engineering components generally demand a lot more than this. Engineering components must perform certain functions completely and effectively, under various conditions in aggressive environments. Engineering environments are normally complex, combining loading with chemical and physical degradation to the surface of the component.

Surface wear is a phenomenon, which effects how a component will last in service. Surface coatings can help to deal with the circumstances such as component working in an aggressive environment. In wear-resistant components, their surface must perform many engineering functions in a variety of complex environments. The behavior of a material is therefore greatly dependent on the surface of a material and the environment under which the material must operate. The surface of these components may require treatment to enhance the surface characteristics. Surface modification techniques such as hardfacing and surface coating may be used enhance the wear resistance. In this paper, surface protection by hardfacing techniques, manual metal arc welding (MMAW) process, and applications of hardfacing are discussed.

2 Wear

Wear is a process of removal of material from one or both of two solid surfaces in solid-state contact, occurring when two solid surfaces are in sliding or rolling motion together (Bhushan and Gupta 1991). The deterioration of surfaces is a very real problem in many industries. Wear is the result of impact, erosion, metal-to-metal contact, abrasion, oxidation, and corrosion, or a combination of these. Figure 1 shows the five main categories of wear and the specific wear mechanisms that occur in each category (Stokes 2008) (Figs. 2, 3, 4, 5).

3 Surface Protection by Surface Modification Techniques

Serviceable engineering components not only rely on their bulk material properties but also on the design and characteristics of their surface (Stokes 2008). Although considerable attention has already been paid by the researchers to develop modern techniques to prevent and control the problems resulting from wear; still, there is a need for further research to reduce the losses incurred. These wear- and corrosion-related problems can be minimized mainly by following two methods (Arulmani and Sunil 2003):

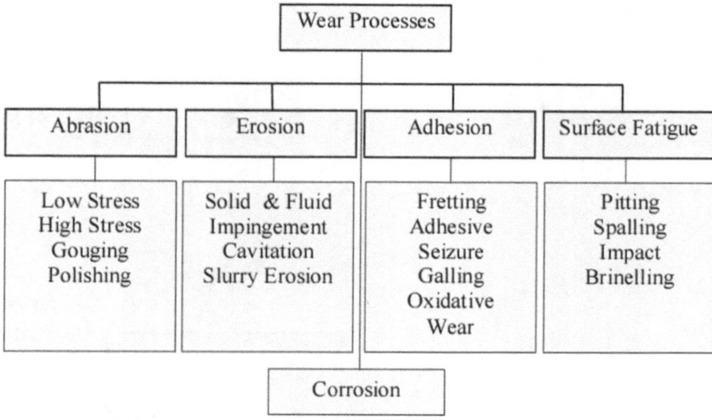

Fig. 1 Flowchart of various wear mechanisms

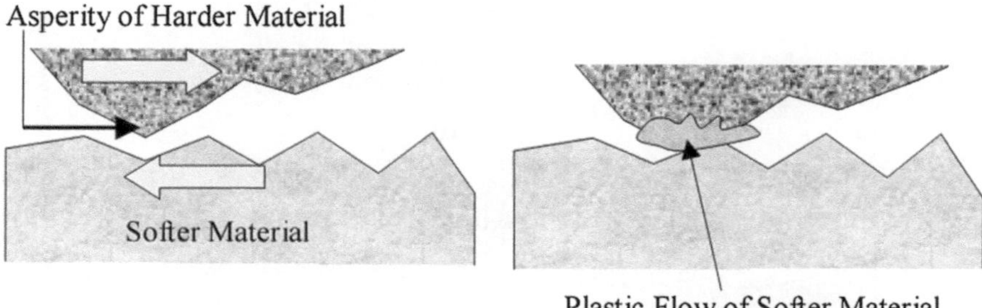

Fig. 2 Abrasion mechanism

- By using high-cost wear-resistant alloys/metals better than the existing low-cost ones.
- By improving the wear resistance of the existing metals and alloys by applying certain modifications to the surface.

Individuals and industry tend to focus on the wearing surface that has the greatest impact on their own economic situation. As the wear is a surface phenomenon and occurs mostly at outer/mating surfaces, it is more appropriate and economical to use the latter method of making surface modifications than using the former one which will not only involve very high cost of the operation but also involve longer time as compared to the second technique. To this end, a host of surface modification techniques can be used such as hardfacing by welding or thermal spraying in which a layer of strong and hard alloys is fused onto the surface of the component for improving its wear resistance (Winkelmann and Badisch 2009).

Fig. 3 Schematic of erosive wear

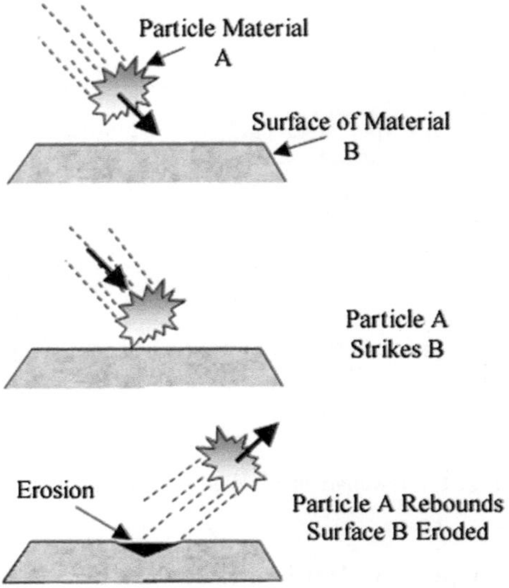

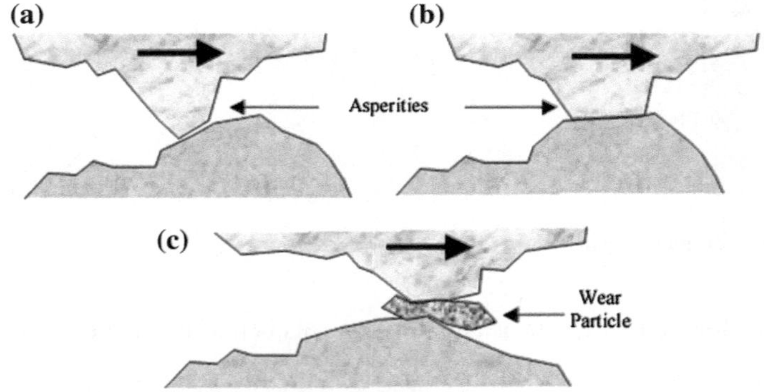

Fig. 4 Schematic of generation of wear particle as a result of adhesive wear

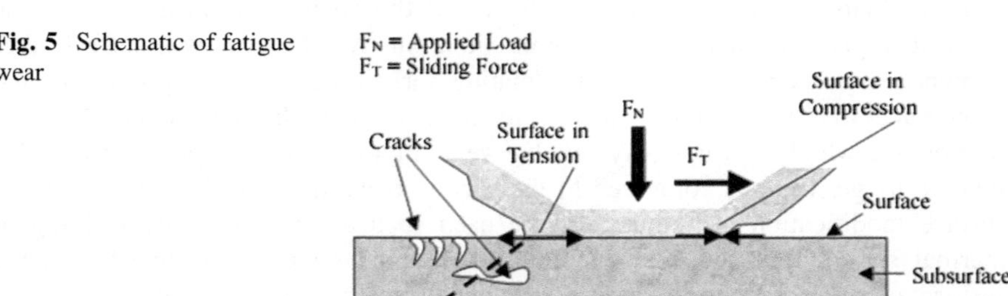

Fig. 5 Schematic of fatigue wear

4 Hardfacing

Hardfacing is one of the versatile techniques that can produce the hard and wear-resistant surface layer of various metals and alloys on metallic substrate. It not only helps them withstand wear, but also helps to prevent corrosion and high-temperature oxidation (Wang and Han 2008). Hardfacing is a commonly employed method to improve surface properties of agricultural tools, components for mining operation, soil preparation equipments, and others. An alloy is homogeneously deposited onto the surface of a soft material (usually low- or medium-carbon steels) by welding with the purpose of increasing hardness and wear resistance without significant loss in ductility and toughness of the substrate (Buchely and Gutierrez 2005). The hardfacing technique has in the mean time grown into a well-accepted industrial technology. Due to continuous rise in the cost of materials as well as increased material requirements, the hardfacing has been into prominence in the last few decades. MMAW process is commonly selected for hardfacing applications, as it is highly versatile and most economical (Selvi et al. 2008).

5 Hardfacing Deposition Techniques

The various types of hardfacing deposition techniques are as under:

- Thermal spraying
- Cladding
- Welding.

5.1 Thermal Spraying

These processes are preferred for applications requiring thin, hard coatings applied with minimal thermal distortion of the work piece and with good process control. These processes are most commonly used the coating material in the powder form, and almost any material capable of being melted without decomposition, vaporization, sublimation, or dissociation can be thermally sprayed.

5.2 Cladding

These processes are used to bond bulk materials in foil, sheet, or plate form to the substrate to provide tribological properties. The cladding processes are used either where coatings by thermal spraying and welding cannot be applied or for applications which require surfaces with bulk-like properties. Since relatively thick sheets can be readily clad to substrate, increased wear protection may be possible compared to thermal spraying and welding. If the coating material is available in sheet form, then cladding may be cheaper alternative to surface protection. It is difficult to clad parts having complex shapes and extremely large sizes.

5.3 Welding

Welding is preferred for applications requiring dense relatively thick coatings (due to extremely deposition rates) with high bond strength. Welding coatings can be applied to substrate which can withstand high temperatures (typically 7,900 °C). Welding processes most commonly use the coating material in the rod or wire form. Thus, materials that can be easily cast in rods or drawn into wires are commonly deposited. In arc welding, the substrate and the coating material must be electrically conductive. Welding processes are most commonly used to deposit primarily various metals and alloys on metallic substrates.

Hardfacing by arc welding is performed using all of the common processes and equipment. From the arc welding group, MMAW or stick welding is the most common and versatile process, although it does not provide the highest deposition rate.

6 Hardfacing Processes

There are various processes for hardfacing. They can be grouped in the following ways (Pradeep et al. 2010):

6.1 Hardfacing by Arc Welding

Shielded metal arc welding, flux-cored arc welding, and submerged arc welding.

6.2 Hardfacing by Gas Welding

Deposition by oxy-acetylene gas welding.

6.3 Hardfacing by Combination of Arc and Gas

Tungsten inert gas welding and gas metal arc welding.

6.4 Powder Spraying

Flame spraying, high-velocity oxy-fuel process, electric arc spraying, and plasma transferred arc.

6.5 Laser Hardfacing

Laser hardfacing (Laser Cladding).

7 Hardfacing Alloys

Different types of hard-facing alloys are available, and they fall into four general categories (Pradeep et al. 2010):

- Low-alloy iron-base alloys,
- High-alloy iron-base alloys,
- The cobalt-base and nickel-base alloys,
- Tungsten carbide materials.

8 Base Materials

Almost 85 % of the metal produced and used is steel. The term steel encompasses many types of metals made principally of iron. The various types of steels used in the industry for making different components for different applications are grouped into the following types (Pradeep et al. 2010):

- Low-carbon steels and low-alloy steels
- Medium-carbon steels
- High-carbon steels
- Other steels are low-nickel chrome steels, low-manganese steels, low-alloy chromium steels, and the electric furnace steels.

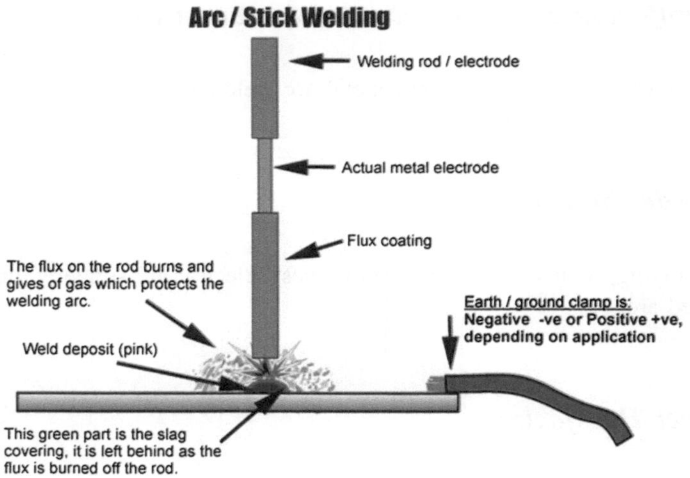

Fig. 6 Manual metal arc welding (MMAW) process

9 Manual Metal Arc Welding (MMAW)

Welding with stick electrodes is called MMAW. In this process, heat required for fusion is generated by the electric arc formed between a metallic electrode and the base metal. The electrode is consumed in the arc and provides the filler metal on the substrate. The extremely high arc temperature of over 5,000 °C permits it to supply a large amount of heat. Among the arc processes, MMAW is the most common, versatile, inexpensive one and has advantages in areas of restricted access and accounts for over 60 % of the total welding in advance countries and over 90 % of the total welding in India (Fig. 6).

10 Benefits of Hardfacing

1. *Most versatile*: Hardfacing is the most versatile process to improve the life of the worn-out component.
2. *Best chosen*: Hardfacing is the best chosen process these days for reducing the cost of replacement.
3. *Reduces downtime*: Hardfacing reduces downtime because parts last longer and fewer shutdowns are required to replace them.
4. *Any steel material*: Hardfacing can be done on any steel material using wide variety of welding processes.

5. *Desired property*: Different alloying elements can be introduced into the base metal in the form of weld consumables to achieve any desired properties such as hardness, wear resistance, abrasive resistance, and crack resistance.
6. *Longer service life*: Fewer replacements of parts are needed when parts are hard faced.
7. *Higher productivity*: Upon improving wear life, this contributes to the equipment working and producing more per hour. This increases the productivity and profits.
8. *Less downtime*: Greater availability of machine, a longer service life means that you will spend less time replacing the tips. This contributes to a reduction in operating costs.
9. *Reduced cost*: As wear resistance and hardness are the required at surface, one can deposit the superior material on the substrate to enhance the surface characteristics at less cost.

11 Some Industrial Applications of Hardfacing

Hardfacing is widely used in agriculture, mining, metallic, pulp and paper, dredging, foundry, petroleum, metal production, cement/concrete, glass, railroad, plastic, steel making, metal forming, brick/clay, and crushing/sizing industry. There are many different items that could potentially benefit from hardfacing on the farm. Primarily, hardfacing was used to restore worn pats, but nowadays, its use in making new components is also increasing.

12 Surface Coating Methods

Surface coating methods are classified as under:

- Thermal spraying (Metal Spraying)
- Chemical vapor deposition (CVD)
- Physical vapor deposition (PVD).

12.1 Thermal Spraying

It is a group of processes wherein a feedstock material is heated and propelled as individual particles or droplets onto a surface. Sprayed particles impinge upon the surface, and they cool and build up, splat by splat, into a laminar structure forming the thermal spray coating. A detonation gun unit mainly consists of a double-walled barrel, a combustion chamber, a powder feeder, and control panels to regulate gas flows and gas operation. The setup also includes an appropriate manipulator to hold the work piece and control its movements. A barrel is filled with a small amount of powder and an explosive oxygen–acetylene mixture. With the use of a spark plug, the mixture is ignited. After ignition, a detonation wave accelerates and heats the entrained powder particles. Instead of a continuous combustion process, it uses an intermittent series of explosions to melt and propel the particles onto the substrate. After each detonation, the barrel is purged with nitrogen. This step is essential to remove all remaining "hot" powder particles from the chamber; otherwise, these can detonate the explosive mixture in an irregular fashion. With this, one detonation cycle is completed. The above procedure is repeated at a particular frequency until the required thickness of coating is deposited. The process produces noise levels that can exceed 140 dB and requires special sound and explosion proof rooms. Depending upon the ratio of the combustion gases, the temperature of the hot gas stream can go up to 3,890 °C and the velocity of the shock wave can reach 3,500 m/s. Depending on the required coating thickness and the type of coating material, the detonation spraying cycle can be repeated at the rate of 1–10 shots per second.

12.2 Chemical Vapor Deposition (CVD)

It involves the dissociation and/or chemical reactions of gaseous reactants in an activated (heat, light, and plasma) environment, followed by the formation of a stable solid product. The deposition involves homogeneous gas-phase reactions, which occur in the gas phase, and/or heterogeneous chemical reactions, which occur on/near the vicinity of a heated surface leading to the formation of powders or films, respectively.

12.3 Physical Vapor Decomposition (PVD)

It is based on separating atoms from surfaces and accumulating (atomic or ionic) them to sub-material surface to be coated, by evaporating or sloping materials under vacuum. Coating material, in PVD method, is transmitted to surface in atomic, molecular, or ionic form, obtaining it not chemically but physically from solid, liquid, and gas sources. Chemical reactions can exist on main material surface too colder than CVD coating (50,500 °C); however, such a reaction formation is not necessary. It is more interesting that PVD operation is performed in relatively lower temperatures. In addition, after completion of coating, microstructure and properties of main material are not affected. PVD method is carried out by three methods called as evaporating, dispersion, and ionic coating.

13 Conclusion

Surface modification techniques improve the life of the worn-out component and reduce the cost of replacement. These techniques reduce downtime by extending the service life, and hence, few shutdowns are required to replace them. The purpose of surface technology is to produce functionally effective surfaces. Hardfacing and surface coating can improve the corrosion, erosion, and wear resistance of materials. Surfacing by hardfacing is an economical tool which can be used to increase the service life of the components used in various types of industries. The economic success of the surface modification techniques depends on selective application of material and its chemical composition for a particular application. Effort should be made for the right selection of surfacing materials and the process to achieve the full advantage of these techniques.

Acknowledgments The authors would like to thank the referees of ICRIME-2013 for their valuable suggestions and comments for modifying the contents of this paper.

References

Archard JF (1980) In: Peterson MB, Winer WO (eds) Wear theory and mechanisms. Wear control handbook. ASME, New York

Arulmani R, Sunil P (2003) Surfacing through plasma enhanced shielded metal arc welding. In: International welding symposium, pp 61–68

Bhushan B, Gupta BK (1991) Material coating and surface treatments. McGraw-Hill, New York

Buchely MF, Gutierrez JC (2005) The effect of microstructure on abrasive wear of hardfacing alloys. Wear 259:52–61

Pradeep GRC, Ramesh A, Durga Prasad B (2010) A review paper on hardfacing processes and materials. Int J Eng, Sci Technol 2(11):6507–6510

Selvi S, Sankaran SP, Srivatsavan R (2008) Comparative study of hardfacing of valve seat ring using MMAW process. Elsevier, J Mater Process Technol 207:356–362

Stokes J (2008) Theory and application of the high velocity oxy-fuel (HVOF), Thermal spray process. Dublin City University, ISBN 1-87232-753-2, ISSN 1649-8232

Wang X, Han F (2008) Microstructure and wear properties of the Fe–Ti–V–Mo–C hardfacing alloy. Wear 265:583–589

Winkelmann H, Badisch E (2009) Wear mechanisms at high temperatures part-1: wear mechanisms of different Fe-based alloys at elevated temperatures. Springer, Tribol Lett, pp 155–166

Reduction in Fuel Consumption and CO_2 Emissions by Elimination of Steam Leak from Steam Traps

Mehboob Singh Gill and Paramjit Singh Bilga

Abstract Steam traps are automatic mechanical valves installed in the steam distribution system (SDS) of various process industries. The steam traps operate to discharge condensate, air, and other impurities that reduce the efficiency of the steam lines. Generally during the maintenance of the SDS, the steam traps are neglected as they are placed in locations which are not easily approachable. But if steam traps are not working properly, they result in huge loss of energy as they are in large quantity in the SDS. With the increasing energy crisis and global warming, there is an urgent need to save each bit of energy and reduce green house gas emissions. The main purpose of the present study is to reduce the energy consumption needed to produce steam by decreasing the losses of the steam traps installed in the SDS. The sight method and temperature measurement method have been employed to detect the losses and working of steam traps. Further, the Masoneilan's formula has been used to quantify the steam saving from a leaking steam trap. The results of the present research for a typical textile industry show that the annual wastage of steam through leaking steam traps amounts to 450 tonnes. Thus, if the wastage of steam is stopped, this would result in saving of 115 tonnes of rice husk (fuel) which amounts to saving of Rs. 4 Lakhs per annum. Due to this savings, there will be a reduction of 189 tonnes of CO_2 emissions per annum to the environment.

Keywords Steam trap · Steam distribution system · Rice husk · CO_2 emissions

M. S. Gill (✉)
Department of Mechanical Engineering, Guru Gobind Singh College of Engineering and Technology, Talwandi Sabo, Punjab, India
e-mail: mehboob.gill@gmail.com

P. S. Bilga
Department of Mechanical Engineering, Guru Nanak Dev Engineering College, Ludhiana 141006, Punjab, India
e-mail: psbilga@gndec.ac.in

S. S. Khangura et al. (eds.), *Proceedings of the International Conference on Research and Innovations in Mechanical Engineering*, Lecture Notes in Mechanical Engineering, DOI: 10.1007/978-81-322-1859-3_45, © Springer India 2014

1 Introduction

The steam distribution system (SDS) is the most important and critical component of most of the process industries. Steam traps are automatic mechanical valves installed through the SDS that open to discharge condensate, air, and other impurities that reduce the efficiency of steam lines. When using steam, at any temperatures and pressures, failure to remove the condensate, the air and non-condensable substances reduce the heat transfer and causes "water hammer", leading to significant energy waste and unsafe working conditions (Bhattacharya 2002).

In most industrial facilities, steam traps are often not applied or installed properly, not inspected often enough, or not replaced when determined to be failed.

Failed steam traps fall into two categories (Kellstrom 2005), with associated consequences:

1. Failed closed or undersized and flooded—this type of steam trap failure obstructs the process, not allowing the condensate to be removed and possibly blocking the flow of steam through the system
2. Failed open, leaking, or blowing through—this type of failure causes steam loss leading to performance inefficiencies and other steam system problems though the final decision to implement the suggestion is left to the management.

Fuhr (2008) specified that the steam traps are like loyal servants, they are usually tucked away in the corner doing their job day in and day out. Forbes (2008) in his study highlighted, good operating practices can determine the success of an energy conservation effort. Every firm should know its specific energy consumption—and monitor and display it daily.

The operation of a mechanical steam trap is driven by the difference in density between condensate and steam. The denser condensate rests on the bottom of any vessel containing the two fluids. As additional condensate is generated, its level in the vessel will rise. This action is transmitted to a valve via either a "free float" or a float and connecting levers in a mechanical steam trap. One common type of mechanical steam trap is the inverted bucket trap, as shown in Fig. 1. Steam entering the submerged bucket causes it to rise upward and seal the valve against the valve seat. As the steam condenses inside the bucket or if condensate is predominately entering the bucket, the weight of the bucket will cause it to sink and pull the valve away from the valve seat. Any air or other non-condensable gases entering the bucket will cause it to float and the valve to close. Thus, the top of the bucket has a small hole to allow non-condensable gases to escape. The hole must be relatively small to avoid excessive steam loss.

As the name implies, the operation of a thermostatic steam trap is driven by the difference in temperature between steam and sub-cooled condensate. Valve actuation is achieved via expansion and contraction of a bimetallic element or a

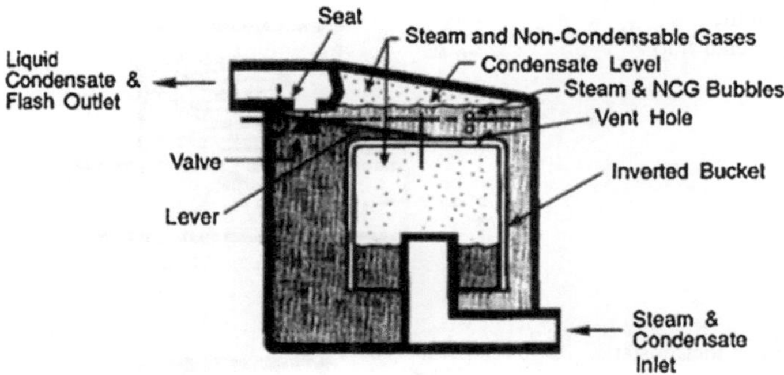

Fig. 1 Inverted bucket steam trap

liquid-filled bellows. Bimetallic and bellows thermostatic traps are shown in Figs. 2 and 3.

The inverted bucket trap allows rapid purging of air at start-up. The inverted bucket trap relies on fluid density differences to actuate its valve. Therefore, it cannot distinguish between air and steam and must purge air (and some steam) through a small hole. Recognition of this deficiency with inverted bucket traps or other simple mechanical traps led to the development of float and thermostatic steam traps. The condensate release valve is driven by the level of condensate inside the trap, while an air release valve is driven by the temperature of the trap.

A float and thermostatic trap is shown in Fig. 4.

Thermodynamic stream trap valves are driven by differences in the pressure applied by steam and condensate, with the presence of steam or condensate within the trap being affected by the design of the trap and its impact on local flow velocity and pressure. Disc, piston, and lever designs are three types of thermodynamic traps with similar operating principles; a disc trap is shown in Fig. 5. When sub-cooled condensate enters the trap, the increase in pressure lifts the disc off its valve seat and allows the condensate to flow into the chamber and out of the trap. The narrow inlet port results in a localized increase in velocity and decrease in pressure as the condensate flows through the trap, following the first law of thermodynamics and the Bernoulli equation. As the condensate entering the trap increases in temperature, it will eventually flash to steam because of the localized pressure drop just described. This increases the velocity and decreases the pressure even further, causing the disc to snap closed against the seating surface. The moderate pressure of the flash steam on top of the disc acts on the entire disc surface, creating a greater force than the higher pressure steam and condensate at the inlet, which acts on a much smaller portion of the opposite side of the disc. Eventually, the disc chamber will cool, the flash steam will condense, and inlet condensate will again have adequate pressure to lift the disc and repeat the cycle.

The present research paper deals with the study carried out in a typical textile industry with the aim to quantify the reduction in fuel consumption and CO_2 emissions by elimination of steam leaks from steam traps.

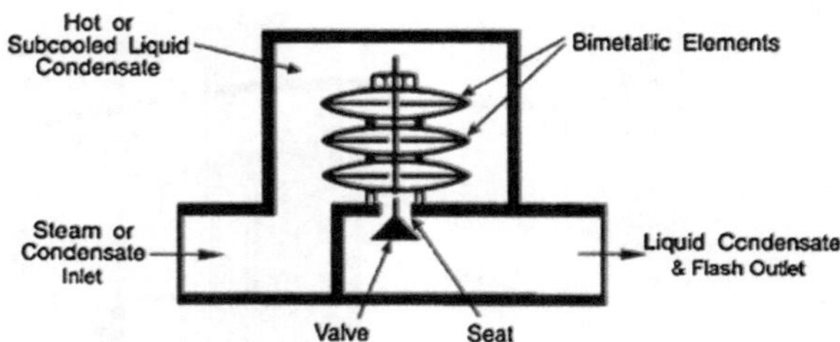

Fig. 2 Bimetallic thermostatic steam trap

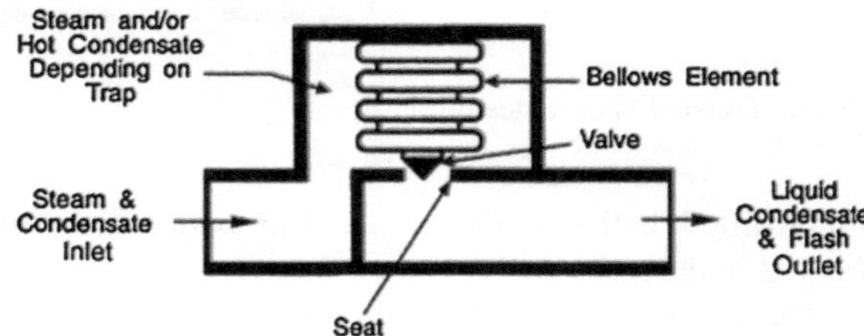

Fig. 3 Bellows thermostatic steam trap

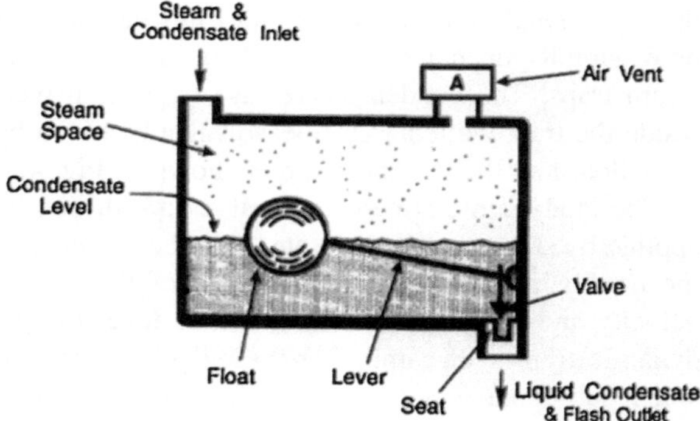

Fig. 4 Float steam trap

2 Methodology

There are three basic methods for evaluating a steam trap that are commonly discussed in the literature: sight, sound, and temperature. Two of the three methods should be used to increase the chances of correctly identifying the condition of a steam trap. A less commonly discussed method is based on fluid conductivity. Although this method should be at least as reliable as sonic-based methods, it is discussed less frequently in the literature, and no general consensus on its relative reliability was evident. Two methods were involved in finding if the steam traps have failed in open or closed position as discussed below.

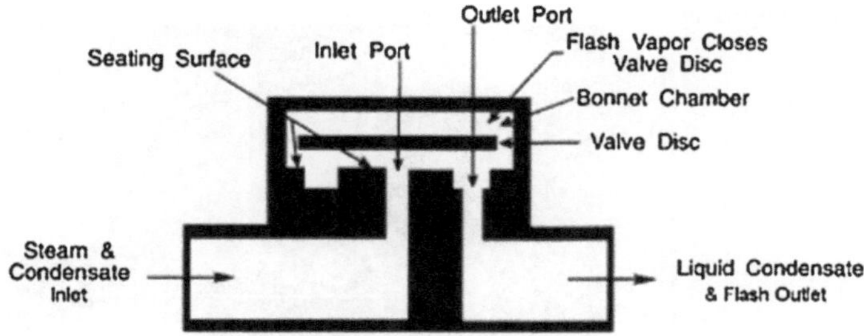

Fig. 5 Disc steam trap

2.1 Temperature Method

Measuring the temperature of the steam trap is generally regarded as a reliable of the three basic evaluation techniques (Crockett 2006). Saturated steam and condensate exist at the same temperature; of course, so it is not possible to distinguish between the two based on temperature. Still, temperature measurement provides important information for evaluation purposes. A cold trap (i.e. one that is significantly cooler than the expected saturated steam temperature) indicates that the trap is flooded with condensate, assuming the trap is in service. As described above for the visual test via a sight glass, a flooded trap could mean several things, but barring measurement during start-up, when flooding can be expected, generally indicates a problem that needs to be addressed. Downstream temperature measurement may also yield useful clues in certain circumstances. For example, the temperature downstream of a trap should drop off relatively quickly if the trap is working properly (mostly condensate immediately past the trap). On the other hand, the temperature downstream of the trap will be nearly constant if significant steam is getting past the trap. Care must be taken not to use this technique where other traps could affect downstream conditions, however (Fig. 6).

Temperature measurement methods, like sound measurement, vary tremendously in the degree of sophistication. At the low end, spitting on the trap and watching the sizzle provide a general indication of temperature. For the more genteel, a squirt bottle filled with water will serve the same purpose.

Alternatively, a glove-covered hand can provide a similar level of accuracy. More sophisticated are various types of temperature-sensitive crayons or tapes designed to change colour in different temperature ranges. Thermometers, thermocouples, and other devices requiring contact with the trap offer better precision. Finally, non-contact (i.e. infrared) temperature measuring devices provide the precision of thermometers and thermocouples without requiring physical contact. Non-contact temperature measurement makes it easier to evaluate traps that are relatively difficult or dangerous to access closely.

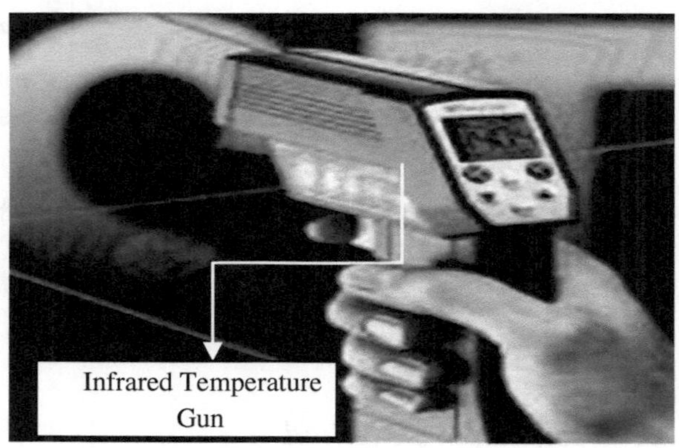

Fig. 6 Infrared temperature gun

2.2 Sight Method

The sight method is usually based on a visual observation of the fluid downstream of the trap. This is possible if there is no condensate recovery system or if test valves have been installed to allow a momentary discharge of the downstream fluid from the condensate recovery system. In either case, the steam trap evaluator must be able to distinguish between "flash" steam, which is characteristic of a properly working trap, and "live" steam, which is characteristic of a trap that has failed open and is leaking or blowing a significant amount of steam. Flash steam is created when a portion of the condensate flashes to vapour upon expansion to atmospheric pressure. Flash steam is characterized by a relatively lazy, billowy plume. Live steam, on the other hand, will form a much sharper, higher velocity plume that may not be immediately visible as it exits the test valve or steam trap. The difference between live steam and flash steam is illustrated in Fig. 7.

The rate of energy loss can be estimated based on the size of the orifice and system steam pressure. For a 3/8-inch-diameter orifice steam trap at 60-psia saturated steam system, steam loss of 2,500 million Btu/year or 630 million kcal/year will occur (refer Fig. 2.3). Further assuming steam trap is 50 % blocked and steam system is energized for 50 % of the year will result in energy loss of 210 million kcal/year at 75 % boiler efficiency.

On the basis of standard sizing practice, as well as some observations from practical installations, steam traps can be sorted into either "process" or "drip and tracer" traps. For these two broad categories of traps, the safety factor and service factor can noted from the following:

Category	Safety factor (SF)	Service factor (FS)
Process traps	1.75	0.9
Drip and tracer traps	3.0	1.4

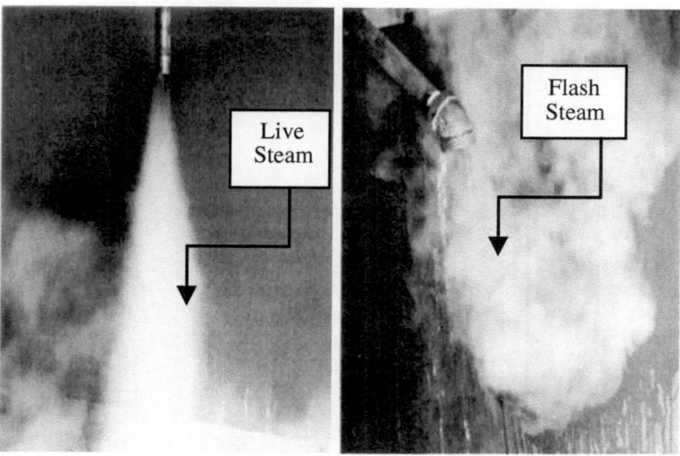

Fig. 7 Live steam versus flash steam (Yarway Inc 1995)

The failure type factor (FT) for various types of failures of steam traps has been considered as follows:

Failure type	FT
Leaking	0.25
Blow Thru	1

Masoneilan's formula used to quantify the steam loss from a steam trap that has been failed open is given as,

$$L_{st} = \text{OP.FT.FS.CV.}\{(P_1 - P_2)(P_1 + P_2)\}^{1/2}, \text{lbs/year}$$

P_1 Upstream pressure, psia

P_2 Downstream pressure, psia, with the limitation that $P_2 \geq (P_1/2)$; therefore if the actual downstream pressure is less than $P_1/2$, the value of $P_1 = P_1/2$ is used in the formula

CV Flow coefficient = 22.1 d^2

d Orifice diameter of trap, inch

OP Operating period, h/year

Using measured data such as steam trap hours of use obtained after conducting a detailed evaluation survey of steam traps in the industry (Table 1), the steam savings from steam trap replacement reduce steam use in the plant and therefore lead to reduced fuel input for steam generation. The steam savings can be directly

Table 1 Steam trap evaluation survey

S. No.	Trap location	Condition of trap	Operation period (h/year)	Orifice size (inch)	Pressure (psia)		$P_{out,t} \geq P_{in,t}/2$, if condition is not satisfied then use $P_{in,t} = P_{in,t}/2$ in formula		Pressure calculation	Failure type factor	Service factor	Flow coefficient	Estimated steam saving per year (kg/year)
		BT/LK / RC /PL/ FL[a]	OP	d	$P_{in,t}$	$P_{out,t}$	$(P_1 - P_2) = (P_{in,t}/2 - P_{out,t})$	$(P_1 + P_2) = (P_{in,t}/2 + P_{out,t})$	$\{(P_1 - P_2)(P_1 + P_2)\}^{1/2}$	FT	FS	$CV = 22.1\ d^2$	L_{st}
1	Boiler header	LK	6,000	0.196	106.68	28.45	24.89	81.79	45.11	0.25	1.4	0.849	80,442.63
2	Hank Dyeing m/c No. 1	LK	6,000	0.196	85.34	14.22	28.45	56.89	40.23	0.25	0.9	0.849	46,110.22
3	Hank Dyeing m/c No. 5	LK	6,000	0.198	85.34	14.22	28.45	56.89	40.23	0.25	0.9	0.866	47,056.05
4	Hank Dyeing m/c No. 6	LK	6,000	0.196	85.34	14.22	28.45	56.89	40.23	0.25	0.9	0.849	46,110.22
5	Hank Dyeing m/c No. 8	LK	6,000	0.118	85.34	14.22	28.45	56.89	40.23	0.25	0.9	0.308	16,712.79
6	Knitting area	BT	6,000	0.118	85.34	14.22	28.45	56.89	40.23	1	0.9	0.308	66,851.19
7	Knitting area	LK	6,000	0.118	42.67	7.11	14.22	28.44	20.11	0.25	0.9	0.308	8,356.39

(continued)

Table 1 (continued)

S. No.	Trap location	Condition of trap	Operation period (h/year)	Orifice size (inch)	Pressure (psia)		$P_{out,t} \geq P_{in,t}/2$, if condition is not satisfied then use $P_{in,t} = P_{in,t}/2$ in formula		Pressure calculation	Failure type factor	Service factor	Flow coefficient	Estimated steam saving per year (kg/year)
		BT/LK / RC /PL/ FL[a]	OP	d	$P_{in,t}$	$P_{out,t}$	$(P_1 - P_2) = (P_{in,t}/2 - P_{out,t})$	$(P_1 + P_2) = (P_{in,t}/2 + P_{out,t})$	$\{(P_1 - P_2)(P_1 + P_2)\}^{1/2}$	FT	FS	$CV = 22.1\ d^2$	L_{st}
8	Washing Range Drum Drier—2	LK	6,000	0.118	85.34	7.11	35.56	49.78	42.07	0.25	0.9	0.308	17,478.26
9	Washing Range Drum Drier—5	LK	6,000	0.118	85.34	14.22	28.45	56.89	40.23	0.25	0.9	0.308	16,712.79
10	Washing Range Drum Drier—8	LK	6,000	0.118	85.34	0	42.67	42.67	42.67	0.25	0.9	0.308	17,726.07
11	Near Fongs Machine No. 3	LK	6,000	0.118	85.34	14.22	28.45	56.89	40.23	0.25	0.9	0.308	16,712.79

(continued)

Table 1 (continued)

S. No.	Trap location	Condition of trap	Operation period (h/year)	Orifice size (inch)	Pressure (psia)		$P_{out,t} \geq P_{in,t}/2$, if condition is not satisfied then use $P_{in,t} = P_{in,t}/2$ in formula		Pressure calculation	Failure type factor	Service factor	Flow coefficient	Estimated steam saving per year (kg/year)
		BT/LK / RC /PL/ FL[a]	OP	d	$P_{in,t}$	$P_{out,t}$	$(P_1 - P_2) = (P_{in,t}/2 - P_{out,t})$	$(P_1 + P_2) = (P_{in,t}/2 + P_{out,t})$	$\{(P_1 - P_2)(P_1 + P_2)\}^{1/2}$	FT	FS	$CV = 22.1\ d^2$	L_{st}
12	Near Fongs Machine No. 6	LK	6,000	0.118	85.34	7.11	35.56	49.78	42.07	0.25	0.9	0.308	17,478.26
13	Near Fongs Machine No. 7	LK	6,000	0.118	85.34	14.22	28.45	56.89	40.23	0.25	0.9	0.308	16,712.79
14	Near Fongs Machine No. 8	LK	6,000	0.118	85.34	7.11	35.56	49.78	42.07	0.25	0.9	0.308	17,478.26
15	Near Fongs Machine No. 9	LK	6,000	0.118	85.34	7.11	35.56	49.78	42.07	0.25	0.9	0.308	17,478.26
Total estimated steam saving (kg/h)													449,417.07

[a] *BT*: Blow Thru; *LK*: leaking; *RC*: rapid cycling; *PL*: plugged; *FL*: flooded

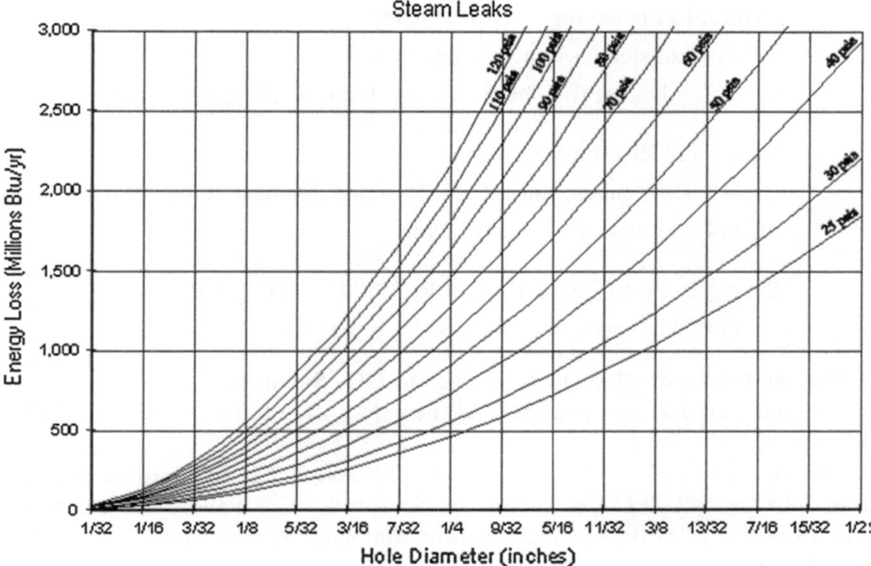

Fig. 8 Energy loss from leaking steam trap (Manczyk 2000)

translated into reduced rice husk (fuel) consumption and the resulting CO_2 emissions reductions. This can be done by calculating how much rice husk would have been required to produce the steam that is now being saved minus rice husk that would have been burned but for the implementation of the project (Fig. 8).

3 Results

The calorific value for rice husk is = 3,300 kcal/kg.

Average boiler efficiency at 75 %

The steam heat value at 662 kcal/kg at 8.5 psig (from steam tables)

Equivalent Rice husk savings (tonnes coal/tonne steam)

$$= \frac{\text{Steam generated (tonne)} \times \text{steam heat value (kcal/kg at given psig)}}{\text{Rice husk heat value (kcal/kg)} \times \text{boiler efficiency (\%)}}$$

$$= \frac{1 \text{ tonne} \times (662-30) \text{ kcal/steam kg}}{3300 \text{ kcal/husk kg} \times 75\ \% \text{ eff}}$$

$$= 0.25535 \text{ tonne husk/ tonne steam}$$

$$\textbf{Annual Fuel savings (tonnes/year)}$$
$$= \text{Annual steam savings (tonnes/year)}$$
$$\times \text{ Equivalent Rice husk savings (tonnes husk/tonne steam)}$$

Equivalent Rice husk Saved:

$$= 449.417 \text{ tonne steam} \times 0.25535 \text{ tonne rice husk/tonne steam}$$
$$= 114.76 \text{ tonnes Rice husk/annum}$$

Saving in Cost of Rice Husk at Rs. $3500/-$ per tonne

$$= 114.76 \times 3500 = \text{Rs. } 401,660/- \text{ per year}$$

Conversion from Rice Husk Savings to CO_2 Savings:

Following are the data required and the values used further in the calculations (Fischer 1995):

Carbon content of rice husk at 27.6 (t C/TJ) and 20.52: TJ/kilotonne rice husk from IPCC = 449.396 t C/kilotonne of rice husk = 0.449396 t C/tonne of rice husk (Revised 1996 IPCC Guidelines for National Greenhouse Gas Inventories, pp. 1.6, 1.16).

Complete Combustion process

$$C + O_2 = CO_2 + \text{Heat}$$

Thus, for each pound-mole of carbon in the rice husk, during the combustion process, there is an equivalent formation of carbon dioxide.

Carbon C	12 moles
Oxygen O_2	$2 \times 16 = 32$ moles
Carbon Dioxide CO_2	$12 + (2 \times 16) = 44$ moles

Mole content of C in CO_2 = 12 / 44 = 0.272727

0.449396 t C/tonne of rice husk $\times$ 44/12 = 1.6477853 t CO_2/tonne of rice husk

Then after implementation of the project,

$$\textbf{Annual } \mathbf{CO_2} \textbf{ Emissions Reduction}:$$
$$= 114.76 \text{ tonne Rice husk} \times 1.6477853 \text{ tonnes } CO_2/\text{tonnes rice husk}$$
$$= \mathbf{189.1 \text{ tonnes } CO_2/\text{year}}.$$

4 Conclusions

The present study conducted in a typical textile unit shows that 450 tonnes of steam is wasted every year from the leaking steam traps which can be avoided to save 115 tonnes of fuel, i.e. rice husk amounting to expenditure of approximately

Rs. 4,00,000/- per year. This would also reduce 189 tonnes of CO_2 emissions to the atmosphere and thus save the environment. Further, it can be easily found that if the same traps are replaced by equivalent new steam traps, then the return on investment will be less than 1 month which is a very viable proposition as it results in good savings to the plant. A proper maintenance schedule should be made to avoid the above losses which can be either routine maintenance covering either replacement of faulty internals of the steam trap or replacement of the steam trap.

References

Bhattacharya DK (2002) Energy balance and accounting for boilers and process house in textile industry through software approach. Northern India Textile Research Association, Ghaziabad—201 002

Crockett M (2006) Leak detection with thermography & ultrasonic acoustics. APP Site visit, Nov 2006

Fischer DW (1995) Assessing the impact of energy losses in steam systems. Plant Eng. Armstrong International Inc, 10 July 1995. www.armstrong-intl.com/university/su.html

Forbes N (2008) In his study highlighted that, "*good operating practices can determine the success of an energy conservation effort*. Every firm should know its specific energy consumption—and monitor and display it daily"

Fuhr P (2008) Easy ways to save energy now—Take care of those steam traps. U.S. Department of Energy, GovEnergy Conference, Nov 2008

Kellstrom C (2005) Problems in steam traps due to corrosion and erosion. In: Procedures of 23rd industrial energy technology conference, Houston, TX, p 253

Manczyk H (2000) Estimating the cost of steam loss through the orifice of a steam trap

Yarway Inc (1995) Simple techniques for surveying steam traps. 231 Rev, July 95 10M998Q

Comparative Study of Thermal Performance of Liquid Metal and Water Flow Through a Channel

Peijie Li and Sarada Kuravi

Abstract The heat transfer performance of a liquid metal flowing inside a 1-mm channel with different lengths was numerically investigated and compared with water. Inlet velocity of the liquid metal was varied to simulate same inlet velocity and pressure drop conditions as water. Pressure drop and heat transfer coefficient for both fluids were compared for different channel lengths and wall heat flux. It was found that the performance of the fluids is influenced by the heat flux, but is strongly dependent on the channel length-to-diameter ratio and the inlet flow rate used.

Keywords Liquid metal flow · Mini/Microchannel · Electronics cooling · Heat transfer enhancement

1 Introduction

Current high-power electronics and semiconductor devices, with their steadily increasing heat loads and heat fluxes, require thermal solutions that go beyond the traditional electronic heat sinks that perform at a higher level. Heat sinks with mini- and microchannels are one of the methodologies that are employed for thermal management of these systems. In a minichannel, the channel dimensions may range between 200 μm and 3 mm and may be between 50 and 200 μm. Microchannels are capable to remove heat fluxes as high as 1,000 W/cm^2 (Tuckerman and Pease 1981). The compact design enables the heat to be

P. Li (✉) · S. Kuravi
Department of Mechanical Aerospace Engineering, Florida Institute of Technology, Melbourne, FL 32901, USA
e-mail: pli2012@fit.edu

S. Kuravi
e-mail: skuravi@fit.edu

S. S. Khangura et al. (eds.), *Proceedings of the International Conference on Research and Innovations in Mechanical Engineering*, Lecture Notes in Mechanical Engineering, DOI: 10.1007/978-81-322-1859-3_46, © Springer India 2014

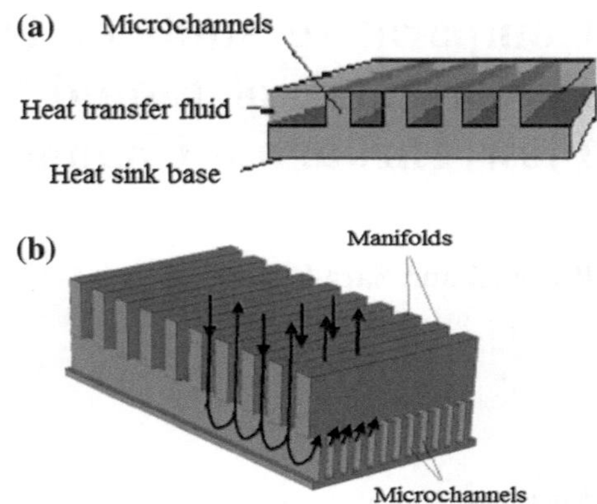

Fig. 1 **a** Traditional microchannel heat sink. **b** Manifold microchannel heat sink (length ~1–2 mm)

efficiently carried from the substrate into the coolant because of its inherently small passageways and a very large surface-to-volume ratio. The typical lengths of traditional microchannels (TM) are 1–2 cm (Fig. 1). However, due to high heat flux conditions, the temperature variation along the heat source may be large. In order to maintain isothermality, the mass flow rates are increased, which leads to considerable pressure drop. In applications where high-temperature variation and large pressure drop are unacceptable, manifold microchannel heat sink (MMC) (Copeland 1995; Copeland et al. 1998) is shown to be an effective way in reducing the pressure drop and significantly reducing the temperature variation. The MMC heat sink features many inlet and outlet channels alternating at a periodic distance (~1 mm) along the length of the MMC (Fig. 1). The flow enters the microchannels from the manifold inlet channel, splits and flows through the microchannels and then exits to the manifold outlet channel. This pattern is repeated along the length of the MMC.

The performance of the heat sinks can be improved by utilizing advanced heat transfer fluids such as phase change slurries (Kuravi et al. 2007, 2009a, b, 2010), nanofluids with metallic particles (Gong et al. 2011a) or by changing surface patterns of walls such as wavy channels (Jung et al. 2009; Gong et al. 2011a, b; Kota et al. 2010). Researchers have also investigated liquid metals to improve the heat sink performance, due to their high thermal conductivity (Sharma et al. 2013; Ma and Liu 2007a, b; Li et al. 2006). They found that liquid metals perform better in minichannels, whereas water performs better in microchannels. The channel sizes were varied to determine the performance comparison.

In this work, the thermal performance of liquid metal and water was analyzed numerically and is compared for varying lengths. A channel width of 1 mm is used for this study, which could be assumed as a minichannel; however, the lengths assumed were up to 1 cm (lengths encountered in MMC and TM heat sinks), in an attempt to understand the effect of length. Several parameters were varied in the study and results are presented.

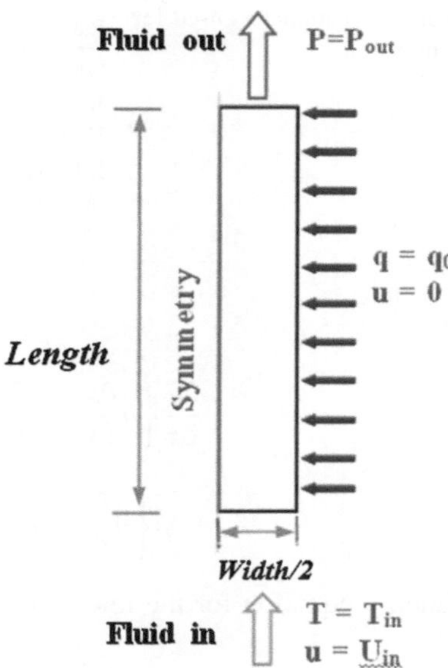

Fig. 2 Schematic of flow domain

2 Numerical Model

For the analysis, a high aspect ratio channel was assumed and hence 2D simulation was performed. The simulation domain along with the boundary conditions is shown in Fig. 2.

2.1 Assumptions

The flow is assumed as laminar and steady state. The properties of fluids were assumed to be constant, and a constant heat flux boundary condition was assumed for the walls. Table 1 shows the parameters assumed for the simulation.

2.2 Governing Equations

The governing equations are as follows:

The mass and momentum equations for the fluid inside the simulation domain can be written as:

$$\frac{\partial u}{\partial x}+\frac{\partial v}{\partial y}=0 \tag{1}$$

Table 1 Parameters used for simulation

Parameter used	Value	Units
width	1	mm
Inlet temperature	293.15	K
Exit pressure	1e5	Pa
Inlet velocity	0.24, 0.13	m/s
Heat flux	25, 50, 75	W/cm^2
Channel length	2, 5, 10	mm

$$\frac{\partial u}{\partial x}\left\{u\frac{\partial u}{\partial x}+v\frac{\partial u}{\partial y}\right\}=-\frac{\partial p}{\partial x}+\mu\frac{\partial^2 u}{\partial x^2}+\mu\frac{\partial^2 v}{\partial y^2} \tag{2}$$

$$\rho\left\{u\frac{\partial v}{\partial x}+v\frac{\partial v}{\partial y}\right\}=\frac{\partial p}{\partial y}+\mu\frac{\partial^2 u}{\partial x^2}+\mu\frac{\partial^2 v}{\partial y^2} \tag{3}$$

Energy equation for the microchannel wall/fin is:

$$\frac{\partial^2 T_w}{\partial x^2}+\frac{\partial^2 T_w}{\partial y^2}=0 \tag{4}$$

2.3 Boundary Conditions

For flow inside the channels:

$$u = 0 \text{ at the wall} \tag{5}$$

$$p = p_0,\ \text{atmospheric pressure at the outlet} \tag{6}$$

$$u = |u_{in}| \text{ at the inlet} \tag{7}$$

The boundary conditions for the heat transfer equations are as follows:

$$q \cdot n = q_o;\ \text{constant heat flux at the base} \tag{8}$$

$$q \cdot n = 0;\ \text{adiabatic at the symmetry} \tag{9}$$

$$T = T_{\text{in}} \text{ at the inlet} \tag{10}$$

$$q \cdot n = (\rho c_p u T) \cdot n;\ \text{which is the convective heat flux boundary condition at the outlet/exit} \tag{11}$$

Table 2 Properties of water and liquid metal

	Water	Galinstan
Density (kg/m^3)	998	6,363.2
Specific heat (J/kg K)	4,180	365.81
Thermal conductivity (W/m K)	0.6	39
Viscosity (Pa s)	1.003×10^{-3}	2.215×10^{-3}

2.4 Fluid Properties

Galinstan was used as the liquid metal and the results were compared with water. The properties of the fluids are shown in Table 2.

3 Results

3.1 Flow and Thermal Performance

Initial simulations were run with constant inlet velocity of 0.24 m/s for both the fluids for a channel length of 10 mm and for a heat flux of 50 W/cm^2. The pressure drop for water and liquid metal was found to be 106 and 178 Pa.

Figure 3 shows the temperature profile for both the fluids inside the channel. It can be observed that the maximum temperature at the wall for liquid metal is lower compared to water. This is due to the resultant larger mass flow rate of liquid metal when the inlet velocity is constant. For all the cases, the global mass and energy balance was checked and it was found that the maximum difference between the theoretical and numerical results was around 4 %, with less than 2 % for most of the cases.

The pressure drop and bulk temperature rise are defined as in Eq. (12), whereas the bulk mean temperature of the fluid at the exit is calculated as in Eq. (13):

$$\Delta P = p_{\text{in}} - p_{\text{out}}; \quad \Delta T = T_{\text{in}} - T_{\text{out}} \tag{12}$$

$$T_{\text{out}} = \frac{\int T(n \cdot \vec{u})\mathrm{d}s}{\int (n \cdot \vec{u})\mathrm{d}s} \tag{13}$$

where ds is the surface area at the outlet for Eq. (13).

3.2 Grid Independency

Unstructured mesh was created using the default mesh option in COMSOL, and the results were checked for grid independency. Three different meshes with

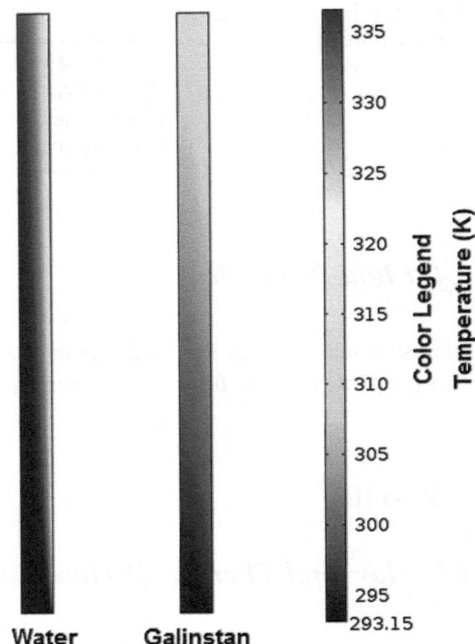

Fig. 3 Comparison of temperature inside the channel for water and galinstan ($U_{\text{in}} = 0.24$ m/s, Heat flux = 50 W/cm^2)

triangular free mesh elements (number of elements 71,608, 34,184, 16,494) were used for grid independency check, and the resultant bulk temperature and maximum temperature were compared. It was found that the difference in the values predicted with the meshes was less than 0.001 K.

4 Parametric Study

4.1 Effect of Length-to-Diameter Ratio

The length of the channel was varied between 2 and 10 mm to analyze the performance of galinstan and water. With these lengths, the length-to-diameter ratio will be 2, 5 and 10, respectively. Figure 4 shows the pressure drop obtained for the two fluids. As expected, the pressure drop increased with increase in channel length. Since the resultant mass flow rate of liquid metal is higher than water for the same inlet velocity, the pressure drop is higher for liquid metal and is higher than water for all the lengths.

Figure 5 shows the heat transfer coefficient obtained for both liquids at a heat flux (q_o) of 50 W/cm^2. The heat transfer performance was evaluated by comparing the heat transfer coefficient defined as:

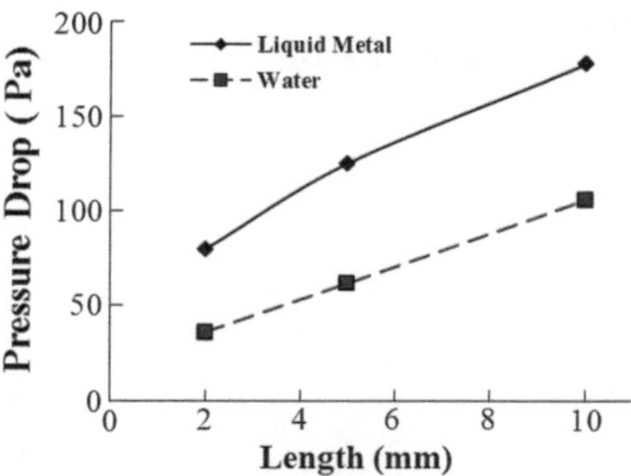

Fig. 4 Pressure drop results, $U_{in} = 0.24$ m/s

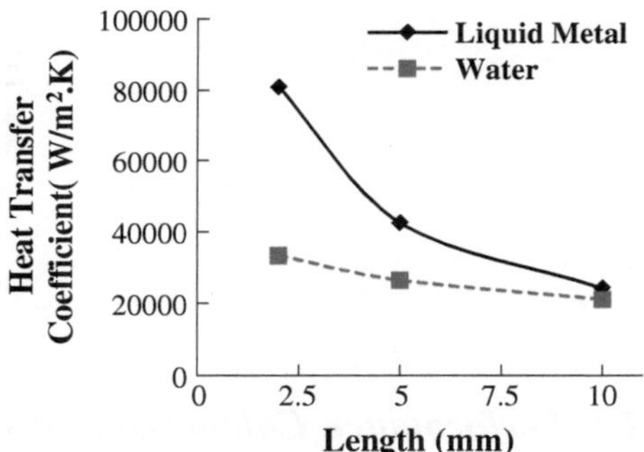

Fig. 5 Heat transfer coefficient, $U_{in} = 0.24$ m/s

$$h = \frac{q_o}{(T_{wall} - T_{inlet})} \quad (14)$$

T_{wall} is the maximum wall temperature at the heater. It can be observed that liquid metal has higher heat transfer coefficient compared to water for all the cases presented in Fig. 5.

4.2 *Effect of Heat Flux*

The base heat flux was varied and also the performance of the fluids for different channel lengths. Figure 6 shows the results. It can be observed that for all the cases considered, the heat transfer coefficient of galinstan is higher compared to water. In can also be observed that the heat transfer coefficient of liquid metal does not vary much for the range of heat fluxes used, but it increased for water at higher heat flux.

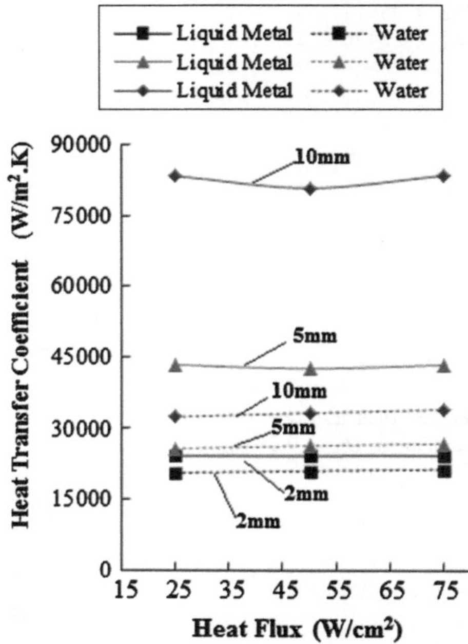

Fig. 6 Heat transfer coefficient for different heat

4.3 Performance Comparison at Constant Pressure Drop

Since the pressure drop in case of liquid metal is higher, simulations were run with velocities that result in constant pressure drop for both the fluids. Figure 7 shows the heat transfer coefficients obtained as a function of length for both the fluids at a constant heat flux of 75 W/cm^2. It can be observed that unlike previous simulations, where liquid metal's thermal performance is always higher compared to water, the thermal performance of galinstan is lower in some cases at constant pressure drop. For channel length of 2 mm, the heat transfer coefficient of liquid metal is around 1.88 times higher compared to water and is around 1.14 times higher in case of 5-mm channel length. With increase in length-to-diameter ratio, the performance of water increases and it was found that at 10 mm length, performance of water is comparable or slightly higher than liquid metal.

Figure 8 shows the Nusselt number comparison for both the fluids. It can be observed that the Nu for liquid metal is lower for all the cases though the heat transfer coefficient might be higher. This suggests that the heat transfer into liquid metal is mostly governed by conduction, whereas the heat transfer into water is mostly governed by convection. This is expected since the thermal conductivity of liquid metal is almost 65 times of water.

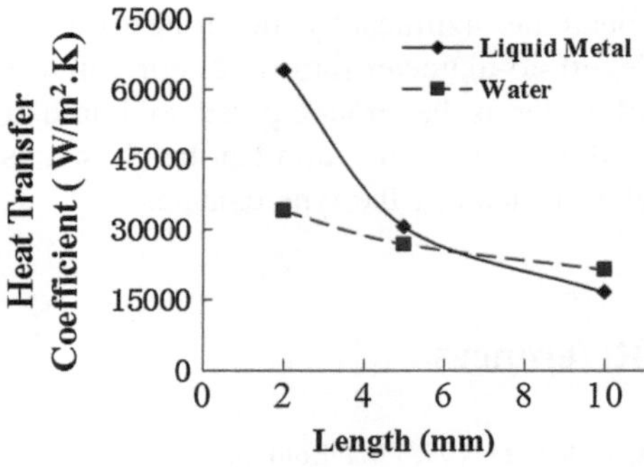

Fig. 7 Heat transfer coefficient for constant pressure drop

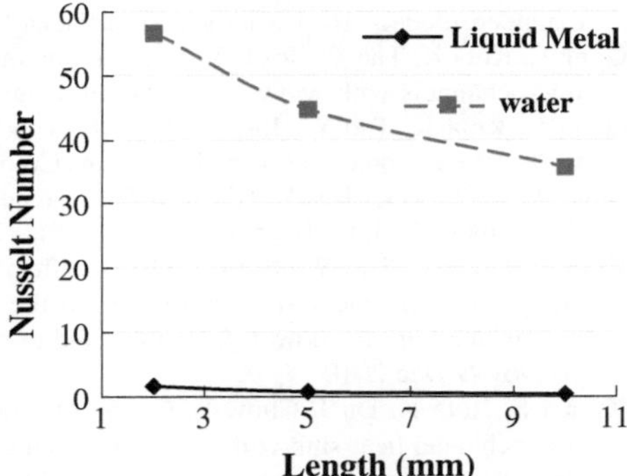

Fig. 8 Nu number comparison

5 Discussion

From the results, it can be assumed that the performance of the fluids is strongly dependent on the flow rate and channel dimensions. It can be assumed that for the flow and heat transfer parameters used in the study, water will perform much better compared to liquid metal if the length-to-diameter ratio is much higher than 10, even when a minichannel diameter is used. This provides an important conclusion that the choice of fluid strongly depends on length-to-diameter ratio of the channels and hence can be used for choosing the fluid instead of purely depending on whether the channel is a minichannel or microchannel.

6 Conclusion

The heat transfer performance of liquid metal was compared with water in 1-mm-wide channels with varying lengths. It was found that the heat transfer coefficient of liquid metal is higher compared to water for same inlet velocity condition. In case of flow rates with constant pressure drop condition, it was found that liquid metal has significantly higher heat transfer performance compared to water, if the length-to-diameter ratio is 2. For length-to-diameter ratio of 10, the performance of water is better compared to liquid metal. This suggests that the length-to-hydraulic diameter ratio can be chosen as one of the parameters for choosing the fluid instead of the type channel.

References

Copeland D (1995) Manifold microchannel heat sinks: analysis and optimization. Therm Sci Eng 3(1):7–12

Copeland D, Behnia M, Nakayama W (1998) Manifold microchannel heat sinks: conjugate and extended models. Int J Microelectron Packaging Mater Technol 1(2):139–152

Gong L, Kota K, Tao W, Joshi Y (2011) Parametric numerical study of flow and heat transfer in microchannels with wavy walls. J Heat Transfer 133(5):051702(1–10)

Gong L, Kota K, Tao W, Joshi Y (2011b) Thermal performance of microchannels with wavy walls for electronics cooling. IEEE Trans Compon Packaging Manuf Technol 1(7):1029–1035

Jung JY, Oh HS, Kwak HY (2009) Forced convective heat transfer of nanofluids in microchannels. Int J Heat Mass Transf 52(1–2):466–472

Kota K, Gong L, Tao W, Joshi Y (2010) Effect of wall roughness on thermal performance of wavy microchannels. In: Presented at the 10th international workshop on micro and nanotechnology for power generation and energy conversion applications, Leuven, Belgium, 30 Nov–3 Dec 2010

Kuravi S, Kota K, Du J, Chow L, Colvin D (2007) Numerical simulation of heat transfer in a microchannel heat sink with micro encapsulated phase change material (MEPCM) slurry. In: Presented at proceedings of ASME-JSME thermal engineering summer heat transfer conference, Vancouver, Canada, 8–12 July 2007

Kuravi S, Kota K, Du J, Chow L (2009) Numerical investigation of flow and heat transfer performance of nano-encapsulated phase change material (NEPCM) Slurry in microchannels. J Heat Transfer 131:062901(1–9)

Kuravi S, Du J, Chow L (2009) Encapsulated phase change material slurry flow in manifold microchannles. In: Presented at 41st AIAA thermophysics conference, San Antonio, Texas, 22–25 June 2009

Kuravi S, Du J, Chow L (2010) Encapsulated phase change material slurry flow in manifold microchannels. J Thermophys Heat Transfer 24(2):364–373

Li T, Lv YG, Liu J, Zhou YX (2006) A powerful way of cooling computer chip using liquid metal with low melting point as the cooling fluid. Forsch Ingenieurwes 70(4):243–251

Ma K, Liu J (2007a) Liquid metal cooling in thermal management of computer chips. Front Energy Power Eng Chin 1(4):384–402

Ma KQ, Liu J (2007b) Nano liquid-metal fluid as ultimate coolant. Phys Lett Sect A 361(3):252–256

Sharma D, Garg H, Singh PP, Karar V (2013) Numerical study on the performance of double layer microchannel with liquid gallium and water. Adv Mech Eng 2013(324578):15 pp

Tuckerman DB, Pease RFW (1981) High-performance heat sinking for VLSI. IEEE Electron Dev Lett EDL-2:126–129

Comparison of Different Agro Residue Biomass for Power Generation in Paper Industry

Arvind Dhingra, Tejinder Singh Saggu and Manpreet Kaur

Abstract Punjab is an agro-rich state. Two crop cycles, namely Rabi and Kharif are followed. Of late, the farmers have been resorting to burning of stubble of the agro residue from wheat and rice. Although the government has put a ban, yet in the absence of economy of scale, the practice continues. This paper is an attempt to find out the agro residue-based mixture which can be a useful source of power generation for one agro-based industry—the paper industry. We have tried to gather information regarding the availability of different agro residues and have tried to project a mixture of residues which would be most efficient in terms of calorific value.

Keywords Agro residue · Calorific value · Ash content · Moisture content

1 Introduction

The problem of stubble burning in an agricultural rich economy as of state of Punjab has assumed alarming proportions due to high cost of labor and uneconomical methods of waste disposal. In this paper, we have tried to work out a solution to this problem by using the agro residue biomass for power generation. To make the study effective, a case study of a paper mill has been taken. Also, this

A. Dhingra (✉)
Guru Nanak Dev Engineering College, Ludhiana 141006 Punjab, India
e-mail: arvinddhingra@gmail.com

T. S. Saggu
PEC University of Technology, Chandigarh, India
e-mail: saggutejinder@gmail.com

M. Kaur
Electrical Engineering Department, Guru Nanak Dev Engineering College, Ludhiana 141006 Punjab, India
e-mail: colmag@gndec.ac.in

S. S. Khangura et al. (eds.), *Proceedings of the International Conference on Research and Innovations in Mechanical Engineering*, Lecture Notes in Mechanical Engineering,
DOI: 10.1007/978-81-322-1859-3_47,

study aims to provide an alternate renewable source of power generation in the power starved state of Punjab.

Various fuels that we have compared include rice husk, maize cob, wheat straw, sugarcane trash, mungra, cow dung cake, wooden chips, and mustard trash. Various parameters have been compared for these fuels namely, moisture content, ash content, and calorific value. To determine the values of above-said parameters of various agro residues considered, various tests were performed using hot air oven, muffle furnace, and bomb calorimeter. Combinations of these fuels in various proportions were considered to arrive at the combination which gave maximum calorific value. Results obtained from the tests reveal that sugarcane trash has minimum and wood chips has maximum moisture content. Mungra has minimum and wood chips has maximum ash content. Cow dung cake has minimum and wood chips has maximum calorific value. In the mixture of two agro residues, the mixture of rice husk and sugarcane trash has minimum and mixture of rice husk and sludge has maximum moisture content. In the mixture of two agro residues, the mixture of wheat straw and wood chips has minimum and mixture of rice husk and sludge has maximum ash content. In the mixture of two agro residues, the mixture of rice husk and cow dung cake has minimum and mixture of wheat straw and wood chips has maximum calorific value. In the mixture of three agro residues, the mixture of rice husk, wood chips, and cotton stalks has minimum and the mixture of rice husk, maize cob, and leaves has maximum moisture content. In the mixture of three agro residues, the mixture of ashoka leaves, grass, and tree shedding has minimum and mixture of rice husk, sugarcane trash, and leaves has maximum ash content. In the mixture of three agro residues, the mixture of rice husk, maize cob, and leaves has minimum and mixture of ashoka leaves, grass, and tree shedding has maximum calorific value. So, each agro residue has its own significant property, but availability of each fuel varies seasonally. Hence, agro residue-based cogeneration could play a substantial role in bridging the gap between the electricity supply and demand.

2 Various Sources of Renewable Energy

Renewable sources of energy are renewed again and again by nature. Due to various environmental problems and depleting fossil fuels, scientists are in search for substitute of fossil fuels for electricity generation. The various renewable sources of energy are as follows:

- Wind energy
- Solar energy
- Geothermal energy
- Hydropower
- Nuclear energy
- Biomass.

Biomass is a renewable source of energy in form of wood, agriculture residue, animal, municipal, and industrial waste. Biomass does not create pollution because it will not give carbon dioxide to surrounding as it absorb the same amount of carbon in growing as it releases when consumed as a fuel. Now a day's biomass is becoming an important source of energy for power generation after coal, oil, and natural gas.

Types of Biomass Resources

The following are types of Biomass Resources:

1. Biomass in traditional solid form
2. Biomass in non-traditional form
3. Biomass in gaseous form.

In traditional form of biomass, we can directly burn it and get energy. In non-traditional form of biomass, biomass is first converted into ethanol and methanol to be used as fuels in liquid form in engines. The third type is biomass in gaseous form called biogas.

Biomass feed stocks include dedicated energy crops, agricultural crops, forestry residues, aquatic crops, biomass processing residues, municipal waste, and animal waste.

2.1 Power Generation from Agro Waste

By the process of photosynthesis, plants can convert the solar energy of pants into biomass which is used as a fuel for power generation.

Process of Photosynthesis

$$CO_2 + \text{Chlorophyll} + \text{Moisture} + \text{Solar energy} \text{ - - - - - } \text{Biomass} + O_2.$$

Biomass carries the energy of sun, and plants absorb that energy in a process which is known as photosynthesis. When the chemical energy of biomass burned is converted into heat energy, that heat energy is used for power generation. Biomass can also be used for in other forms like methane gas or transportation fuels like ethanol and biodiesel.

Types of Agro Residues

Following are some of the agro residues which are available from the agriculture fields:

1. Rice—Stalks, Straw, Husk
2. Wheat—Stalks, Straw
3. Cotton—Stalks, Kernels

4. Maize—Stalks, Cobs
5. Sunflower—Stalks, Kernels
6. Sugarcane—Trash, Bagasse
7. Wood—Popular, Eucalyptus, Saw dust
8. Mustard—Husk, Stalks
9. Soyabean.

Electrical energy can be generated from agro residue by constructing small agro residue-based power plants in rural areas; one plant can easily satisfy the needs of cluster of villages. The main problem with agro residue is their collection from fields.

2.2 Agro Residue Waste Scenario

In rural areas, agro residue is a valuable source of heat energy in all over the world. Due to gradual depletion of fossil fuels agro residue are playing a leading role in developing countries. Various processes have to be done for adequate power generation like as agro residue collection, briquetting, combustion, gasification, and then cogeneration. From the environmental point of view, agro residue is a serious threat to the surrounding which disturbs the ecology, creates problems for human health because some of the agro waste is burnt in the fields, which results in environmental pollution and also harms the fertility of the fields. If these are to be used for power generation, then it will automatically help to solve the pollution problem. We collected the agriculture waste manually or in mechanized manner. In the manual manner, threshers were used for harvesting rice and wheat which in turn gave us rice straw or wheat straw but in mechanized manner, harvesting combines were used which gave us output in form of grains, stalks, and straw.

The prominent factors that influence the storage of agro waste include:

- Physical properties like size, shape, nature, density, etc.
- Conversion methods and electric power generation operations.
- Area availability for agro waste storage.
- Weather conditions and seasonal conditions.
- Backup storage capacity.

3 Case Study

We chose one paper mill in Sangrur district of state of Punjab for carrying out this study. The steps followed in evaluating the efficiency of the mixtures were:

Finding out the moisture content of agro residue
Finding out ash content of agro residue

Finding out calorific values of agro residue

We shall now list the moisture content as found out for three major agro residues:

Moisture content for rice husk

S. no.	Rice husk sample	Weight W1 (g)	Weight W2 (g)	Moisture content (%)
1	RH1	55.68	47.718	14.3
2	RH2	55.62	46.44	16.5

Average moisture content: 15.4 %

Moisture content for wheat

S. no.	Wheat sample	Weight W1 (g)	Weight W2 (g)	Moisture content (%)
1	WS1	34.68	31.11	9.5
2	WS2	34.40	31.61	8.1

Average moisture content: 8.8 %

Moisture content for maize cob

S. no.	Wheat sample	Weight W1 (g)	Weight W2 (g)	Moisture content (%)
1	MC1	39.92	32.934	17.5
2	MC2	39.83	32.939	17.3

Average moisture content: 17.4 %

We shall now list the ash content of three major crops:

Ash content for rice

S. no.	Weight of ash (g)	Initial weight of rice (g)	Ash content (%)
1	2.10	12.90	16.27
2	2.15	12.80	16.79

Average ash content: 16.53 %

Ash content for wheat

S. no.	Weight of ash (g)	Initial weight of wheat (g)	Ash content (%)
1	0.30	3.01	9.96
2	0.32	3.03	10.56

Average ash content for wheat: 10.26 %

Ash content for maize cob

S. no.	Weight of ash (g)	Initial weight of maize cob	Ash content (%)
1	0.59	7.89	7.47
2	0.55	7.87	6.98

Average ash content: 7.22 %

Now, we shall list the calorific values of the fuels under investigation:

S. no.	Sample	Wt. of thread	Wt. of wire	Wt. of tablet	Temp. reading	Cal value (cal/g)
1	Rice	0.02	0.005	0.939	1.39	3688.38
2	Wheat	0.03	0.008	0.887	1.48	4115.4
3	Maize cob	0.03	0.007	0.865	1.46	4161.42

Then, a mixture of any two of the above three was taken. Then, their moisture content, ash content, and calorific values were found out. The calorific values are presented below:

S. no.	Sample 80:20	Wt. of thread	Wt. of wire	Wt. of tablet	Temp. reading	Cal value (cal/g)
1	Rice and wheat	0.02	0.005	0.875	1.36	3870.63
2	Rice and maize cob	0.03	0.004	0.968	1.50	3825.16

Then, a mixture of two of these with dried leaves was taken. The calorific value is listed below:

S. no.	Sample 55:25:20	Wt. of thread	Wt. of wire	Wt. of tablet	Temp. reading	Cal value (cal/g)
1	Rice wheat and leaves	0.03	0.007	0.864	1.50	4284.44

In addition to the above results, various other samples and in various proportions were also considered and the findings are presented below:

- Sugarcane trash has minimum and wood chips has maximum moisture content.
- Mungra has minimum and wood chips has maximum ash content.

- Cow dung cake has minimum and wood chips has maximum calorific value.
- In the mixture of two agro residues, the mixture of rice husk and sugarcane trash has minimum and mixture of rice husk and sludge has maximum moisture content.
- In the mixture of two agro residues, the mixture of wheat straw and wood chips has minimum and mixture of rice husk and sludge has maximum ash content.
- In the mixture of two agro residues, the mixture of rice husk and cow dung cake has minimum and mixture of wheat straw and wood chips has maximum calorific value.
- In the mixture of three agro residues, the mixture of rice husk, wood chips, and cotton stalks has minimum and the mixture of rice husk, maize cob, and leaves has maximum moisture content.
- In the mixture of three agro residues, the mixture of ashoka leaves, grass, and tree shedding has minimum and mixture of rice husk, sugarcane trash, and leaves has maximum ash content.
- In the mixture of three agro residues, the mixture of rice husk, maize cob, and leaves has minimum and mixture of ashoka leaves, grass, and tree shedding has maximum calorific value.

4 Conclusion

We can see from the above observations that agri residue biomass offers a viable solution for power generation. It is more advantageous to have the biomass-based plant in conjunction with agro-based industry such as paper. In addition to solving the problem of power shortage, it also helps in keeping the environment pollution free.

Thermal Management of Low Volume Complex Electronic Systems

Rajesh Kanda and Karan Jain

Abstract Advances in the field of electronics have resulted in a significant increase in density integration, clock rates, and emerging trend of miniaturization of modern electronics. This resulted in dissipation of high heat flux at the chip level. In order to satisfy the junction temperature requirements in terms of performance and reliability, improvements in cooling technologies are required. The task of maintaining acceptable junction temperature by dissipating the heat from the integrated circuit chips is a significant challenge to thermal engineers. Much work has been done on cooling one hot spot with one heat sink, but there has not been as much investigation into cooling multiple hot spots with a single heat sink. A circuit board with a specific geometry and chip arrangement will be cooled using a liquid-cooled cold plate.

Keywords Thermal management · Cold plate · Junction temperature · Pressure drop · Thermal resistance

1 Introduction

The authors concluded that thermal management must be taken into account early in the design process. Room must be made for the cold plate to fit the electronics, and components that do not need to be cooled should not interfere with the cold plate.

R. Kanda (✉)
Department of Mechanical Engineering, PEC University of Technology, Chandigarh, India
e-mail: kandarajesh@yahoo.co.in

K. Jain
Larson & Toubro Constructions, Mumbai, India

S. S. Khangura et al. (eds.), *Proceedings of the International Conference on Research and Innovations in Mechanical Engineering*, Lecture Notes in Mechanical Engineering,
DOI: 10.1007/978-81-322-1859-3_48, © Springer India 2014

Electronics cooling is viewed in three levels, which are non-separable. First, the maintenance of chip temperature at a relatively low level despite high local heat density. Second, this heat flux must be handled at system or module level. Finally, the thermal management of the computer machine room, office space, or telecommunication enclosure. The thermal design of the system is influenced by the key drivers such as chip size, power dissipation, junction temperature, and ambient air temperature.

The high chip temperature results in thermal failures such as mechanical stresses, thermal de-bonding, and thermal fracture. The failure in electronics during operation occurs mainly due to temperature. Therefore, thermal management is a key enabling technology in the development of advance electronics. The main constraint for any thermal management is the cost. Therefore, the cooling technology must be cost-effective and keep pace with the reduction in overall package and system cost per function. The cost of cooling is also recognized as a factor playing important role in maintaining competitiveness.

2 Earlier Research

There has been a dramatic shift in cooling high-power devices in the industry during the past decade. Although it is quite difficult to make a distinction based on total power dissipation, it seems that beyond a range of about 1,500 W dissipation, there are many physical and design constraints that may dictate a shift toward liquid as the preferred medium.

Nottage (1945) suggested that the heat sink fin and channel might be thought of as a type of heat exchanger. The solid fin is considered as a hot stream. The flow stream direction relative to heat flow direction plays a significant role in determining the heat transfer effectiveness of a fin-fluid arrangement. The counterflow arrangement has the greatest potential to achieve high effectiveness [Evolution of cold plate design].

Jones and Smith (1970) performed a similar study for rectangular fin assemblies facing upward and downward in relation to the gravity vector. In both studies, the correlation equations were restricted to a fixed range of geometric and flow conditions limiting their use as general-purpose design tools [Evolution of cold plate design].

The need for liquid cooling has resulted in a major paradigm shift for many electrical system designers and manufacturers. This brought the realization that liquid cooling offers a better, and often the only feasible, solution. The compactness of the cold plates and the supply and return lines, lower power consumption, and reduction in noise levels are some of the attractive features.

The use of microchannels as a viable cooling solution was first proposed in 1981 by Tuckerman and Pease, who designed and tested an integral, water-cooled

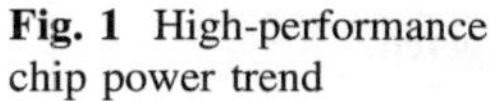

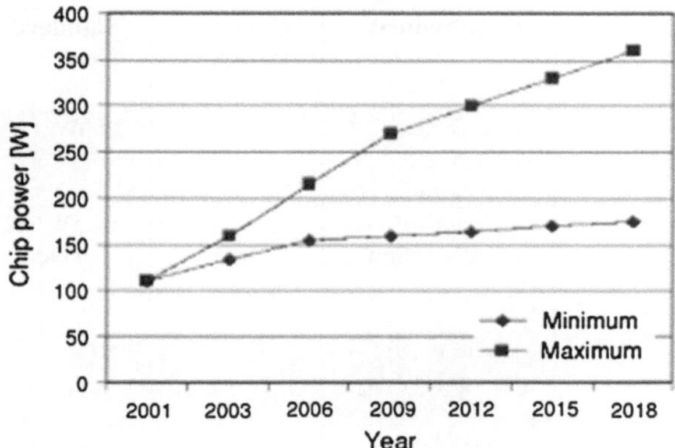

Fig. 1 High-performance chip power trend

heat sink and achieved a high heat flux of 790 W/cm^2 with a temperature rise of 71 °C above the inlet water temperature. Microchannel heat sinks are one of the effective cooling methods for high-power density and compact electronic devices according to Zhao. Further confirmation was provided by studies done by Chien-Hsin.

The available literature on the cold plate design is primarily focused on relatively low power dissipation. A number of thermal management solutions in use for cooling power electronic modules in automotive applications are reviewed in Nakayama et al. (1984) and Garg and Velusamy (1986). The coolant in such cooling must be accomplished with a low temperature difference between the semiconductor and the coolant (Fig. 1).

3 Research Objectives

The objective of this project is to design and fabricate a highly effective, liquid-cooled cold plate for high heat flux cooling. This project is unique in that the main objective was to accommodate chips of different heights and power densities.

The cold plate requirements are listed below. The cold plate must work in low-temperature environments, as well as environments where both the magnitude and direction of the gravity force are constantly changing, while remaining inexpensive.

The cold plate must accommodate the complex geometry of the multi-chip circuit board, and different chip heights must be accommodated while maintaining good conduction contact.

Requirement	Current Design	Standard Goal	Stretch Goal
Power Dissipated by the module	16.6W	200W	400W
Highest Power Concentration	2.96W/in2 (0.46W/cm2)	64.5W/in2 (10W/cm2)	258 W/in2 (40 W/cm2)
Highest Single Power Device	2.7W	60W	150W
Fluid Used	N/A	50/50 Propylene Glycol/Water	PAO
Pressure Drop	N/A	20 psi (138 kPa)	2.5psi (18 kPa)
$T_{max,module}$- $T_{fluid,inlet}$	30 C	10 C	3 C
Maximum Module Thickness	.125 in(.318cm)	.25 in (.635 cm)	.1 in. (.25 cm)
Inlet Fluid Temperature Range	N/A	0 C to 4 C	40 C to 60 C

Fig. 2 Project objectives

4 Experimental Setup

A test section was designed and constructed to evaluate the capabilities of the cold plate. The test section was designed to emulate the geometry and thermal characteristics of the circuit board.

It can be seen that Chips 1–4 have the highest heat flux. Attention in this project was focused on finding a thermal solution capable of cooling those chips. Anything with a smaller heat flux should be cooled just as effectively by the same cold plate (Fig. 2).

4.1 Test Section Description

The test section consists of a rectangular G10 fiber glass composite board 88 mm wide, 225 mm long, and 2.50 mm thick. G10 is a fiberglass that is often used in low-temperature applications. Four copper squares were made. Three of the

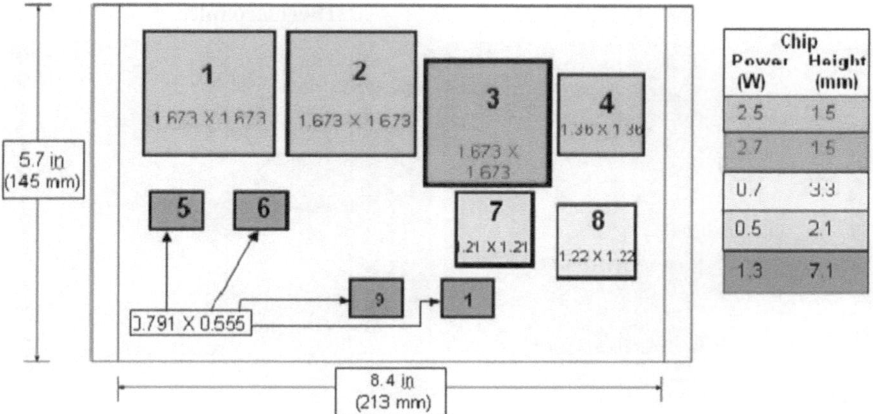

Fig. 3 Test section

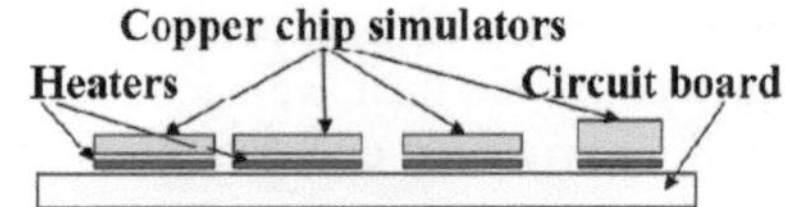

Fig. 4 Test section schematic

squares are 42.5 mm on a side and 2 mm thick. The fourth square is 34.5 mm on a side and 3.5 mm thick. These copper squares are the same size and shape as Chips 1–4 in Fig. 8. Each copper square has a heater attached to the back of it for heat generation (Fig. 3).

4.2 Test Section Drawings and Pictures

A schematic drawing of the test section is shown in Fig. 4.

A picture of the fabricated test section is shown in Fig. 5, including the thermocouples on the surface of the copper chip simulators. These thermocouples measured the temperature of the copper chip simulators at the interface between the copper chip simulators and the cold plate.

5 Experimental Procedure

The experimental procedure for the two cold plates is discussed in this section. The procedures do not vary much because most of the data, except for the pressure drop, were taken from the test section. The same test section was used for each experiment.

The test section and cold plate were then placed in contact. The chiller water was circulated through the cold plate at 5 °C for fifteen minutes. Once the chiller

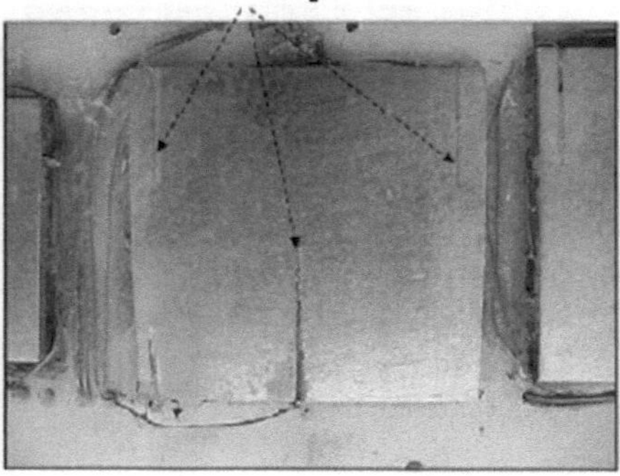

Fig. 5 Test section

water reached steady state and flow rate, power was applied to the heaters. The pressure drop across the cold plate and the temperatures of each thermocouple were recorded both by computer and by hand. The three temperatures recorded for each chip were then averaged to get an average chip temperature. Each steady state reading produced one data point for each copper chip simulator.

5.1 Cold Plate Setup

The cold plate is a U-type cold plate with pressure ports on the inlet and outlet of the cold plate to measure pressure drop (Fig. 6).

5.2 Channels

The cold plate consists of 12 parallel channels fed by the inlet header. Figure 7 shows cross-sectional views of the cold plate revealing the channel geometry.

The channel was compressed to the point where the fins which extended only partway into the channel became joined and the channel walls bent. The direction of the bending of the walls in the compression was random and uncontrollable with the process used to create the compression.

Figure 7 shows the final cold plate. The cold plate has a compression to accommodate a taller chip and pressure ports on both the inlet and outlet.

5.3 Modifications Based on Setup 1: Setup 2 Cold Plate

The first-generation cold plate was constructed and tested. A great non-uniformity was found in the thermal resistances of the chips.

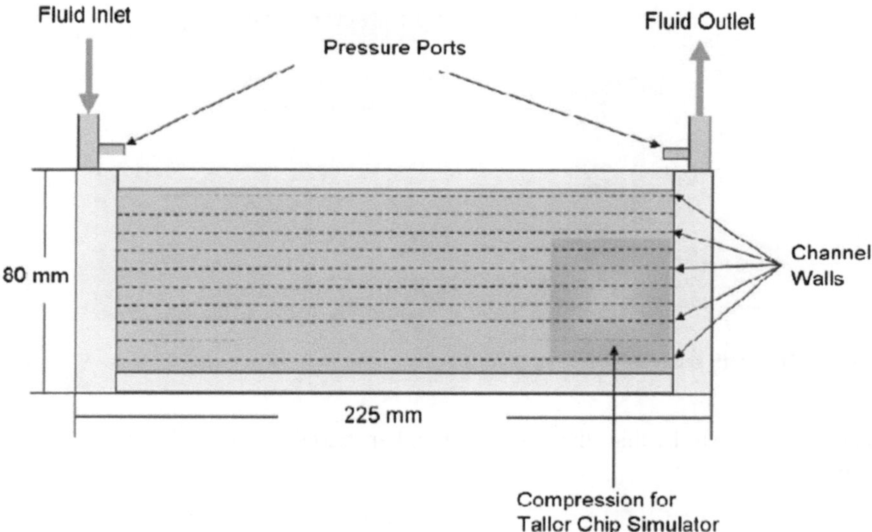

Fig. 6 Cold plate setup 1

Fig. 7 Aluminum section of cold plate

The second-generation cold plate was designed with long headers and short channels, both to take advantage of the entrance effect while also decreasing the pressure drop. The channels of the second-generation cold plate had to be smaller in order to satisfy the thermal resistance stretch goal. Since the pressure drop through a closed channel increases as the hydraulic diameter decreases, the pressure drop through the second-generation cold plate channels was expected to increase.

The modified design reduced the flow length through the channels and thus lowered the pressure drop. Figure 8 illustrates the second-generation cold plate.

The second generation had a smaller hydraulic diameter than the channels of the first-generation cold plate. Smaller channels were chosen for several reasons.

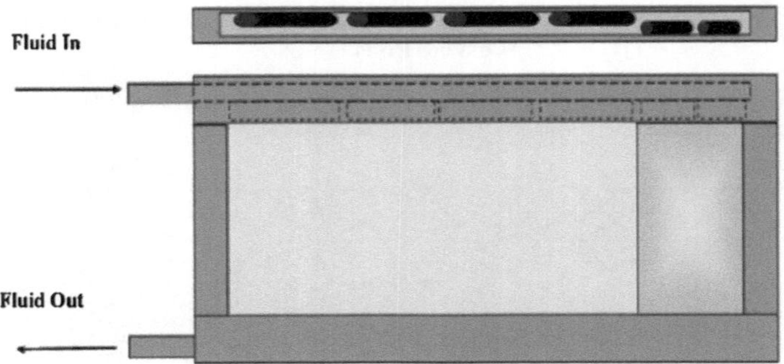

Fig. 8 Fluid path setup 2

The first reason is that the Nusselt number based on the channel hydraulic diameter,

$$\mathrm{Nu}_d = \mathrm{hd}_h/k$$

converges to a constant value, 4.36, in fully developed, laminar, closed channel flows exposed to a constant heat flux. This means that channels with a smaller hydraulic diameter have larger convection coefficients.

The second reason is that channels were smaller in the second-generation cold plate inorder to meet the goal for maximum allowed cold plate thickness. The first-generation cold plate met the standard goal for cold plate thickness, but did not satisfy the stretch goal. As the channel dimension shrinks, the pressure drop in the channels increases.

The channel flow length also had to be shortened to help meet the stretch goal for pressure drop. Since the channels were smaller than the previous cold plate, a new technique was used to accommodate the taller chip height.

Compressing the channels would not have been possible because the channels were not thick enough to accommodate the required depth of the compression. Instead, a common header and fluid inlet and outlet paths were used, but to accommodate the taller chip height, the connecting slot and the slots for fluid distribution were offset within the header. This eliminated the need to compress the channels while still accommodating the taller chip height and remaining easy to design and manufacture.

5.4 Cold Plate Setup 3

5.4.1 Modifications from earlier design

The cold plate designs so far were full contact type. This forced us to keep the design bulky and required extra rigidity from the test section to hold it together.

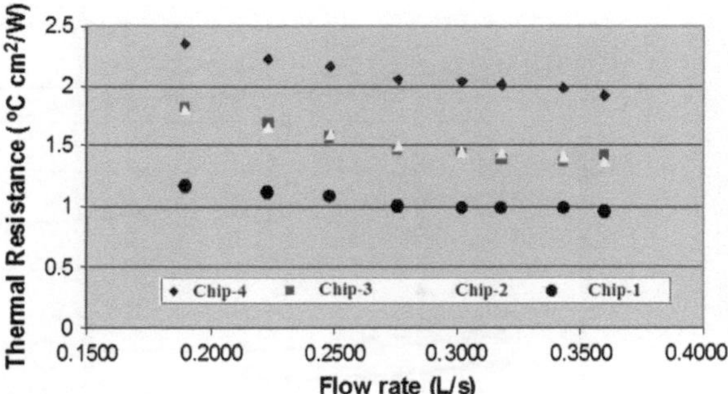

Fig. 9 First-generation cold plate thermal resistance

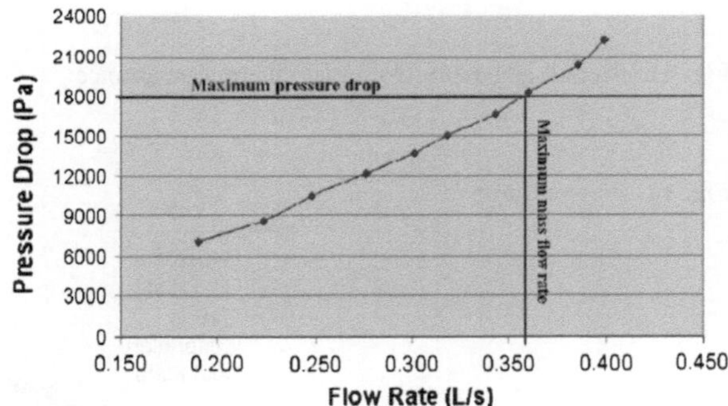

Fig. 10 First-generation cold plate pressure drop

Material of construction remained aluminum as in the earlier designs. At the same time, selective coating was provided to the external structure of the cold plate to reduce emissivity and justify the cooling aspect.

In the new cold plate, flow length was increased slightly, allowing some pressure drop across the section.

6 Results and Discussions

6.1 First-generation heat transfer

Figure 9 is a graph showing the first-generation cold plate thermal resistance as a function of the flow rate through the cold plate.

Figure 9 shows that the first-generation cold plate meets the standard goals for thermal resistance while adhering to the pressure drop restriction. The most important feature of the graph is the uneven distribution of the thermal resistances of each of the copper chip simulators.

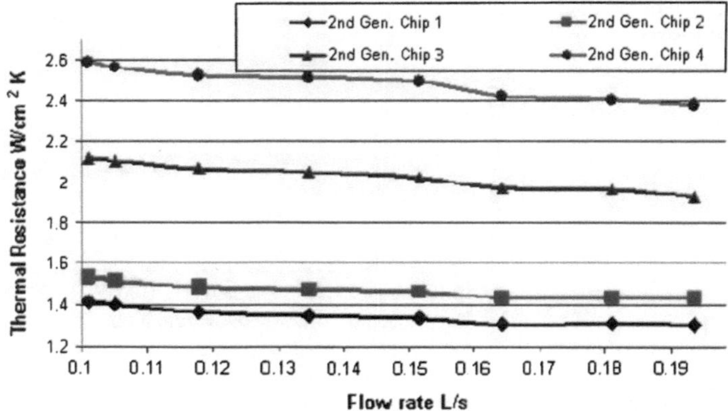

Fig. 11 Second-generation cold plate thermal resistance

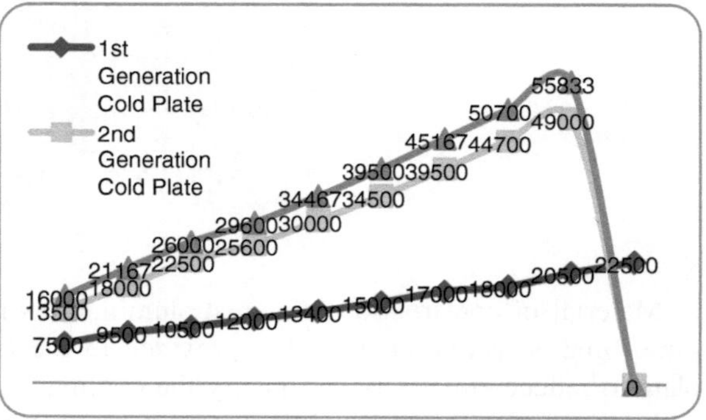

Fig. 12 Pressure drop

The maximum pressure drop allowed by the standard goal is shown along with the corresponding flow rate. The thermal resistance goal was satisfied while also satisfying the pressure drop standard goal (Fig. 10).

When compared to the first-generation results, several differences present themselves. The first difference is that the thermal resistance non-uniformity has become greater. Some thermal resistances have increased (Chip 1 and Chip 3), some have decreased (Chip 2), and the thermal resistance for Chip 4 has not changed (Fig. 11).

6.2 Pressure Drop

The pressure drop from the second-generation and the third-generation cold plates is measured as shown in Fig. 12.

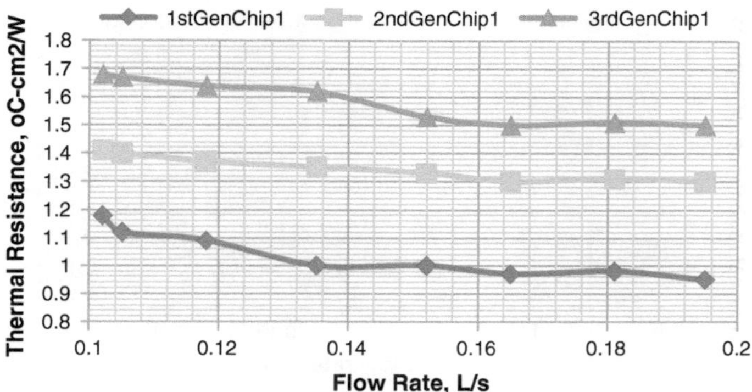

Fig. 13 Thermal resistance comparison for cold plates (Chip 1)

Figure 13 shows that the pressure drop for the second-generation cold plate is greater than the pressure drop for the first-generation cold plate and that for the third-generation cold plate is greater than both previous versions. This is a result of the second- and third-generation cold plates having smaller channels than the first-generation cold plate and the third-generation cold plate having smaller channels than the second.

6.3 Comparisons

There are two important differences in heat transfer between the different setups of cold plates. The first difference is that the cold plates cooled different chips differently.

Switching from one generation of cold plate to another caused the thermal resistances of some chips to increase while causing the thermal resistances of other chips to decrease. The second important difference is that while the range of thermal resistances was about the same for the three cold plates, the range of the second- and third-generation cold plates had higher average temperatures than the first-generation cold plate.

Overall, the second and third setups of cold plates were less effective at cooling the entire test section.

Figures 13, 14, 15, 16 are comparisons of the thermal resistances of Chip 1 and Chip 4. From these figures, the trend is clear that at a given flow rate, the first-generation cold plate is a better heat exchanger than the second-generation cold plate.

The thermal resistance data presented from each chip are composed of three separate thermal resistances: thermal interface resistance, conduction resistance through the cold plate, and convection resistance from the cold plate to the water.

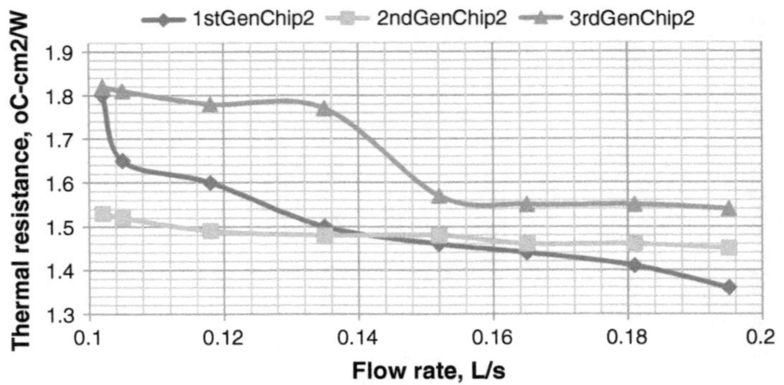

Fig. 14 Thermal resistance comparison cold plates (Chip 2)

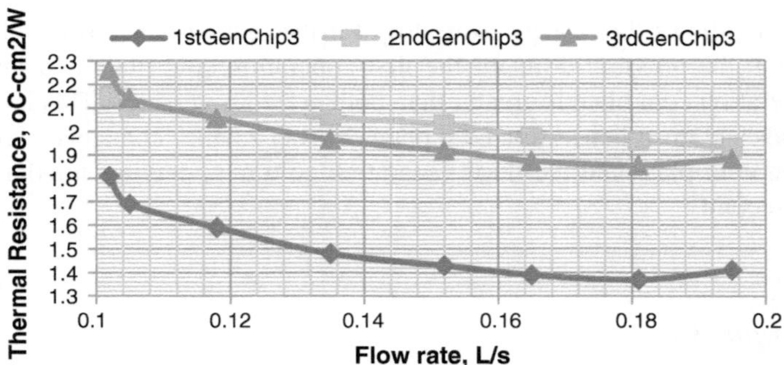

Fig. 15 Thermal resistance comparison cold plates (Chip 3)

Fig. 16 Thermal resistance comparison for cold plates (Chip 4)

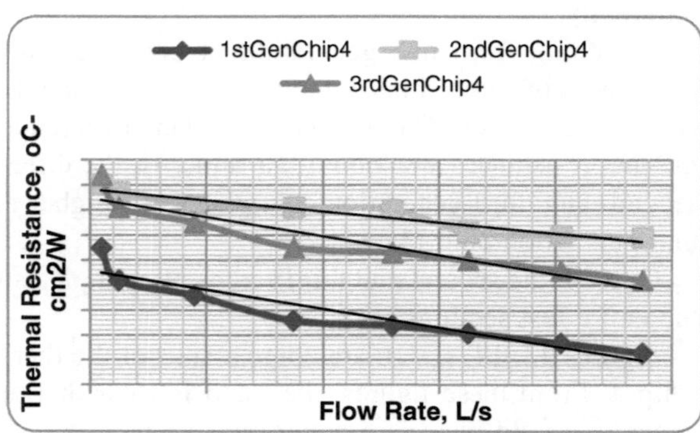

Of these resistances, the only one that changes with flow rate are the convection resistance. Therefore, it must be the convection resistance that causes both the high initial thermal resistance and causes it to fall more rapidly in the first-generation cold plate. The source of this trend is believed to be flow maldistribution in the header. Numerical simulations were performed on this cold plate, which showed that there was a jet effect in the header due to the small diameter and high velocity. This jet effect caused fluid to skip over the first few channels in the cold plate resulting in more fluid in the other channels, thus giving the other channels a higher mass flow rate.

It is believed that Chip 2 benefited from this effect because it was cooled by the channels in the middle of the cold plate. As the jet effect increased, more fluid was delivered to the channels cooling Chip 2 in the first-generation cold plate, resulting in a drop in the total thermal resistance.

7 Conclusions

The experimental boundary conditions are well characterized and specified, and the setup is verified to produce good results via a verification of the energy balance achieve and a comparison of single phase measurements to predictions. We observe that:

- The heat transfer coefficient peaks at a thermodynamic vapor quality of approximately 20 % and decreases significantly with increasing vapor quality.
- The heat transfer coefficient shows a strong increase with increasing heat flux.
- The influence of the saturation pressure in the investigated range is negligible.

The header of the second-generation is an elongated version of the first-generation cold plate header. The fluid inlet and outlet paths are the same diameter; however, the paths in the second-generation header are longer.

The difference in pressure drop between the two cold plates has to do with both the header design and the channels. The second-generation cold plate was designed so that most of the pressure drop would take place in the channels.

Due to the high flow rate and the design of the fluid distribution slots, the flow through the channels was not even, and a large amount of the pressure drop is believed to have occurred due to the high velocity in the fluid inlet and outlet paths. This increased the flow maldistribution in the cold plate.

7.1 Heat Transfer Performance

The heat transfer differences are also believed to be strongly dependent on the flow distribution in the header and the channels. The entrance effect alone could not explain the difference in thermal resistance for the first-generation cold plate.

This was further proven by the second-generation cold plate. If the entrance effect were the cause of the improved heat transfer, all the chips in the second-generation cold plate would have benefited from it.

Uneven flow rate in the channels would account for the uneven cooling performance. Channels receiving little or no flow would not cool the chip as effectively. The average temperature for each chip was not very far from the individual thermocouple readings.

This suggests that it is a flow problem and that many channels are affected by it, not just one or two scattered channels.

Today's advanced microelectronic technology is pushing semiconductor devices to their thermal limits, underscoring the importance of a well-designed reliable heat sink. Unlike fuses and semiconductors that can be standardized in size and performance, high-power thermal management is always customized for each individual application.

To select the thermal management method that best meets the needs of application at hand, the OEM should collaborate closely with the heat sink manufacturer at the earliest possible stages of the design project. After carefully considering the application's mechanical, hydraulic, power, and cost requirements, the OEM and heat sink manufacturer will together be able to design a solution that delivers the optimum combination of performance, size, and cost.

Once the fluid flow problem had been dealt with, the solution of the heat transfer problem was undertaken. In recognition of the fact that the needed computer resources for a cold plate whose stream wise dimension is very large are not generally available, a simplified solution strategy was developed and implemented. This strategy takes full advantage of the periodic nature of the problem and, at first, focuses attention on the solution for one of the periodic modules that is representative of the others in the array.

That solution is necessarily conjugate. From the solution to the conjugate problem, heat transfer coefficients are extracted, which facilitate the subsequent heat conduction solution for the temperature distribution in the bounding walls of the entire array. These heat transfer coefficients were subjected to an accuracy check that ensures the quality of the final results.

To illustrate the use of the method, a problem that typifies the thermal management of electronic equipment by means of a cold plate was formulated and solved numerically. The results of the numerical solution were interpreted from the standpoint of the quality of the overall performance.

References

Garg VK, Velusamy K (1986) Heat transfer characteristics for a plate fin. J Heat Transf 108(1):224–226

Jones CD, Smith LF (1970) Optimum arrangement of rectangular fins on horizontal surfaces for free convection heat transfer. J Heat Trans 92:6–10

Nakayama W, Nakajima T, Hirasawa S (1984) Heat sink studs having enhanced boiling surfaces for cooling microelectronic components. ASME paper, 84-WA/HT-89
Nottage HB (1945) Efficiency of extended surface. Trans ASME 67:621–631

Productivity Improvement Through 5S Implementation in Indian Manufacturing Industries

Abhishek Jain, Rajbir Bhatti and Harwinder Singh

Abstract 5S is a systematic technique used by organizations, and this comes from five Japanese words: seiri (sort), seiton (set in order), seiso (shine), seiketsu (standardize), and shitsuke (sustain). This system helps to organize a workplace for increasing efficiency and decrease wasting and optimize quality and productivity via monitoring an organized environment. There is a real need for the studies in field of new management systems and their impact on company's performance. Aim of this paper is to determine the impact of 5S in industrial organizations and identifying the effectiveness of 5S implementation on organizational performance as well. The target organization (Medium size organization) is chosen for implementation. The results of this research obtained from a comparative measurement of organizational performance before and after 5S implementation. The results show that 5S is an effective tool for improvement of organizational performance, work culture, productivity, etc. Consequently, 5S techniques would strongly support the objectives of organization to achieve continuous improvement and also in implementation of total productive maintenance (TPM).

Keywords 5S · Performance · Quality · Productivity · Efficiency · Implementation · Organization

A. Jain (✉)
Punjab Technical University, Jalandhar, Punjab, India
e-mail: abhi_mpct@rediffmail.com

R. Bhatti
Mechanical Engineering Department, SBSSTC, Firozepur, Punjab, India
e-mail: rajbirbhatti@gmail.com

H. Singh
Mechanical Engineering Department, Guru Nanak Dev Engineering College, Ludhiana 141006 Punjab, India
e-mail: harwin75@rediffmail.com

S. S. Khangura et al. (eds.), *Proceedings of the International Conference on Research and Innovations in Mechanical Engineering*, Lecture Notes in Mechanical Engineering, DOI: 10.1007/978-81-322-1859-3_49, © Springer India 2014

1 Introduction

In Japan, 5S was first used in the manufacturing sector in the mid-1950s (Gapp et al. 2008). The Toyota production system (TPS) is a well-known example of its application throughout this period (Sawada 1995). This technique has been applied in Japanese organizations for several years (Ho and Cicmil 1996), forming an integral part of other Japanese approaches to improvement. The importance of 5S lies in the fact that it can help to reduce costs by maximizing process efficiency, effectiveness, and performance through the establishment and maintenance of a high-quality, clean working environment (Liker 2004; Liker and Hoseus 2008). Ablanedo-Rosas et al. (2010) shed some light on 5S implementation in Mexico and conclude that 5S plays a leading role in improvement practices or approaches. More specifically, we seek to answer the main research question of the study, namely How does 5S implementation take place in Indian organization (Table 1).

2 Literature Review

5S is not only useful for improving the working environment but also they raise process and product quality standards, reduce and optimize lead time, and also reduce operating costs and enhance process performance (Monden 1998; Liker 2004; Liker and Hoseus 2008). Schonberger (2007) states that 5S is one of element of the TPS that forms part of Japanese Production Management. Recently, 5S has even started to expand into forming part of environmental management systems (Tice et al. 2005; Bicheno and Holweg 2009). Result of the study indicates that 5S technique is an effective way to improve health and safety standards, environmental performance, and housekeeping (Ab Rahman 2010). Even though the 5S housekeeping program aids production (Eckhardh 2001), the 5S technique is one of the most known in industrial and business environment, and there are few proofs about its adoption in organizations (Bayo-Moriones et al. 2010).

5S is a useful method for founding an organization and spread out a design and can improve communication and help employees to develop their characteristics to decrease downtime, lead time, inventory, defect, injury, and associated costs (Van Patten 2006). 5S is a method for development of companies, change, and training. According to the Japanese organizations, 5S has two components, a high level of management and organizational system and the other one is management provision tools position (Gapp et al. 2008). Survey of one factory in Iran supports pervious findings in this field and indicates that 5S execution provides better condition for implementation of total productive maintenance (TPM) (Moradi et al. 2011). As it is mentioned in several studies, the 5S method is recorded as a way for improving health and safety standard and performance in a holistic operation with high level of efficiency (Khamis 2009) and also helps improving data management system in factories (Ananthanarayanan 2006).

Table 1 Details of 5S

Japanese	English	Japanese meaning	English meaning	Typical example
Seiri	Structure	To clearly separate necessary things from unnecessary ones and abandon the latter	Organization	Discarding any material that is not of use (cardboard, boxes, etc.)
Seiton	Systematize	To neatly arrange and identify things for ease of use	Neatness	Keeping shelves and filing cabinets tidy and organized
Seiso	Sanitize	To always clean up: to maintain tidiness and cleanliness	Cleaning	Establishing specific individual responsibility for corrective and preventive cleanliness
Seiketsu	Standardize	To constantly maintain the three Ss mentioned above (seiri, seiton, seiso). Includes the individual well-being and physical cleanliness of each person	Standardization	Setting standards and maintenance plans for the above three Ss. Also as a plan or program for physical well-being (exercise, etc.)
Shitsuke	Self-discipline	To have workers make a habit of always conforming to rules	Discipline	Executing the standards set in a disciplined fashion on a day-to-day basis

Sources Adapted from Ho (1999a, b), Sawada (1995) and Monden (1998)

We need the 5S at our workplace because many people do their things without thinking about it. In addition, 5S can be a reflection of our behavior. If we have a consideration to 5S, the majority of our routine problems that we face in everyday works could be solved (Hamzah and Ho 1994). Hubbard in 1999 showed that orderliness which is one of the five pillars of the visual workplace intends to eliminate three types of waste: searching waste, difficulty-of-use waste, and the waste of returning items to their proper place (Hubbard 1999). He also believes Japanese 5S fundamentals can be used for continuous quality improvement (Ho 1997). A useful environmental tool for organization management is 5S, which comes from lean manufacturing process. The 5S practice simplifies the workplace and maintenance system's procedures, decrease waste and non-value-added activities. It also improves quality, efficiency, and safety.

From the quality management point of view, order and cleanliness have been considered as a part for continuous improvement (Yusuf and Aspinwall 2001). 5S can link with TPM (Ahuja and Khamba 2008) and Japanese management approaches such as TPM, JIT, and TQM (Gapp et al. 2008). Generally, the success of 5S implementation depends on organizational characteristics (Sousa and Voss 2008). The 5S practice is beneficial for every organization, because it helps everyone having a better life (Da Silveira 2006). In fact, many successful organizations in the world have already included some aspects of the 5S in their daily activities without complete awareness of its benefits. More investigation is needed on implementation of the 5S as improving tools in a business section and some items such as company size and structure could affect application of the 5S and its effectiveness (Oh Eocha 2000). This system is beneficial for any workplace to make high-quality products and services (Sui-PPheng and Khoo 2001). 5S is applied in most of the factories in manufacturing sections with priority compare to other sections with different ways which can be attributed to the maturity of the 5S programme (Ho and Cicmil 1996).

3 Introduction of Company

The Bapuna Group is based on Central India with its liquor and beer manufacturing facilities at Gwalior (Madhya Pradesh) and blending and bottling units at Nagpur (Maharashtra). The group's main business is the manufacture of rectified spirit (RS), extra neutral alcohol (ENA), and branded alcoholic beverages. Gwalior Distillers Limited (GDL), incorporated in 1987, is the flagship company of the Bapuna Group, a group with a rich history and strong track record in the alcohol industry. GDL is engaged in the manufacture of RS and ENA from molasses and grain. Apart from the company's own brands, which are today well accepted and strongly entrenched in the market, the company also blends and bottles popular brands of IMFL, for the UB Group, Seagram and other industry leaders. Following the success of GDL, Tripti Alcobrew Limited was set up at Morena (M.P.). As a modern unit for manufacturing beer, it manufactures its own brands as well as for leading labels such as SKOL, Kingfisher, and Sandpiper.

4 Study Area

This case study was conducted in a leading wine manufacturing industry "Gwalior Distillers Ltd. Rairu" of Madhya Pradesh, India. *Bottling Section* was chosen for the study. There are six conveyor lines in bottling section, and there are four machines on each conveyor line, i.e., one filling machine, two sealing machines, and one tag machine. Filling machine is used to fill the prepared wine form holding tank. There is a blending section in which inspection have already done that wine is out of any poison. Sealing machines at each two sealing station are used to seal the bottles running on conveyor. Inspection is also done on before and after the sealing machines. There is a tag machine, which is used to tag the sticker on bottles after that there is an inspection station then packing station at last (Fig. 1).

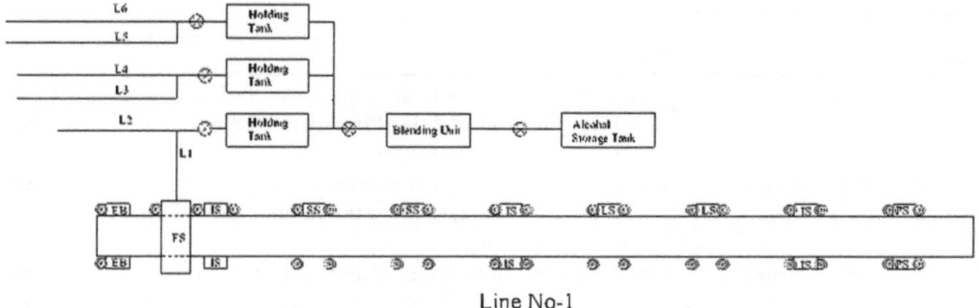

Fig. 1 Line diagram of bottling section

5 Objective of the Study

There are many practices in the world to improve the quality of products/services and performance of organizations, but generally, companies looking for the best practice to implement and utilize for achieving their organizational goals and objectives easier, sooner, and with less expenses. Available techniques are different from each other in terms of their specific characteristics, factors, and ways that consider for reaching their purpose. This study aims to investigate the impact of 5S practices to enhance productivity of the plant as well as employees and also improve the work culture or environment of the organization. Accordingly, the main objective of the research is to measure and compare the production and rejection of bottling section of a wine manufacturing plant before and after implementing 5S practice.

6 Methodology

The purpose of this study was to assist bottling section to improve the process flow with 5S methods in order to reduce possible wastes and enhance the efficiency of operations so that bottling section can reduce costs. 5S can help the company to improve the working environment; this study has provided the evidence that 5S is a great way to assist this organization. 5S provides a method to rearrange the layout and enhance the discipline. This study focused on how to use 5S in each step to help bottling section to rearrange the workplace and improve the efficiency. It was found that low capacity of sealing machine was due to conveyor running at low speed. We have suggested that another one sealing machine starts at each conveyor line so that conveyor can run at its full speed and production will increase. According to Jeffery liker, there are eight types of waste as mentioned in Table 2.

Some companies think that they are too busy to rearrange the workplace because it will take too much time to keep the workplace neat and clean. On the other hand, it means they do not want to keep the work environment clean and neat

Table 2 The eight wastes according to Jeffery liker

Waste	Waste description
1. Over production	Producing items for which there are no orders comes under overproduction and resulting in wasted inventory space and excess transportation time
2. Waiting time (time in hand)	Workers wait for the next processing step due to not having tools, lot processing, equipment downtime, lack of space, bottlenecks, etc
3. Unnecessary transport	Carrying work in progress at long distances, creating inefficient transport, or moving materials, parts, or finished goods into or out of storage or between processes
4. Over processing or incorrect processing	Taking unneeded steps to process the parts, inefficiently processing due to poor tools and product design, causing unnecessary motion and producing defects
5. Excess inventory	Excess raw material, WIP, or finished goods causing longer lead times, damaged goods, increase transportation and storage costs, and delay. Also, extra inventory hides problems such as production imbalances, late deliveries from suppliers
6. Unnecessary movement	Any wasted motion employees have to perform during the course of their work, such as looking for, reaching for, or stacking parts, tools, also, walking is waste
7. Defects or rejections	Production of defective parts or correction is defect. Repair or rework, scrap, replacement production, etc. are also defect
8. Unused employee creativity	Losing time, ideas, skills, improvements, and learning opportunities by not engaging or listening to your employees

(Hirano 1995). Running 5S can be divided into three sections which create a structured process for the project, make a clean environment, and create a clear method of management for the project (Rowlinson 2004). If companies do 5S in the right way, it will help the company to have a smooth operation; hence, all the employees will be happy to remain with the new process in order to have a better environment (Olofsson 2010). The following are some important benefits from implementing a 5S process:

- Orderliness (seiri and seiton)—by using the simple way to maximize the company's efficiency and reduce defects;
- Cleanliness (seiso and seiketsu)—once they have better environment, they can improve the healthier life, safety and transparency; and
- Discipline (shitsuke)—enhances the quality control of work life and work criteria due to training, and education improves the level of morale (Gapp et al. 2008).

We have also suggested implementing cleaning of machines and shop floor area before starting their lines so that rejection or rework can reduce and keep everything at its own place, and there is a place for every item (Fig. 2).

Sort is the first step in 5S to clean a place for better operation. Set is the way to make the items look neat and reduce the time for searching the tools. If the company does not set all items in order, it will be very difficult to look for them.

Fig. 2 Relation of 5S with waste elimination

The general idea is to make everything in the workplace clean, shiny, and neat under the shine. After first three steps, the managers will ask all workers to keep everything the same as it was before. Standardization is the way to help the company to set the normal, and it is possible to change these rules if it can be done better when they truly run first three Ss. Sustain is the final step for 5S implementation and asks the executives to keep everything going every day. The company should make a checklist to help the manager to make sure whether everyone follows the rule to sustain.

7 Result

We find a messy and dirty working place in bottling section. The purpose of this study was to enhance the company's efficiency and reduce different waste by using the 5S concept resulting in a better working place so that employees in the company can feel comfortable and be productive. The results of the data will be provided on how the company used 5S to enhance the production efficiency and reduce waste and keep the working place neat and clean. We have considered only one waste out of eight, i.e., defects or rejections. The defects in the wine factories are not discarded but reworked. In the final inspection, if any bottle or cork is found defective, then they are reworked if possible otherwise contents of bottle can be poured in a new bottle. Sometimes less defective is also sold at lower price. In general, the defects or rejections are likely to be considered as waste which should be disposed of. The goal should be zero defects to provide better product right the

Table 3 Before 5S implementation

Items/date		1/6	2/6	3/6	4/6	5/6
Production (nos.)		13,674	12,983	14,017	13,743	13,162
Rejection (nos.)	LC	1,025	1,047	986	1,121	942
	LP	276	212	221	162	198
	DB	191	154	202	161	172
	Other	198	160	76	132	125
	Total	1,690	1,573	1,485	1,576	1,437
Final goods (nos.)		11,984	11,410	12,532	12,167	11,725

LC loose cap; *LP* level pasting; *DB* damaged bottle

Table 4 After 5S implementation

Items/date		16/6	17/6	18/6	19/6	20/6
Production (nos.)		14,212	13,971	14,743	14,642	13,987
Rejection (nos.)	LC	925	997	854	911	862
	LP	157	182	132	111	126
	DB	158	151	168	107	151
	Other	114	148	121	88	112
	Total	1,354	1,478	1,275	1,217	1,251
Final goods (nos.)		12,858	12,493	13,468	13,425	12,736

LC loose cap; *LP* level pasting; *DB* damaged bottle

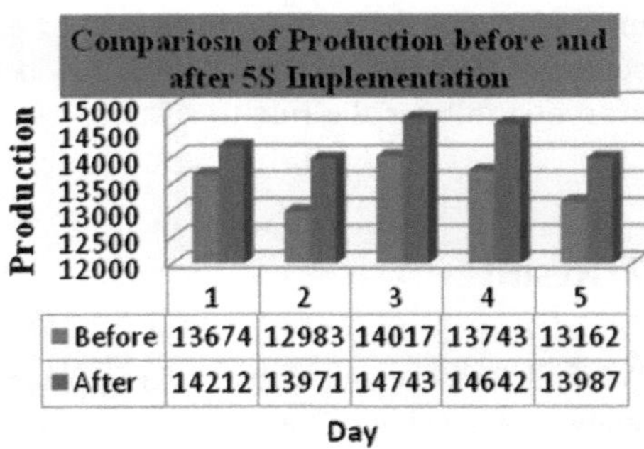

Fig. 3 Comparison of production

first time and every time. The data were collected by recording the production and various rejections as loose cap, level pasting defect, damaged bottle, etc. before and after implementing 5S. The comparison between before and after 5S will be fruitful and represented in Tables 3 and 4 as well as represented by bar charts as shown in Figs. 3, 4 and 5.

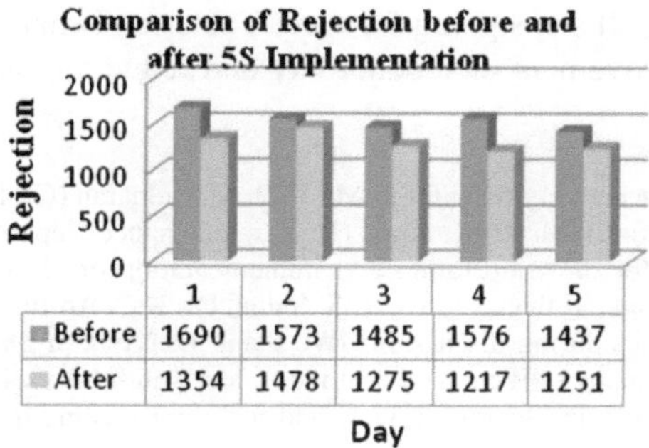

	1	2	3	4	5
Before	1690	1573	1485	1576	1437
After	1354	1478	1275	1217	1251

Fig. 4 Comparison of rejection

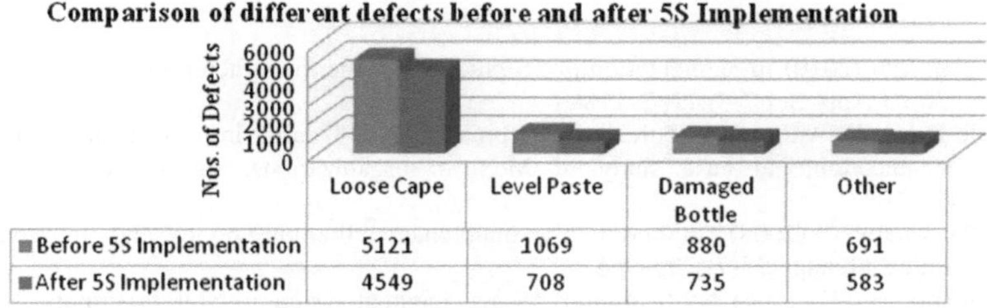

	Loose Cape	Level Paste	Damaged Bottle	Other
Before 5S Implementation	5121	1069	880	691
After 5S Implementation	4549	708	735	583

Fig. 5 Comparison of rejections before and after 5S implementation

8 Conclusion

During our case study, we have observed that most of the medium- and small-scale industries in India do not have well-defined policies and long-term vision. Once they achieve a mile stone, they prefer to act in monotonous manner. Moreover, researcher feels to successfully implement 5S concept in a bottling section of a wine manufacturing plant to improve the overall performance of that section. 5S is also the strongest foundation of TPM. Lack of sense of belongingness in employees and pure profit-making attitude of employer can never yield fruitful result in the long run. In India, employees in most of the cases treat their job as a meal to earn livelihood only. Poor productivity and high rejections in the view of researcher is a mere result of negligence, lack of motivation, mismanagement, and some personal conflicts. Therefore, this is an urgent need to improve work environment of employees by implementing 5S concept. However, data collected before and after 5S implementation; it showed that 5S helped the company to improve. Bottling section thought 5S could only improve the result in the working environment, but it also proved to help the section to gain better productivity and reduce rejections. Sort helped bottling section to decide between used and unused items; in addition, the section got more space from that. Shine made the working environment look better than the previous situation. Standardized and sustain are working for keeping all steps going forward. Tables and charts are clearly showed the remarkable improvement of productivity and reduction on different rejections of bottling section.

Acknowledgement We are very thankful to Mr. P. B. Murlidharan (G.M.), Mr. Saurabh Sharma (In-charge Bottling Section), Mr. S. K. Khan (Head maintenance Department), Mr. J. P .Tiwari (Production Engineer) for the cooperation and implementation of 5S and the changes we had suggested. We also extend our thanks to Dr. T. S. Sidhu, Dr. K.C. Arora, and Dr. A.K. Saxena for their valuable suggestions during discussions. We thank Dr. Avdesh Sharma (MITS, Gwalior), Dr. Dheeraj Gupta (Graphics University, Deharadun), Mr. Rahul Sikarwar (Research scholar IIT Chennai) for the assistance in literature review and for proof reading the manuscript.

References

Ab Rahman MN (2010) Implementation of 5S practices in the manufacturing companies: a case study. Am J Appl Sci 7(8):1182–1189

Ablanedo-Rosas H, Alidaee B, Moreno JC, Urbina J (2010) Quality improvement supported by the 5S, an empirical case study of Mexican organizations. Int J Prod Res 48(23/24):7063–7087

Ahuja IPS, Khamba J (2008) Total productive maintenance: literature review and directions. Int J Qual Reliab Manage 25(7):709–756

Ananthanarayanan K (2006) Application of 5S management system in NDE laboratory. National seminar on non-destructive evaluation

Bayo-Moriones A, Bello-Pintado A, de Cerio JMD (2010) 5S use in manufacturing plants: contextual factors and impact on operating performance. Int J Qual Reliab Manage 27(2):217–230

Bicheno J, Holweg M (2009) The lean toolbox, 4th edn. Picsie Books, Buckingham

Da Silveira GJC (2006) Effects of simplicity and discipline on operational flexibility: an empirical reexamination of the rigid flexibility model. J Oper Manage 24(6):932–947

Eckhardt B (2001) The 5S housekeeping program aids production. Concrete products, vol 104(11), p 56

Gapp R, Fisher R, Kobayashi K (2008) Implementing 5S within a Japanese context: an integrated management system. Manage Decis 46(4):565–579

Hamzah A, Ho S (1994) TQM training for small and medium industries in Malaysia. Training Qual 2(2):27–35

Hirano H (1995) 5 pillars of the visual workplace: the sourcebook for 5S implementation

Ho SKM (1997) Workplace learning: the 5-S way. J Workplace Learn 9(6):185–191

Ho S (1999a) 5-S practice: the first step towards total quality management. Total Qual Manage 10(3):345–356

Ho SKM (1999b) Japanese 5-S—where TQM begins. TQM Mag 11(5):311–321

Ho S, Cicmil S (1996) Japanese 5-S practice. TQM Mag 8(1):45–53

Hubbard R (1999) Case study on the 5S program: the five pillars of the visual workplace. Hosp Materiel Manage Q 20(4):24

Khamis N (2009) Development of 5S practice checklist for manufacturing industry

Liker J (2004) The Toyota way. Simon & Schuster Inc, New York

Liker J, Hoseus M (2008) The Toyota culture: the heart and soul of the Toyota way. McGraw-Hill, New York

Monden Y (1998) Toyota production system: an integrated approached to just in time. Engineering and Management Press, Norcross

Moradi M, Abdollahzadeh M, Vakili A (2011) Effects of implementing 5S on total productive maintenance: a case study in Iran. IEEE

OhEocha M (2000) A study of the influence of company culture, communications and employee attitudes on the use of 5Ss for environmental management at Cooke Brothers Ltd. TQM Mag 12(5):321–330

Olofsson O (2010) 5S implementation. Retrieved from http://world-classmanufacturing.com/5S/5S_3.html

Rowlinson Stephen M (2004) Construction safety management systems. Taylor & Francis Inc, USA and Canada

Sawada N (1995) The Kaizen in Toyota production system. CHU-SAN-REN Quality Control Course, Nagoya, pp 1–38

Schonberger RJ (2007) Japanese production management: an evolution—with mixed success. J Oper Manage 25(2):403–419

Sousa R, Voss CA (2008) Contingency research in operations management practices. J Oper Manage 26(6):697–713

Sui-PPheng L, Khoo SD (2001) Team performance management: enhancement through Japanese 5-S principles. TQM Mag 7(7/8):105–111

Tice J, Ahouse L, Larson T (2005) Lean production and EMSs: aligning environmental management with business priorities. Environ Qual Manage 15(2):1–12

Van Patten J (2006) A second look at 5S. Qual Prog 39(10):55

Yusof SM, Aspinwall E (2001) Case studies on the implementation of TQM in the UK automotive SMEs. Int J Qual Reliab Manage 18(7):722–744

A Multiple Linear Regression Approach in Modeling Traffic Noise

R. Panchal, M. Dahiya, P. K. Saini and N. Garg

Abstract An analytical model is developed to predict road traffic noise for busy roads of Delhi, India. Equivalent continuous sound pressure level, L_{AeqT}, is analyzed at eight different busy road locations of Delhi. A multiple linear regression analysis is conducted to predict the single noise metrics L_{Aeq} in terms of traffic flow rate (Q), percentage of heavy vehicles (H), and average traffic speeds (V). The model so developed is validated with actual experimental data. The coefficient of regression for test data set is observed to be 0.74 between predicted and experimental L_{Aeq} values. The work thus shows that a validated analytical model can be useful for predicting noise levels and conducting the noise impact assessment studies in Delhi. The accuracy of model so developed can be further enhanced by feeding more real-time data from different locations with varied traffic density.

Keywords Equivalent continuous sound pressure level · L_{eq} · Multiple linear regression · Traffic noise

1 Introduction

In Indian scenario, the vehicular population is increasing at an enormous rate as compared to road infrastructure. In National Capital Region, the vehicular population is increasing at a rate of 135.6 %, while the road network has increased by 16.5 %. The number of vehicles both cars and two wheelers are rising at a rate of

R. Panchal (✉) · M. Dahiya · P. K. Saini
Department of Mechanical Engineering, National Institute of Technology, Kurukshetra 136119, Haryana, India
e-mail: rajan8abc@gmail.com

N. Garg
CSIR-National Physical Laboratory, New Delhi 110012, India

S. S. Khangura et al. (eds.), *Proceedings of the International Conference on Research and Innovations in Mechanical Engineering*, Lecture Notes in Mechanical Engineering, DOI: 10.1007/978-81-322-1859-3_50,

8 % per annum. For a city of 17 million, Delhi has close to 8 million registered vehicles more than Mumbai, Kolkata put together (Hindustan Times 2013). Although, the introduction of mass rapid transit system (MRTS) and bus transit system (BTS) has proven to be a convenient means of transport, yet the noise and vibration induced due to these sources has to be controlled for their betterment and mass use by community. Thus, it is imperative to conduct studies pertaining to the increased noise levels due to heavy vehicular density for planning suitable effective measures to control it. The present work is an attempt to focus on this critical issue and tries to use multiple linear regression approach in modeling the sound pressure level generated by vehicular traffic correlating the parameters, viz. number of vehicles, percentage heavies, average speed of vehicles, etc. The accuracy of model so developed is ascertained by the coefficient of determination (R^2). The accuracy of the model can be further enhanced by feeding more real-time data.

2 Studies Done in India

The objective is to develop a regression-based model for traffic noise prediction and forecasting with definite accuracy so that it could be helpful in carrying out noise impact assessments in metropolitan cities like Delhi. It can also be helpful for planning and control especially for urban development bodies. Various models have been developed for different Indian cities in past few years. Rao et al. developed a regression equation for modeling L_{A10} as a function of traffic density (Rao and Rao 1991). In urban areas, most of the traffic flow is often interrupted by traffic signals, and thus, interrupted traffic flow conditions on urban roads create substantially different noise characteristics from highways to expressways (Nirjar et al. 2009; Rajakumar and Gowda 2009). Rajakumar et al. (2009) developed a regression noise prediction model for both acceleration and deceleration lanes. Agarwal and Swami (2011) introduced equivalent number of light and heavy vehicles for the calculation of L_{eq} values. Light motor vehicles have been analyzed to be the major culprit in noise pollution. The recent investigations of Kalaiselvi and Ramachandraiah (2012) introduce horn noise component into account. It has been observed that horn noise events with frequency of 16/min raises L_{eq} by 12 dB (A). There have been various studies reported in India so far, and few have been focused for Delhi city. The heterogeneous traffic conditions and flow characteristics have been also studied by various researchers in the past (Kumar and Jain 1999; Mishra et al. 2010). It has been previously discussed that conventional noise prediction models were developed under uninterrupted traffic flow conditions, do not compromise Indian traffic characteristics, road geometries, and environmental conditions (Rajakumar and Gowda 2009). There have been some studies that have reported a new factor, i.e., tendency to blow horn in conventional federal highway administration (FHWA) model, which could predict the equivalent noise level, L_{eq}, to an accuracy of ±3 dB(A) (Agarwal et al. 2009). In developed nations, there has been an extensive research in this area and every nation has developed its own scientific and validated model such as CORTN for the UK, RLS 90 for Germany, and ASJ-RTN 2008 for Japan (Givargis and Mahmoodi 2008; Campbell 2001; Yamamoto 2010). The present study reports observations from a limited study carried out at eight locations in Delhi city. The selection of sites is done on the basis of type of traffic, varied traffic density, and type of area. Short-term measurements were conducted using calibrated precision digital sound level meter (NOR118) and calibrated industrial sound level meter (Lutron SL-4030).

3 Experimental Data Acquisition

The precision digital sound level meter used was kept at a distance of 0.4 m from the body and at a height of 1.4 m from the ground level and 3 m from the road for avoiding any reflections from roadside barriers. The measurement of sound pressure in dB (A) was done along with the monitoring of the average speed of vehicles with the speed gun (Bushnell) and numbers of vehicles were counted manually. Precision digital sound level meter helped in measuring L_{eq} and statistical parameters, e.g., L_{10}, L_{50}, and L_{90}. Short-term, L_{AeqT}, measurements ranging from 15 to 30 min were undertaken at 8 different busy road locations of Delhi. While measurements, it was ensured that there are no reflections from the adjoining building facades or wall. It may be noted here that the recent studies conducted by Maruyama et al. (2013) show that a minimum 170 number of vehicles are enough for obtaining a reliable L_{AeqT} during measurement time interval T. The equivalent hourly A-weighted continuous sound pressure level can be calculated as

$$L_{eq} = 10\text{Log}_{10}\frac{1}{T}\int_{0}^{T}\left(\frac{p_A(t)}{p_0(t)}\right)^2 \mathrm{d}t \tag{1}$$

where $p_A(t)$ is instantaneous A-weighted sound pressure level (in Pa). A basic empirical relation is objective function based upon the three parameters:

$$L_{eq} = A + B \times \text{Log}_{10}Q + C \times \text{Log}_{10}V + D \times H \tag{2}$$

where L_{eq} is the predicted equivalent continuous sound pressure level, and Q, V, and H are total traffic volume, average traffic speed, and % of heavy vehicles, respectively. A, B, C, and D are the constants whose value is to be computed.

4 Multiple Linear Regression

Multiple linear regression is an extension of simple linear regression. It is used to predict the value of a variable based on the value of two or more other variables. The variable desired to be predicted is called the dependent variable. Additionally, multiple linear regression approach also facilitates the determination of the overall fit of the model and the relative contribution of each of the predictors to the total variance explained. Standardized regression coefficients (constants) are a measure of how strongly each predictor variable influences the criterion variable. It can be generalized as

$$y = \beta_0 + \sum_{i=1}^{k} \beta_i x_i + \varepsilon \tag{3}$$

where β is coefficient of each term, k is number of independent variables, and ε is error. The coefficient of determination is calculated as

$$R^2 = 1 - \frac{\mathrm{SS_e}}{\mathrm{SS_T}} \tag{4}$$

where $\mathrm{SS_e}$ is sum-squared error, and $\mathrm{SS_T}$ is sum-squared total error (Kuram and Ozcelik 2013). The analysis of experimental data shows that sound pressure level $L_{\mathrm{Aeq}T}$ varied from 72 to 81 dB (A) at various locations. Figure 1 shows the typical spectrum of road traffic noise in frequency domain measured at Pusa Road. The analysis of spectrum shows that in range 2.5–4 kHz, the peak in noise levels may be attributed to horn noise. Predicted values were computed by the empirical formula using software tool (SPSS downloaded version 20).

The equation suggests the dependence of predicted L_{eq} upon three basic parameters. The parameter estimates observed in the regression analysis in SPSS downloaded version are shown in Table 1. These analytical investigations thus suggest that vehicular density plays a pivotal role in controlling the traffic noise. If traffic flow is considered as line source, it could be observed that with doubling of the distance, there shall be at least a decrement 3 dB in free-field conditions. The difference between the maximum $L_{\mathrm{Aeq}T}$ and long-term $L_{\mathrm{Aeq}T}$ increases as percentage of heavy vehicles increases (Maruyama et al. 2013). It is imperative to correlate the geometrical divergence of sound waves or distance effect along with the average vehicular speed, traffic density, and percentage heavy vehicles plying on the roads. Some studies (Naka et al. 1999) show that the change in the vehicle speed influences $L_{\mathrm{Aeq}T}$ and greatly influences the percentage accedence level L_{AFNT} through change in the traffic volume. It may be noted that the short-term levels can be very helpful in analyzing the long-term noise indices such as day–night average sound level L_{dn} and equivalent continuous sound pressure level $L_{\mathrm{Aeq,24h}}$. There have been some studies (Tang and Chan 2003) wherein energy-based noise indices are observed to be more sensitive to change in population density,

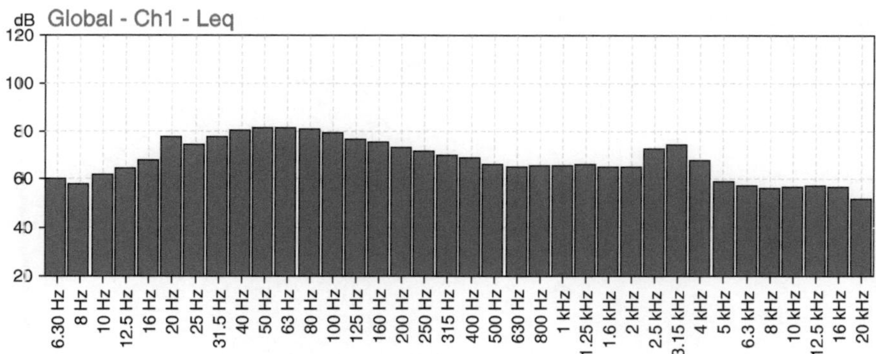

Fig. 1 Typical traffic noise spectrum at Pusa Road, New Delhi, during peak hours of the day

Table 1 Parameter estimates in SPSS software for L_{AeqT}

Parameters	Estimate	Std. error	95 % Confidence interval	
			Lower bound	Upper bound
A	60.332	3.274	53.839	66.826
B	6.898	0.851	5.211	8.585
C	−5.328	1.490	−8.283	−2.374
D	0.079	0.080	−0.080	0.239

and consequently, such investigations can be pursued for National Capital Region to develop a regression-based model.

It may be noted there are various sources of uncertainty that cause an error in prediction of accurate average sound pressure levels, L_{AeqT}. Horn noise, error is average density and speed measurements, meteorological conditions, diffraction effect of sound waves, ground effects, etc., are such parameters to be considered for improving the prediction accuracy.

The correlation between predicted L_{eq} and measured L_{eq} for test data set is shown in Fig. 2. The coefficient of determination (R^2) comes out be 0.74. A similar set of data was also collected by Manish et al. in 2012 at Delhi. The regression value of 0.82 and mean-squared error of 0.00019 have been reported using artificial neural networks (ANN) technique (Sharma et al. 2012). The total traffic flow and number of heavy vehicles have been identified as most significant factors of urban traffic noise in a study conducted in Hong Kong using multiple regression model (To et al. 2002). The paired *t* test for a test data set of few observations only yields a *t* statistic value of −0.036 indicating statistically significance at 5 % level (Table 2).

The overall significance of the multiple regression is tested by the hypothesis that all the slope coefficients are simultaneously zero as reported by Kumar et al. (1998). The computed F-ratio = 150 as shown in Table 3 is very large as compared to critical value of F for 51 degrees of freedom at 5 % level of significance.

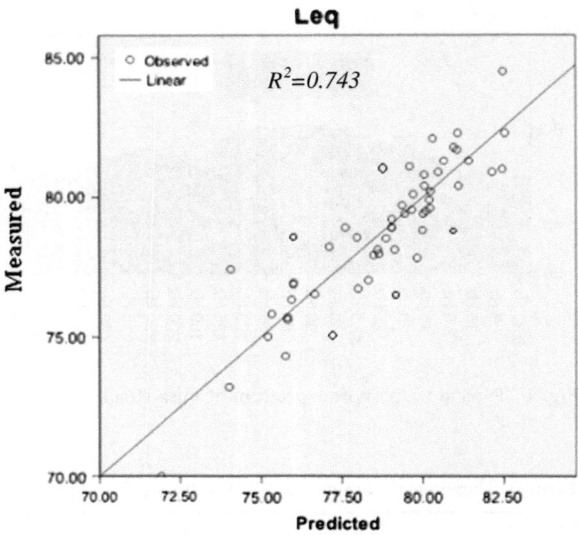

Fig. 2 Measured L_{eq} vs predicted L_{eq}

Table 2 Paired t test for measured and predicted $L_{Aeq,\ T}$

	Measured	Predicted
Mean	78.7	81.3
Variance	6.78	5.71
Observations	51	51
Pearson correlation	0.80	
Hypothesized mean difference	0	
df	50	
t stat	−0.036	
$P\ (T \leq t)$ one-tail	0.49	
t critical one-tail	1.68	
$P\ (T \leq t)$ two-tail	0.97	
t critical two-tail	2.01	

This indicates that null hypothesis is rejected. Although the value of R^2 and the *F*-ratio are fairly high, the coefficients of *A* and *C* have high standard errors.

It may be noted that the type of zone, geographic features, landscape, topography are features on which noise emission and transmission depend (Banerjee et al. 2008). Noise maps prove to be beneficial for impact analysis and formulation of noise risk zones. It is imperative to develop such noise maps for Delhi city for Delhi city for devising suitable measures for abatement. The regression-based models can be thus integrated with these noise maps for impact analysis (Akhtar et al. 2012). Also, the measurement accuracy associated with average speed and density measurements can be further enhanced by using automated systems, i.e., video processing of traffic for tracking the density and speed measurements (Sen et al. 2013).

Table 3 Analysis of variance for the regression model ($R^2 = 0.753$, standard error = 2.31, R^2 adjusted = 0.74)

Source	dof	Sum of squares	Mean sum of squares	*F*-ratio	*P* value
Model	1	255.59	255.58	150.03	0.000
Error	49	83.47	1.70		
Total	50	339.06			

5 Conclusions

A traffic noise survey is conducted on various sites in Delhi, and a multiple linear regression approach is used to fit the data for development of model. The experimental data were collected from eight locations at various times during the day. The traffic density, average speed of vehicles, and percentage heavies were simultaneously monitored along with the traffic noise. A multiple linear regression approach was utilized to develop a model using software tool SPSS (Downloaded free version 20). A correlation analysis ascertained between various variables, i.e., traffic volume, traffic noise, percentage heavies, and equivalent sound pressure level (L_{eq}). The coefficient R^2 is observed to be 0.74. The present investigations are conducted for short-term L_{eq} measurements of around 15–30 min, which is required to develop reliable estimate of L_{AeqT} as enunciated by Maruyama et al. However, the regression model can be further upgraded with more experimental database for traffic noise predictions. The present models also follow a generalized approach of modeling traffic noise without considering the geometrical propagation of sound waves and receiver locations, which may induce uncertainty in predictions. Also, it requires a further in-depth investigation of the contribution of horn noise component, reflections through building facades, characteristics of traffic flow, etc., which have to be included for enhancing the accuracy and precision associated with developing a model. It is imperative to further extend these studies for developing a validated model, which can be very helpful in predicting the noise levels rampant at various locations so as to devise suitable measures to control the traffic noise in India.

References

Agarwal S, Swami BL (2011) Comprehensive approach for the development of traffic noise prediction model for Jaipur city. Environ Monit Assess 172(1–4):113–120

Agarwal S, Swami BL, Gupta AB (2009) Development of a noise prediction model under interrupted traffic flow conditions: a case study for Jaipur city. Noise Health 11(45):189–193

Akhtar N, Ahmad K, Gangopadhyay S (2012) Road traffic noise mapping and a case study for Delhi region. Int J Appl Eng Technol 2(4):39–45

Banerjee D, Chakraborty SK, Bhattacharyya S, Gangopadhyay A (2008) Evaluation and analysis of road traffic noise in Asansol: an industrial town of Eastern India. Int J Environ Res Public Health 5(3):165–171

Campbell S (2001) A critical review of some traffic noise prediction models. Appl Acoust 62(3):271–287

Givargis Sh, Mahmoodi M (2008) Converting the UK calculation of road traffic Noise (CORTN) to a model capable of calculating L_{Aeq}, 1 h for the Tehran's roads. Appl Acoust 69:1108–1113

Hindustan Times (2013) New Delhi, where all roads lead to chaos, 7 May 2013

Kalaiselvi R, Ramachandraiah A (2012) A model for traffic noise prediction in heterogeneous traffic conditions. Int J Curr Res 4:180–184

Kumar K, Jain VK (1999) Autoregressive integrated moving averages (ARIMA) modelling of a traffic noise time series. Appl Acoust 58(3):283–294

Kumar K, Jain VK, Rao DN (1998) A predictive model of noise for Delhi. J Acoust Soc Am 103(3):1677–1679

Kuram E, Ozcelik B (2013) Multi-objective optimization using Taguchi based grey relational analysis for micro-milling of Al 7075 material with ball nose end mill. Measurement 46(6):849–864

Maruyama M, Kuno K, Sone T (2013) The minimum measurement time for estimating L_{AeqT} of road traffic noise from the number of vehicle pass-bys. Appl Acoust 74:317–324

Mishra RK, Parida M, Rangnekar S (2010) Evaluation and analysis of traffic noise along bus rapid transit system corridor. Int J Environ Sci Technol 7(4):737–750

Naka I, Naro Y, Kuno K (1999) Relation between road traffic condition and L_{Aeq}—a study of L_{eq} based on Q–V curve. J Acoust Soc Jpn 55:467–473

Nirjar RS, Jain SS, Parida M, Katiyar VS, Metal N (2009) A study of transport related noise pollution in Delhi. J Inst Eng (India) Environ 84(1):6–15

Rajakumar HN, Gowda RMM (2009) Road traffic noise pollution model under interrupted traffic flow condition. Environ Monit Assess 14(2):251–257

Rao P, Rao S (1991) Prediction of LAT traffic noise levels in the city of Visakhapatnam, India. Appl Acoust 101–110

Sen R, Cross A, Vashistha A, Padmanabhan VR, Cutrell E, Theis W (2013) Accurate speed and density measurements for road traffic noise in India. http://research.microsoft.com/pubs/189809/Sen-dev2013-traffic.pdf

Sharma M, Dahiya M, Saini PK, Garg N (2012) Application of artificial neural network for modelling of traffic noise on roads in Delhi. IOSR J Mech Civ Eng 31–36

Tang SK, Chan WY (2003) Predictability of noise-indices in a high rise residential environment. J Acoust Soc Am 114(3):1222–1225

To WM, Ip RCW, Lam GCK, Yau CTH (2002) A multiple regression model for urban traffic noise in Hong Kong. Acoust Soc Am 112:551–556

Yamamoto K (2010) Road traffic noise prediction model, "ASJ RTN- model 2008": report of the research committee in road traffic noise. Acoust Sci Technol 31:2–55

Factor Analysis of Sourcing Flexibility Among Supply Chain Partners in Indian Industries

Gaurav Tejpal, R. K. Garg and Anish Sachdeva

Abstract Primary purpose of this paper is to apply factor analysis on the data collected for sourcing flexibility; second to reduce data collected to a few factors; and third and finally to propose name of factors. A total of 117 responses of questionnaires were collected and analyzed, and a number of key findings emerged. The field is relatively "new" one for supply chain management as consensus is lacking on the definition of the term. Factors loading on sourcing flexibility emphasize to focus on two factors, and the proposed name for the key factors of sourcing flexibility are dynamic capability of the partners and range capability of partners.

Keywords Factor analysis · Literature · Supply chain management · Sourcing flexibility

1 Introduction

Nowadays, many companies are facing a highly volatile and uncertain environment: short product life cycles, frequent and unpredictable changes in demand (in volume and mix), and global logistics issues. This has led companies to put more pressure on their ability to change or react to environmental uncertainty with little

G. Tejpal (✉)
Amritsar College of Engineering and Technology, 12 km stone on Amritsar–Jallandhar Road, NH1, Amritsar, Punjab, India
e-mail: gaurav_tejpal@acetedu.in

R. K. Garg · A. Sachdeva
Dr. B. R. Ambedkar National Institute of Technology, Jallandhar, Punjab, India
e-mail: gargrk@nitj.ac.in

A. Sachdeva
e-mail: asachdeva@nitj.ac.in

S. S. Khangura et al. (eds.), *Proceedings of the International Conference on Research and Innovations in Mechanical Engineering*, Lecture Notes in Mechanical Engineering,
DOI: 10.1007/978-81-322-1859-3_51, © Springer India 2014

penalty in their performance. In other words, companies have become more aware of the need to be flexible. This is evident by marked increases in practitioner and academic publications, conferences, professional development programs, and university courses in the area.

This paper contributes to this debate by focusing on sourcing flexibility, which we define as "the ability of the purchasing function to respond in timely and cost effective manner to changing requirements of purchased components, in terms of volume, mix and delivery date." Although some studies have recognized the influence of sourcing practices on manufacturing flexibility (e.g., Narasimhan and Das 2000; Jack and Raturi 2002), sourcing flexibility has rarely been studied as a separate construct

Therefore, the four objectives of this paper are the following: one, to examine and develop a context to comprehend the concept called sourcing flexibility; second, to apply factor analysis on the data collected for sourcing flexibility; third, to reduce data collected to a few factors; and finally, to propose name of factors. The next section provides literature review and attempts to explain concept of flexibility and sourcing flexibility. Section three discusses methodology. Section four discusses the results. Finally, the paper concludes by proposing the names for factors extracted on sourcing flexibility.

2 Literature Review

2.1 Concept of Flexibility/Sourcing Flexibility

Based on strategic objections, the multi-dimensional nature of manufacturing flexibility indicates that supply chain organizations may require different types and levels of flexibility. Supply chain organizations must look beyond manufacturing flexibility, to achieve the level of flexibility that adds value to the customers. From the perspective of filling customer orders, no single part of the value chain working above can significantly reduce customer lead time, (Zhang et al. 2002; Bhatt and Emdad 2001). According to Zhang et al. (2002), flexibility is "the organization's ability to meet an increasing variety of customer expectations without excessive costs, time, organizational disruption, or performance losses."

In previous studies, the terms supply and sourcing flexibility have been used indistinctively to designate similar constructs. For simplicity reasons, we will use the term sourcing flexibility. Swafford et al. (2000) defined sourcing flexibility as "the ability of the purchasing function to respond in a timely and cost effective manner to changing requirements of the purchased components."

3 Methodology

3.1 Research Model

This research draws upon the work of "Pujawan (2004)" that evaluated the measurement of sourcing flexibility because this study focuses on the buyer–supplier relationship; adaptations have been made by focusing on inter-organizational relationships.

3.2 Research Methods

Questionnaires were used to collect the required data from manufacturing companies located in the northern region of India over a period of three months. A total of 685 questionnaires were distributed through e-mail and postal mail and were directed to the purchasing directors, managers, executives, and buyers in manufacturing companies as they had direct contact with suppliers. The lists of possible companies were generated from CII directory, 2012. A total of 127 questionnaires were collected from respondents, a response rate of approximately 18.5 %. However, of the 127 questionnaires, only 117 questionnaires could be used. Four respondents answered the questionnaires incompletely and another six questionnaires were rejected due to improper answers.

3.3 Measurement and Validation

The questionnaire in this study was derived from the literature, either through adoption or slight modification to make it relevant to this study. A pilot test was performed by distributing the questionnaires to purchasing managers from four organizations and two from academic organizations to solicit feedback on questionnaire design. The pilot test was conducted to achieve an understanding among purchasers in order to improve the overall quality of the questionnaire. Based on their feedback, several minor changes were made to tailor the questionnaires to the target audience. The questionnaires were deemed ready for distribution after these modifications.

3.4 Research Instruments

A total of seven questions were constructed for sourcing flexibility. Questions for sourcing flexibility were adapted from "Pujawan (2004)". A 5-point rating scale, ranging from "1 = strongly disagree" to "5 = strongly agree", was adopted to measure the sourcing flexibility by both organizations and their suppliers to facilitate the supply chain.

4 Results

4.1 Respondent Profile

Summary of the respondents is as follows: 46.0 % of respondents are senior executives, while 26.6 % of respondents hold middle-level managerial positions; another 25.64 % of respondents hold higher-level managerial posts.

4.2 Respondent Organization Profile

Majority of the respondents come from original equipment manufacturers (OEM) with 77.56 and 22.44 % from the supporting industry. 43.10 % of the organizations have been operating for more than 20 years; furthermore, 62.9 % of companies employ more than 1,000 personnel.

5 Factor Analysis

Factor analysis was conducted to group the items related to one another under sourcing flexibility. Total seven questions were framed out of the literature and put to FA. Following questions were considered for factor analysis. Q1: The cost of switching from one supplier to another is low. Q2: Most suppliers have the capability of producing/supplying various different types of items. Q3: There is a large extra total supply capacity for most items. Q4: Most suppliers are capable of producing a large quantity of items in a relatively short time. Q5: Most suppliers are capable of supplying even the small quantities whenever required, keeping the same price. Q6: Our suppliers have multiple modes of transportation available for delivery of goods. Q7: Our suppliers are able to deliver mix supply of items in a single delivery load.

The Pearson correlation coefficient has been checked between all pairs of questions. Since most of the values are greater than 0.3, no problem due to multicolinearity may occur. Determinant value has also been found greater than the necessary value of 0.00001.

Table 1 shows several important parts of the output: the Kaiser–Mayer Olkin measure of sampling adequacy which is $0.662 > 0.5$. This is specified that factor loading may be applied on the present data. Also Bartlett's test of sphericity shows large value of 252.56 at 21 degree of freedom.

The communality before and after extraction has been checked. Communality is the proportion of common variance within variables. Principal component analysis works on the initial assumption that all variance is common; therefore, before extraction, the communalities are all 1. In effect, all of the variance associated with

Table 1 Kmo and Bartlett's test for sourcing flexibility

Kaiser–Meyer–Olkin measure of sampling adequacy	0.662
Bartlett's test of approx. chi-sphericity square	252.56
Df	5
Significant	21
	0.000

a variable is assumed to be common variance. Once factors have been extracted, we have a better idea of how much variance is, in reality, common. The communalities in the column-labeled extraction reflect this common variance. However, after extraction, the amount of variance in each variable which can be explained by the retained factors is represented by the communalities.

The component matrix before rotation has also been checked. The matrix contained the loadings of each variable onto each other. The matrix is given three components on which loading can be distributed.

5.1 Factor Extraction

Table 2 lists the eigenvalues associated with each linear component (factor) before extraction, after extraction, and after rotation. The eigenvalues associated with each other factor represent the variance explained by that particular linear component. Also it has been displayed through the eigenvalues in terms of the percentage of variance explained. This extracts all factors with eigenvalues greater than 1 and that which leaves us with three factors. In the final part of the table (labeled rotation sums of squared loadings), the eigenvalues of factors after rotation are displayed. Rotation has the effect of optimizing the factor structure, and one consequence for those data is that the relative importance of the four factors is equalized.

Before rotation, factor 1 is accounted for considerably more variance than the remaining two (38.811 % compared to 21.343.). But after extraction, it accounts for only 34035 % of variance.

Figure 1 shows scree plot. Curve is difficult to interpret because it steeply drops to tail off after two factors. Therefore, we could probably justify retaining either two or three factors.

Table 3 shows the rotated component matrix (also called the "rotated factor matrix" in factor analysis); it is a matrix of the factor loading for each variable onto each factor. There are two things to consider about the format of this matrix. First, factor loadings less than 0.4 have not been displayed. Second, the variables are listed in the order of size of their factor loadings.

The next step is to look at the content of questions that load onto the same factor to try to identify common themes. We see from the table number 3, first factor contains the questions 3, 4, 5, and 6. On seeing it more closely, we can

Table 2 Total variance explained for sourcing flexibility

Component	Initial eigenvalues			Extraction sums of squared loadings			Rotation sums of squared loadings		
	Total	% of variance	Cumulative (%)	Total	% of variance	Cumulative (%)	Total	% of variance	Cumulative (%)
1	2.717	38.811	38.811	2.717	38.811	38.811	2.382	34.035	34.035
2	1.494	21.343	60.154	1.494	21.343	60.154	1.828	26.118	60.154
3	0.959	13.694	73.848						
4	0.725	10.355	84.204						
5	0.524	7.479	91.683						
6	0.402	5.741	97.424						
7	0.180	2.576	100.000						

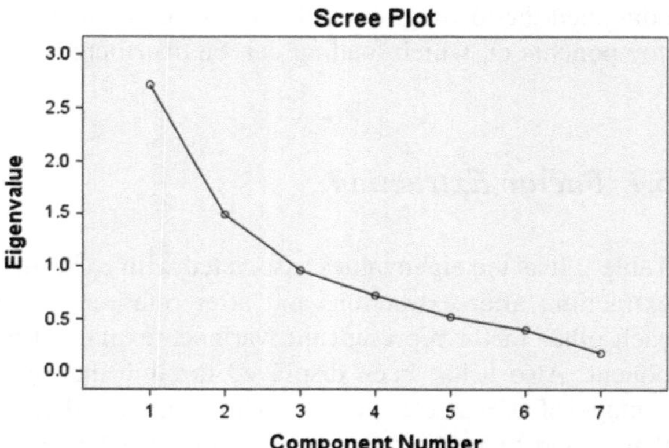

Fig. 1 Scree plot for sourcing flexibility

Table 3 Rotated component matrix[a] for sourcing flexibility component

	1	2
SF5	0.826	
SF4	0.780	
SF3	0.621	
SF6	0.620	
SF1		0.811
SF7		0.809
SF2	0.478	0.582

Extraction method: principal component analysis
Rotation method: varimax with Kaiser normalization

[a] Rotation converged in five iterations

observe that the orientations of the questions are on the dynamic capability of the partners; so subsequently, we may refer it as the same. On same lines, questions 1, 2, and 7 are grouped under range capability of partners or simply as range.

6 Conclusions

Factor analysis was performed on 07 items for sourcing flexibility in the partner relationship scale. The results are shown in various tables mentioned above. The KMO and Bartlett's test of sphericity was significant at the 0.01 level. The anti-image correlation matrix ranged >0.50; so there were sufficient correlations among the items. Two factors were extracted, and these are named as dynamic capability of the partners and range capability of partners.

References

Bhatt GD, Emdad AF (2001) An analysis of virtual value chain in electronic commerce. Logistics Inf Manag 14(1/2):78–91

Jack EP, Raturi A (2002) Sources of volume flexibility and their impact on performance. J Oper Manag 20(5):519–548

Narasimhan R, Das J (2000) An empirical examination of sourcing's role in developing manufacturing flexibility. Int J Prod Res 38(4):875–893

Pujawan N (2004) Assessing supply chain flexibility: a conceptual framework and a case study. Int J Integr Supply Manag 1(1):79–97

Swafford P, Ghosh S, Murthy N (2000) A model of global supply chain agility and its impact on competitive performance. In: Proceedings of the 31st national DSI meeting, Orlando, Florida, pp 1037–1039

Zhang Q, Vonderembse MA, Lim J (2002) Value chain flexibility: a dichotomy of capability. Int J Prod 40(3):561–583

Performance Enhancement of Manufacturing Unit Using Six Sigma DMAIC Approach: A Case Study

Jagdeep Singh and Harwinder Singh

Abstract Six Sigma is the management approach of achieving major enhancement in the process by implementing DMAIC cycle through elimination of waste. Manufacturing units are implementing such strategies to increase productivity of their manufacturing system processes. The purpose of this paper is to reduce quality rejection by implementing DMAIC approach in a systematic manner on the shop floor of the manufacturing unit of northern India. Results indicated the decrease in quality rejection level from 11,179.87 to 28.69 ppm, which results in net savings of 17. 66 lakhs per year.

Keywords Six Sigma · DMAIC · Manufacturing system

1 Introduction

In the mid- to late-1980s, Six Sigma techniques were formally introduced by Motorola as a means to reduce the variability observed in the development of their manufactured products (Montgomery 2009). Within a span of few years, it proved to be a successful business improvement strategy for corporate growth and breakthrough quality improvement. Now, it has been felt that perhaps Six Sigma is the most successful business improvement strategy developed during the last 50 years. It is the process management tool that has yielded the greatest results (Dusharme 2006). Six Sigma is a business performance improvement strategy that

J. Singh (✉)
Punjab Technical University, Kapurthala, Punjab, India
e-mail: jagdhoor605@yahoo.com

H. Singh
Mechanical Engineering Department, Guru Nanak Dev Engineering College, Ludhiana 141006, Punjab, India
e-mail: harwin75@rediffmail.com

S. S. Khangura et al. (eds.), *Proceedings of the International Conference on Research and Innovations in Mechanical Engineering*, Lecture Notes in Mechanical Engineering, DOI: 10.1007/978-81-322-1859-3_52, © Springer India 2014

aims to reduce the number of mistakes/defects to as low as 3.4 occasions per million opportunities. Sigma is a measure of "variation about the average" in a process, which could be in manufacturing or service industry (Snee 2004).

Literature regarding the theory and implementation of Six Sigma has been reported by researchers from many developed countries. Similar kind of studies in other parts of the world needs to be carried out to understand various issues affecting Six Sigma implementation. Moreover, the report on implementation of Six Sigma and its benefits in a developing country like India are rather scanty (Antony and Banuelas 2002). Six Sigma is a business improvement strategy used to improve profitability, to drive out waste, to reduce quality costs, and to improve the effectiveness and efficiency of all operational processes that meet or exceed customers' needs and expectations. The study deals with an application of Six Sigma DMAIC (define–measure–analyze–improve–control) methodology in a manufacturing industry, which provides a framework to identify, quantify, and eliminate quality rejections in their manufacturing system process. The data collection before and after implementing Six Sigma has been analyzed in a quantitative manner to ascertain the important benefits of implementing this approach.

2 Implementation of DMAIC Cycle

2.1 Define Phase or Identification of Critical Areas

Project team of concerned persons has been constituted to define and measure the problem efficiently. The leader selected is a master black belt, and team contains one black belt and three green belt members in it. A schedule chart for effective execution of respective DMAIC phases has been made by Six Sigma team. The rejections of various crankshafts (P19, P90, KZAA) coming from the customer have been analyzed. The crankshaft P19 contributes 70 % of the total rejections of month. The monthly rejection for the various crankshafts manufactured by the company has been analyzed at different levels of production. The defective parts rejected are the sum of carrying in defects + bought out semi-finished defects + bought out defects parts. Figure 1 shows various quality rejections of different crankshafts manufactured. So, crankshaft P19 is considered more critical product to carry out continuous improvement in its production processes.

Figure 2 shows the operation-wise rejection of the crankshaft P19 as determined at the operation level. The 52 out of 95 rejections are caused due to ball height undersize at facing and centering operation. Further, this rejection analysis shows that common defects in facing and centering include ball height undersize; drill broken; double drill and center out; and total length undersize.

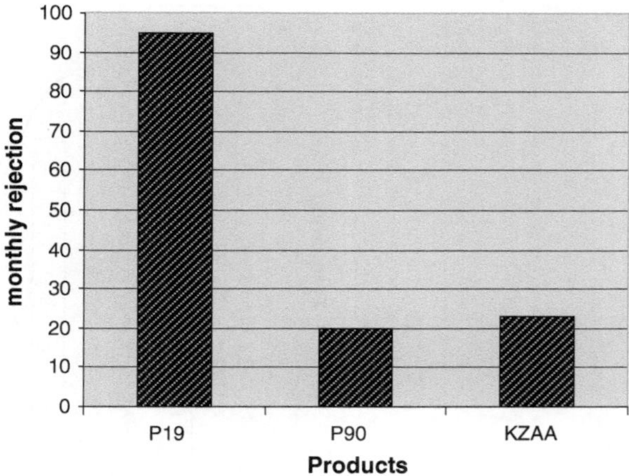

Fig. 1 Monthly rejection of crankshafts manufactured

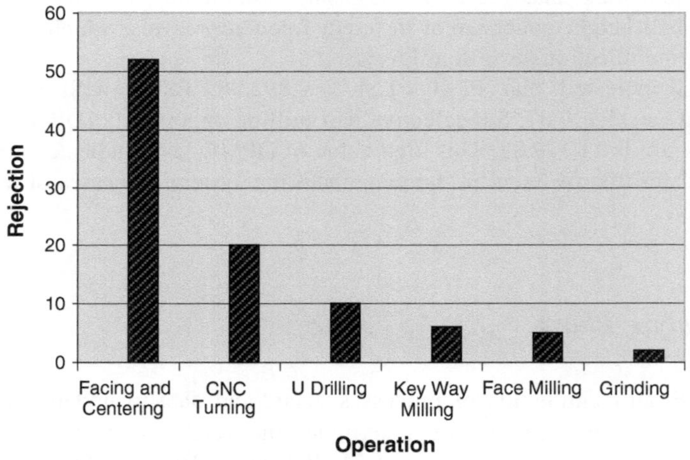

Fig. 2 Operation-wise rejection of the crankshaft P19

2.2 Measure Phase

The goal of the measure phase of a Six Sigma DMAIC project is to get as much information as possible, both how it works and how well it works (Ibrahim 2011). In order to verify the variation in the ball height ranging from 0.00 to 0.10 mm (LSL = 0.00; USL = 0.10), the histogram has been prepared showing variation in different readings of ball height undersize. The sample of 25 observations has been

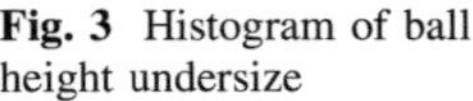

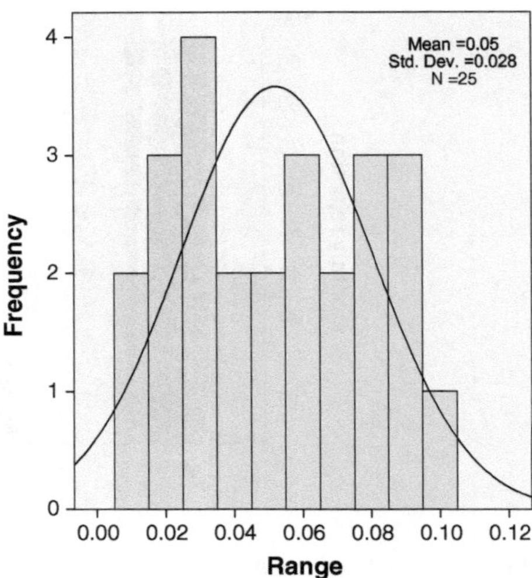

Fig. 3 Histogram of ball height undersize

taken randomly and analyzed using SPSS software. Figure 3 indicated the different range of ball height undersize at different frequency levels, which is helpful in depicting the rejection at million levels.

Results indicated that PPM $<$ LSL = 9,892.98; PPM $>$ USL = 12,906.89; PPM total = 11,179.87. So, defective part million opportunity (DPMO) level for the histogram is 11,179.87. This high value of DPMO level indicates that there is high need for improvement in the manufacturing system processes of the present case study.

2.3 Analyze Phase

It refers to an examination of processes, facts, and data to understand the root cause of given problem. It also determines the various opportunities to bring improvements through Six Sigma DMAIC cycle (Babbar 2005). The brainstorming session has been held including team of Six Sigma experts, HODs of different departments, and managers. After critical analysis and expert's suggestions, fishbone diagram (as shown in Fig. 4) has been prepared, depicting probable causes of ball height undersize at different levels of production. Root cause of the problem has been analyzed by taking appropriate preventive and corrective action through continuous improvement in their manufacturing system processes.

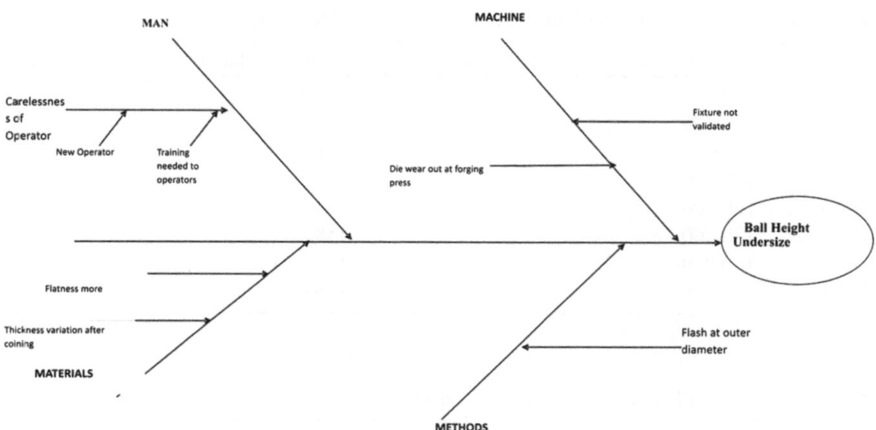

Fig. 4 Causes of ball height undersize

2.4 Improve Phase

2.4.1 Fixture Validation Data of Crankshaft P19

It includes monitoring of processes, analysis of results, and accordingly taking corrective actions when necessary to maintain the business process in central or stable conditions (Johnson 2002). In improvement phase, various improvements have been done to ascertain the problems identified in the analyze phase of implementing DMAIC cycle in a systematic manner. Actual production on various fixtures has been identified along with the validation frequency of these fixtures. Clearly, it has been found that actual production on various fixtures exceeds the validation frequency of the fixtures. This may lead to cracks, wear, and tear of the fixtures used for the process. So, validation of fixtures is needed after the number of pieces manufactured. Table 1 indicated the fixture validation data for crankshaft P19 line.

Table 2 shows the fixture validation of facing and centering according to 04 fixtures attached to this machine. Flatness of V Clamp is improved, and depression in pin-type locator has been validated.

Table 3 shows the various problems identified as root cause of the problem (flatness more, new and unskilled labor, maximum die runs) and steps taken to improve the process.

2.5 Control Phase

To control the variation in the ball height, the standard operating procedure (SOP), skilled matrix cum training need identification (providing on job and off the job training as per recommendation of the different experts from various departments)

Table 1 Fixture validation data before implementing DMAIC cycle

Fixture no.	Validation frequency (production in pcs)	Actual production (production in pcs)
FC-CL-03	160,000	275,277
FC-CR-01	160,000	253,328
FC-CL-01	160,000	259,679
FC-CR-02	160,000	291,450
UD-CR-02	150,000	258,202
UD-CL-02	150,000	158,463
UD-CR-01	150,000	190,230
UD-CL-04	150,000	195,075

Table 2 Fixture validation of facing and centering

Fixture no.	Flatness specified	Actual flatness	Validation
FX-CS-0108	V Clamp > 0.02 mm	0.34	OK
FX-CS-0115	Depression in pin > 0.3 mm	0.40	OK
FX-CS-0116	Depression in pin > 0.3 mm	0.25	Not OK
V BLOCK	Depression in V clamp < 0.30	0.5	Not OK

Table 3 Root cause analysis

Problem	Route cause of the problem	Activity made to improve the process
Flatness more at coining	Pressure not proper at coining press	New punch with proper pressure has been introduced
	Punch wear out at coining	
Flatness more at forging	Radius of outside ring diameter of flange face (0.4 mm)	Radius of outside ring reduced to 0.15
	Die wear out at forging	New die has been introduced
New operator	Skill matrix to be prepared for providing on job and off the job training	Skilled matrix has been prepared for the crankshaft line
Die runs	Maximum die run at facing and centering	Only one or two die run at facing and centering

and check sheet as shown in Table 4 (to be filled by the operator daily). Table 4 has been prepared to eliminate the roadblocks in facing and centering operation of the crankshaft P19.

SOP for facing and centering has been discussed below:

- Firstly, the operator will check the zero of special gauge (ball height on surface plate).
- Operator will clean the machine and then fill the checklist.
- Keep the master piece of special gauges as per running component and remove the other model gauge from the machine.

Table 4 Check sheet

Checkpoint	Day 1	Day 2	Day 3
Cleaning the area			
Cleaning the workstation			
Lubrication of slide			
Coolant level as per level indicator			
Hydraulic pressure of clamping (15–20 kgf/cm^2)			
Guards			
Gauge color code			
Gauge identification			
Locator			
Resting pads			
Vertical and horizontal clamp			
Process operation sheet available			
Five piece strand should be available			
Tool consumption report			

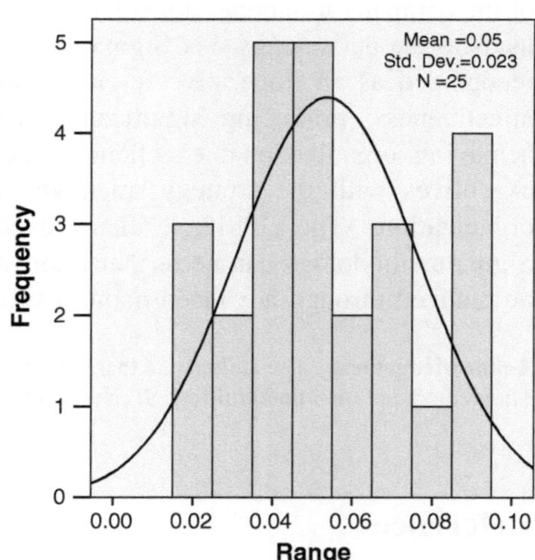

Fig. 5 Histogram after implementing Six Sigma DMAIC cycle

- Whenever the setting changes, gauges should also be changed accordingly to the component.
- Setting of tool by tool setter.
- After first five pieces, inspection by operator should be done.
- Operator will verify the length variation in center drill, and if variation found in length, return the drill to supervisor.
- Only one or two dies run at facing and centering.
- Maintain 5S around your machine.

Figure 5 indicated that PPM < LSL = 10.22; PPM > USL = 18.47; PPM total = 28.69, So, DPMO level for the histogram is reduced from 11,179.87 to 28.69.

- PPM before Six Sigma = 11,179.87
- Achieved PPM = 28.69
- Cost per piece = rupees 110
- PPM reduction = 11,151.18
- Monthly production = 120,000
- Savings per month = 11,151.18 × 120,000/ 10^6 = 1,338.1416 pcs
- Yearly savings = 1,338.1416 × 110 * 12 = rupees 1,766,346.384/17. 66 lakhs.

3 Conclusions and Scope of Future Work

From the results of the study, it is concluded that Six Sigma improvement methodology, viz. DMAIC cycle, is highly helpful in improving the performance of the company to increased level as regards to quality improvement which further leads to the net savings. Six Sigma as a powerful business strategy has been well recognized as an imperative for achieving and sustaining operational (process) effectiveness, producing significant savings to the bottom line, and thereby achieving organizational excellence. Six Sigma provides business leaders and executives with the strategy, methods, tools, and techniques to change their organizations. The DPMO level achieved after implementation of DMAIC Six Sigma methodology can be further improved, and new performance standards can be realized through advanced manufacturing technologies.

Acknowledgement The authors are thankful to the reviewers for improving the quality of paper. The authors are also thankful to DST New Delhi for providing funding under SERC Category.

References

Antony J, Banuelas R (2002) Key ingredients for the effective implementation of Six Sigma program. Measuring Bus Excellence 6(4):20–27

Babbar S (2005) Service quality and business ethics. Int J Serv Oper Manage 1(3):203–219

Dusharme D (2006) Survey: Six Sigma packs a punch. Qual Dig 103(1/2):85–97. Accessed 14 Feb 2006

Ibrahim RA (2011) Waste elimination using quality function deployment. Int J Serv Oper Manage 10(2):216–238

Johnson A (2002) Six Sigma in R&D. Res Technol Manage 45(2):12–16

Montgomery DC (2009) Introduction to statistical quality control, 6th edn. Wiley, Hoboken, New Jersey

Snee R (2004) Six Sigma: the evolution of 100 years of business improvement methodology. Int J Six Sigma Competitive Adv 1(1):4–20

Thermal Energy Saving Using Total Quality Management

Sudip Banerjee and Gurinder Singh Brar

Abstract Milk is considered as an essential and complete food. Hence, dairy industries are getting more importance nowadays. The profitability of the organization thereby depends not only on efficient processing and product handling, but also on saving of energy as much as possible. The later one bears lot of importance in the present era where energy saving becoming essential not only due to scarcity of natural resources but also for the survival of the humanity. The present study was conducted at one of the premier milk industries of northern India in reference to the thermal energy savings. After the implementation of total quality management tools, it was observed that the cost of heavy petroleum stock (HPS) used as boiler fuel got reduced by 20 % by improving the milk handling per kg of fuel from 39 to 47 kg up to the month of May 2013 resulting into a total amount of Rs. 4,347,972 in 6 months.

Keywords TQM · Dairy industry · Thermal energy saving

1 Introduction

The milk plant established in 1976 with a capital cost of 110 crore and initial milk handling capacity of 50,000 LPD. Presently with handling capacity of 3 lakh liter

S. Banerjee (✉)
Guru Nanak Dev Engineering College, Ludhiana, India
e-mail: dairysolutions1@rediff.com

S. Banerjee
MilkFed Milk Plant Chandigarh, The Punjab State Cooperative Milk Producers' Federation Limited, SCO-153-155, Sector 34A, Chandigarh, India

G. S. Brar
Mechanical Engineering Department, Guru Nanak Dev Engineering College, Ludhiana 141006, Punjab, India
e-mail: brar.gurinder@gmail.com

S. S. Khangura et al. (eds.), *Proceedings of the International Conference on Research and Innovations in Mechanical Engineering*, Lecture Notes in Mechanical Engineering, DOI: 10.1007/978-81-322-1859-3_53, © Springer India 2014

per day, it was producing liquid milk, panjiri, ghee, curd, paneer, lassi, kheer, special pinni, etc. The plant is making net profits for the last 5 years and annual growth of 25 % from 2010 onward. It was observed that out of total annual expenditure of Rs. 222,005,206 for the year 2011–2012, the fuel and electrical expenditure was Rs. 30,445,357, which was about 14 % of the total cost and fuel cost contributed 67 % to the total cost of fuel and electricity. Since electricity always was in short supply in Punjab, reduction in fuel cost became utmost necessity in the present era.

Figures 1 and 2 depict the percentage contribution of the major cost factors and percentage contribution of electricity and fuel cost in total cost.

In the year 2011–2012, milk handled in kg per kg of fuel was 39, which was in very lower side and could be increased. Accordingly, this project was taken to reduce the fuel usage cost by 20 % by improving the milk handling per kg of fuel from 39 to 47 kg up to the month of May 2013. Thereby, a saving of approximately 28 lakhs per annum was expected.

2 Literature Review

Rodgers and Vries (1992) suggested that the most efficient way of problem solving was in groups. A visual tool for problem-solving "CEDAC" (which stands for cause and effect diagram with the addition of cards) was demonstrated which created a virtual workplace. How the tool was used in group working led by a facilitator was also explained. Moreover, an advantage of using CEDAC over other methods was considered, and the conditions which needed to be met in order for it to be used successfully were listed.

Kunene and Petkov (2002) described that for brainstorming tasks, the process support of the system had been restricted to ensuring member anonymity and allowing the simultaneous entry of ideas. Little work has been done investigating other process improvements for idea-generating groups. Authors investigated the effect on idea quality and quantity of decomposing a brainstorming task with a South African setting. Using team expert choice, an analytic hierarchy process (AHP)-based group support system. Authors conducted an experiment with two groups. Authors hypothesize that task decomposition will generate more and better quality ideas. Findings show that task decomposition resulted in 40 % more ideas than no decomposition; the effect on decision quality is statistically significant only when decision quality is measured as the number of good ideas.

Kumar et al. (2011) tried to learn about the quality and efficiency of US healthcare services. It seeks to examine the impact of quality and efficiency on various stakeholders to achieve the best value for each dollar spent for healthcare. The US healthcare system was examined utilizing various data on recent trends in spending, budgetary implications, economic indicators, i.e., GDP, inflation, wage, and population growth. Process maps, cause and effect diagrams, and descriptive data statistics were utilized to understand the various drivers that influence the

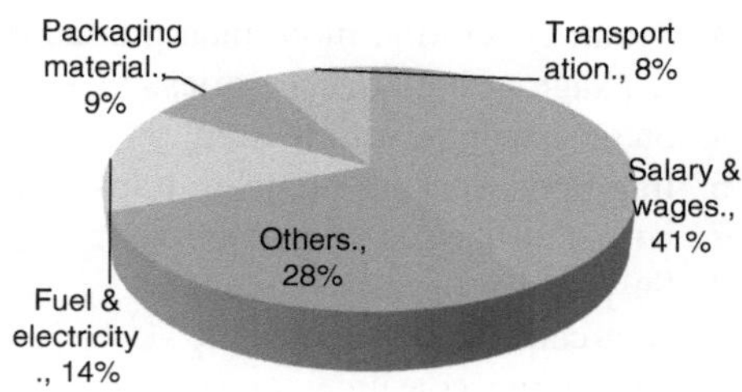

Fig. 1 Major cost factors with percentage contribution

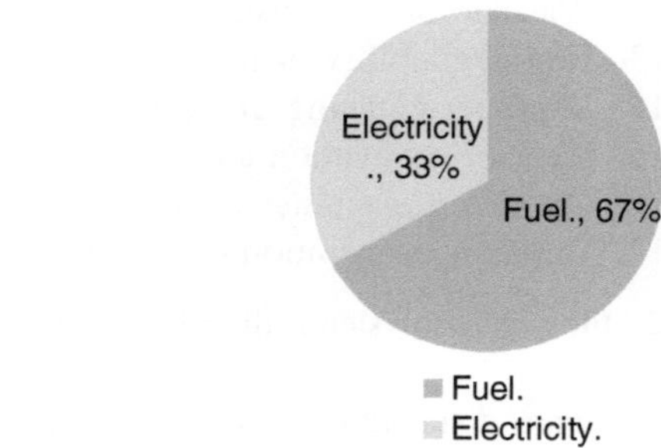

Fig. 2 Contribution of electricity to fuel in total cost

rising healthcare cost. A proposed cause and effect diagram is presented to offer potential solutions, for significant improvement in US healthcare. The US healthcare system was characterized as the world's most expensive yet least effective compared with other nations. Growing healthcare costs have made millions of citizens vulnerable. Major drivers of the healthcare costs were institutionalized medical practices and reimbursement policies, technology-induced costs, and consumer behavior.

3 Methodology

To solve the above problem, TQM tool was used which comprised of the following systematic studies.

1. Brainstorming

To know the probable causes of thermal energy wastage, employees from different categories were invited for getting ideas as each human being is a store house of logical and innovative ideas. Accordingly, 27 persons gathered. During brainstorming, 34 ideas came forward that were recorded and then scrutinized and then 17 valid ideas were identified.

1. Inefficient boiler.
2. Scattered running of production activities.

3. Steam generation more than production requirement.
4. Leakage at boiler/accessories.
5. Inconsistent steam pressure.
6. Improper working of steam traps and valves.
7. Under utilization of machines.
8. Suspension of milk processing.
9. Non-calibrated instruments.
10. Pasteurizer consumes more steam.
11. Long header and uninsulated steam pipe lines.
12. Wastage of condensate.
13. Non-availability of task force.
14. Improper cleaning of equipments using steam.
15. Lack of motivation and belongingness among staff.
16. Out-dated machines using more steam.
17. Unplanned operation of boiler.

2. Flowchart of steam distribution before project starts up

Steam distribution system flowchart was prepared first. From the flowchart, it was observed that the unnecessary lengthy steam lines were laid and processes were staggered due to which heavy losses were occurring. With the help of flowchart, unnecessary steam lines were plugged/rerouted and staggered processes such as paneer section and kheer section were shifted near the processing which ultimately saved thermal energy (Fig. 3).

3. Cause and Effect Diagram

To have in-depth knowledge insight into the causes of thermal energy wastage, cause and effect diagram was drawn in order to have probable causes contributing toward the wastage of thermal energy (Fig. 4).

4. Validation of causes

Main causes of thermal energy wastage which were drawn in the form of fish bone diagram helped to validate the causes of thermal wastage. From which, main probable root causes of thermal energy wastage were identified (Table 1).

4 Data Collection and Root Cause Analysis

Data collected and respective root cause were analyzed, and accordingly, action taken to save thermal energy is as shown in Table 2.

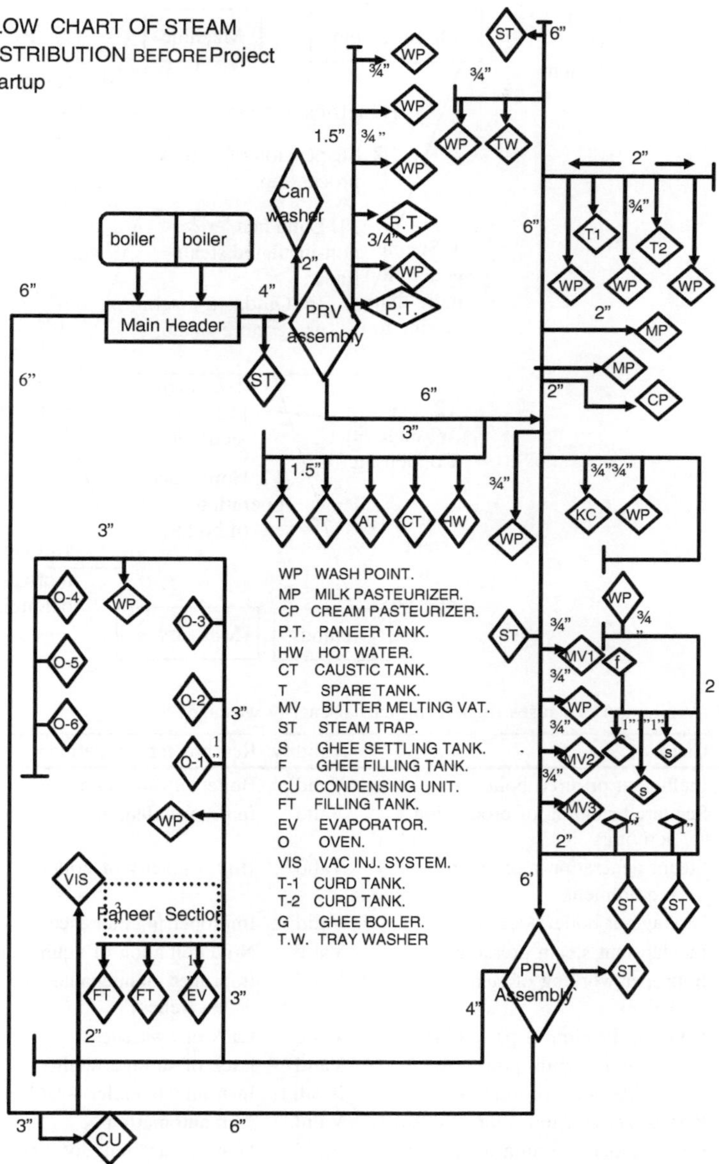

Fig. 3 Flowchart of steam distribution before starting the project

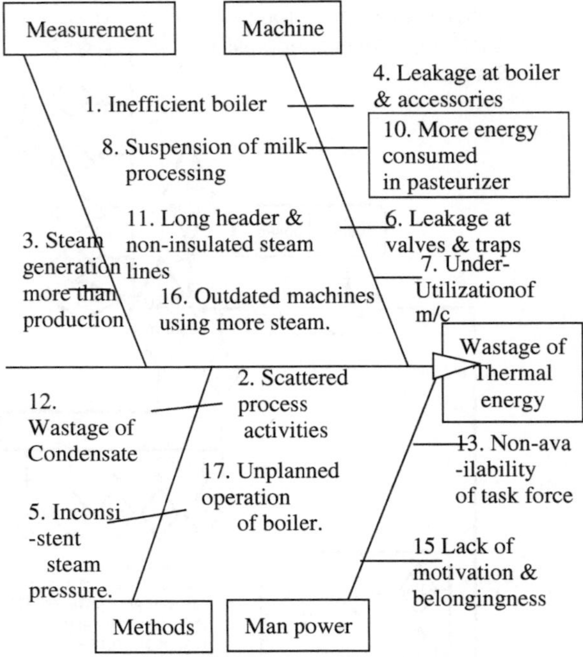

Fig. 4 Cause and effect diagram to find out the probable causes of thermal energy wastages

Table 1 Validation of causes related to thermal energy wastage

S. No.	Causes	Validity	Reasons for validation
1.	Inefficient oil-fired boiler	Valid	Boiler of old technology
2.	Scattered running of production activities	Valid	Improper planning
3.	Steam generation more than requirement	Valid	High capacity of boiler
4.	Leakage at boiler/accessories	Valid	Improper maintenance
5.	Inconsistent steam pressure	Valid	No automation of equipments
6.	Improper working of steam traps and valves	Valid	Improper maintenance and lack of awareness
7.	Under utilization of pasteurizer	Valid	Lack of awareness
8.	Suspension of milk processing	Valid	Lack of storage facility
9.	Non-calibrated instruments	Invalid	Instruments under AMC
10.	Pasteurizer consumes more steam	Valid	Non-automation
11.	Long header and uninsulated steam pipe lines	Valid	Unnecessary lengthy pipe lines
12.	Wastage of condensate	Valid	Lack of infrastructure/awareness
13.	Non-availability of task force	Valid	Lack of awareness
14.	Improper cleaning of equipments using steam	Invalid	Effective CIP being done
15.	Lack of motivation among floor staff	Valid	Lack of proper guidance
16.	Outdated m/c using more steam	Valid	Tray washer of old/out-dated design
17.	Unplanned operation of boiler	Valid	Lack of coordination among producer and user

Table 2 Collection of data against each brainstorming points and modified action taken

S. No.	Root causes	Steam losses (kg/day)	Action taken	Steam saving (kg/day)
1.	Inefficient boiler	2,000	New boiler of 3-ton capacity under trials	Under trial
2.	Scattered running of production activities	2,500	Production activities clubbed together	2,000
3.	Steam generation more than requirement	200	New boiler of desired capacity of 3 ton on trials	Under trial
4.	Leakage at boiler/ accessories	300	Maintenance work done	200
5.	Inconsistent steam pressure	600	Work under progress	–
6.	Improper working of steam traps and valves	1,500	Maintenance done and old traps replaced	1,300
7.	Under utilization of machines	2,500	Both the pasteurizers run together	2,000
8.	Suspension of milk processing	1,000	Extra storage tank installed	800
9.	Invalid cause			
10.	Pasteurizers consume more steam	1,600	Automation under progress	Under assessment
11.	Long header and uninsulated pipelines	300	Unused pipelines plugged	200
12.	Wastage of condensate	1,600	Condensate water reused in CIP system	1,300
13.	Non-availability of task force	Intangible	Training imparted to the staff	Intangible
14.	Invalid cause			
15.	Lack of motivation and belongingness among staff	Intangible	Frequent session at floor level staff held for energy saving	Intangible
16.	Out-dated machine	2,500	New tray washer installed	2,500
17.	Unplanned operation of boiler	1,300	Coordination strengthened	1,000

4.1 Flowchart of Steam Distribution After Modification

Flowchart of steam distribution was prepared after modification to see further scope of improvement to save thermal energy (Fig. 5).

4.2 Schedule of Actions Taken

After data collection and root cause analysis, action plan was formed; i.e., firstly, those actions were taken that could be implemented easily and required less time

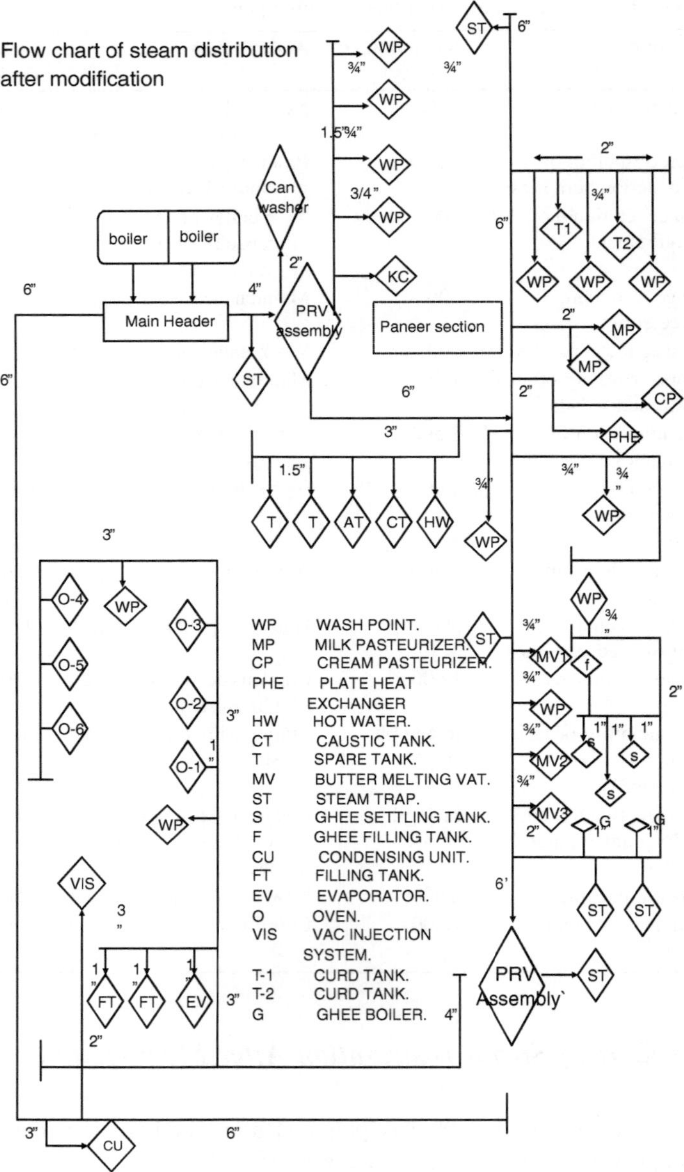

Fig. 5 Flowchart of steam distribution after taking modified action

Table 3 Monthwise action taken for implementation

S. No.	Action taken	Month of implementation					
		Dec–May	Jan	Feb	March	Apr	May
1.	New 3-ton capacity more economical boiler purchased to save energy						
2.	Scattered production activities clubbed together to reduce running hours	√	√	√	√	√	√
3.	New boiler of desired steam generation capacity of 3 ton is purchased and under trials to save energy						
4.	Maintenance work done at boiler and accessories to avoids team losses	√	√				
5.	Rectification work of inconsistent steam pressure is under progress					√	√
6.	Maintenance/replacement of valves and traps to avoid steam losses	√	√				
7.	Both the pasteurizers run together to curtail running hours, which saves energy		√				
8.	To stop suspension of milk processing due to lack of storage facility extra storage tank installed			√	√		
9.	Automation of pasteurizers under progress to stop more consumption of steam					√	√
10.	Unused pipelines plugged to avoid thermal losses			√	√	√	
11.	Condensate water reused in CIP system to save energy and water					√	√
12.	Task force constituted to monitor daily energy consumption		√	√	√	√	√
13.	Training imparted to the staff regarding importance of energy conservation		√	√	√	√	√
14.	Out-dated m/c like tray washer, which consumes more steam, replaced with new tray, which runs without steam					√	√
15.	Boiler run with the consent of user to avoid idles running of boiler	√	√				

and money and could give us maximum saving in thermal energy. Schedule of action taken is as under monthwise (Table 3).

4.3 Data of Saving

The comparative details of milk handled and running hours of tube well in the year 2011–2012 and year 2012–2013 and the respective saving of electrical unit during that period with amount are shown in Tables 4, 5, 6.

Table 4 Calculation of net savings of HPS

Month	Milk handled (in kg) 2011–2012	Milk handled (kg) 2012–2013	HPS consumed (kg) 2011–2012	Equivalent HPS (kg) to be consumed in 2012–2013	Actual HPS (in kg) consumption in 2012–2013	Net saving of HPS (in kg)
Dec	2,703,438	2,899,721	7,7850	83,493	65,800	17,693
Jan	2,966,548	3,249,305	79,000	86,533	63,650	22,883
Feb	2,828,444	3,900,301	86,250	118,947	73,493	45,454
Mar	3,121,951	3,636,843	91,850	106,999	69,900	37,099
April	2,781,223	3,177,382	74,550	85,161	62,000	23,161
May	26,54,947	3,007,156	69,650	78,886	47,100	31,786
Total				560,019	381,943	178,076
Cost of saved HPS in (Rs.)						3,258,003

Table 5 Electrical cost calculation

Month	Milk handled (in kg) 2011–2012	Milk handled (in kg) 2012–2013	Electrical unit consumed in 2011–2012	Equivalent electrical unit to be consumed in 2012–2013	Actual electrical unit consumption in 2012–2013	Net saving of electrical unit
Dec	2,703,438	2,899,721	146,970	157,679	150,910	6,769
Jan	2,966,548	3,249,305	155,890	170,836	150,470	20,366
Feb	2,828,444	3,900,301	153,625	211,857	177,933	33,924
Mar	3,121,951	3,636,843	178,270	207,700	165,900	41,800
April	2,781,223	3,177,382	172,970	197,721	165,170	32,551
May	2,654,947	3,007,156	170,700	193,386	170,470	22,916
			Total	1,139,179	980,853	1,58,326

Table 6 Monthwise collection of data with net saving of tube well running hour

Month	Milk handled (in kg) 2011–2012	Milk handled (in kg) 2012–2013	Running hours of tube well 2011–2012	Equivalent running hour to be consumed in 2012–2013	Actual running hours of tube well 2012–2013	Net saving of tube well running hour
Dec	2,703,438	2,899,721	268	287	197	90
Jan	2,966,548	3,249,305	264	289	191	98
Feb	2,828,444	3,900,301	201	277	179	98
Mar	3,121,951	3,636,843	246	287	182	105
Apr	2,781,223	3,177,382	231	264	186	78
May	2,654,947	3,007,156	248	280	182	98
			Total	1,684	1,117	567
Equivalent electric unit saved at submersible pump 567 × rating of motor 25 × 0.746 = 10,574 unit. Amount saved in Rs.						51,812
Equivalent amount of unit at disposable pump = 10,574 unit. Amount saved in Rs.						51,812
Electrical unit saved at ETP. (40 HP × 2 no. motor × 0.746 × 30 days × 6 months × 4 h daily saving) Total electric unit saved = 42,969. Amount saved in Rs.						210,548
Total saving in Rs.						314,172

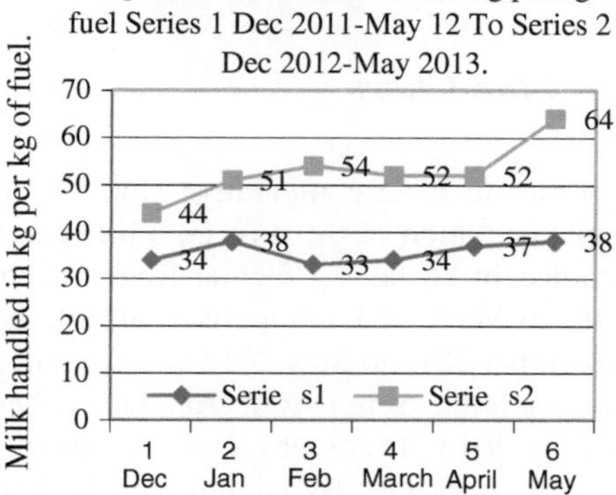

Fig. 6 Comparison of milk handled in kg per kg of fuel in consecutive 2 years

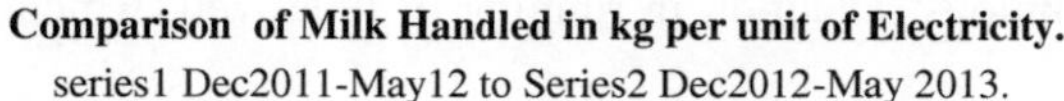

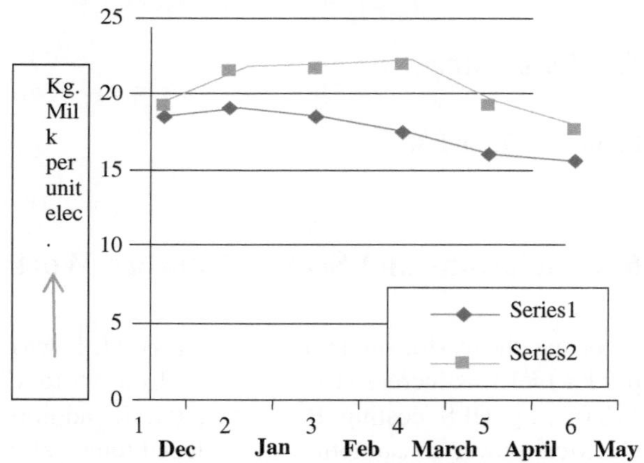

Fig. 7 Comparison of milk handled in kg per unit of electricity in consecutive 2 years

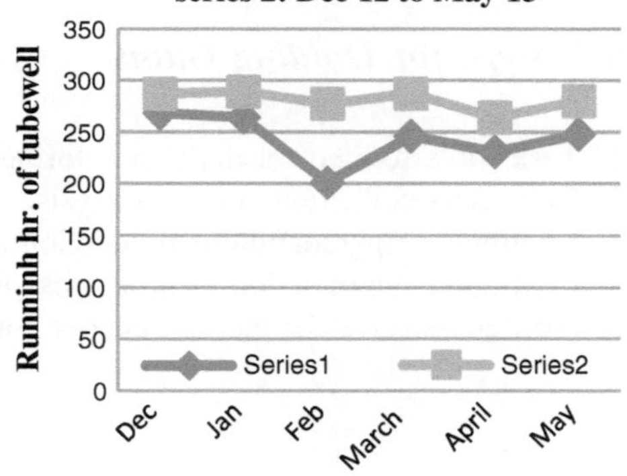

Fig. 8 Comparison of monthwise running hour of tube well in consecutive 2 years

5 Result and Discussions

5.1 Line Chart

In order to have comparative study of the losses of the thermal energy during implementation of the project, a line chart was drawn for the comparison of milk handled in kg per kg of fuel from December 2011 to May 2012 and December 2012 to May 2013, comparison of milk handled in kg per unit of electricity from December 2011 to May 2012 and December 2012 to May 2013, and comparison of running hours of tube well from December 2011 to May 2012 and December 2012 to May 2013. There was a clear contrast in the comparison in line chart as compared to bar chart. By adoption of the TQM project, there was continues increase in handling of milk per liter of fuel and per unit electricity and there was a continuous reduction in running of tube well, which shows saving of water.

5.2 Line Chart

Figures 6, 7, and 8.

6 Conclusions and Scope of Future Work

After the successful implementation of TQM, it was observed that milk handled per kg fuel was increased @20 %, i.e., from 39 to 47 kg resulting into saving of 178,076 kg HPS costing Rs. 3,258,003. In addition to that amount saved in electrical power consumption and reduced tube well running hour per annum was Rs. 775,797 and Rs. 314,172 respectively, amounted to be Rs. 4,347,972 as sum total.

6.1 Steps for Holding Gains

1. Task force formed for daily monitoring of production activities.
2. Daily data collection and its analysis.
3. Continuous upgradation of production activities.
4. Continuous upgradation of processes and procedures.
5. Awareness drive for thermal energy conservation.

References

Kumar S, Ghildayal NS, Shah RN (2011) Examining quality and efficiency of the US healthcare system. Int J Health Care Q Assur 24(5):366–388

Kunene K, Petkov D (2002) Task structuring a brainstorming group activity with an AHP-based group support system. In: 35th annual Hawaii international conference on system sciences (HICSS'02), vol 8, p 221

Rodgers L, Vries JD (1992) A visible solution. TQM Mag 4(6):369–372

An Artificial Neural Network Model for Traffic Noise Predictions

M. Dahiya, R. Panchal, P. K. Saini and N. Garg

Abstract The major environmental challenge encountered by metropolitan city today other than air pollution is traffic noise. So urban planning needs methods to aid in designing, planning, and forecasting in order to accommodate the increasing population and increasing traffic noise levels. Since the problem of traffic noise is nonlinear in nature, a model based on backpropagation neural network to counter this problem is suggested and examined. In order to have a clear and distinct insight on the magnitude of this problem, single-noise metrics L_{Aeq} is modeled. It is observed that the model is accurate in predictions and can be employed efficiently to predict traffic noise levels and conduct sensitivity analysis of factors affecting the noise levels. It can serve as an important tool for urban planning and development.

Keywords Equivalent continuous sound pressure level · L_{Aeq} · Mean-squared error · MSE · Coefficient of determination · R^2 · Multilayer perceptron (MLP)

1 Introduction

Technological development has led to urbanization which has also given rise to the problem of increasing traffic noise. The ever-increasing transportation infrastructure such as roads, railways, and airports has accentuated the problem. It has been scientifically proven that high noise levels seriously affect the health of the people

M. Dahiya (✉) · R. Panchal · P. K. Saini
Department of Mechanical Engineering, National Institute of Technology, Kurukshetra 136119, Haryana, India
e-mail: manojdahiya4865@yahoo.com

N. Garg
CSIR-National Physical Laboratory, New Delhi 110012, India

S. S. Khangura et al. (eds.), *Proceedings of the International Conference on Research and Innovations in Mechanical Engineering*, Lecture Notes in Mechanical Engineering, DOI: 10.1007/978-81-322-1859-3_54, © Springer India 2014

exposed. In the developed countries, stricter norms have been laid to encounter this problem. There have been numerous studies conducted in order to generate a model which can predict the noise levels with a definite accuracy and precision. In Indian context, there have been some studies reported for different cities. Since Delhi has been facing a lot of traffic-related problems, a model based on Delhi's traffic noise would be sufficient to generalize this problem and can be used to predict the traffic noise in other parts of country as well. The previous studies in Indian context are generally focused on regression-based approach (Rao and Rao 1991; Nirjar et al. 2003; Rajakumar and Gowda 2009; Agarwal and Swami 2011; Kalaiselvi and Ramachandraiah 2012; Kumar and Jain 1999). As such, the dependence of noise levels with traffic density, average vehicular speed is non-linear in nature, so other approaches especially the soft computing algorithms should also be tried to ascertain their compatibility in comparison with regression-based approach, the present study focuses to use the artificial neural network (ANN) approach in modeling traffic noise levels. Several studies have used this technique to predict traffic noise. Givaris and Karimi (2009) have made a comparative study analyzing both a linear regression model and ANN-based model for the Tehran-Karaj (Iran) express train for predicting $L_{\max}$. Parabat and Nagarnaik (2007) developed an ANN model to predict the sound pressure level for continuous traffic flow conditions. Givaris and Karimi (2010) have also developed an ANN-based model to predict L_{eq} for Tehran's roads. Most of the results have indicated a high degree of accuracy for environmental noise prediction. Nucara et al.'s (2002) investigations reveal that dynamic behavior of neural networks that allow complete and detailed description of the involved phenomena suggests increasing applications in research field. Based on study of 25 previously selected input variables, Genaro et al. (2010) developed an ANN-based model to predict urban environmental noise (L_{Aeq}). The recent studies thus have shown that ANN can be efficiently utilized for traffic noise predictions with definite accuracy and precision. In present study, a limited set of noise measurements carried out at eight locations in Delhi is utilized to ascertain the applicability of ANN in traffic noise predictions and forecasting. A multilayer perceptron (MLP) is implemented in MATLAB language.

2 Network Architecture and Methodology

Neural networks are similar to linear and nonlinear least squares regression and can be viewed as alternative statistical approach in solving the least squares problem (White et al. 1992). ANN architecture is based on biological neural network and it consists of interconnected artificial neurons which are grouped under the input, hidden, and output layers, where the number of input and output decides the number of neuron in the respective layer. The most significant part is the hidden layer influences the final output. There is no thumb rule for determining the optimum number of neurons. Some studies (Kaastra and Boyd 1996) recommend that all neural networks should start with preferably one or almost two hidden layers. The hidden layers provide the network with its ability to generalize. In practice, neural networks with one and occasionally two hidden layers are used widely and have performed well in many studies. For backpropagation ANN architecture, following steps should be considered:

- A network is classified for a particular problem.
- Network topology, i.e., number of neurons, hidden layers, output neurons, and error function to be designed.
- Selection of training parameters, i.e., learning rate, epoch size, momentum constant, size of training, testing, and validation data set.
- Data are divided randomly for training, testing, and validation.
- Implementation and holdout validation for test data set.
- Characterization and validation by performance criteria, i.e., MSE, mean absolute percentage error (MAPE), etc.

Figure 1 shows a pictorial example of such a network of 3 neurons in input layer and 4 in output layer with two hidden layers each having 4 neurons. The backpropagation algorithm is used for the present study. Backpropagation networks are a class of feed forward neural networks with supervised learning rules. Supervised learning is a process of comparing each of the network's forecasts with the known correct answer and adjusting the weights based on the resulting forecast error to minimize the error function (Kaastra and Boyd 1996). Optimization of weights is made by backward propagation of error during learning phase. The ANN model reads the input and output values in the training data set, and network weights and biases are refined until the difference between the predicted and output value is minimum (Taghavifar and Mardari 2013). A complete round of forward–backward passes and weight adjustments using all input–output pairs is called an epoch or iteration. If a network is left for training too long, network can be overtrained also and can lose ability to generalize. Figure 2 shows the flowchart of backpropagation algorithm used for developing traffic noise model.

3 Experimental Data Acquisition

The precision digital sound level meter used was kept at a distance of 0.4 m from the body and at a height of 1.4 m from the ground level and 3 m from the road for avoiding any reflections from roadside barriers. The measurement of sound pressure in dB(A) was done along with the monitoring of the average speed of vehicles with the speed gun (Make: Bushnell), and numbers of vehicles were counted manually. Precision digital sound level meter helped in measuring L_{eq} and statistical parameters, e.g., L_{10}, L_{50}, L_{90}, etc. Short-term L_{AeqT} measurements ranging from 15 to 30 min were undertaken at 8 different busy road locations of Delhi. While measurements, it was ensured that there is no reflections from the

Fig. 1 Network architecture

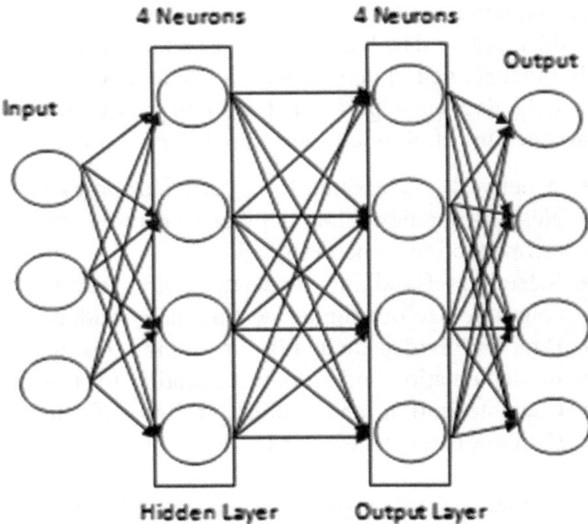

adjoining building facades or wall. It may be noted here that the recent studies conducted by Maruyama et al. (2013) shows that a minimum of 170 vehicles are enough for obtaining a reliable $L_{\mathrm{Aeq}T}$ during measurement time interval T. The output variables L_{eq}, L_{10}, L_{50}, and L_{90} are also measured with the help of sound level meter in A-weighing mode.

4 Model Development

The various parameters considered which provided the best results of the measured model have employed the learning function "learngdm." The training function is "trainlm" and the data are divided by the "dividerand" function. The data are divided into training data (70 %), testing data (15 %), and validation data (15 %). The network is designed considering the parameters as discussed previously. ANN has the advantage to work out with all the variables at the same time. After the segregation of data set, the network is trained and run for different no. of neurons in the hidden layer. The main input variables for the model are total no. of vehicles, average speed, and percentage of heavy vehicles. The selection of number of hidden neurons was done based on hit and trial approach. Initially, one hidden neuron was selected and gradually the number of neurons was increased until a network of minimum mean-squared error for test data set is attained. Further increase in hidden neurons produces higher error. Thus, an MLP feed forward network consisting of one hidden layer with four neurons was built with three inputs and four outputs (3:3:4). The architecture of MLP network used in the study is shown in Fig. 3. The sigmoidal function is used as transfer function for the hidden and output layer

Fig. 2 Flowchart for backpropagation algorithm

nodes. Trainlm is a network training function that updates weights and bias values according to Levenberg Marquardt optimization (Taghavifar and Mardari 2013). Since the range of input variables was different, in order to achieve faster convergence to minimal MSE, each of the input variables was normalized in the range −1 to 1.

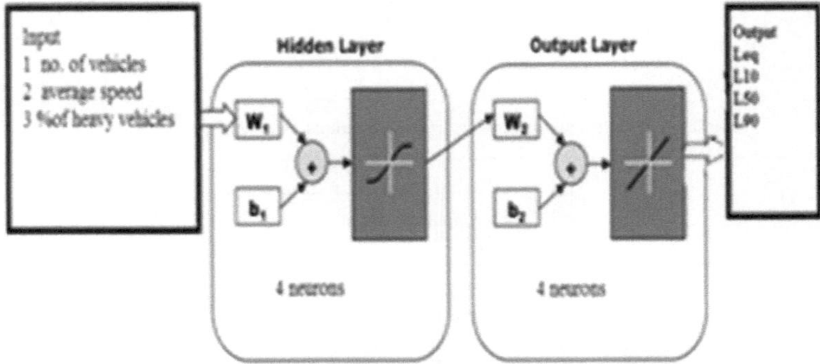

Fig. 3 Model structure

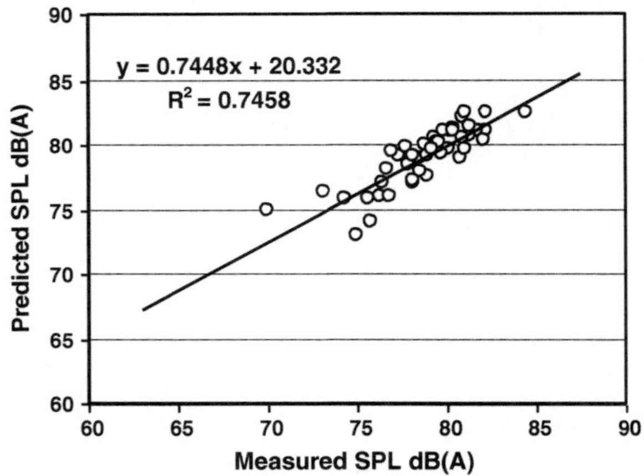

Fig. 4 Predicted (output) versus measured in dB(A)

5 Result and Discussion

The model implements training function "trainlm" and the learning function "learngdm" to establish a relationship between the measured variables, i.e., L_{eq}, L_{10}, L_{50}, L_{90}. The predicted values have been calculated for different no. of neurons in the hidden layer, and the least MSE has been achieved for 3 neurons.

It may be noted that while training the network, R^2 value greater than 0.80 is obtained for network topology of 3:3:4. The value of R^2 is 0.75 for the test data set used for holdout validation as shown in Fig. 4, and the MSE for 3-neuron configuration is 0.22. A graph has been plotted for MSE w.r.t. no. of neurons in the hidden layer, and with increasing the no. of neurons, the MSE increases initially, but after 6 neurons, the MSE decreases. Best curve is obtained for 3 neurons.

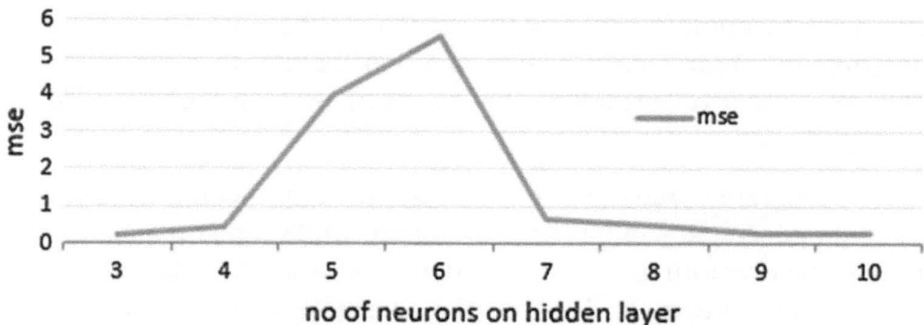

Fig. 5 Mean-squared error (MSE) in 10^{-3} versus no. of neurons

Table 1 Variation in MSE (in 10^{-3}) with number of neurons

No. of neurons	MSE	Epochs	Mu value	Gradient	Performance
3	0.22	46	0.0001	0.165	4.94
4	0.41	31	0.0001	1.65	2.47
5	4	19	0.0001	0.469	3.11
6	5.6	193	0.0001	3.80	3.04
7	0.65	389	0.0001	0.883	5.48
8	0.46	176	0.0001	0.408	2.54
9	0.25	12	0.0001	0.573	2.76
10	0.29	417	0.0001	2.57	2.51

It can be observed (Fig. 5) that at 3 hidden neurons, the MSE is 0.22×10^{-3} and the value of correlation coefficient R is 0.97. Although value of R is same for other configuration as enlisted in Table 1, the MSE is minimum for 3 neurons, so MSE can be considered a good criterion for the selection as well as validation of the model. Moreover, value of R described the fitness of the curve, which does not explain the deviation from the measured values.

Thus, it can be observed that ANN can be suitable for traffic noise predictions with a definite accuracy. Previous studies (Sharma et al. 2012; Garg and Maji 2011) in this respect with a network topology of 7:6:1 also confirm the suitability of ANN in modeling traffic noise comparable to analytical formulations.

6 Conclusion

The value of R^2 is 0.75 for the model developed whose input variables are no. of vehicles, average speed, and percentage of heavy vehicles. The model has predicted values which are very close to the measured values, and the MSE is 0.22×10^{-3}. The training function used “trainlm,” and the learning function is “learngdm”. The final structure of the model has 3 neurons in the hidden layer having transfer function (transig) and 4 neurons in the output layer with transfer function (purelin). From these results, we can deduce that the model aids in predicting the noise level accurately. Following conclusions can be made from the analysis as follows:

- Model is accurate in predicting traffic noise levels of medium to high intensity.
- Information generated can be applied directly while urban planning.
- MSE is a good performance testing criterion. However, if the model so developed is trained with more data set, the accuracy of the ANN model can be further enhanced.
- The model can be improved with more no. of readings at greater no. of sites and considering more no. of variables including road characteristics, vehicle characteristics, weather, environmental conditions, vegetation, type of locality, presence of industries nearby, etc.

The study reveals that backpropagation algorithm has good capability in data modeling. However, despite many advantages, there are some disadvantages too. Construction of ANN model is time-consuming and depends on size of training data and network structure. Also it is sometimes like a black box wherein one cannot adjudge the weights and biases developed while training the network. However, in spite of these shortcomings, ANN can serve as vital substitute for analytical models in traffic noise predictions and forecasting.

References

Agarwal S, Swami BL (2011) Comprehensive approach for the development of traffic noise prediction model for Jaipur city. Environ Monit Assess 172:113–120

Garg N, Maji S (2011) Comparison of principal traffic noise models. In: Presented at international conference on traffic noise, Dresden, 28th Sept to 1st Oct 2011

Genaro N, Torija A, Ramos-Ridao A, Requena I, Ruiz DP, Zamorano M (2010) A neural network model for urban traffic noise prediction. J Acoust Soc Am 128(4):1738–1746

Givargis S, Karimi H (2009) Mathematical, statistical and neural models capable of predicting L_{max} for the Tehran-Karaj express train. Appl Acoust 70(7):1015–1020

Givargis S, Karimi H (2010) A basic neural traffic noise prediction model for Tehran's roads. J Environ Management 91(12):2529–2534

Kaastra I, Boyd M (1996) Designing a neural network for forecasting financial and economic time series. Neurocomputing 10:136–215

Kalaiselvi R, Ramachandraiah A (2012) A model for traffic noise prediction in heterogeneous traffic conditions. Int J Curr Res 4:180–184

Kumar K, Jain VK (1999) Autoregressive integrated moving averages (ARIMA) modelling of a traffic noise time series. Appl Acoust 58:283–294

Maruyama M, Kuno K, Sone T (2013) The minimum measurement time for estimating L_{AeqT} of road traffic noise from the number of vehicle pass-bys. Appl Acoust 74:317–324

Nirjar RS, Jain SS, Parida M, Katiyar VS, Mittal NA (2003) A Study of transport related noise pollution in Delhi. J Inst Eng (I) Environ 84:6–15

Nucara A, Pietrafera M, Scacciance G, Stalixi G (2002) A comparison between analytical models and Artificial neural networks in the evaluation of traffic noise levels. In: Proceedings of ICA, Rome

Parabat K, Nagarnaik PB (2007) Assessment and ANN modeling of noise levels at major road intersections in an Indian intermediate city. J Res Sci Comput Eng 4(3):39–49

Rajakumar HN, Gowda RMM (2009) Road traffic noise pollution model under interrupted traffic flow condition. Environ Monit Assess 14:251–257

Rao P, Rao S (1991) Prediction of L_{AT} traffic noise levels in the city of Visakhapatnam, India. Appl Acoust 34:101–110

Sharma M, Dahiya M, Saini PK, Garg N (2012) Application of artificial neural network for modeling of traffic noise on roads in Delhi. IOSR J Mech Civil Eng 1:31–36

Taghavifar H, Mardari A (2013) Application of artificial neural networks for the prediction of traction parameters. J Saudi Soc Agri Sci 13:35–43

White H, Gallant AR, Harnik K, Stinchcombe M, Wooldridge J (1992) Artificial neural networks: approximation and learning theory. Blackwell, Cambridge

Growth Through Engineering Six Sigma

Rupinder Singh Kanwar, Gurinder Singh Brar and Vikas Ucharia

Abstract Indian organizations are losing edge to China low-cost manufacturing practices and are facing turbulent marketing conditions. Emulation of Chinese practices is rare possibility because of vibrant Indian democracy. Indians have other avenues and opportunities to learn and apply Six Sigma methodology in their manufacturing endeavors to become resilient and profit making. Tremendous work has been done in lean manufacturing. Lean with Six Sigma amalgamation with customer centricity has great potential. It is evident that significant control on defect rates improves financial condition of the company. It further improves image of the company and enhances status among organization on the attribute of quality parameters. It is proven fact now that defect rates have direct impact on the success or failure of the company. It must be the predisposition of each organization to achieve Six Sigma levels and achieve zero defects. Research work is carried out to reduce the defect rate in an automobile parts manufacturing company of Himachal Pradesh, by implementing DMAIC methodology. It is found that lean with Six Sigma implementation defects is reduced to 0.9 % from 3.6 % with better surface finish by implementing right burnishing process. This is not the end of the story, further improvements are possible to reduce it below 0.25 %.

Keywords Resilient · Defect rate · DMAIC · Profit · Burnishing

R. S. Kanwar (✉) · G. S. Brar
Department of Mechanical Engineering, Guru Nanak Dev Engineering College, Ludhiana, 141006, Punjab, India
e-mail: rupinder.singhkanwar@gmail.com

G. S. Brar
e-mail: brar.gurinder@gmail.com

V. Ucharia
Department of Mechanical Engineering, Bahra University, Solan, Himachal Pradesh, India

S. S. Khangura et al. (eds.), *Proceedings of the International Conference on Research and Innovations in Mechanical Engineering*, Lecture Notes in Mechanical Engineering, DOI: 10.1007/978-81-322-1859-3_55, © Springer India 2014

1 Introduction

Lean through Six Sigma with customer centricity is an amalgamation of "lean" thinking in which business attempts to cut waste and Six Sigma, a quality management strategy that strives to eradicate defects and provide impeccable products and services created through customer centricity process. Six Sigma is one of the methodologies, launched by Motorola in 1987. The same concept is applied by various companies of the world in different areas to improve quality of their products. The health care and other service sectors are successfully experimented with Six Sigma to reduce cost and to improve their systems, thereby improving their status in their sphere of their specialties.

Two basic requirements to survive in a competitive environment are growth of business and profitability.

- Business growth is possible only if the organization is "customer focused" and is able to satisfy all the stated and implied needs of the customer with respect to quality, delivery, and price (Desai 2006).
- Profitability can be achieved by increasing price and/or reducing the manufacturing cost. Since price is decided by the competition, the surest way to increase profit is to cut down the manufacturing cost by achieving total excellence in design and
- Lean Six Sigma is a business strategy and methodology that increases process performance, resulting in enhanced customer satisfaction and improved bottom-line results. It is also being widely recognized that Lean Six Sigma is an effective leadership development tool. Welch and Welch (2005) points out that "Perhaps the biggest but most unheralded benefit of Six Sigma is its capacity to develop a cadre of great leaders."

2 Case Study

2.1 Define Phase

Define: Is the process of ascertaining the root cause for the problem in line with smart goals associated with customer centric initiatives in the company. Moreover, this phase encourages developing mechanism for collecting customer satisfaction/dis-satisfaction data or consumer's feedback for preventive and corrective actions. Our study emphasis on the pivot oil pump (POP) manufacturing is manufactured for various automobile manufacturers and is facing major problem of rejection.

Research Case: Microturner Barotiwala is leading manufacturer of automobile parts. POP is the major part manufactured as shown in Fig. 1, which was facing a problem of rejection. It is the goal of manufacturer to reduce the rejections.

Problem Statement: Manufacturing organization is facing a rejection rate of 36,000 ppm.

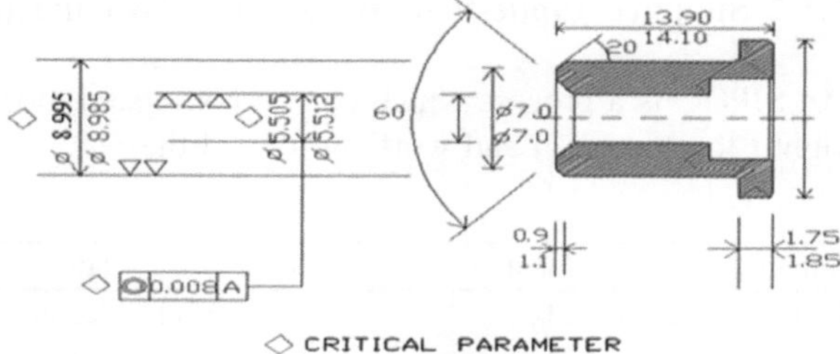

Fig. 1 Drawing of part (POP)

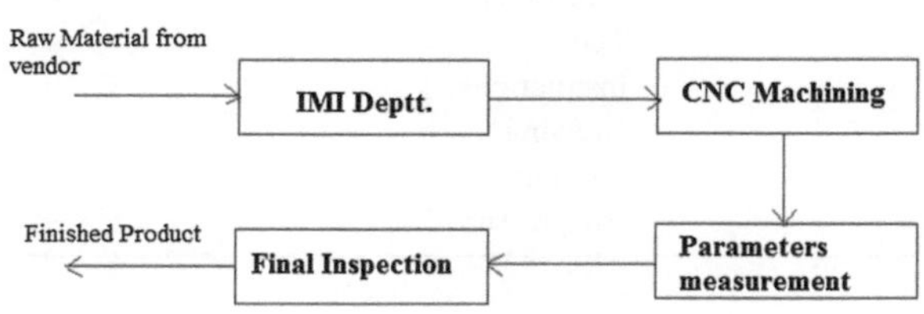

Fig. 2 High-level process map

Goal statement: Increase productivity and reduce defect percentage to minimum level.

Following parameters are responsible for rejection of POP

1. Internal diameter
2. Outer diameter
3. Length of pivot
4. Collar diameter
5. Concentricity
6. Miscellaneous (raw material not provided as per drawing)

To identify the scope of the project, a high-level process map is used as shown in Fig. 2. The map shows the main production operations that affect the performance of the production process.

2.1.1 Brainstorming

A brainstorming session was organized with the team of 8 members and following conclusive points came to notice

1. Material from vendor.
2. Handling of machined part.
3. Unskilled workers with less knowledge of work, not following instruction properly.
4. Jaws of machine play in slides and improper butting.
5. Air leakage and air fluctuation for measurement.
6. Voltage fluctuation.
7. Vibration in machine or due to other machines.

2.1.2 Supplier–Input–Process–Output–Customer (SIPOC)

An SIPOC is a process which can be completed most easily by starting from the right ("customers") and working toward the left.

Supplier	Inputs	Process	Output	Customer
POP from vendor	Man Machine Electricity Tools Instructions Drawing Program Air pressure	CNC machining Measurement	Manufacture part (POP) Worn tool rejection	Inspection Rework process Scrap

2.1.3 Critical to Quality (CTQ) Identification

CTQs are derived from customer needs. For cost considerations, scrap plays major role, as scrap is directly related with rejection as shown in Fig. 3.

2.2 Measure Phase

Measurement system analysis (MSA) is used as pilot tool for checking the gauge repeatability and reproducibility. A measurement systems analysis evaluates the test method, measuring instruments, and the entire process of obtaining measurements to ensure the integrity of data used for analysis.

To start the measurement phase, a data collection plan is prepared and distributed to the team, for gathering necessary information which is analyzed in the next phase. After set of pilot run data sheet is prepared in the form of gauge repeatability and reproducibility (R&R). Operators are advised to enhance collaboration and to spread out the knowledge among new operators. And it is found that both man and gauges are working properly. A sample from the collected data for the gauge R&R is shown in Table 1.

2.3 Analyze Phase

Analyze phase is finding root causes responsible for customer dissatisfaction. It deals with finding the critical factor in the system and current operation which could be improved to minimize the variation. Before starting this phase, different studies are performed.

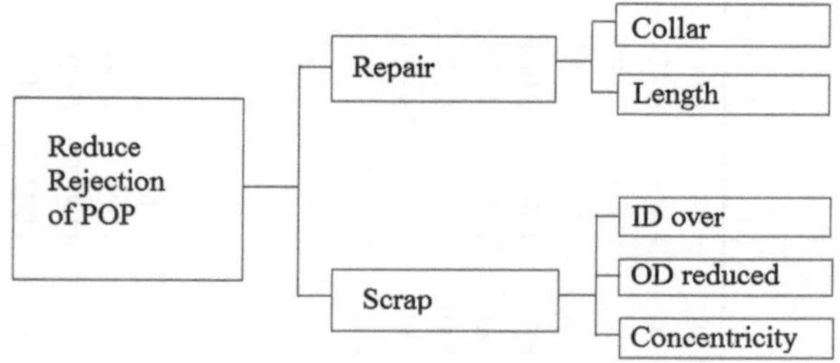

Fig. 3 CTQ tree for POP

To check the machine for its vibration and disturbance due to neighboring machines, environment of the organization, trainings are provided to the operators, air pressure of air pressure gauge (APG) is checked, and leakage in air pipe lines is checked.

2.3.1 Scatter Diagram

It gives the correlation of OD and ID and how both are dependent on each other; a scatter plot is drawn as shown in Fig. 4, and correlation shows that if ID decreases OD decreases vice versa.

2.4 Improvement Phase

This is the phase of originating and implementing right solutions for achieving our predetermined goals in line. Having identified and verified main types of wastes, the team proposed actions to minimize the probability of reoccurrence for these problems. Brainstorming sessions are held to generate solution alternatives, and it has been decided by the keeping the previous experience in the mind of the team members to shrink the tolerance of outer diameter so that concentration can be made to maintain the OD parameter only which may be assumed to follow scatter plot to meet the customer demand to achieve the customer centricity to be resilient organization.

A total of 750 pieces of said component are machined, and measurement of OD and ID performed 750 reading are divided into 15 sets, group of 50. Correlation of all 15 sets is determined, and on the basis correlation, data process capability chart is prepared as shown in Fig. 5.

It is found that after doing this practice value of process ($C_{pk)}$ comes to be nearly 15,000 ppm as the cost of each part to the organization is Rs. 4/- which has saved the rejection cost.

Table 1 Gauge repeatability and reproducibility data sheet

Operator name		1	2	3	4	5	6	7	8	9	10	Average
Dilip Kr	1	8.982	8.987	8.965	8.966	8.992	8.994	8.989	8.980	8.998	8.983	8.988
	2	8.982	8.987	8.965	8.988	8.993	8.995	8.990	8.980	8.999	8.983	8.988
	3	8.983	8.963	8.986	8.988	8.994	8.995	8.990	8.981	8.999	8.983	8.989
	Average	8.9823	8.9873	8.9857	8.9873	8.9930	8.9947	8.9897	8.9803	8.9987	8.9830	8.9852
	Range	0.001	0.001	0.001	0.002	0.002	0.001	0.001	0.001	0.001	0.000	0.001
Mukesh Kumar	1	8.982	8.966	8.980	8.988	8.993	8.994	8.989	8.980	8.990	8.981	8.9863
	2	8.982	8.967	8.986	8.988	8.994	8.994	8.989	8.980	8.990	8.983	8.9873
	3	8.983	8.967	8.987	8.988	8.996	8.996	8.989	8.981	8.990	8.986	8.9883
	Average	8.9823	8.9867	8.9843	8.9880	8.9943	8.9947	8.9890	8.9803	8.9900	8.9833	8.9873
	Range	0.001	0.001	0.007	0.000	0.003	0.002	0.000	0.001	0.000	0.005	0.002
Vikas	1	8.980	8.965	8.981	8.987	8.993	8.993	8.987	8.979	8.987	8.981	8.9853
	2	8.981	8.965	8.985	8.988	8.993	8.993	8.988	8.979	8.988	8.982	8.9862
	3	8.981	8.966	8.985	8.988	8.994	8.994	8.989	8.980	8.988	8.982	8.9867
	Average	8.9807	8.9653	8.9837	8.9877	8.9933	8.9933	8.9880	8.9793	8.9877	8.9817	8.9861
	Range	0.001	0.001	0.004	0.001	0.001	0.001	0.002	0.001	0.001	0.001	0.001
Part	Average	8.9818	8.9864	8.9846	8.9877	8.9936	8.9942	8.9889	8.9800	8.9921	8.9827	0.002
	Range	0.001	0.001	0.004	0.001	0.002	0.001	0.001	0.001	0.001	0.002	0.012
Measurement unit analysis				*Percentage process variation*								
Reproducibility (Av)		0.0046		% Ev	21.851							
Reproducibility		0.0057		% Av	27.220							
R&R		0.0073		% R & R	34.906							
Part variation		0.01962		% PV	93.710							
Total variation		0.02094										

If GRR < 10, gauge is acceptable
If 10 < GRR < 30, gauge is conditionally acceptable
If 30 < GRR, gauge is unacceptable and must be replaced/modified

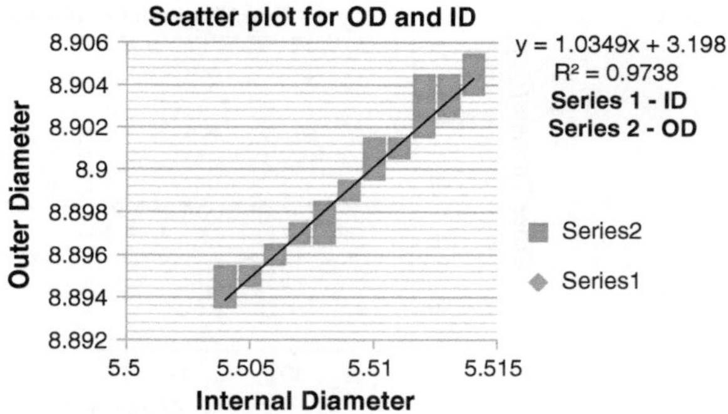

Fig. 4 Scatter plot for correlation of OD and ID

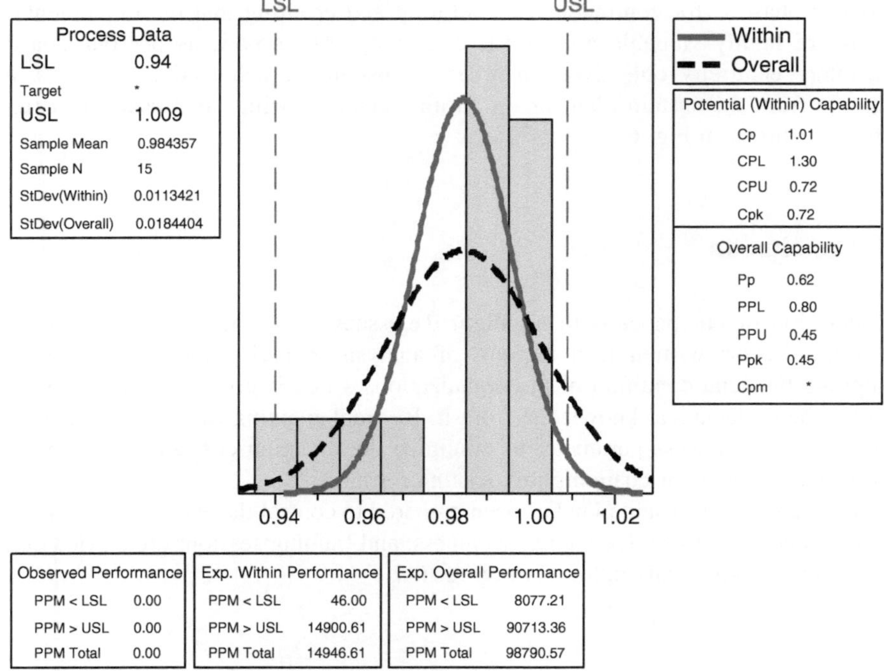

Fig. 5 Process capability for correlation of OD and ID

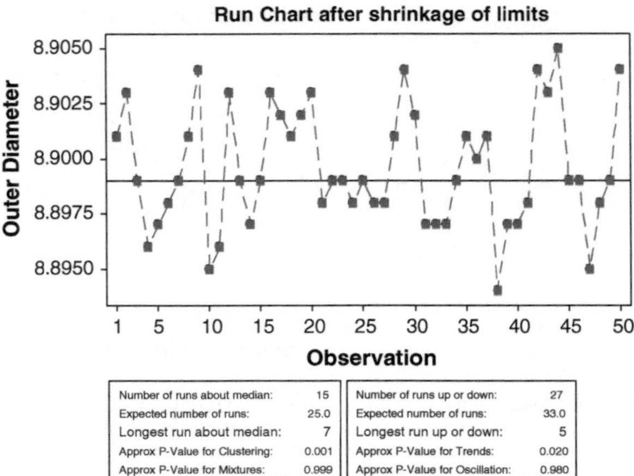

Fig. 6 Run chart after shrinkage of limits

2.5 Control Phase

Control phase is for continuously monitoring and ensuring that the implementations are highly valuable and effective, serving the purpose as per our goals/ customer centricity objectives. In order to sustain the achieved goal/results, a sample from the sustain plan for the turning center machine is presented in run chart as shown in Fig. 6.

3 Conclusion

Conclusion of our paper is to highlight the issues arising in the manufacturing process. Report written is a mixture of analysis, experimental experience, to improve financial condition of the organization, which is good experience for us and gained a technical knowledge from it. By implementing Six Sigma, we have optimized the process parameter to minimize the rejection cost and increase the profit margin without changing the existing resources.

Furthermore, a sustain plan has been prepared to control the achieved results for future benefits of Lean Six Sigma awareness, and training sessions are carried out, focusing on individuals rather than on results.

References

Desai DA (2006) Improving customer delivery commitments the Six Sigma way: case study of an Indian small scale Industry. Int J Six Sigma Competitive Advantage 2(1):23–47

Welch J, Welch S (2005) Winning. Harper Business, New York

Minimizing the Make Span and Tardiness of Identical Parallel Machines by Genetic Algorithm

Kamaljeet Bhambri

Abstract Recent customer demand for high-variety products has contributed to an increase in product complexity that emphasizes the need for improved scheduling. The scheduling leads to increase in capacity utilization efficiency and hence thereby reducing the time required to complete jobs and consequently increasing the profitability of an organization in present competitive environment. Job shop scheduling is generally considered to be the one of the most significant issues in the planning and operation of a manufacturing system. Job shop scheduling (JSS) with make span criteria is non-polynomial hard. A genetic algorithm has been proposed which generates improved make span time and reduce tardiness, earliness costs for *m* number of identical machines with *n* given number of jobs in Sartaj Industry, Chandigarh. A Java-based program has been developed. The proposed algorithm is found to generate improved results, viz, cost saving for an annum and the total cost reduction.

Keywords Non-polynomial (NP) · Genetic algorithm (GA) · Job shop scheduling (JSS)

1 Introduction

Machine scheduling is process in which loading a work piece on machine from starting time of machine to completion time of jobs. With optimal scheduling, the completion time of last job to leave the machine is improved. When a last job leaves earlier and make span is improved which will make production cost lesser and better utilize resource.

K. Bhambri (✉)
Department of Mechanical Engineering, SUSCET, Tangori, Punjab, India
e-mail: bhambri2002@yahoo.co.in

S. S. Khangura et al. (eds.), *Proceedings of the International Conference on Research and Innovations in Mechanical Engineering*, Lecture Notes in Mechanical Engineering,
DOI: 10.1007/978-81-322-1859-3_56, © Springer India 2014

Traditional scheduling is done through experience of foreman and experience of operator, but it is very difficult to solve because in practical scheduling, problems are NP-hard. That scheduling done are not optimal and consume much more time. The genetic algorithm with sectional codes is most effective method to solve these NP hard problem.

1.1 Identical Parallel Machine

It is defined as that every workpiece can machine on same type of machine, where machines are set up in parallel as in production system and manufacturing system. In identical system jobs are represented by $J = 1, 2, 3, 4, \ldots, n$ and machine are represented by M_1, M_2, M_3, M_4, … , M.

1.2 Nonidentical Parallel Machine

It is defined as that every work piece can machine on different type of machine, where machines are set up in non-parallel as in production and manufacturing system. In nonidentical parallel system, jobs are represented by $J = 1, 2, 3, 4, \ldots,$ n and machines are represented by M_1, M_2, M_3, M_4, … , M.

2 Methods

As the name suggests, GA is the process of evolution and uses the idea of natural selection, recombination, and mutation to control the search process. The set of solutions is called population. A single solution of this population is called chromosome. Chromosomes are made of genes. Genetic algorithms are an optimization technique based on natural evolution. They include the survival of the fittest idea into a search algorithm, which provides a method of searching. In nature, the fittest individuals are most likely to survive. Therefore, the next generation should be fitter and healthier because they were bred from healthy parents. This same idea is applied to a problem by GA. An initial population is created, and subsequent generations are generated according to a prespecified breeding and mutation methods inspired by nature. The methods for creating an initial population are varied feasible only, randomized, using heuristics, etc. Due to the large search space in job shop scheduling, it is expected that random generation of initial solutions provides relatively strong results. In the present work, the GA has been proposed for generations of initial chromosome.

The proposed GA is described in the following:

Step 1 Set a tournament size $K = 4$
Step 2 Generate a random permutation of the chromosomes in the current population.
Step 3 Compare the fitness value of the first k chromosomes listed in the permutation and copy the best one into the next generation. Discard the strings compared.
Step 4 If the permutation is exhausted, generate another permutation.
Step 5 Repeat steps 3 and 4 until no more selections are required for the next generation.

2.1 Total Tardiness

It is defined as the due dates of jobs to be scheduled as considered for minimization of lateness of jobs for its delivery.

$$\begin{aligned} T_J &= C_k - D_k \quad \text{if } C_k - D_k > 0 \\ &= 0 \quad \text{otherwise.} \end{aligned}$$

2.2 Total Earliness

It is defined as that the jobs to be produced before its due dates which include inventory cost.

$$\begin{aligned} E_j &= D_k - C_k \quad \text{if } D_k - C_k > 0 \\ &= 0 \quad \text{otherwise.} \end{aligned}$$

2.3 The Objective Function

It is including all the above three performance measures that sum of total tardiness, total earliness, and make span simultaneously for job shop scheduling with sequence-dependent setup times and represented as

$$\textbf{Min}\,\mathbf{X} = \alpha_k \sum_{(j=1)}^{k} T_k + \beta_k \sum_{(j=1)}^{k} E_k + \gamma\,\mathbf{C}_{\max},$$

2.4 Assumptions

1. Machines have not been idle.
2. Splitting of job or job cancelation has not been allowed.
3. All process time on the machine has been known, i.e., deterministic and finite.
4. Each machine has been continuously available for assignment, without any breakdown or maintenance.
5. The first machine has been assumed to be ready whichever job has been to be processed on it first.
6. All the jobs and machines have been available at zero time.
7. Setup times for operation has been sequence dependent and included in processing time.

2.5 Nomenclature

α_k	Weight of tardiness or penalty for order
β_k	Weight of earliness or penalty for order
γ_k	Weight of make span
E_k	Earliness of jobs, J
T_k	Tardiness of jobs, J
C_{max}	Make span time
D_k	Due dates of jobs
M	No. of machines = 1, … , m
N	No. of jobs $J = 1, \ldots, n$
P	Production order
C_k	Completion time of jobs, j
S_t	Starting time of order
M_k	Machine type
P_s	Population size
P_m	Probability of machine selection
Q	No. of operations.

A chromosome representation has been developed. Each chromosome has been a sequence of genes whose length is equal to the number of machines to which operations are assigned. In a chromosome, each gene represents a machine and the value of each gene represents the operation numbers and of which the corresponding machine processes. Each part represents one type of machines. Each operation can only be assigned to the machines that can handle it. Table 2 shows the representation that considers a problem with 20 operations to be assigned to one type. These machines have been categorized into four types: type 1 includes machines no. 1–4, type 2 includes machines no. 5–8, type 3 includes machines no. 9–12, and type 4 includes machines no. 13–16. Operations 1, 2, 3, 4, … , must be processed on these types of machines.

2.6 Algorithm

Step 1 Choose a coding to represent problem parameters, a selection operator, a crossover operator, and a mutation operator. Choose population size n and crossover probability P_{m}. Initialize a random population of strings of size 1. Choose a maximum allowable generation number, t = 0 tmax.
Step 2 Evaluate each string in the population.
Step 3 If **t > tmax** or other termination criteria were satisfied, terminate.
Step 4 Perform **reproductions** on the population.
Step 5 Perform **crossovers** on the random pairs of string.
Step 6 Perform **mutations** on every string.
Step 7 Evaluate strings in the new population. Set **t = t + 1** and go to **step 3.**

The algorithm was straightforward with repeated application of three operators **(step 4–7)** to a population of points (string).

2.7 The Initialization Process

1. Index $j = 1$, a population size $P_{\text{s},}$ Q has been no. of operations which each machine can process or handle.
2. For each machine equal chance of its selection. This has been decided on the probability.
3. Random selected machine m which can process operation. The machine's selections depend upon probability P_{m}. The machines which have higher value of P_{m} have been selected first. If value of P_{m} is zero, then machine has been not selected.
4. Assign all operation from start to end from 1 to n. $P_{\text{m}} = P - 1/Q$ For every machine $P = 1$, Q has been the no. of operations.
5. Assign all operation from $I = i + 1 \text{ if } I > n$.
6. If all machines have been assigned, stop the process.

A genetic optimization process based on GA has been developed, which includes a chromosome representation, a heuristic initialization process, and modified crossover and mutation operators.

The algorithm has been then be used to solve the job shop scheduling (JSS) problems using data from Sartaj industry Chandigarh, respectively.

Population size = 200, maximum number of generation = 60, and crossover probability = 0.6.

Mutation probability = 0.01.

The GA algorithm was coded in Java. The various experiments were executed on a personal computer (PC) equipped with a Intel core i3 processor working at a speed of 3.40 GHz with RAM 3 GB and 320 hard disk. The GA heuristic required an execution time of program about 5 s to 1 min.

Table 1 Chromosome representation

Machine processing operation

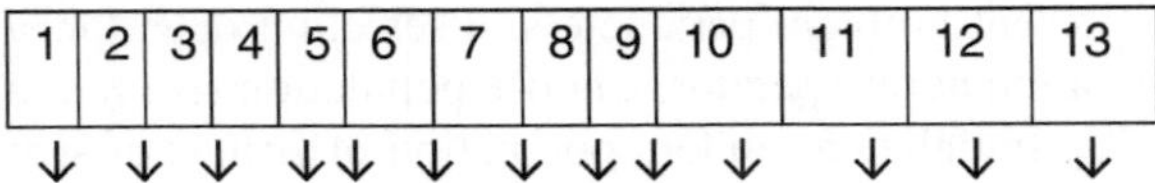

1	2	3	4	5	6	7	8	9	10	11	12	13

Machine types

Type 1	Type 2	Type3	Type 4
Highly Skilled Worker	Semi Skilled Worker	Skilled Worker	Unskilled Worker

3 Results

3.1 Classification of Workers

In this work, the workers have been categorized into different levels of skill and type of work: (1) CNC machines have been operated by highly skilled worker, (2) general-purpose machines which have been operated by semi-skilled worker, (3) special-purpose machines which have been operated by skilled worker, and (4) simple machines that have been operated by unskilled worker Table 1.

3.2 Data Collection

The data have been collected from the Sartaj Industry Chandigarh which is Vendor of PTL, Indo Farm, Swaraj Mazda. The machine shop in this is responsible for production planning and control for different machines which are depicted in Table 2 (10 jobs and 6).

3.3 Operation Assignment for ALL Categories

The scheduling results in which 20 operations have been assigned to four workstations with types 1–4 in which only one type has been selected in the result of Java.

Table 2 Processing time of 10 jobs, 6 machines (min)

Job no.	CNC turning center	Lathe machine	Milling machine	Drilling machine	Tapping machine	Surface grinder
1	20	17	16	10	9	19
2	17	27	18	9	8	18
3	25	18	18	8	7	14
4	27	19	17	11	11	17
5	29	11	19	12	10	18
6	35	19	28	8	14	17
7	19	17	24	9	11	21
8	20	10	29	9	9	22
9	22	14	21	11	8	26
10	27	12	18	10	9	27

3.4 Cost Benefits

Cost Saving for month = Rs. 14,450
Cost Saving for one year = 14,450 × 12 = Rs. 173,400
Total time save per month = 49.33 h
Total time save per year = 599.91 h
No of days saves per year = 24.99 days
The total cost reduction achieved by 23 %.

4 Conclusions

This paper develops a model of JSS problem for the machine shop. The job shop operation of various machines where scheduling and line balancing are optimized in terms of GA. The scheduling objective in this paper will meet due dates and reduction in cost which will minimize inventory.

Our future research will focus on demand fluctuation, unbalanced orders, and material shortage, because these factors were assumed constant while in assumption.

Implementation of Quality Control Tools in an Automobile Organization to Reduce the Rejection of Casting Components

Jatinder Pal

Abstract The importance of quality has been long recognized in the business environment in order to obtain or manufacture higher quality products. In manufacturing environment, quality improves reliability, increases productivity, and customer satisfaction. In order to survive in a competitive market, improving quality and productivity of product or process is a must for any company. Quality control (QC) tools are essential to improve the quality of manufacturing process. These tools are used to solve 95 % of the problems faced by the organizations for improving its effectiveness. A case study has been conducted in an automobile organization using QC Tools. It has been found that the defects of casting components have 9.79 % per month which have been finally reduced to 8.52 % per month by using QC Tools. The net saving due to reduction in casting components rejections has 2.31 lakhs per month and 27.7 lakhs per year.

Keywords Casting defects · Rejection · Quality · Quality control tools

1 Introduction

The continuous quality improvement process assumes and even demands that team of experts in field as well as company leadership actively use quality tools in their improvement activities and decision-making process. Quality tools can be used in all stages of production process, from the beginning of product development up to product marketing and customer support. The main goals of the quality tools are

J. Pal (✉)
Department of Mechanical Engineering, Guru Nanak Dev Engineering College, Ludhiana 141006 Punjab, India
e-mail: pal.jatinder74@gmail.com

S. S. Khangura et al. (eds.), *Proceedings of the International Conference on Research and Innovations in Mechanical Engineering*, Lecture Notes in Mechanical Engineering,
DOI: 10.1007/978-81-322-1859-3_57,

customer satisfaction by delivery of defect-free products at quality cost. The primary objective of quality control (QC) in any organization is to reduce the cost of its operation. If control efforts do not lead to any saving in costs or cost reduction, then in principle there is no need of QC. The conducted research has investigated possibilities of successful application of QC tools. Improving quality with basic statistical process control tools to significantly improve the monthly defect quality from 13.49 to 7.4 %. Rs.12,677.57 saved per year for an operation of control valve cylinder head. About 75–80 % US companies are practicing QC tools for managing quality. Regarding this, survey has been conducted to see the extent to which they used quality management practice.

2 Literature Review

The researcher described the current situation of implementation of statistical process control tools in twenty-six Indian bicycle industries. Research has identified the motives for implementation of statistical process control tools and difficulties encountered during implementation of statistical process control tools. It has been revealed that histogram and check sheets are the most popular statistical process tools in Indian bicycle industries (Grewal and Gill 2007). The research has presented the reduction in the claims of the cement bag by practical applications of (QC) tools in process industry (Rahman et al. 2009). The results of eight case studies performed as part of a research project in development of statistical process control (SPC) in Malaysian manufacturing companies (Paveletic et al. 2008). Automotive industries experts' opinion in Malaysia and Indonesia investigates the difficulties associated with the use of quality engineering tools and techniques (Putri and Yusuf 2009). QC tools in three companies of Malaysia were implemented and discussed how practitioners and policy makers could select QC techniques and overcome some difficulties and problems in its implementations (Judi et al. 2009). The case study was conducted to reduce the cost of the component using the applications of QC tools in "Automotive Industries in Central India." By improving quality, the method of optimization reduces process operational costs, cost saving was the main objectives of applying QC tools. Rs. 12,677.57 saved per year (Mandavgade and Jaju 2009).

3 Methodology

Identification of the problem: Critical casting defects: Sand crush, mold shift, cold shut, pin hole, slag, shrinkage, mold broken, fettling defect, core shift, vermicular defects, hard spot, and miscellaneous (machining) have been found in the

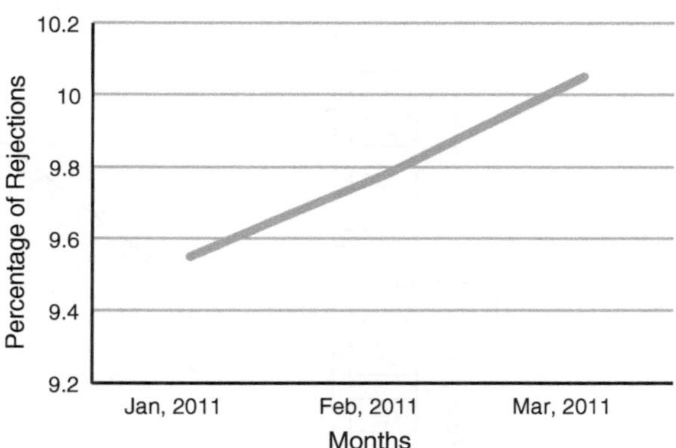

Fig. 1 Rejection trend of casting components

SHBC133, 241, 640, 124, 569, 562, and 563 casting components in G.S auto, Ludhiana. The monthly rejection of the casting components has been alarming, i.e., 10 %. Temperature, time, pressure, and design parameters have the greatest effect on the rejection. The casting defects has to be reduced by implementation of QC tools, data collection: The rejection data have been collected from the casting shop from January 2011 to March 2011 from the daily inspection reports and rearrange the data defect wise.

Data analysis: The Pareto chart, fishbone diagram, and flow diagram have been used for data analysis in this study. Pareto chart allows the user to focus attention on a few important factors in a process. Fishbone diagram was used to associate possible causes with a single effect. Flowcharts have used in identifying where errors are likely to be found in the system.

Remedies/actions: Corrective and preventive action have been taken in the casting shop to reduce the rejection of the casting components.

Implementation of QC Tools: On the basis of outcome of previous steps results, QC Tools have been implemented.

Standardization: Standard operating procedure was revised after analyzing the results.

Rejection trend of casting components: Fig. 1 shows the rejection trend of the casting components from January 2011 to March 2011. The average rejection of the casting components has approximate 10 % which has alarming.

Figure 2 shows the Pareto analysis of the casting defects. Casting defects (a–i) represent along x-axis, and cumulative rejection represents along y-axis. First seven defects contribute more than 80 % in rejection. Sand crush defect contributes maximum 24.03 %, and hard spot defect contributes least, i.e., 1.39 % in rejection (Table 1).

Fishbone diagram for sand crush defect (Fig. 3).

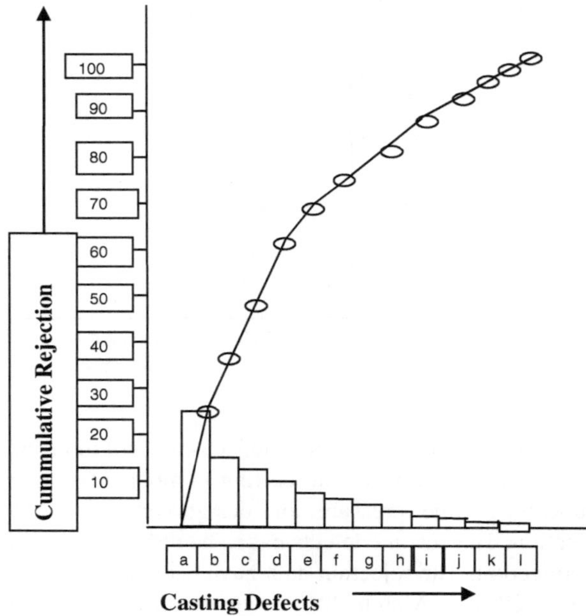

Fig. 2 Pareto analysis

Table 1 Cumulative casting rejection

S. No.	Casting defect	Rejection	Cumulative rejection	Cumulative %
a	Sand crush	3,424	3,424	24.03
b	Mold shift	1,831	5,255	36.88
c	Cold shut	1,776	7,031	49.35
d	Pin hole	1,544	8,575	60.19
e	Slag	1,361	9,936	69.74
f	Miscellaneous	940	10,876	76.34
g	Shrinkage	902	11,778	82.67
h	Mold broken	883	12,661	88.87
i	Fettling defect	704	13,365	93.81
j	Core shift	348	13,713	96.25
k	Vermicular defect	336	14,049	98.61
l	Hard spot	198	14,227	100

4 Counter Measure for Man

Sufficient training was provided to operators regarding the causes and cures of sand crush defect. Specific work instructions were provided to operators, regarding the sand crush defect. The pattern plate condition (fitment of screws, dowels, and cleanliness) was properly checked before loading on machine. Before making the

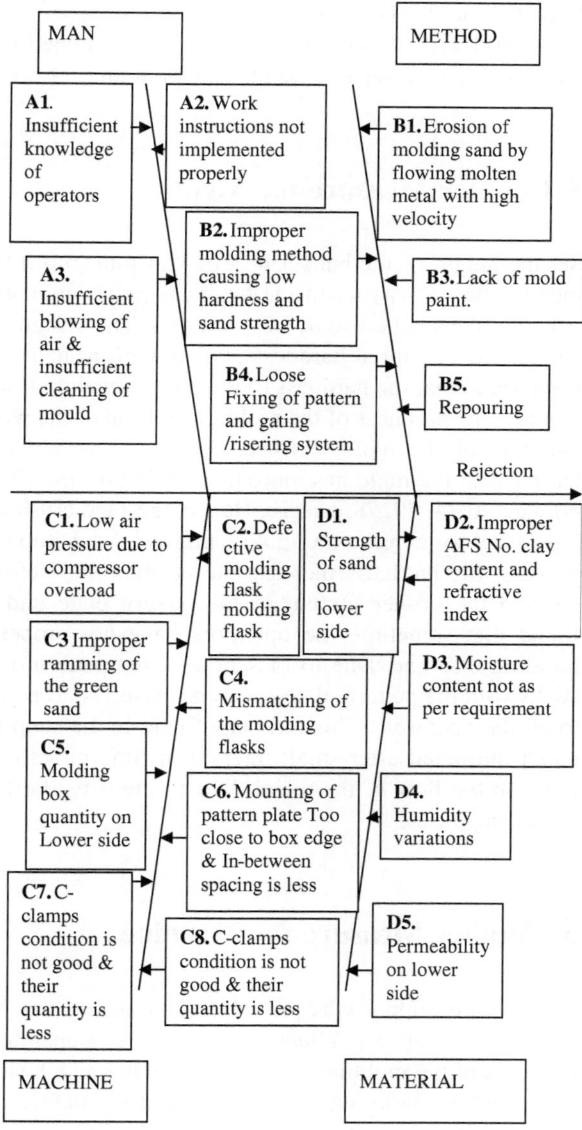

Fig. 3 Fishbone diagram for possible causes

molds, the pressure was maintained to 90–120 psi. The condition of the machine was checked before loading the pattern plate. In case of any discrepancy, the maintenance incharge was informed. The mold was cleaned with air pressure before closing. The mold hardness was maintained to more than 80. Strength of sand is checked twice in a day. OK C-clamps are used, and rejected C-clamps are separated and sent to the nonconforming area. The mold was filled to less height buckets of up to six inches. (152.4 mm). The range of temperature was maintained at 1,380–1,420 °C. The mold boxes were clamped diagonally, and the defective mold boxes were separated out. Sufficient air was provided for cleaning the mold.

5 Counter Measure for Method

Set the height of the bucket up to 152.4 mm (6 in.) to control the velocity of the metal. $V = \sqrt{(2gh)}$ when the height is more, velocity is also more. Due to high velocity, the sand crush defect increases. The design of the wooden plate has been changed for required hardness and strength near the edges of the mold. This has been found that the hardness of the mold found on lower side, i.e., less than 80. To increase the hardness of the mold, the design of the wooden plate has changed. The hardness of the mold has been found not as per company plan. The required hardness of the mold has more than 80 but during checking it was found less than 80, i.e., 76.74, 79, 78, and 70. Hence, the sand crush defects more. Mold paint has been used to increase the mold hardness. After implementation of the mold paint process, the hardness increased more than 80. Before loading, the pattern plate verified the proper fitment of the pattern plate and runner/riser. This has been found that sometime the operators have not properly fitted the pattern on the molding machine. Due to looseness of the pattern on the machine, molding sand sticks with the pattern plate and hence crushes. The proper instructions have given to all the operators. This has been found in the shop that the operators will report about the metal after small interval of time in some cases, which causes irregularity in the flow of the metal. Due to the irregularity in the flow, the sand crush defect increases.

6 Counter Measure for Machine

Separate air supply was provided for molding machines with required range of pressure 90–120 psi. When the machine has operated under 90 psi pressure, the hardness of the molding sand was found below 80 which causes sand crush defect. The molding flasks were inspected, and the defective molding flasks were separated out. The fitment of the bush, GO, NO GO size with plug gauge and alignment of the flasks were checked. This has been found that some operators ignore these points so proper training has been provided with the help of supervisors to those operators in the company. The matching of the molding flasks was checked and found that some molding flask has been mismatching. The mismatching molding flasks that cause sand crush defect was separated out. Molding box quantity increased from 190 to 204 to reduce the time for pouring. The modification in design of pattern plate was needed to increase the size of spacing between the molding boxes and pattern plate edges to reduce the sand crush. The defective over

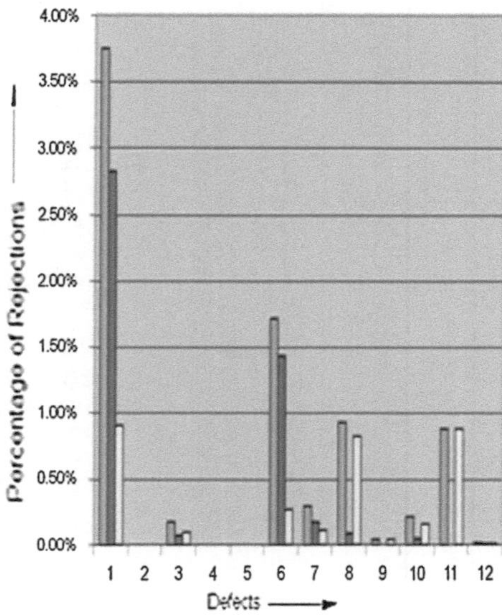

Fig. 4 Reduction in rejection SHBC124

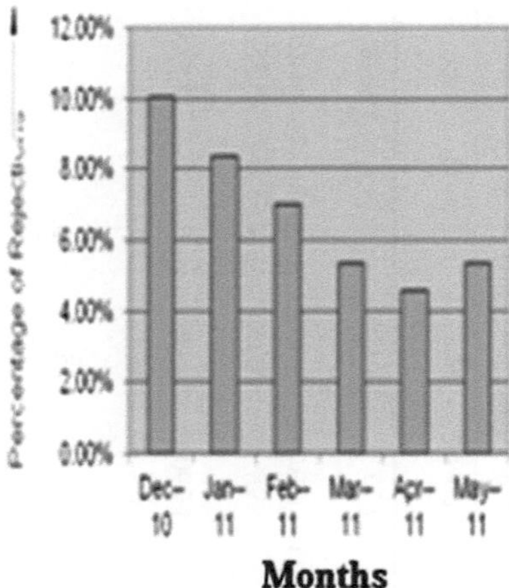

Fig. 5 Monthly rejection of SHBC124

size bushes and under size pins of molding boxes clamping were changed which causes sand crush. Damaged C-clamps without thread have been used in the shop which causes not to clamp the molding boxes properly. Due to the looseness of the C-clamps on the molding box, the sand crushed when track has moved. Damaged threaded C-clamps have been separated out, and their quantity has increased from 150 to 200 (Fig. 4).

The range of the moisture was set to 3–3.5 % for greater strength. In case of humidity on higher side, the moisture content was set near 3 %.This has to depend upon the climate condition. Permeability has found on lower side in some cases due to over-ramming of the mold. Hence, back pressure has been generated in the

Table 2 Rejection analysis of casting component SHBC124

Rejection analysis	Dec-10	Jan-11	Feb-11	Mar-11	Apr-11	May-11	Rejection % (Dec–Mar)	Rejection % (Apr–May)	Reduction in rejection
Monthly rejection (%)	8.56	9.54	9.77	10.05	8.29	8.78			
Item	SHBC124	SHBC124	SHBC124	SHBC124	SHBC124	SHBC124			
Percentage of rejection	10.02	8.34	6.98	5.35	4.58	5.35			
OK	4,187	3,838	4,290	2,602	2,582	2,885			
Not Ok	466	350	322	147	124	163			
REJ. PTS									
Sand crush	148	164	163	85	53	102	3.75	2.83	0.91
Pin holes									
Mold shift	6	16	3	2	3	1	0.18	0.07	0.10
Slag									
Miscellaneous									
Shrinkage	66	85	70	37	45	34	1.72	1.44	0.27 %
Mold broken	13	11	16	6	2	8	0.30	0.18	0.11 %
Fettling defect	68	50	22		4	1	0.93	0.09	0.83 %
Hard spot	4			2			0.04		0.04 %
Core shift	10	13	9	2		3	0.22	0.05	0.16 %
Vermicular Graphite	132						0.88		0.88 %
Cold shut	1	1	1	1	1	0	0.02	0.01	0.01 %
Total pieces (Dec–Mar)	14,927								
Reject (Dec–Mar)	1,285								

(continued)

Table 2 (continued)

Rejection analysis	Dec-10	Jan-11	Feb-11	Mar-11	Apr-11	May-11	Rejection % (Dec–Mar)	Rejection % (Apr–May)	Reduction in rejection
Rejection % (Dec–Mar)	8.6								
Total pieces (Apr–May)	5,467			Reduction in rejection = 8.60 − 5.24 = 3.35 = 183 pieces/2 months = 92					
Reject (Apr–May)	287			Cost of one piece = Rs. 411.18					
Rejection % (Apr–May)	5.24			Saving in 2 months = 183 × 411.18 = Rs. 75,245.94					
				Saving in 1 year = 75,245.94 × 6 = Rs. 451,476.64					

Shot blast charges
Runner/riser wt/pc for item SHBC 124 = 1.61 kg/pc
Runner/riser shot blast cost/kg for item
SHBC 124 = Rs.0.12
Runner/riser shot blast cost/pc for item
SHBC 124 = 0.12 × 1.61 = Rs.0.19
Number of pcs produced from April to May = 5,467
Runner/riser shot blast cost for item
SHBC 124 (April–May) = 0.19 × 5,427 = Rs. 1,031.13
Mold paint charges
Mold paint charges for 1 pc = Rs. 2.5
Mold paint charges for 5,467 pcs in two months = Rs. 13,667.75
Net Saving for SHBC 124 IN Two months = 75,245.94 – 14,698.88 = Rs. 60,547.06

mold that causes sand crush. The range of the permeability was set to 90–150 mm. Similar techniques and methodology were used to minimize the others defects (Fig. 5).

Rejection analysis of casting component SHBC124 is shown in Table 2.

Similar rejection analysis method has been used for other casting components

7 Conclusions and Future Scope of Work

Rejection has been reduced by 1.27 % (9.79–8.52 %) per month for all the casting components, the saving per month is Rs. 2.31 lakhs and the saving per year is Rs. 27.7 lakhs. The pouring time has been reduced from 10 to 7 min. There is a large scope for this research in the nonferrous foundry and machine shop for the same organization. The study can also be extended to other manufacturing companies in India or abroad. Automation solution can be introduced to reduce the cycle time for pouring in foundry shop. Also analysis will be done on customer complaints in any organization using QC tools.

References

Grewal C, Gill S (2007) Measuring SPC implementation in Indian bicycle Industry. Udyog Pragati 31(3):30–37

Judi H, Jenal R, Genasans D (2009) Some experiences of quality control

Mandavgade K, Jaju B (2009) Optimization of cost by using 7 QC tools in India. Int J Eng Stud 1(3):149–160

Paveletic D, Skovic M, Paliska G (2008) Practical application of quality tools. Int J Qual Res 2(3):199–205

Putri Nilda, Yusuf SM (2009) Critical success factors for implementing quality engineering tools and techniques in Malaysian and Indonesian's automotive industries. Int Multiconf Eng Comput Sci 2:978–988

Rahman M, Zail R, Nopiah Z, Ghani A, Deros B, Mohammad N, Ismail A (2009) The implementation of SPC in Malaysian manufacturing companies. Eur J Sci Res 26(3):453–464

Traffic Noise Modeling Using Artificial Neural Network: A Case Study

Raman Kumar, Arun Kumar, Mahakdeep Singh and Jagdeep Singh

Abstract The heterogeneous features of traffic noise, together with the characteristics of environmental noise, with their great spatial, temporal, and spectral variability makes the matter of modeling and prediction a very complex problem. A need is being felt to develop a traffic noise prediction model suitable for the Indian condition. The present work represents a traffic noise prediction model taking Patiala–Sangrur highway as a representative/demonstrative site. All the measurements of noise levels were made at selected points around the highway at different time on number of days in a staggered manner in order to account for statistical and temporal variations in traffic flow conditions. The noise measurement parameters recorded were traffic volume, i.e., number of vehicles passing through in a particular time period, vehicle speed, and the noise descriptors recorded were the equivalent noise level (L_{eq}) and percentile noise level (L_{10}). Artificial neural network (ANN) approach has been applied for traffic noise modeling in the present study. After training and testing of the ANN, it was found that the values of correlation coefficient (R) were 0.9486, 0.9577, and 0.9255 for the training, validation, and testing samples, respectively, and the percentage error varied from −0.19 to 0.64 and 0.54 to 0.99 for L_{eq} and L_{10}. Therefore, a good correlation coefficient and less percentage error between experimental and predicted output obtained is an indication of prediction capability of neural network.

Keywords ANN · Traffic noise model

R. Kumar (✉)
Punjab Technical University, Jalandhar, Punjab, India
e-mail: ramankakkar@gmail.com

R. Kumar · A. Kumar · M. Singh
Department of Mechanical Engineering, Chandigarh University, Gharuan, Punjab, India

J. Singh
Department of Mechanical Engineering, MMU University, Mullana, Haryana, India

S. S. Khangura et al. (eds.), *Proceedings of the International Conference on Research and Innovations in Mechanical Engineering*, Lecture Notes in Mechanical Engineering,
DOI: 10.1007/978-81-322-1859-3_58,

1 Introduction

Highway traffic noise has been a federal, state, and local problem. Emanating from vehicle engines, exhaust systems, and tires interacting with pavement, traffic noise affects the quality of life for nearby residents and businesses by drowning out conversations, disrupting sleep, and discouraging outdoor activities. Over the years, community and motorist concerns have fueled the push to improve noise measurement and modeling tools that help transportation agencies address the highway traffic noise problem. Traffic noise prediction models are required means in designing new highways and other roads or redesigning traffic flow in existing roads to have comfortable traffic noise conditions. In the last few years, a number of prediction tools have been created to estimate noise levels in several countries. For instance, Steele (2001) has reviewed commonly used models such as CORTN, STAMINA, and FHWA. A GIS-based road traffic noise prediction model was developed for use in China by Lia et al. (2002). Pichai and Prakob (2002) formulated a model of highway traffic noise based on vehicle types in Thailand; Calixto et al. (2003) from Brazil developed a statistical model to estimate road traffic noise in an urban setting.

2 Methodology

This study is aimed to develop a more relevant and accurate free-flow traffic noise prediction model for a 2-lane highway in India, taking Patiala (Punjab) city as a representative city, based on L_{eq} and L_{10} as noise descriptors and traffic flow, percentage of heavy vehicle, and average speed as traffic noise parameters. Data were recorded for a total of 133 h duration on different dates and timings in a random or staggered manner in order to account for the statistical and temporal variations in traffic flow characteristics. To fulfill the above objective, artificial neural network (ANN) approach was applied in the present study. The measured parameters were divided into two classes, i.e., output parameters (L_{10}, L_{eq}) and input parameters (vehicle volume/h., percentage of heavy vehicles and average vehicle speed). The input parameters were randomly subdivided into three sets:

Training These are presented to the network during training, and the network is adjusted according to the error.

Validation These are used to measure network generalization and to halt training when generalization stops improving.

Testing These have no effect on training and so provide an independent measure of network performance during and after training.

A comparison was also made between experimental and the ANN output parameters.

Fig. 1 Sound level meter on a tripod with windscreen

2.1 Measurement Procedure

For traffic noise measurements at a suitable site, systematic noise monitoring was done during April–May 2012 using sound level meters (Cesva SC-310) (Fig. 1).

The sound level meter was suitably calibrated, and microphone mounted on a tripod was suitably leveled.

The SLM was mounted at a height of 1.2 m above the ground level and was located at a distance of 10 m from the center of the road lane. Continuous data were recorded with sound level meter during daytime from 9.00 a.m. to 5.00 p.m. at the selected site. The noise descriptors L_{eq} and L_{10} were measured for 15 min. duration in dB (A) weighting with slow response.

Unusually, high values of L_{eq} represent the cases of vehicles honking continuously or the vehicles are without proper silencers, etc.

2.2 Measurements

The data were measured at the selected site as per the described procedure (Fig. 2).

The measured parameters were traffic volume, average speed of vehicles, and noise descriptors (L_{eq} and L_{10}). Traffic volume was measured during the measurement period, and it was varied from 1,155 to 1,380 Veh/h out of which range

Fig. 2 Measurement site

Table 1 Measured data

Time	Traffic vol. Q (Veh/h)	Heavy vehicles *P* (%)	Avg. vehicle speed *V* (km/h)	Sound pressure level dB (A)			
				L_{eq}	L_{10}	L_{max}	L_{min}
Date: 31/03/12 (Wednesday)							
09:00–10:00 a.m.	1,218	32.67	48.6	76.7	79.9	92.5	52.8
10:00–11:00 a.m.	1,221	34.31	49.2	75.6	80.4	91.8	56.3
11:00–12:00 a.m.	1,270	32.12	48.6	75.2	80.5	92.6	55.1
12:00–01:00 p.m.	1,227	33.82	47.1	75.8	80.4	88.9	54.6
02:00–03:00 p.m.	1,261	32.51	49.3	74.9	79.6	89.6	57.5
03:00–04:00 p.m.	1,205	31.12	46.1	76.3	80.8	96.3	54.6
Date: 01/04/12 (Thursday)							
09:00–10:00 a.m.	1,285	37.19	50.3	75.8	78.9	98.0	56.3
10:00–11:00 a.m.	1,255	33.69	51.9	76.3	79.2	96.7	54.4
11:00–12:00 a.m.	1,320	34.57	47.1	75.6	79.6	98.8	55.8
12:00–01:00 p.m.	1,278	31.84	48.0	74.8	79.4	96.3	56.6
02:00–03:00 p.m.	1,246	35.79	50.6	75.4	79.2	98.5	54.8
03:00–04:00 p.m.	1,265	36.04	52.4	75.5	78.8	96.4	53.3

of percentage of heavy vehicle is 27.45–39.73 %, and average vehicle speed varied from 45.0 to 55.0 km/h. Vehicles are categorized into seven different types as heavy trucks, medium trucks, buses, cars, tractors, three wheelers, and two wheelers considering the Indian conditions. The noise descriptors L_{eq} and L_{10} were displayed on the SLM screen and were recorded in its memory. Representative data for 2 days, and similar data were recorded for 24 days spread over 133 h of measurement time (Table 1).

3 Modeling

In the present study, the multilayer feedforward neural network was trained by the backpropagation learning algorithm which is based on gradient descent with momentum weight and bias learning function. The Levenberg–Marquardt optimization technique was used in backpropagation algorithm. The training of a network by backpropagation involves the feedforward of the input training pattern, calculation of backpropagation of the associated error and adjustment of weights to minimize the error. The hyperbolic tangent sigmoid transfer function $F(x) = [(ex - e\text{-}x)/(ex + e\text{-}x)]$ was chosen for hidden layer and linear transfers function $[f(x) = x]$ for output layer. The performance of the neural network was evaluated in terms of mean square error (MSE) between the targeted output and predicted output for given samples size.

In order to develop a traffic noise prediction model, steps given below are followed:

Step 1 The MATLAB neural network tool box was used for ANN analysis which includes neural network training, testing, performance evaluation, and comparison.

Step 2 Total of 133 hours data was randomly divided into training, validation, and testing data. Data Distribution:

Training = 60 %

Validation = 15 %
Testing = 25 %

Step 3 ANN Architecture

The prediction accuracy of any neural network is dependent on the number of hidden layer and the numbers of neurons in each layer. So to find out the optimal neural network architecture, in the present case, a number of neural networks architecture have been trained and tested by varying number of neurons in hidden layer as shown in Table 2. One-layered neural network architecture 3–12–2 (3—input neurons, 12—neurons in hidden layer, and 2—output neurons) was found to be optimum because of better performance in terms of MSE during training, validation, and testing in both the highway noise descriptors.

Step 4 Training and Testing ANN Architecture

A total of 24 days experimental data sets (samples), including vehicle volume/h. (Log Q), percentage of heavy vehicles (P), average vehicle speed (Log V), L_{10}, and L_{eq}, are randomly distributed for training and testing the ANN model. Initial weights and biases were generated and updated depending upon the error between

Table 2 Comparison of training, validation, and testing data for ANN architecture (3–N–2)

No. of hidden layer neurons (N)	Correlation coefficient (R)			Mean square error (MSE)		
	Training	Validation	Testing	Training	Validation	Testing
5	0.9342	0.9220	0.9343	0.7081	0.9888	0.6937
6	0.9454	0.9482	0.9202	0.5411	0.5333	0.8232
7	0.9580	0.9653	0.8985	0.4246	0.3325	1.1219
8	0.9560	0.9055	0.9189	0.4316	0.9453	0.7756
9	0.9635	0.8868	0.8903	0.3818	1.0612	1.0562
10	0.9586	0.9427	0.8970	0.4454	0.5580	1.0417
11	0.9377	0.8985	0.9229	0.9473	0.9654	0.9285
12	0.9486	0.9577	0.9255	0.5641	0.4058	0.6638
13	0.9579	0.8777	0.9064	0.4220	1.2652	0.9572
14	0.9402	0.9093	0.9439	0.5725	0.9093	0.5917
15	0.9723	0.9301	0.8826	0.2805	0.7242	1.1575
16	0.9611	0.9179	0.8746	0.3806	0.8303	1.2017
17	0.9519	0.9326	0.8845	0.4978	0.6580	1.2343
18	0.9697	0.9050	0.8669	0.3105	1.0356	1.3347
19	0.9719	0.9151	0.8855	0.2750	1.0117	1.1885
20	0.9711	0.9164	0.8475	0.2964	0.7862	1.5105

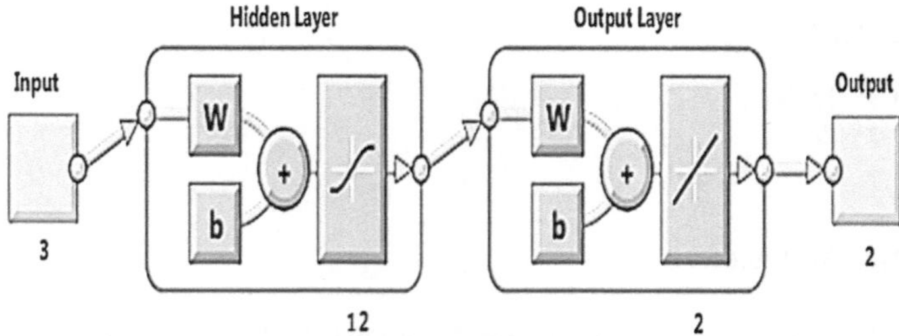

Fig. 3 ANN architecture (3–12–2). *W* weights of connections, *b* bias

predicted and targeted output. The program was terminating the training process if any one of the following condition was achieved:

Maximum number of epochs
Error goal achieved
Minimum gradient reached.

The output equation from the ANN was obtained as: Output = 0.89 × Target + 8.4. (Fig. 3).

4 Results and Discussion

Since the values of correlation coefficient (R) were 0.9486, 0.9577, and 0.9255 for the training, validation, and testing samples, respectively, and the percentage error varied from −0.19 to 0.64 and 0.54 to 0.99 for L_{eq} and L_{10}; therefore, a good correlation coefficient and less percentage error between experimental and predicted output is an indication of better prediction capability of neural network.

5 Conclusions

Multilayer feedforward backpropagation (BP) neural networks were trained and tested by Levenberg–Marquardt (LM) optimization algorithm to predict L_{10} and L_{eq}, highway noise descriptors with different number of neurons in the hidden layer of the neural network, and among all the neural networks tested, one-layered neural network architecture 3–12–2 (3—input neurons, 12—neurons in hidden layer, and 2—output neurons) was found to be optimum because of better performance in terms of MSE during training, validation, and testing in both highway noise descriptors. The values of L_{eq}, L_{10}, and L_{max} from the highway were observed to be high as per Indian standards; therefore, remedial measures are essential to reduce its harmful effect.

References

Calixto A, Diniz FB, Zannin PHT (2003) The statistical modeling of road traffic noise in an urban setting. Cities 20(1):23–29

Lia B, Tao S, Dawson RW, Cao J, Lam K (2002) A GIS based road traffic noise prediction model. Appl Acoust 63:679–691

Pichai P, Prakob V (2002) Noise prediction for highways in Thailand. Transp Res Part D 7:441–449

Steele C (2001) A critical review of some traffic noise prediction models. Appl Acoust 62(3):271–287

A Study on Effect of Operating Parameters on Tire/Road Interaction Noise

Raman Kumar, Jagdeep Singh, Kamaldeep Singh, Arun Kumar and Mahakdeep Singh

Abstract Tire/road interaction noise is a dominant source of noise in almost all types of vehicles running above 50 km/h. Tire/road noise is generated by certain mechanisms such as aerodynamical, mechanical, frictional, or propagational. All these mechanisms depend upon certain operating parameters such as vehicle speed, load on the vehicle, tread pattern type, and road surface. Effects of some of the operating parameters on tire/road noise levels are examined experimentally in this paper. An experimental rig was fabricated, background noise level was measured before taking the measurements, and it was reduced with the help of shielding being done by making a wooden box and a wooden partition having a layer of an absorbing material. Then, the effect of parameters such as vehicle speed, tread pattern design, road surface, and tread wear on tire/road noise was examined. Close proximity method was used for taking the measurements. A-weighted sound pressure level (L_{eq}) and 1-1 octave band frequency spectrum measurements were made under the combined effect of different operating parameters, and their effect was studied.

Keywords Tire noise · CPX · L_{eq}

R. Kumar (✉)
Punjab Technical University, Jalandhar, Punjab, India
e-mail: ramankakkar@gmail.com

R. Kumar · A. Kumar · M. Singh
Department of Mechanical Engineering, Chandigarh University, Gharuan, Punjab, India

J. Singh
Department of Mechanical Engineering, MMU University, Mullana, Haryana, India

K. Singh
Department of Mechanical Engineering, Surya World College of Engineering, Rajpura, Punjab, India

S. S. Khangura et al. (eds.), *Proceedings of the International Conference on Research and Innovations in Mechanical Engineering*, Lecture Notes in Mechanical Engineering, DOI: 10.1007/978-81-322-1859-3_59,

1 Introduction

Nowadays, traffic noise is one of the major problems related to noise pollutions. It is increasing day by day with the increase in the number of vehicles on the road. The sources of noise from a vehicle can be separated into four main areas viz. the engine, air intake, exhaust, and the noise produced as a result of the tires rolling in contact with the road surface. Reductions in the levels of noise from the different sources have been significant, and it is now recognized that the tire/road interaction noise is dominant for constant speeds in almost all types of vehicle. It is also a significant contributor to the overall noise during acceleration and is the dominant noise source above a vehicle speed of 50 km/h. For reducing the noise generation from tires and eliminating the ill effects caused by noise pollution to environment and society, noise produced form the tire/road interaction and the mechanisms involved in noise production have to be studied carefully (Fig. 1).

2 Literature Review

The basic mechanisms of tire/road noise generation were investigated largely from theoretical viewpoint, and the relative importance of noise to the overall vehicle noise has been examined (Leasure and Bender 1975). The mechanisms involved in tire/road noise generation was explained which contribute very much to the vehicle exterior noise by dividing the factors into exciting force, vibration characteristics, and acoustic radiation characteristics (Keijiro and Ichiro 1996). The effect of road roughness in the generation of tire vibration noise has been investigated. The detailed effects of road roughness parameters on tire tread vibration were estimated using a tire/road contact model (Tatsuo et al. 2005). The behavior of sound generated from porous surface has been investigated. For this analysis of sound generated from the interaction of tire and road surface near the contact patch in close proximity and as a function of vehicle speed in acoustic range was done (Douglas et al. 2005). The dynamical tire/road contact for noise prediction was experimentally studied. In situ measurements of contact forces and close proximity noise levels were carried out for as lick tire rolling on six different road surfaces between 30 and 50 km/h. The results show that the macro-texture generates contact forces linearly around 800 Hz, and consequently, noise levels between 500 and 1,000 Hz via the vibrations transmitted to the tire (Bueno et al. 2009). The effect of adhesion forces in the tire/road contact interface was examined, which significantly influence the contact dynamics and resulting noise generation for certain tire/road combinations. For that, an experimental setup was built where tire tread samples were loaded and rapidly unloaded.

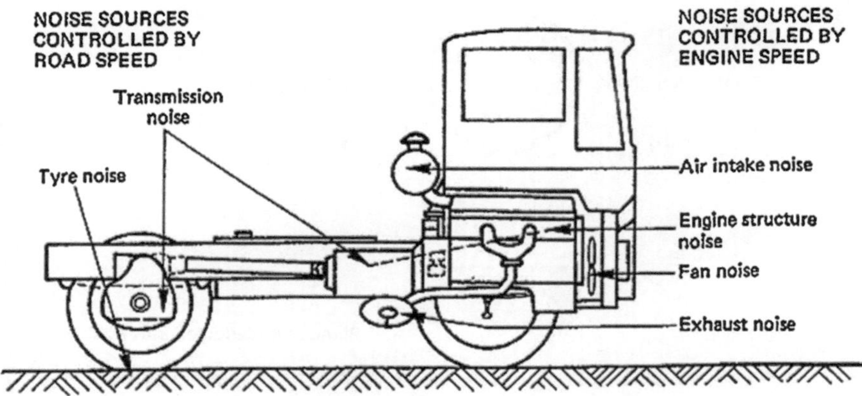

Fig. 1 Main sources of vehicle noise (Goel 2009)

3 Problem Formulation

Several studies indicate that tire noise is primarily produced by the interaction of the tire with the roadway, rather than by other mechanisms such as aerodynamic flow over the tire. Accordingly, tread design and road texture are the major factors in the production of tire noise. Quite clearly, other parameters which influence the tire/roadway interaction, such as tread wear, speed, load, inflation pressure, carcass design, and tire temperature, also can have an influence on noise generation. Tire/road interaction is influenced majorly by speed of the vehicle, tread patterns of different designs, pavement surfaces of different texture, and varying conditions such as wet, dry, and oily. Identifying the effect of these operating parameters on tire/road noise is important for better providing the optimal combined conditions having minimum noise generation from tires during running. In the present work, an experimental rig was prepared having provision for running two-wheeler tires of three different tread patterns at four different speeds on two pavement surfaces viz. asphalt and concrete under three conditions each, i.e., wet, dry, oily surfaces.

4 Methodology

The CPX method (ISO/CD 11819-2, 2000) for measuring tire/road noise has been actively pursued in Europe. Near field or close proximity methods consist of measuring the sound levels at or near the tire/pavement interface. In CPX method, sound pressure or sound intensity is measured by locating the microphones near the road surface. Microphones are placed near the tire/pavement interface to directly measure the tire/pavement noise levels and not the whole vehicle noise.

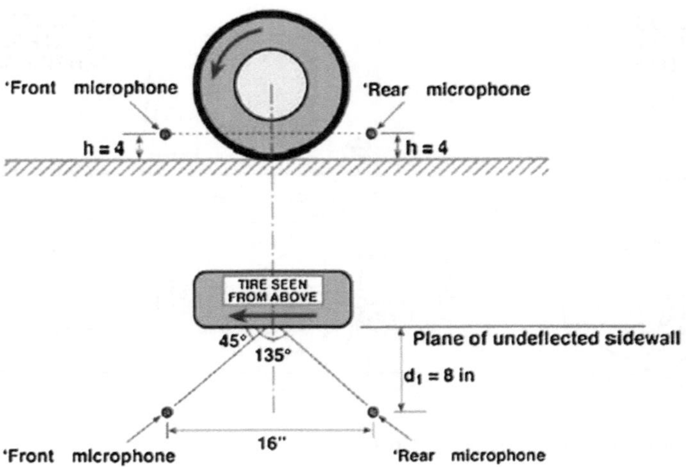

Fig. 2 Microphone positions in close proximity method

According to the ISO standards, microphones are mounted 8 inches from the center of the tire and 4 inches above the surface of the pavement as shown in the Fig. 2.

5 Result and Discussion

5.1 Effect of Tread Pattern on Sound Pressure Level

Measurements were taken for studying the effect of tread pattern on sound pressure level. Sound pressure levels were recorded for three types of tread pattern on two pavement surfaces at constant speed of 750 rpm (58 km/h) (Fig. 3). The tires with different tread pattern that were used are shown in Fig. 4.

Figure 4 shows that the noise generated for all the three tires with asphalt surface was more than that for concrete surface. For concrete surface, tire C was having the maximum sound pressure level, and for asphalt surface, tire A had maximum sound pressure level.

5.2 Effect of Tire Wear on Sound Pressure Level

Measurements had been taken for tread depth and corresponding sound pressure levels for different tires at constant speed 58 km/h and concrete dry pavement surface (Fig. 5).

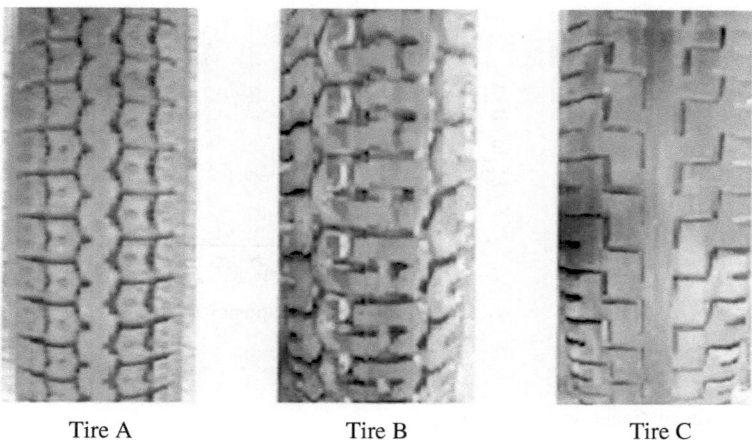

Fig. 3 Three tires of different tread pattern

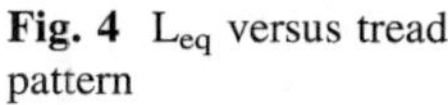

Fig. 4 L_{eq} versus tread pattern

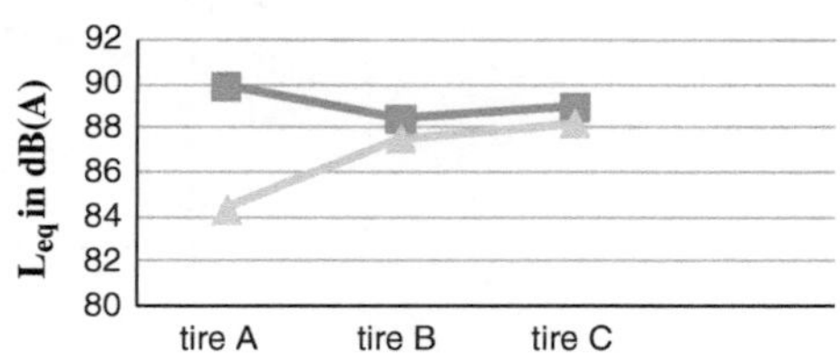

Fig. 5 L_{eq} versus tread depth

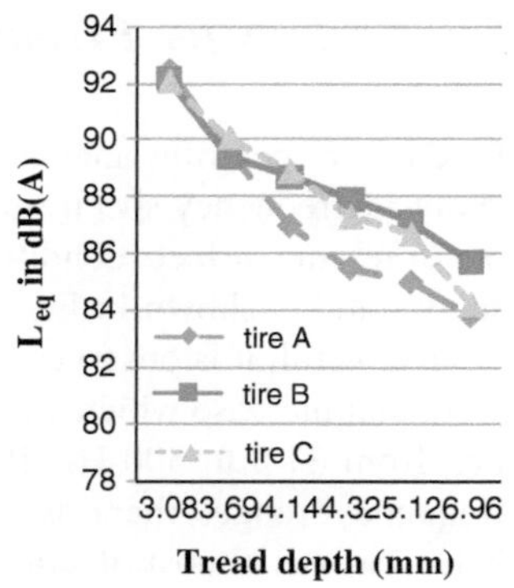

It had been found that with the approximate 50 % wear of tread depth, sound pressure level increases from approximately (84–86) dB (A) to (92–93) dB (A) which is about 8 dB (A); this could be due to the increase in contact area because of wear of tread.

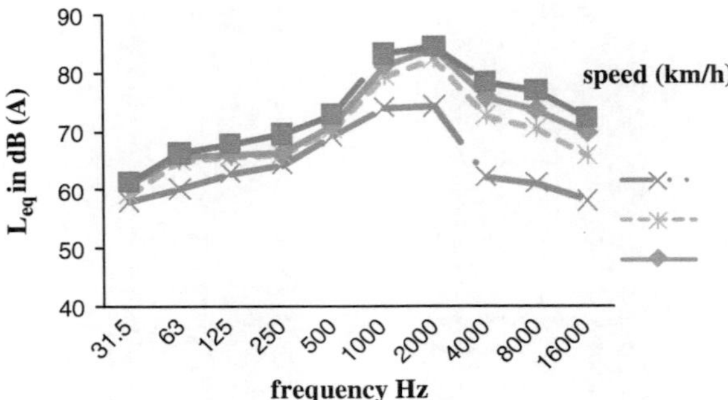

Fig. 6 L_{eq} versus frequency

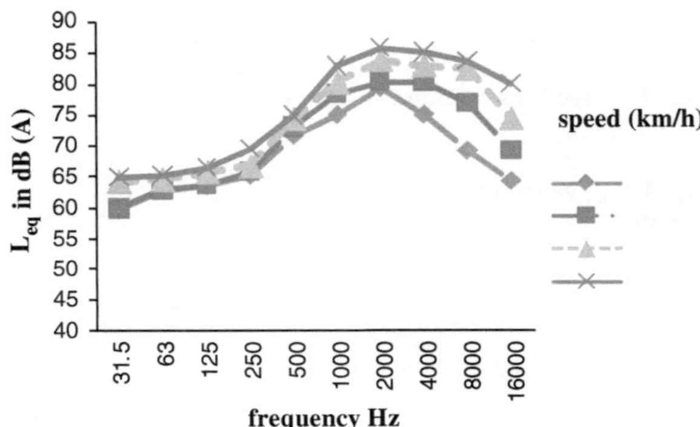

Fig. 7 L_{eq} versus frequency

5.3 Analysis for Frequency Spectrum in 1-1 Octave Band

Frequency spectrum analysis for tire A on asphalt and concrete surfaces. In this analysis, frequency spectrum was recorded at one-one octave band mode for tire A on different surface conditions of asphalt and concrete pavements at varying speeds and is shown in Figs. 6, 7, and 8.

In general, it is observed from the 1-1 octave spectra that in all cases, there was a gradual increase which was almost constant in the levels in low frequency range, i.e., from 31.5 to 500 Hz. Peaks are observed in 2–4 K/8 K range. In the higher frequency range, there is a sharp decline in the levels. Frequency spectrum behavior was almost of same kind for asphalt dry and wet surface, but for asphalt oily surface, it showed a different behavior at speed of 12 km/h, where there was sharp decrease in the levels beyond 4,000 Hz, which was not so in case of other two conditions. A sharp decrease in the L_{eq} levels was observed at 2,000 Hz frequency for concrete wet and oily surfaces which was not there in case of concrete dry surface.

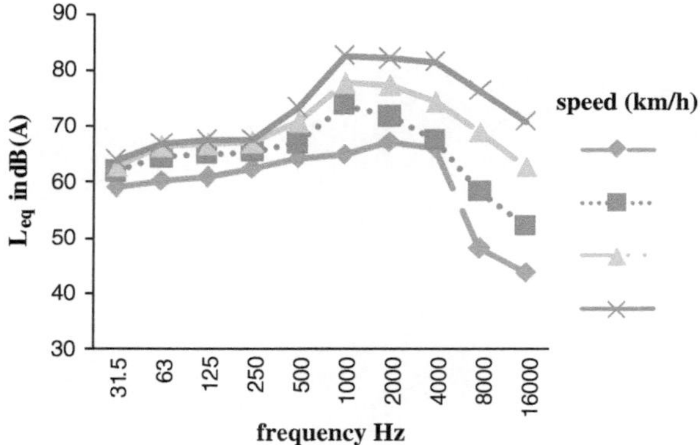

Fig. 8 L_{eq} versus frequency

6 Conclusion

Effect of different operating parameters on tire/road noise in case of two wheelers was studied for equivalent sound level and frequency band spectra in 1-1 octave band mode. An experimental rig was designed for taking the measurements according to CPX method. It had been found that sound pressure level varies with the change in the tread pattern design change for same sized tire and increases with the increase in speed of the vehicle on all type of pavements. Sound pressure level was more for asphalt rough surface than concrete surface. Moreover, different surface conditions of the same surface also affect the sound pressure level as it had been found that wet condition of a surface was noisier than dry and oily. Oily surface produces least noise than other two. Tire wear also had an effect on sound pressure level as with the increase in wear of tire, i.e., decrease in tread depth, sound pressure level increases.

References

Bueno M, Viñuela U, Terán F, Paje SE (2009) Toward the acoustical characterization of asphalt pavements: analysis of the tire/road sound from a porous surface. J Acoust Soc Am 125:5–7

Douglas LH, Roberet S, Brian W (2005) Kansas tire/pavement noise study. Hot Mix asphalt Technology

Goel VK (2009) Vibration and noise control. Document by Indian Institute of Technology, Roorkee

Keijiro I, Ichiro Y (1996) A study on mechanisms of tire/road noise. JSAE Rev 17:39–144

Leasure WA Jr, Bender EK (1975) Tire-road interaction noise. J Acoust Soc Am 58:1–12

Tatsuo F, Hiroshi K, Yasuo O, Hideki T (2005) Definition of road roughness parameters for tire vibration noise control. Appl Acoust 66:501–512

Evaluating the Contributions of Just-in-Time Manufacturing on Manufacturing Performance in Indian Manufacturing Industry

Raminderpal Kaur and Inderpreet Singh Ahuja

Abstract The purpose of this paper is to evaluate the contributions of just-in-time (JIT) manufacturing initiatives toward building core competencies in Indian manufacturing industry. This paper critically examines the implications of JIT implementation initiatives in Indian manufacturing organizations. The study reveals that organization culture and management commitment, employee's involvement and workplace organization, and quality improvements of JIT can significantly contribute toward accruing core competencies in the organization.

Keywords Just-in-time manufacturing · Manufacturing industries · Competencies · Competitive advantage

1 Introduction

Just-in-time production is defined as a philosophy that focuses attention on eliminating waste by purchasing or manufacturing just enough of the right items just in time. Yasin et al. (2004) have stated that after World War II, Japanese manufacturers were faced with the dilemma of vast shortages of material, financial, and human resources. The problems that Japanese manufacturers were faced with differed from those of their western counterparts. These conditions resulted in the birth of the "JIT" production system concept. In order to make a move toward improvement early, Japanese leaders such as Toyoda Kiichiro, Shigeo Shingo, and TAIICHI OHNO devised a new, disciplined, process-oriented system, which is known today as the "Toyota production system," or "Just-in-Time manufacturing" (Voss and Robinson 1987). The JIT advocates the elimination of waste by simplifying production processes, reductions in setup times, controlling material

R. Kaur (✉) · I. S. Ahuja
Department of Mechanical Engineering, University College of Engineering, Punjabi University, Patiala, Punjab, India
e-mail: raminderbatth@gmail.com

S. S. Khangura et al. (eds.), *Proceedings of the International Conference on Research and Innovations in Mechanical Engineering*, Lecture Notes in Mechanical Engineering,
DOI: 10.1007/978-81-322-1859-3_60, © Springer India 2014

flows, and emphasizing preventive maintenance are seen as ways by which excess inventories can be reduced or eliminated, and resources utilized more efficiently (Kannan and Tan 2005). JIT is just not a technique or set of techniques of manufacturing, but is an advanced approach or philosophy which embraces both new and old techniques and provides a wide range of benefits by renovation of existing manufacturing systems.

The basic underlying idea of this system is to minimize the consumption of resources that add no value to a product (Salaheldin 2005). In order to compete in today's fiercely competitive market, many manufacturers have come to realize that the traditional mass production concept has to be adapted to the new ideas of JIT manufacturing (Panchal et al. 2013). Just-in-time manufacturing is closely associated with the principles of pull production control. A pull system is control by downstream information and is inherently made to stock. For example, closed lines are pull systems because buffer spaces act as stock voids to trigger release (Berkley 1992; Brown and Mitchell 1991).

2 Literature Review

The notion of JIT production was described by TAIICHI OHNO, the godfather of Toyota production system, as Liker (2004) states that "all we are doing at the time line from the moment the customer gives us an order to the point when we collect the cash, and we are reducing that time line by removing the non-value-added wastes" (McKone et al. 2001). According to Cheng and Podolsk (1996), one motivating reason for developing JIT and other better production techniques was that after World War II, Japanese people had a very strong incentive to develop good manufacturing techniques to help them rebuild the economy (Hallihan et al. 1997). Kinney and Wempe (2002) have deployed a matched-pair research design to investigate the profitability of JIT and non-JIT firms. Inconsistent with the earlier studies, their results indicate that the ROA of the JIT firms fell significantly less compared to the non-JIT firms when tested after three post-JIT adoption years. A more recent study by Fullerton et al. (2003) provides empirical support for the relationship between the degree of JIT practices used and profitability.

3 Methodology

The methodology deployed for the research work is depicted in Fig. 1.

4 Research Work

In the present study, the key organizational initiatives have been classified as organization culture and management commitment (X1), employee's involvement and workplace organization (X2), inventory control and purchasing (X3),

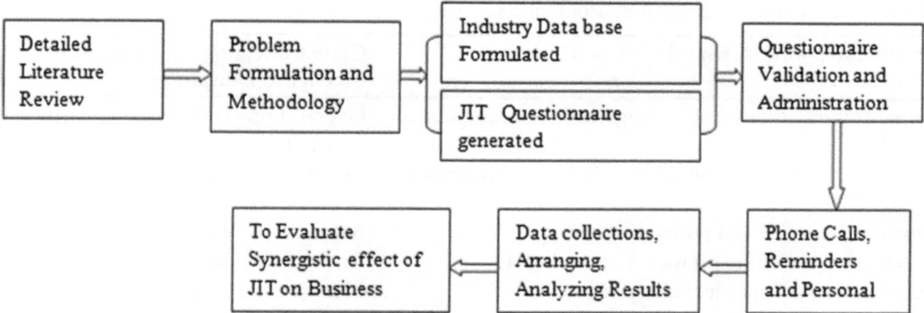

Fig. 1 Block diagram of methodology

production system and process control (X4), Kanban and pull production system (X5), and customer satisfaction (X6). The manufacturing excellence indicators have been classified into six categories namely, overall organization achievements (Y1), employee competencies (Y2), supplier coordination (Y3), production capabilities (Y4), plant performance (Y5), delivery compliance (Y6), and quality improvement (Y7). The following inputs and outputs have been deployed in the study (Table 1).

The Cronbach α values for all the input and output categories, as depicted in Table 2, in excess of 0.625 indicate the significantly high reliability of data for various input and output categories.

As an initial step, the Pearson correlations values are calculated to find the level of inter correlation among performance parameters (Y) and factors of implementation (X) variables (factors). The correlation coefficients (r) are found to be high and significant at $p = 0.05$ significance level in most of the cases. This indicates that most factors are significantly related to improvement in performance parameter, and this confirms the high correlation between various JIT implementation dimensions and performance parameters. The correlation (r) values through exploratory method using SPSS 13.0 are shown below for each dependent variable.

Y1 – X3 (0.606**), X4 (0.562**), Y2 – X1 (0.497**), X2 (0.602**), X3 (0.436**), X4 (0.607**), X5 (0.468**), Y3 – X2 (0.515**), Y4 – X1 (0.481**), X2 (0.626**), X3 (0.637**), X4 (0.626**), X5 (0.491**), Y5 – X1 (0.514**), X2 (0.561**), X3 (0.526**), X4 (0.512**), X5 (0.577**), X6 (0.487**), Y6 – X2 (0.429**), X5 (0.424**), Y7 – X1 (0.563**), X2 (0.577**), X3 (0.534**), X4 (0.576**), X5 (0.651**).

**Correlation is significant at the 0.01 level (2-tailed).

The correlation values indicate that there is a strong correlation of X3—inventory control and purchasing with Y1—overall organization achievement (0.606**), Y2—employee competencies (0.436**), Y4—production capabilities (0.637**), Y5—plant performance (0.526**), Y7—quality improvement (0.563**). Also, X4—organization culture and management commitment are found to be closely related to Y1—overall organization achievements (0.562**),

Table 1 Description of input and output categories

Inputs categories (organizational initiatives)	Output categories (manufacturing excellence indicators)
Organization culture and management commitment (X1)	Overall organization achievements (Y1)
Employee's involvement and workplace organization (X2)	Employee competencies (Y2)
Inventory control and purchasing (X3)	Supplier coordination (Y3)
Production system and process control (X4)	Production capabilities (Y4)
Kanban and pull production system (X5)	Plant performance (Y5)
Customer satisfaction (X6)	Delivery compliance (Y6)
	Quality improvement (Y7)

Table 2 Calculation of Cronbach's α for input and output category of JIT companies

Various input dimensions	X1	X2	X3	X4	X5	X6	
Cronbach's α value	0.717	0.827	0.710	0.841	0.900	0.625	
Output performance parameters	Y1	Y2	Y3	Y4	Y5	Y6	Y7
Cronbach's α value	0.789	0.784	0.770	0.822	0.829	0.805	0.899

Y2—employees competencies (0.607**), Y4—production capabilities (0.626**) and Y5—plant performance (0.512**), and Y7—quality improvement (0.576**). Here, X1—organization culture and management commitment-related to Y2—employee competencies (0.497**), Y4—production capabilities (0.481**), Y5—plant performance (0.514**), and Y7—quality improvement (0.563**). Also, X2—employee's involvement and workplace organization have strong relation with Y2—employees competencies (0.602**), Y3—supplier coordination (0.0515**), Y5—plant performance (0.561**), Y6—delivery compliance (0.429**), and Y7—quality improvement (0.577**). In this test, X5—Kanban and pull production system also related to the Y2—employees competencies (0.468**), Y4—production capabilities (0.491**), Y5—plant performance (0.577**), Y6—delivery compliance (0.424**), and Y7—quality improvement (0.651**). Also, X6—customer satisfaction related to the Y5—plant performance (0.487**). Further, multiple regression analysis test has been conducted for identification of significant implementation dimensions contributing toward realization of performance parameters for JIT companies. The results of this test have been depicted in Table 3. The significant factors with (p) significance level, R/R^2, β value, t value, and F values for each performance parameter are given.

T Tests for comparative effectiveness of JIT approach on performance parameters

The effect of period of implementation of JIT performance is assessed by many researchers with the help of an empirical study on a manufacturing industry. The study for JIT strategy is carried out for three time periods or phases to examine the

Table 3 Results of multiple regressions between performance parameters and implementation factors for JIT companies

Performance factor	Significance I/P factor	Beta value	*t*- value	*p* value	R/R^2 value	*F* value
Y1	X3	0.606	4.444	0.000	0.606/0.367	19.751
Y2	X4	0.607	2.197	0.000	0.607/0.369	19.850
Y3	X2	0.515	3.505	0.001	0.515/0.265	12.287
Y4	X3	0.552	4.493	0.000	0.637/0.406	23.237
	X5	0.361	2.940	0.006	0.728/0.529	18.554
Y5	X5	0.479	3.760	0.001	0.577/0.332	16.927
	X3	0.414	3.246	0.003	0.703/0.494	16.106
Y6	X2	0.351	2.311	0.027	0.429/0.184	7.657
	X6	0.323	2.124	0.041	0.531/0.282	6.479
Y7	X5	0.556	4.769	0.000	0.651/0.423	24.946
	X6	0.404	3.464	0.001	0.760/0.577	22.508

short-, medium-, and long-term effects on performance as depicted in Table 4. The justification for each of these time periods is explained below.

Period of transition (up to three years of implementation): This is a period during which initial investment and efforts are made to overcome initial resistance and to orient organization as per requirements of the strategies. In this phase, the organization experiences changes in work culture, work practices, and policies. The period of transition normally goes for three years since beginning of implementation.

Period of stability (more than three years but up to five years of implementation): This is the period during which the improvement drives get stabilized and organization starts realizing benefits of JIT. The period of stability is taken as more than three years but up to five years of implementation.

Period of maturity (more than five years of implementation): Over a long period, the benefits accrued from improvement drives give strategic and competitive edge in terms of cost, delivery, flexibility, and customer satisfaction in comparison with competitors. This phase represents long-term experiences with these two improvement drives in terms of strategic leverage over other companies.

The detail summary of comparative analyses for effectiveness of JIT strategy with the help of *t* tests is shown in Table 5. The values of mean and standard deviations obtained for JIT strategy in three different phases, that is, transition, stability, and maturity phase indicate considerable improvements on all performance parameters. The results of two-tailed *t* tests at 5 % significance level for each parameter of performance for three phases are given in Table 5.

To investigate the effect of JIT implementation program on the manufacturing performance over a period of time, the responses obtained from various manufacturing organizations have been classified into three categories depending upon the experience gained over an extended period. The tests *t*(II/I), *t*(III/II), *t*(III/I)

Table 4 Details of *t* tests used for comparative analyses

Test	Phase	Purpose
t test	Phase-I period of transition	Comparative effectiveness of performance parameters in combined approach and JIT of phase A
t test	Phase-II period of stability	Comparative effectiveness of performance parameters in combined approach and JIT of phase B
t test	Phase-III period of maturity	Comparative effectiveness of performance parameters in combined approach and JIT of phase C

Table 5 Test results of two-tailed *t* tests for Phase-I, Phase-II, and Phase-III

Performance parameters	Phase-I		Phase-II		Phase-III		II/I	III/II	III/I
	Introductory phase		Stability phase		Maturity phase		*t* value at $p < 0.05$	*t* value at $p < 0.05$	*t* value at $p < 0.05$
	Less than 3 years		3–5 years		>5 years				
	$N = 13$		$N = 14$		$N = 10$				
	Mean	SD	Mean	SD	Mean	SD			
Y1	2.57	0.518412	3.15	0.520321	3.21	0.730248	2.876	0.199	2.331
Y2	2.84	0.413939	3.24	0.592466	3.53	0.48919	2.016	1.319	3.540
Y3	2.52	0.320908	2.70	0.696143	3.36	0.505964	0.866	2.690	4.543
Y4	2.57	0.398482	2.66	0.658336	3.28	0.491709	0.430	2.653	3.688
Y5	2.67	0.429235	3.09	0.506659	3.52	0.491709	2.283	2.106	4.292
Y6	2.40	0.723627	2.60	0.748331	3.28	0.526624	0.692	2.613	3.294
Y7	2.73	0.534846	3.17	0.726878	3.52	0.559365	1.765	1.327	3.350

have been conducted to ascertain the statistical difference in effectiveness of various manufacturing performance improvements accrued as a result of successful JIT implementation initiatives.

Table 5 depicts the average and standard deviations of gains accrued for various manufacturing performance parameters as a result of adaptation of effective JIT initiatives in the organizations. The significant values of *t*(II/I) and *t*(III/II) have clearly demonstrated the mean values of the manufacturing excellence indicators accrued in Phase-II and Phase-III are significantly higher than those obtained in Phase-I, while the significant values of *t*(III/I) have also demonstrated that the mean values of manufacturing excellence indicators accrued in Phase-III have been higher than those obtained in Phase-II. The reason behind is the sincere efforts made by Indian entrepreneur over period of time, toward adopting strategic JIT initiatives at the organizations. Thus, in the context of the present study, it can be concluded that usually 3–5 years of whole hearted JIT implementation in needed to realize significant results.

5 Conclusions

The research has investigated some significant revelations about various facts of JIT paradigm in Indian industry. The study reveals that Indian entrepreneurs have been reasonably successful in imitating their western counterparts toward implementation of JIT paradigm. The study provides a careful and systematic evaluation of contributions of JIT in manufacturing organizations. The exploits of Indian entrepreneurs with traditional and contemporary production rate improvement initiatives have been investigated in the research work, for developing an understanding of production rate competencies accrued by Indian manufacturing organizations. The contributions JIT in manufacturing improvement initiatives toward enhancing manufacturing competencies of the organizations have been ascertained by employing various statistical, graphical, and through modeling techniques.

The study also highlights the critical success factors necessary for realization of various manufacturing excellence attributes. It has been observed that overall organization achievements (Y1) are closely associated with inventory control and purchasing (X3) and production system and process control (X4), employees competency (Y2) and production capabilities (Y4) with organization culture and management commitment (X1), employee involvement and workplace organization (X2), inventory control and purchasing (X3), production system and process control (X4), supplier coordination (Y3) with employee involvement and workplace organization (X2), plant performance (Y5) with organization culture and management commitment (X1), employee involvement and workplace organization (X2), inventory control and purchasing (X3), production system and process control (X4), Kanban and pull production system (X5), customer satisfaction (X6), delivery compliance (Y6) with employee involvement and workplace organization (X2), Kanban and pull production system (X5), quality improvement (Y7) with organization culture and management commitment (X1), employee involvement and Workplace organization (X2), inventory control and purchasing (X3), production system and process control (X4), and Kanban and pull production system (X5). The results show that quality-oriented tools like JIT paradigm in the enhancement of manufacturing excellence along with competitive advantage of manufacturing organization. Hence, we can see that to have a total JIT manufacturing system, a company-wide commitment, proper materials, quality, people, and equipment must always be made available when needed. In addition; the policies and procedures developed for an internal JIT structure should also be extended into the company's supplier and customer base to establish the identification of duplication of effort and performance feedback review to continuously reduced wastage and improve quality. The research has revealed improvements in manufacturing performance over time extending up to five years and beyond.

References

Berkley BJ (1992) A review of Kanban production control research literature. Prod Oper Manage 1(4):393–411

Brown K, Mitchell R (1991) A comparison of just-in-time and batch manufacturing: the role of performance obstacles. Acad Manuf J 34(4):907–917

Cheng TCE, Podolsk S (1996) Just-in-time manufacturing, an introduction. Chapman and Hall, London

Fullerton RR, McWatters CS, Fawson C (2003) An examination of the relationships between JIT and financial performance. J Oper Manage 21(4):383–404

Hallihan A, Sackett P, Williams GM (1997) JIT manufacturing: the evolution to an implementation model founded in current practice. Int J Prod Res 35(4):901–920

Kannan V, Tan K (2005) Just in time, total quality management, and supply chain management: understanding their linkages and impact on business performance. Int J Manage Sci 33(2):153–162

Kinney M, Wempe W (2002) Further evidence on the extent and origins of JIT's profitability effects. Account Rev 77(1):203–225

Liker J (2004) The Toyota Way: 14 management principles from the world's greatest manufacturer. McGraw-Hill, New York

McKone KE, Schroeder RG, Cua KO (2001) The impact of just in time practices on manufacturing performance. J Oper Manage 19(1):39–58

Panchal V, Gupta A, Tiwari PC, Rathi N (2013) Evaluation of just in time (JIT) elements in banking sector using ANOVAs technique. Int J Innovative Res Sci, Eng Technol 2(2):460–468

Salaheldin IS (2005) JIT implementation in Egyptian manufacturing firms: some empirical evidence. Int J Oper Prod Manage 25(4):354–370

Voss CA, Robinson SJ (1987) Application of just-in-time manufacturing techniques in the UK. Int J Oper Prod Manage 7(4):46–52

Yasin MM, Wafa M, Small MH (2004) JIT: an analysis of JIT implementations in the manufacturing service and public sectors. Benchmarking Int J 11(1):74–92

Quantitative Assessment of Hospital Waste Generation to Recover Energy: A Case Study

Amrinder Singh and Jasvir Singh

Abstract Today, the production of energy from waste is emerging as a new technology; however, its implementation in healthcare sector is still missing in developing countries, despite of abundance of valuable hospital waste in the urban areas of these countries. This paper analyzed the energy potential of hospital waste generated by healthcare facilities of Moga city. For this study, generation rates from typical city hospitals were collected and selected cotton, bandages, plastic syringe, and glucose bottles as the main source of energy recovery from hospital waste since their quantity is always on larger side. Five samples of each selected waste (cotton, bandages, plastic syringe, and glucose bottles) were collected. The collected samples were tested in laboratory, where proximate analysis and calorific value tests were conducted. The experimentation of selected wastes (cotton, bandages, plastic syringe, and glucose bottles) in laboratory highlighted that plastic syringe has the highest calorific value which was 42,749 kJ/kg, while cotton has the least 15,190 kJ/kg. Glucose bottles have the highest percentage in terms of mass with 46 %, bandages second highest have 16 %, and cotton has the least with 6 %. The total energy that can be produced from these wastes (cotton, bandages, plastic syringe, and glucose bottles) amounts to 9.03 MW/day.

Keywords Calorific value · Energy recovery · Gasification · Hospital waste · Incineration · Proximate analysis

A. Singh (✉)
Mechanical Engineering Department, Lala Lajpat Rai Institute of Engineering and Technology, Moga 142001 Punjab, India
e-mail: amrindersingh.sethi@gmail.com

J. Singh
Mechanical Engineering Department, Guru Nanak Dev Engineering College, Ludhiana 141006 Punjab, India
e-mail: jasvir44@yahoo.co.in

S. S. Khangura et al. (eds.), *Proceedings of the International Conference on Research and Innovations in Mechanical Engineering*, Lecture Notes in Mechanical Engineering,
DOI: 10.1007/978-81-322-1859-3_61, © Springer India 2014

1 Introduction

Proper management of solid waste especially hospital waste is critical to the health and well-being of urban residents. Land filling is a threat to all basic needs for humans, e.g., clean air to breathe, clean water to drink and non-toxic and healthy food to eat as well as shelter for those living close to the landfill. For instance, land filling pollutes the air with methane, carbon dioxide, and other gases, depending on the chemical and biological activities of the material in the landfill. Leachate and storm water from the landfill contaminate ground and surface waters, transporting the pollutants into the food chain.

Dumping of hospital waste can also affect the shelter of people living near the landfill. For instance, avalanches and waste slides can occur during rainy periods if the waste is not properly compacted.

According to the World Health Organization (WHO), around 85 % of hospital wastes are non-hazardous, 10 % are infectious, and the remaining 5 % are toxic chemicals, pharmaceutical wastes, and radioactive wastes. Hospital waste in India till recently was not being managed, but it was simply 'disposed off'. The disposal of hospital waste can be very hazardous, particularly when it gets mixed with municipal solid waste and is dumped in uncontrolled or illegal landfills. Hospital waste contains organic as well as inorganic matter. The latent energy present in its organic fraction can be recovered for gainful utilization through adoption of suitable waste processing and treatment technologies. The recovery of energy from wastes also offers a few additional

1. Demand for land, which is already scarce in cities, for land filling can reduce.
2. The cost of transportation of waste to far-away landfill sites can be reduced proportionally.
3. Net reduction in environmental pollution.

The main technologies for this are as follows:
1. Incineration and
2. Gasification.

Incineration process has some concerns due to formation of toxic dioxins, especially from polyvinyl chloride (PVC) containing plastics and other materials that form toxic dioxins when they burn, which are harmful for health and environment.

Gasification is significantly different and cleaner than incineration. Dioxins need sufficient oxygen to form, and the oxygen-deficient atmosphere in a gasifier does not provide the environment needed for dioxins to form. The main parameters that determine the potential of recovery of energy from wastes are as follows:

1. Quantity of waste and
2. Physical and chemical characteristics (quality) of the waste.

The actual production of energy will depend on specific treatment process employed, the selection of which is also critically dependent on the above two parameters.

The important physical parameters requiring consideration include the following:

1. Size of constituents
2. Density
3. Moisture content.

2 Literature Review

Mustafa et al. (2013) studied both quantity and volume of municipal solid waste in Kirkuk city. Kirkuk's municipal solid waste poses serious effects on environment and citizen health. The daily waste generation is expected to grow with the rise in city population. The project would lead to improved electricity supply and efficient managing waste. Landfill technology is the most preferred one to generate 5 MW of electricity.

Abdulla et al. (2008) investigated the medical waste management practices used by hospitals in northern Jordan. The results reported focus on the level of medical waste segregation, treatment, and disposal options practiced in the hospitals of the study area. The most frequently used treatment practice for solid medical waste was incineration. Only 48 % hospitals had incinerators, and none of those incinerators met the Ministry of Health regulations. The results indicated that the medical waste generation rate ranges from approximately 0.5 to 2.2 kg/bed day, which is comprised of 90 % of infectious waste and 10 % sharps.

Alamgir and Ahsan (2007) analyzed the daily generation rates of municipal solid waste at six major cities of Bangladesh. A total of 7,690 tons of municipal solid waste is generated daily. Sampling was done at different waste generation sources such as residential, commercial, institutional, and open areas, in different seasons. The potential for waste recovery and reduction based on the waste characteristics were evaluated, and it is predicted that 21.64 million US$/year can be earned from recycling and composting of municipal solid waste.

Azage and Kumie (2010) examined healthcare waste type, generation rate, and its management system in health centers in West Gojjam Zone. The daily mean healthcare waste generation rate was 1.79 ± 0.54 kg, which was equivalent to 0.035 ± 0.05 kg/outpatient/day. About 0.93 ± 0.3 kg/day (52.0 %) was general waste, and 0.86 ± 0.33 kg/day (48.0 %) was hazardous waste. Only four out of ten health centers used local type of incinerators, while others used open burning for the final handling of healthcare wastes. The unit generation rate was relatively small in magnitude when compared with similar health facilities that are found in developing countries.

Bulucea et al. (2009) examined the biomedical waste management practices, followed by the hospitals of Dolj District, Romania. The assessment of biomedical waste situation was taken into account, and the quantities of biomedical waste generated in eleven hospitals of Dolj District were presented. Also, following the rules and legislation of both Romania and European Union, the methods for segregation, packaging, and labeling and the treatment techniques for reduction in volume, neutralization, and final disposal of the biomedical waste were analyzed. It was observed that the Dolj District hospitals have been properly managing their biomedical waste. The hospitals have been segregating the biomedical waste every day, in accordance with the biomedical waste categories, collected in the appropriate type of container and specified color coding, in accordance with the legislation.

Felicia et al. (2008) studied the management practices of hospital solid waste in Limpopo Province of South Africa. The generated hospital waste was weighed to compute the generation rates and was followed through various management practices to the final disposal. The findings revealed a major policy implementation gap between the national government and the hospitals. While modern practices such as landfill and incineration are used, their daily operations were not carried according to minimum standards. Incinerator ash is openly dumped, and wastes are burned on landfills instead of being covered with soil. The incinerators used are also not environmentally friendly as they use old technology. The findings further revealed that there is no proper separation of wastes according to their classification as demanded by the national government. The mean percentage composition of the waste was found in the following decreasing order: general waste (60.74 %) > medical waste (30.32 %) > sharps (8.94 %). The mean generation rates were found to be 0.60 kg per patient per day.

Khambalkar et al. (2008) assessed the bioenergy available in a rural village for self-sustainable development. The biomass consumption of the village for domestic as well as for all the activities has been collected. The study also entailed the collection of all bioenergy sources available in the village. The bioenergy sources, such as biomass available through forestry, agriculture waste, and residues , and animal waste (animal dung), have been collected for the exact quantification of the bioenergy generation capacity of the village. From this study, it has been found that the village has considerable bioenergy potential. A suitable renewable energy generation system in the studied village has been recommended.

Ashekuzzaman et al. (2010) studied the potential of energy recovery from rurally available agro- and household organic wastes and thus the possible impact on supplementing energy demand, reducing deforestation, and replacing fossil fuel as well as avoiding greenhouse gases. Results show that codigestion of a wide range of manure, crop residues, and household wastes with cow manure was successful to produce increased gas yield than what would be if cow dung is digested separately and the energy value from this can supplement 57–79 % of the rural demand, depending on methane yield from organic waste mixtures. It has been suggested that implementation of codigestion in the centralized plant could be a viable solution to produce decentralized energy for the rural households in terms of sustainable waste management, reducing deforestation as well as replacing fossil.

3 Objective

From the literature, it has been found that a lot of emphases are paid on municipal waste and agriculture to electric energy, but no attention has been given on hospital waste to energy, although hospital waste is far more dangerous than municipal and agriculture waste and at the same time the quantity of hospital waste is increasing at a higher rate. Therefore, the present study deals with experimental assessment of amount of energy that can be recovered from hospital waste of Moga city, which is as follows:

1. To determine the generation rates of various types of hospital waste.
2. To quantify the amount of energy that can be recovered from different types of hospital waste.

4 Proposed Experiments

The amount of energy that can be recovered from different types of medical wastes can be obtained by the following tests:

1. Proximate analysis: It has been done to find the following:
a. Moisture content
b. Volatile matter
c. Fixed carbon
d. Ash content.
2. Calorific value test: Calorific value is the amount of potential energy that can be converted into actual heating ability. This heating ability gives the value of energy that can be recovered. The calorific value was found by using apparatus known as bomb calorimeter.

5 Results and Discussion

5.1 Hospital Waste Generation Rates in Surveyed Hospitals

Solid waste generated by each hospital has been weighed, and the average quantity of waste is determined. The highest generation rate of 1.38 kg/patient/day was found in one of the surveyed hospital, followed by 1.35 kg/patient/day. The lowest rate observed was 1.11 kg/patient/day. The average medical waste generated per patient per day was 1.26 kg.

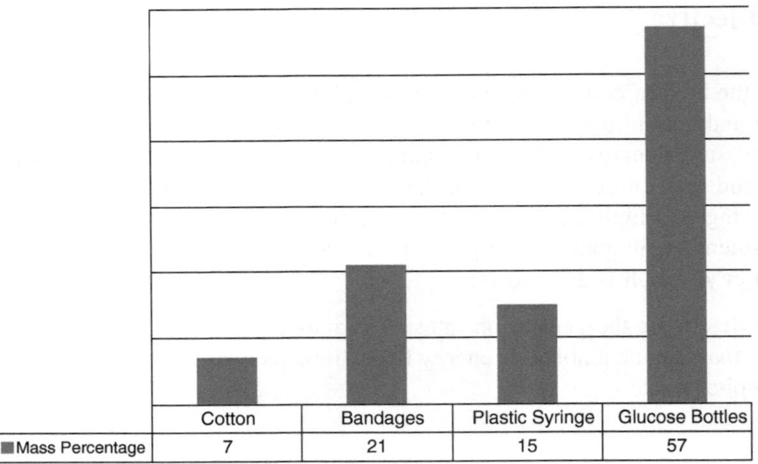

Fig. 1 Classification of hospital waste in surveyed hospitals (mass %)

5.2 Classification of Hospital Waste in Surveyed Hospitals Both in Mass and in Mass%

Quantitative analysis of hospital waste indicates that total cotton waste of surveyed hospital was 66.6 kg/day, total bandages waste was 176.6 kg/day, total plastic syringe waste was 139.2 kg/day, and total glucose bottles waste was 556.4 kg/day.

Average waste generation rate of cotton was 3.33 kg/hospital/day, for bandages 9.83 kg/hospital/day, for plastic syringe 6.96 kg/hospital/day, and glucose bottles 27.82 kg/hospital/day.

Qualitative analysis of hospital waste indicates that glucose bottles have the highest percentage in terms of mass with 57 %, bandages next highest have 21 %, plastic syringe have 15 %, and cotton have 7 % in terms of mass (Fig. 1).

5.3 Experimental Analysis of Collected Samples from Surveyed Hospitals

1. Proximate Analysis

The selected samples (cotton, bandages, plastic syringe, and glucose bottles) were collected and then taken to laboratory where proximate analysis and calorific value tests were conducted. The result of proximate analysis (Table 1) indicates that cotton has the highest moisture content with 16.60 % and plastic syringe has the lowest with 12.80 %. Also, cotton has the highest volatile matter with 47.80 % and plastic syringe has the lowest with 32.80 %. Bandages have the highest percentage of fixed carbon with 24.80 %, and the lowest percentage was found in case

Table 1 Proximate analysis of collected samples from surveyed hospitals

S. no.	Moisture content (%)	Volatile matter (%)	Fixed carbon (%)	Ash (%)
Cotton	16.60	47.80	21.20	14.40
Bandage	14.60	45.20	24.80	15.40
Plastic syringe	12.80	32.80	23.80	30.60
Glucose bottles	14.20	36.20	21.60	28.00

Table 2 Average calorific value content in selected hospital waste

S. no.	Type of waste	Average calorific value (kJ/kg)
1.	Cotton	15,190
2.	Bandages	18,030
3.	Plastic syringe	42,740
4.	Glucose Bottles	39,840

of cotton with 21.20 %, whereas in case of ash, plastic syringe has the highest percentage with 30.60 % and cotton has the lowest percentage with 14.40 %.

2. Calorific Value Test

The experiment of calorific value on selected hospital wastes indicates that calorific value of plastic syringe was among the highest calorific value with 42,740 kJ/kg, glucose bottles have 39,840 kJ/kg, bandages contain 18,030 kJ/kg, and cotton contains the least with 15,190 kJ/kg of energy (Table 2).

6 Calculations

6.1 Calculation of Energy Content from Entire District Hospital Waste

To calculate the energy potential from entire district hospital waste, average calorific value of a particular waste was multiplied by total waste generated by the hospitals and using the energy to power conversion relation 1 MW = 3,600 MJ; hence, energy potential from hospital waste has been calculated (Table 3).

6.2 Payback Period for Waste Treatment Plant

Payback period of waste treatment plant (gasifier) has been obtained by comparing factors such as cost of gasifier, electricity that can be generated per day by energy recovery process, and price of per kilo watt of electricity. Since 9.03 MW (9,030 kW) can be generated per day (Table 4) by process of gasification (excluding installation charges, labor cost, etc.). Therefore, 1-MW gasifier has

Table 3 Calculation of energy content from entire district hospital waste

Waste	Generation rate (kg/day)	Average calorific value (MJ/kg)	Energy (MJ)	Power (MW)
Cotton	66.6	15.19	1017.73	0.28
Bandage	196.6	18.03	3551.91	0.98
Plastic syringe	139	42.74	5940.86	1.64
Glucose bottles	556	39.84	22151.04	6.13
Total			32661.54	9.03

Table 4 Payback period for waste treatment plant (gasifier)

Cost of gasifier (rupees)	Electricity generated per day (kW)	Price of 1 kW (rupees)	Payback time (months)
10 million	9,030	6	31

been suggested sufficient. The cost of 1-MW gasifier was 10 million Indian rupees (Ankur Gasifier, Gujarat). Price of 1 kW electricity was six Indian rupees in the surveyed region. Therefore, payback time of 31 months has been obtained, as shown in Table 4.

7 Conclusions

Average waste generation rate obtained in this study lies within the range of values (1.11–1.38 kg/patient/day). In this study, it has been observed that the plastic contributes 72 % (15 % plastic syringe and 57 % glucose bottles) in the total hospital waste. At the same time, both plastic syringe and glucose bottles contain high calorific value content. Such high calorific value content should be used for gainful purposes such as for generating electricity. By quantitative assessment of hospital waste generation and by performing experiments on samples of hospital waste, it has been observed that 9,030 kW of electricity per day can be generated by energy recovery process. From the assessment of waste generation rate, a 1-MW-capacity gasifier should be installed. The recovered energy (electricity) can be sold, and the investment on gasifier can be recovered within 31 months.

Acknowledgement Authors are thankful to various hospital managements for the collection of data. As per the requirement of management of hospitals, the name of hospitals has been kept secret for public information.

References

Abdulla F, Qdais H, Rabi A (2008) Site investigation on medical waste management practices in northern Jordan. Waste Manage 28(2):450–458

Alamgir M, Ahsan A (2007) Municipal solid waste and recovery potential: Bangladesh perspective. J Iran J Environ Health Sci Eng 4(2):67–76

Ashekuzzaman SM, Badruzzaman ABM, Rafiqul Hoque ATM, Rudra S (2010) Supplementing energy demand of rural households in Bangladesh through appropriate biogas technology. International conference on environmental aspects of Bangladesh, (2010)

Azage M, Kumie A (2010) Healthcare waste generation and its management system: the case of health centres in West Gojjam Zone, Amhara Region, Ethiopia. Ethiop J Health Dev 24(2):119–126

Bulucea CAV, Popescu MC, Cornea E, Olteanu V (2009) Characterizing biomedical waste in hospitals of Dolj District. J Environ Probl Dev 23(9):97–104

Felicia N, Sally M, Luke C (2008) Hospital solid waste management practices in Limpopo Province. Waste Manage 28:1236–1245

Khambalkar VP, Karale DS, Gadge SR, Dahatonde SB (2008) Assessment of bioresources potential pf a rural village for self energy generation. A review article, BioResources 3(2):566–575

Mustafa SS, Mustafa SS, Mutlag AH (2013) Kirkuk municipal waste to electrical energy. J Electr Power Energy Syst 44:506–513

Optimization of Fractal Tree Dipole Antenna at 2.4 GHz Using SuperNEC

Gurpreet Kaur and Munish Rattan

Abstract The aim of this paper is to design multi-band and broadband fractal antennas with fractal tree geometry. The antennas are designed with the wire fractal tree structure and optimized to get the desired frequency response. In this design, the fractal antenna is a broadband antenna with the bandwidth of 350 MHz. This antenna is optimized to operate in the range of 2,400–2,750 MHz frequency band with acceptable return loss <-10 dB (VSWR < 2). The antenna design and simulation is performed using SuperNEC software which is based on the method of moments.

Keywords Fractal antenna · Fractal tree antenna · Multi-band · Fractal geometry · Dipoles

1 Introduction

As part of an effort to improve modern communication system technology, researchers are now studying many different approaches for creating new and innovative antennas. One technique that has received much recent attention involves combining aspects of the modern theory of fractal geometry with antenna design (Petko and Werner 2004). Fractals were first defined by Mandelbrot (Falconer 1990) as a way of classifying complex geometric structures that have non-integer dimensionality and which possess inherent self-similarity or self-affinity within their geometrical structure. While Euclidean geometries are limited to

G. Kaur (✉) · M. Rattan
Guru Nanak Dev Engineering College, Ludhiana 141006 Punjab, India
e-mail: gurpreetsaini3224@gmail.com

M. Rattan
e-mail: dr.munishrattan@gmail.com

S. S. Khangura et al. (eds.), *Proceedings of the International Conference on Research and Innovations in Mechanical Engineering*, Lecture Notes in Mechanical Engineering,
DOI: 10.1007/978-81-322-1859-3_62, © Springer India 2014

points, lines, sheets and volumes of integer dimensionality, fractal structures fall between these Euclidean classifications having non-integer dimensionality. Fractal geometries accurately characterize many non-Euclidean features of the natural including the length of coastline, density of clouds and the branching of trees (Werner and Ganguly 2003) and find application in many areas of science and engineering including antenna design. Nowadays, the design of multi-band and small size antennas is still of major importance as an engineer topic. The growths of the telecommunication systems are driving the engineering efforts to develop multi-service (multi-band) and compact (portable) systems which require such kind of antennas. In case of cellular systems, not only the handset antenna is important, but also those on base stations. It is in this framework where fractal technology appears potentially as powerful tool to meet the telecommunication operator requirements (Wemer and Mitra 2000).

Fractal-shaped antennas are becoming a useful way to design advanced antennas such as multi-band antennas with approximately the same input or radiation characteristics for different frequency bands. This comes out from the fact that most of the fractals are self-similar objects. That is roughly speaking fractals composed many copies of themselves at different scales. The global fractal form is repeated at different sizes as many times as desired within the object structure such that the global object and its parts become identical (Wemer and Mitra 2000). In addition to the simplicity and self-similarity, fractal curves have the additional property of approximately filling a plane which makes them the attractive candidates for use in the design of antennas (Gianvitorio and Rahmat 2002). An example of plane-filling Hilbert curve had been demonstrated in Vinoy et al. (2001); also, the Koch monopole and dipole had been demonstrated in Zainud-Deen et al. (2004).

This paper presents the design and simulation of a fractal tree dipole wire antenna using SuperNEC 2.9 software. The first part of antenna design deals with the implementation of antenna to resonate at desired frequency. In second part, the relevant antenna performance of the proposed design viz. return loss, gain, SWR, 3-D and 2-D radiation pattern are reported and discussed.

2 Proposed Antenna Geometry

Fractals are objects which have a self-similar structure repeated throughout their geometry (Mandelbrot 1983; Peitgen et al. 1992). This self-similar structure may be produced by the repeated application of a generator, and in the case of fractal trees, the generator is defined as a junction from which several smaller branches, called child branches, split from a parent branch. Every branch, with the exception of the first and final branches, has a generator connected to it at each end: one from which is a child and the other to which is the parent.

Figure 1 shows the configuration of fractal tree dipole antenna proposed in this paper.

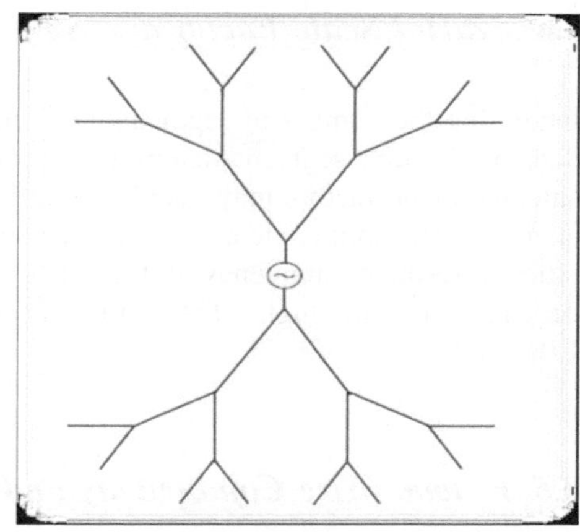

Fig. 1 Configuration of 3rd iterated fractal tree dipole antenna

3 Proposed Antenna

3.1 Selection of Operating Frequency

The operating frequency of 2,450 MHz was chosen for the fractal tree dipole antenna. This frequency band is used for wireless local area network (WLAN, 2.45 GHz for wireless operation).

3.2 Length of the Antenna

The length of half-wave dipole is given by the relation (Balanis 2003):

$L = \lambda/2$
$L =$ length of the antenna

Therefore, length of the fractal tree dipole antenna is 0.06 m according to above relation.

3.3 Fractal Geometry

Fractal geometry means order of the iteration. The iteration used in design of fractal tree dipole antenna is 3, although 1–5 iterations can be used.

3.4 Fractal Scale Factor and Split Angle (Branching Angle)

Increasing the number of segments may increase the coupling between branches. Size of the first segment determines the one of the resonant frequency of the antenna. Scale factors may decide the ratio between the successive resonant frequencies. The split angle also affects the coupling. However, it does not affect the ratio of resonant frequency if the lengths and widths of the branches are not dependent on the angle. The scaling is by a factor of 1 and branch angle is 1.047 rad.

3.5 Fractal Wire Conductivity and Radius

The wire conductivity of all the conductors is assumed to be 5.7E7. The radius of the fractal wire is 1 mm.

3.6 Voltage Source and Input Impedance

The voltage source is of 1 V and input impedance is of 50 Ω.

4 Antenna Design

This section presents the design of the fractal tree dipole antenna. The purpose of building this dipole is to produce a more space-efficient half-wave dipole design. The initial segment is divided by a scale factor, moved at an angle and placed at the top of the initial segment. The same pattern is repeated to construct the tree of any order. After some order, depending on the scale factor and angle, the branches start overlapping each other. Such an antenna can be thought of as a virtual combination of capacitors and inductors, loading the previous structure. This makes the antenna so that it has many different resonances which can be chosen and adjusted by choosing the proper fractal design.

Figure 1 shows the proposed antenna design. The antenna is designed using SuperNEC 2.9 software. SuperNEC is a Method of Moments electromagnetic (EM) simulation package for Windows or Linux platforms. The easy to use 3-D input GUI, making use of multi-level assemblies, provides the easiest ever structure input and model creation tool. The output viewer provides the design engineer with all the necessary information for proper antenna analysis. SuperNEC has a number of predefined antenna assemblies making it easy and quick to design and analyse antennas.

Table 1 Proposed antenna design specifications

Location	[0 0 0]
Segment length	0.1
Orientation	[0 0 0]
Frequency scaling	1
Type of fractal antenna	Dipole 2D
The order of fractal	3
The limit of the fractal curve (m)	[0 0.06]
The radius of the wire (m)	0.001
The split angle	1.047

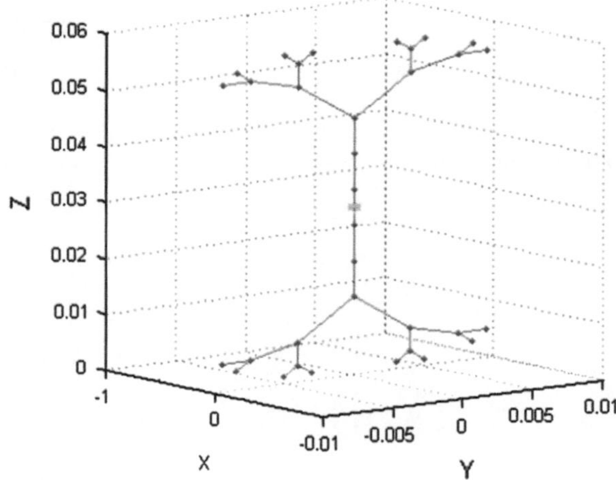

Fig. 2 Proposed antenna design

The fractal tree design structure has the following specifications mentioned in Table 1.

The resultant antenna design is shown in the Fig. 2.

5 Results and Discussion

Figure 3 shows the VSWR of antenna design, and Fig. 4 shows the gain of the antenna at different frequencies. The graph shows that there are two resonant frequencies for this antenna. The table for resonant frequency, gain and VSWR is given in the Table 2.

Some modifications of dipole antenna structure have been done in order to obtain the specification of the designed antenna. To achieve the design, the parameters of the fractal tree dipole antenna have been changed. The length of the antenna is changed from 6 to 5 mm. The scaling factor is increased to 1.7, and branch angle is changed from 1.047 to 0.14836 rad. Initially, the wire radius is selected 1 mm, and to get the desired results, the wire radius is changed to

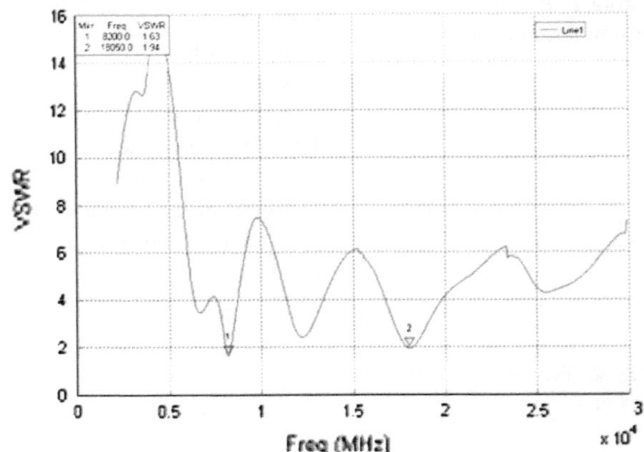

Fig. 3 VSWR plotter

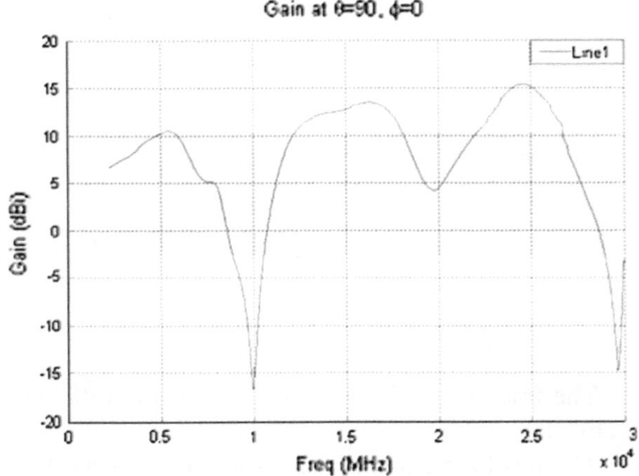

Fig. 4 Gain plotter

Table 2 Gain and VSWR of proposed antenna

Resonant frequency (MHz)	Gain (dBi)	VSWR
8,200	3.6	1.63
18,050	10.3	1.94

1.2 mm. The antenna is designed and simulated. The simulation results are shown in the Figs. 5 and 6.

Figure 5 represents the VSWR, and Fig. 6 represents the gain of the proposed antenna.

From the results, it can be seen that the proposed antenna works at the frequency band of 2,400–2,750 GHz with the bandwidth of 350 MHz. At this frequency band, the VSWR is less than 2. It is found that the antenna has wideband

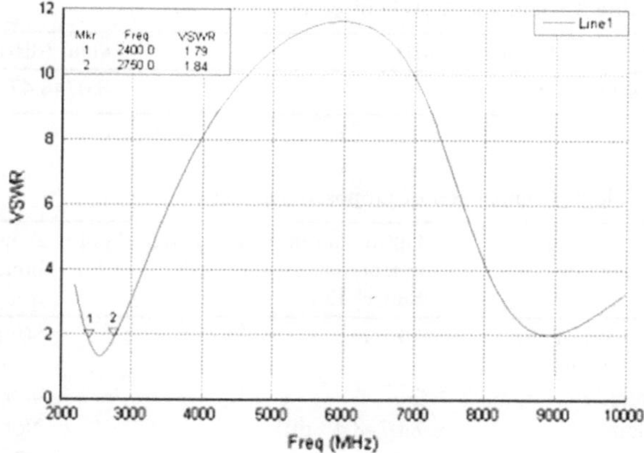

Fig. 5 VSWR plotter at frequency band 2,400–2,750 MHz

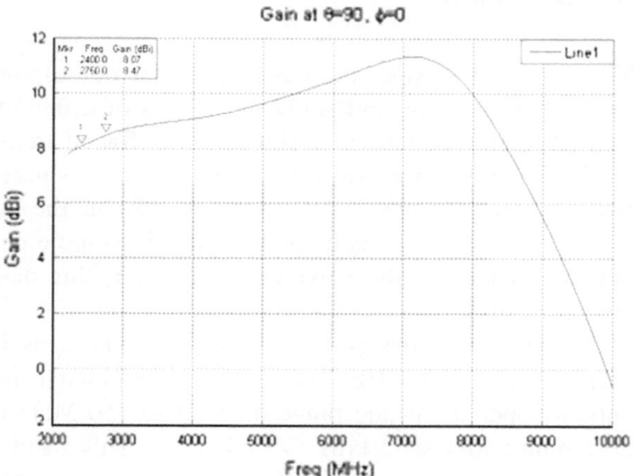

Fig. 6 Gain plotter at frequency band 2,400–2,750 MHz

behaviour. The gain at this frequency band is between 8.07 and 8.47 dBi. Table 3 shows the results of simulated antenna.

The comparison of this paper with the paper titled "A Novel Broadband Quasi-Fractal Binary Tree Dipole" (Xing 2007) reveals the following differences in the results given in Table 4.

As given in Table 4, the proposed antenna is more compact and using the iteration factor 3 as compared to other design in which iteration factor 4 and 5 was used. Therefore, the proposed antenna is compact in size and easy to design.

Table 3 Gain and VSWR of proposed antenna

Resonant frequency (MHz)	Gain (dBi)	VSWR
2,400–2,750	8.07–8.47	<2

Table 4 Comparison of proposed antenna

Name of the parameter	Optimization of fractal tree dipole antenna at 2.4 GHz using SuperNEC	A novel broadband quasi-fractal binary tree dipole
Length of the antenna	50 mm	80 mm
Iteration	3	4–5
Gain	8.07–8.47 dBi	Not given
VSWR	<2	<2

6 Conclusion

The aim of this paper is to design fractal tree dipole antenna for wireless communication systems and study the effects of different parameters. In this, a tree-shaped fractal antenna is designed using fractal tree geometry. This antenna can work in two different ways. In the first case, it is used as multi-band antenna with two resonant frequencies. So for this design, the antenna can be a multi-band antenna. This antenna is implemented using dipole geometry, i.e. the length of the antenna is half of the wavelength. Further, this design is optimized to get the desired results.

In the second design, the fractal antenna is as broadband antenna with the bandwidth of 350 MHz. This antenna is simulated for 2.45 GHz frequency. This antenna operates in the range of 2,400–2,750 MHz frequency band with acceptable return loss <-10 dB (VSWR < 2). This band covers the 2.4 GHz WLAN band (frequency range 2.4–2.483 GHz) and the 2.5 GHz mobile WiMAX operating band (frequency range 2.5–2.7 GHz).

References

Balanis CA (2003) Antenna theory: analysis and design, 2nd edn. Wiley, New York

Falconer K (1990) Fractal geometry: mathematical foundation and applications, 2nd edn. Wiley, Chichester

Gianvitorio J, Rahmat Y (2002) Fractal antennas: a novel antenna miniaturization technique and applications. IEEE Antennas Propag Mag 44:20–36

Mandelbrot BB (1983) The fractal geometry of nature. Freeman, New York

Peitgen HO, Jurgens H, Saupe D (1992) Chaos and fractal. New frontiers of science. Springer, New York

Petko SJ, Werner D (2004) Miniature reconfigurable three-dimensional fractal tree antennas. IEEE Trans Antennas Propag 52:1945–1956

Vinoy KJ, Jose KA, Varadan VK, Varadan VV (2001) Hilbert curve fractal antenna: a small resonant antenna for VHF/UHF applications. Microwave Opt Technol Lett 29:215–219

Werner DH, Ganguly S (2003) An overview of fractal antennas engineering research. IEEE Antennas Propag Mag 45:38–56

Wemer D, Mitra R (2000) Frontiers in electromagnetics, 2nd edn. IEEE Press, New York (IEEE Press Series on Microwave Technology and RF)

Xing C (2007) A novel broadband quasi-fractal binary tree dipole. PIERS online, vol 3, no. 1

Zainud-Deen SH, Awadalla KH, Khamis SA, El-shalaby ND (2004) Radiation and scattering from koch fractal antennas. In: Proceedings of 21st national radio science conference (NRSC), 16–18 Mar 2004, pp B8–B19

A Comprehensive Performance Analysis of MANET Protocols

Rajni Puri and Munish Rattan

Abstract Mobile ad hoc networks are characterized by wireless connectivity, continuous changing topology, distributed operations, and ease of deployment. We compare two reactive routing protocols, dynamic source routing (DSR) and ad hoc on-demand distance-vector routing (AODV), and one proactive destination-sequenced distance-vector routing (DSDV) protocol by using Manhattan mobility model. We have analyzed the performance of protocols by varying network load, mobility, and type of traffic (CBR, TCP). A detailed simulation has been done using NS2. We consider packet delivery fraction, routing overhead, normalized routing load (NRL), and end-end delay as metrics for performance analysis of these protocols.

Keywords MANET · AODV · DSR · DSDV important · Manhattan mobility model

1 Introduction

Ad hoc network is a collection of wireless mobile nodes forming a temporary network without any existing wire-line infrastructure. Communication between nodes is based on radio to radio multi-hoping.

R. Puri (✉) · M. Rattan
Guru Nanak Dev Engineering College, Ludhiana 141006 Punjab, India
e-mail: rajni_puri71@yahoo.com

M. Rattan
e-mail: rattanmunish@gndec.ac.in

S. S. Khangura et al. (eds.), *Proceedings of the International Conference on Research and Innovations in Mechanical Engineering*, Lecture Notes in Mechanical Engineering,
DOI: 10.1007/978-81-322-1859-3_63, © Springer India 2014

Due to the infrastructure less (Bing Lin et al. 2000; Sarkar et al. 2008), self-configuring network property, and absence of a central governing authority, MANET[1] has wide application in industrial and commercial field involving cooperative mobile data exchange, inexpensive alternatives, or enhancement to cellular-based mobile network infrastructures. MANET has potential applications in the locations where setting of infrastructure networks is not possible. Military ad hoc networks detect and gain as much information as possible about enemy movements, explosions, and other phenomena of interest. Such kind of network also has applications in emergency disaster relief operations after natural hazards like hurricane or earthquake.

Some of the wireless traffic sensor networks monitor vehicle traffic on highways or in congested parts of a city. Wireless surveillance sensor networks may be deployed for providing security in shopping malls, parking garages, and many such other areas where direct or wired communication cannot be made.

Ad hoc networks are also characterized by frequent topology change due to mobility of nodes. Nodes may join and leave the network at any time. All nodes in such networks behave as routers and take part in discovery and maintenance of routes to the other nodes in the network. Ad hoc networks have to deal with many challenges and the one that is most important is route selection.[2] So the routing algorithm must be dynamic and must be adaptive to the frequent topology changes due to node mobility. Many algorithms have been proposed in the literature that can be used in ad hoc networks for finding routes.

The ad hoc routing protocols are divided into three categories:

Proactive routing protocol: These are also known as table-driven protocols and will actively determine the layout of the network. Through a regular exchange of network topology packets between the nodes of the network, at every single node, an absolute picture of the network is maintained.

Reactive routing protocol: These are also called on-demand routing protocols and start to set up routes on-demand. The routing protocol will try to establish such a route, whenever any node wants to initiate communication with another node to which it has no route. Unlike proactive routing protocols, reactive protocols do not generate sustained routing overhead.

Hybrid routing protocol: A hybrid routing protocol is the one that combines the best features of proactive and reactive protocols. It reduces the control overhead of proactive routing protocols and decreases the latency caused by route discovery in reactive routing protocols.

Several performance evaluations of MANET routing protocols using CBR and TCP traffic have been done in the literature (Harminder et al. 2010; Kumar et al. 2009) by considering various parameters such as mobility, network load, and

---

[1] IETF Working Group: Mobile Ad hoc Networks (MANET). Available at http://www.ietf.org/html.charters/manet-charter.html.

[2] Routing_basic_from http://www.cisco.com/en/US/docs/internetworking/technology/handbook/Routing-Basics.html. Mar 2010.

pause time. Bindra et al. (2010) compared the two reactive routing protocols, ad hoc on-demand distance-vector routing (AODV) and dynamic source routing (DSR) by using group mobility model with CBR and TCP traffic sources and observed that AODV gives better performance in CBR traffic and real-time delivery of packet. DSR gives better results in TCP traffic and under restricted bandwidth condition. Also Kumar et al. (2009) investigated AODV and DSR routing protocols under random way point mobility model with CBR and TCP traffic sources. They concluded that AODV outperforms DSR in high-load and/or high-mobility situations. Jayakumar and Gopinath (2008) have observed that in dense networks of Manhattan grid model with CBR traffic, the packet delivery ratio for AODV and DSR is relatively near to one another. DSR, however, has very bad results in network latency making AODV favorable choice in more dense networks.

In this paper, we present performance comparison of two reactive routing protocols DSR, AODV and destination-sequenced distance-vector routing (DSDV) (Jayakumar and Gopinath 2008; Jeya Kumar and Rajesh 2009; Feng et al. 2009) to bring out their relative merits. Two are on-demand protocols, and they initiate their routing activities only when required. The motivation behind this comparison is to understand their internal working mechanism and bring out situations where one is preferred than the other and the other is table driven.

There are various mobility models such as random way point (Kumar et al. 2009), reference point group mobility model (RPGM) (Bindra et al. 2010), Manhattan mobility model (Jayakumar and Gopinath 2008), freeway mobility model, Gauss–Markov mobility model that have been proposed for evaluation. Many previous studies have used random way point or reference point group mobility as reference model. In this study, we will make a detailed study based on Manhattan mobility model.

2 Routing Protocols for MANETs

2.1 Dynamic Source Routing

The main feature of DSR is the use of source routing. That is, the sender knows the complete route from source to destination including all intermediate hops. These routes are stored in a route cache. The data packets carry the source route in the packet header. When a node in the ad hoc network attempts to send a data packet to a destination for which it does not already know the route, it uses a route discovery process to dynamically determine such a route. Route discovery works by flooding the network with route request (RREQ) packets. Each node receiving a RREQ, rebroadcasts it, unless it is the destination or it has a route to the destination in its route cache. Such a node replies to the RREQ with a route reply

(RREP) packet that is routed back to the original source. RREQ and RREP packets are also source routed. The RREQ builds up the path traversed so far.

The RREP routes itself back to the source by traversing this path backwards. The route carried back by the RREP packet is cached at the source for future use.

If any link on a source route is broken, the source node is notified using a route error (RERR) packet. The source removes any route using this link from its cache. A new route discovery process must be initiated by the source, if this route is still needed. DSR makes very aggressive use of source routing and route caching. No special mechanism to detect routing loops is needed. Also, any forwarding node caches the source route in a packet it forwards for possible future use.

Promiscuous listening: When a node overhears a packet not addressed to itself, it checks whether the packet could be routed via itself to gain a shorter route. If so, the node sends a gratuitous RREP to the source of the route with this new, better route. Aside from this, promiscuous listening helps a node to learn different routes without directly participating in the routing process.

The DSR protocol is composed of two main mechanisms that work together to allow discovery and maintenance of source routes in MANET.

Route Discovery: When a source node S wishes to send a packet to the destination node D, it obtains a route to D. This is called route discovery. Route discovery is used only when S attempts to send a packet to D and has no information of a route to D.

Route Maintenance: When there is a change in the network topology, the existing routes can no longer be used. In such a scenario, the source S can use an alternative route to the destination D, if it knows one, or invoke route discovery. This is called route maintenance.

2.2 Ad Hoc on-Demand Distance-Vector Routing

AODV inherits the property of both DSR and DSDV protocols. It borrows the basic on-demand mechanisms of route discovery and route maintenance from DSR, plus the use of hop-by-hop routing, sequence numbers, and periodic principal of DSDV. It provides loop-free path and avoids Bellman-Ford count to infinity, and the convergence is also fast when the topology changes. RRREQ, RREP, and RERR are the message types defined by AODV. When a route to a new destination is needed, the node broadcasts a RREQ to find a route to the destination. In case of link break in an active route, a RERR message is used to notify other nodes that the link failure has occurred. AODV is a reactive protocol, and it deals with route table management.

2.3 *Destination-Sequenced Distance-Vector Routing*

DSDV protocol is a table-driven algorithm based on the classical Bellman-Ford routing mechanism. The improvements made to the Bellman-Ford algorithm include freedom from loops in routing tables. Every mobile node in the network maintains a routing table in which all of the possible destinations within the network and the number of hops to each destination are recorded. DSDV tags each route with a sequence number and considers a route r more favorable than r' if r has a greater sequence number or if both have the same sequence number but as a lower metric (hop count).

Each entry is marked with a sequence number assigned by the destination node. The sequence numbers enable the mobile nodes to distinguish stale routes from new ones, thereby avoiding the formation of routing loops. Mobile nodes maintain an additional table where they store the data sent in the incremental routing information packets.

New route broadcasts contain the address of the destination, the number of hops to reach the destination, the sequence number of the information received regarding the destination, as well as a new sequence number unique to the broadcast. The route labeled with the most recent sequence number is always used. In the event that two updates have the same sequence number, the route with the smaller metric is used in order to optimize (shorten) the path. Mobiles node also keep track of the settling time of routes, or the weighted average time that routes to a destination will fluctuate before the route with the best metric is received. By delaying the broadcast of a routing update by the length of the settling time, mobiles node can reduce network traffic and optimize routes by eliminating those broadcasts that would occur if a better route was discovered in the very near future

3 Manhattan Grid Mobility Model

Fan Bai, Narayanan, and Ahmed Helmy introduced the Manhattan model to emulate the movement pattern of mobile nodes on streets (Jayakumar and Gopinath 2008). It can be useful in modeling movement in an urban area where a pervasive computing service between portable devices is provided. The scenario is composed of a number of horizontal and vertical streets. Given below is topography showing the movement of seventeen nodes for Manhattan mobility model.

1. *Applications*: It can be useful in modeling movement in an urban area where a pervasive computing service between portable devices is provided.
2. *Important Characteristics*: Maps are used in this model too. However, the map is composed of a number of horizontal and vertical streets. Each street has two lanes for each direction (north and south direction for vertical streets, east and west for horizontal streets). The mobile node is allowed to move along the grid of horizontal and vertical streets on the map. At an intersection of a horizontal

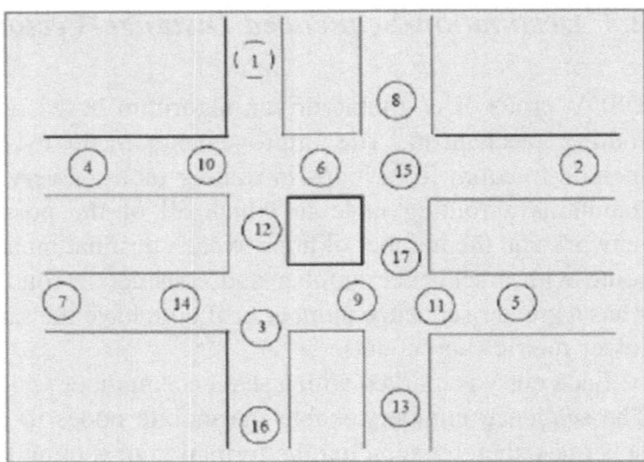

Fig. 1 Topography showing the movement of nodes for Manhattan mobility model

and a vertical street, the mobile node can turn left and right or go straight with certain probability.

Thus, the Manhattan mobility model is also expected to have high spatial dependence and high temporal dependence. It too imposes geographic restrictions on node mobility. However, it differs from the freeway model in giving a node some freedom to change its direction. Most of the mobility models mentioned above are parameterized, e.g., SDR and ADR are some of the parameters used in RPGM, while maps are important parameters in the freeway and Manhattan models. There are other prevalent mobility models: Gauss–Markov mobility model, random walk mobility model, etc. (Ariyakhajorn 2006) (Fig. 1).

4 Simulation Environment

Traffic models: Random traffic connections of CBR and TCP can be set up between mobile nodes using a traffic-scenario generator script. This traffic generator script is available under ~ns/indep-utils/cmu-scen-gen and is called cbrgen.tcl. It can be used to create CBR and TCP traffic connections between wireless mobile nodes. So the command line looks like the following:

- ns cbrgen.tcl [-type cbr|tcp] [-nn nodes] [-seed seed] [-mc connections] [-rate rate]

For the simulations carried out, traffic models were generated for 50 nodes with CBR traffic sources, with maximum connections of 10 or 40 at a rate of 4 packets per sec, and the packet size is 512 bytes.

Mobility model: The movement of nodes in the Manhattan grid mobility model is generated by the software called Mobility Generator which is based on a frame

Table 1 Simulation parameters for Manhattan grid mobility model

Parameter	Value
Mobility model	Manhattan grid mobility model
Routing protocols	AODV, DSR, DSDV
Terrain size	1000 m × 1000 m
No. of horizontal streets	3
No. of vertical streets	3
Packet size	512 bytes
Traffic type	CBR, TCP
No. of sources	10, 30
Packet rate	4 packets/second
Simulation time	900 s
No of nodes	50
Maximum speed	10, 20, 30, 40, 50, 60 m/s

work called IMPORTANT (Impact of Mobility Patterns on Routing in Ad hoc NeTwork, from University of Southern California).

Implementation: We have used network simulator (NS)-2 in our evaluation. The NS-2 is a discrete event driven simulator[3] (Altman and Jimenez 2012) developed at UC Berkeley. We have used Red Hat Linux environment with version NS-2.34 of network simulator. NS-2 is suitable for designing new protocols, comparing different protocols and traffic evaluations. It is an object-oriented simulation written in C++, with an OTcl interpreter as a frontend. NS uses two languages because simulator got to deal with two things: (1) detailed simulation of protocols which require a system programming language which can efficiently manipulate bytes, packet headers, and implement algorithms, (2) research involving slightly varying parameters or quickly exploring a number of scenarios.

Nam is the basic visualization tool used for ns-2 simulations. Using Java program, we analyze the trace file generated, and using MS Excel, we draw the graph.

We have used four traffic patterns with varying number of sources for each type of traffic (TCP and CBR). The goal of our simulation is to evaluate the performance differences of these two on-demand routing protocols. The type of traffic (CBR and TCP) and the maximum number of sources are generated by inbuilt tool of NS2 (M Greis 2010). The parameters used for carrying out simulation are given in the Table 1.

Performance Metrics: RFC2501 (Corson and Macker 1999) describe a number of quantitative metrics that can be used for evaluating the performance of MANET routing protocols. We have used the following metrics for evaluating the performance of two on-demand reactive routing protocols (AODV and DSR):

Packet Delivery Fraction: It is the ratio of data packets delivered to the destination to those generated by the sources. It is calculated by dividing the number of packet received by destination through the number packet originated from source.

[3] Network Simulator-NS-2. Available at http://www.isi.edu/nsnam/ns/.

$$\text{PDF} = (\text{Pr}/\text{Ps}) * 100$$

where Pr is total packet received and Ps is the total packet sent.

Routing Overhead: It is the total number of control or routing (RTR) packets generated by routing protocol during the simulation. All packets sent or forwarded at network layer is considered routing overhead.

$$\text{Overhead} = \text{Number of RTR packets}$$

Normalized Routing Load: Number of routing packets "transmitted" per data packet "delivered" at destination. Each hop-wise transmission of a routing is counted as one transmission. It is the sum of all control packet sent by all node in network to discover and maintain route.

$$\text{NRL} = \text{Routing Packet}/\text{Received Packets}$$

Average End-to-End Delay (second): This includes all possible delay caused by buffering during route discovery latency, queuing at the interface queue, retransmission delay at the MAC, propagation, and transfer time. It is defined as the time taken for a data packet to be transmitted across an MANET from source to destination.

$$\text{D} = (\text{Tr} - \text{Ts})$$

where Tr is receiving time and Ts is sending time.

5 Result and Discussion

5.1 Packet Delivery Fraction

In case of CBR traffic, AODV performs well and delivers almost 80 % packets irrespective of high and low load. DSDV delivers around 70 % irrespective of load. It declines slightly with the increasing speed in case of both AODV and DSDV. DSR uses source routing and also caches some routing entries. It is observed that such caching provides a significant benefit up to a certain extent. But at high loads and increasing speed, the PDF declines as shown in Fig. 2.

For TCP traffic, it is investigated that DSDV performs better than AODV in high or low network load. PDF ratio is nearly 100 % when number of sources is low, the PDF ratio for AODV and DSR is comparable as shown in Fig. 3. Under high load conditions (say 30 sources), when mobility is low, DSR shows better PDF than AODV. This is because DSR uses caching; hence, it is more likely to find a route in cache and perform the route discovery less frequently than with AODV. With the increase in speed, the routes change more frequently and there is

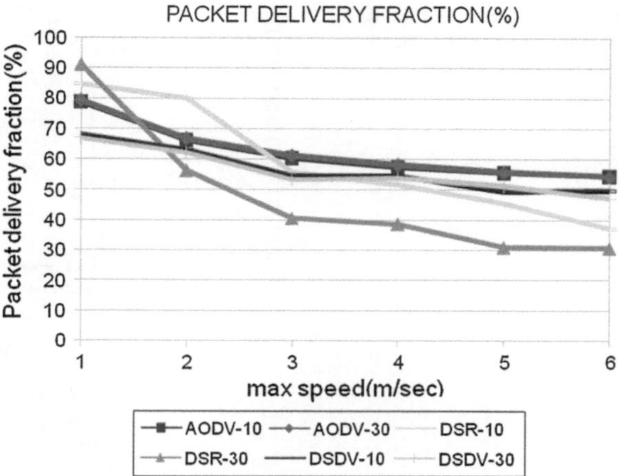

Fig. 2 PDF versus speed (CBR Traffic)

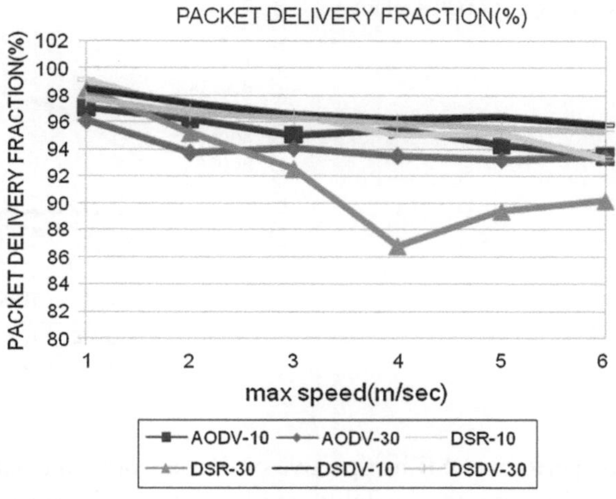

Fig. 3 PDF versus speed (TCP Traffic)

a strong need of finding new routes in DSR as well. Over all, the packet delivery fraction is around (87–99 %) in TCP traffic which is better than CBR which is around (30–90 %).

5.2 Routing Overhead

In case of CBR traffic, the routing load of DSDV is almost similar in high or low network load at all speeds and is low as compared to other two routing protocols as shown in Fig. 4. AODV is comparable at low network traffic, but with more mobility nodes the routing overhead also increases. AODV and DSDV

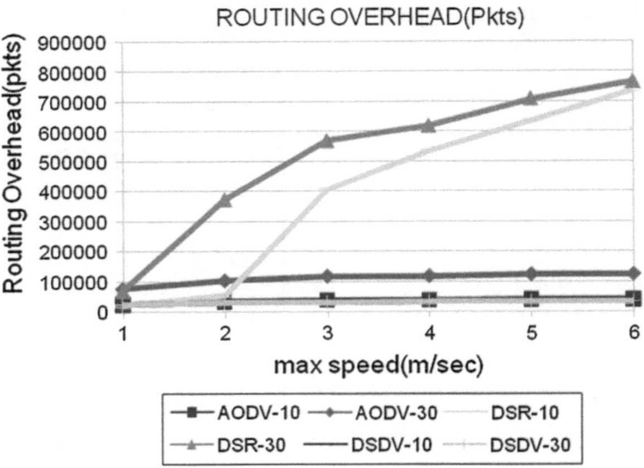

Fig. 4 Routing overhead versus speed (CBR Traffic)

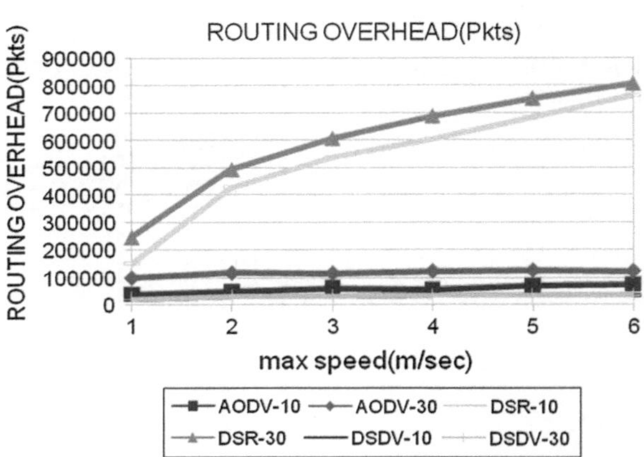

Fig. 5 Routing overhead versus speed (TCP Traffic)

outperforms DSR irrespective of the network load and speed of nodes. Simulation results show that mobility and load affects the performance of AODV and DSR differently. In the presence of high mobility, link failures can happen very frequently. Link failure triggers new route discoveries in AODV since it has almost one route per destination in its routing table. The reaction of DSR to link failures in comparison is mild and causes route discovery less often. The reason is the abundance of cached routes at each node. Thus, the route discovery is delayed in DSR until all cached routes fail. But with high mobility, chances of the caches being stale are quite high in DSR. Eventually when a route discovery is initiated, the large numbers of replies received in response are associated with high overhead. Hence, the cache staleness and high overhead together result in significant degradation in performance of DSR in high-mobility scenarios.

In TCP traffic, DSDV gives lower routing overheads than AODV at all speeds as shown in Fig. 5. AODV performs better at low network load, but the routing overhead increases at high speed.

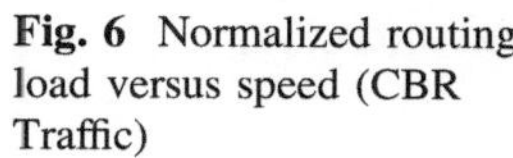

Fig. 6 Normalized routing load versus speed (CBR Traffic)

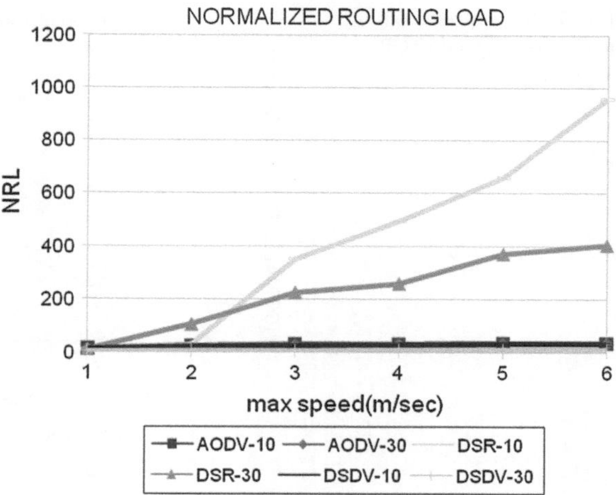

Performance of DSR deteriorates with increasing speed. When nodes are moving fast, there is higher rate of disconnections, which produces more route errors and frequent needs for re-initialization of route discovery process. At high load and at high speeds, performance of DSDV is slightly better than AODV. Over all, the routing overhead in TCP traffic is low as compared to CBR traffic.

5.3 Normalized Routing Load

In case of CBR traffic, AODV and DSDV perform better than DSR irrespective of the network load and speed of nodes as shown in Fig. 6. DSR at low network load substantially increases with increase in speed.

In TCP traffic also, AODV and DSDV outperforms DSR irrespective of network load and speed of nodes as shown in Fig. 7. Normalized routing load of AODV and DSDV is comparable. At high network load, performance of DSR is less as compared to low network load. NRL in CBR traffic is high as compared to TCP traffic.

5.4 Average End-to-End Delay

In CBR traffic, average delay of AODV is very low as compared to DSR and DSDV as shown in Fig. 8. The delay of DSR has a significant order of magnitude difference. This is highly undesirable for delay-sensitive applications.

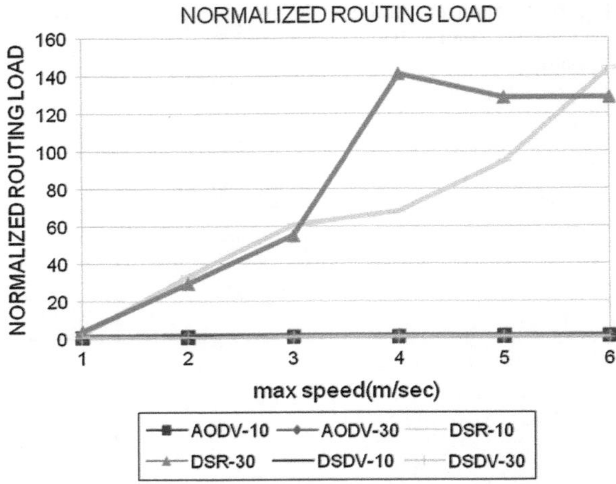

Fig. 7 Normalized routing load versus speed (TCP Traffic)

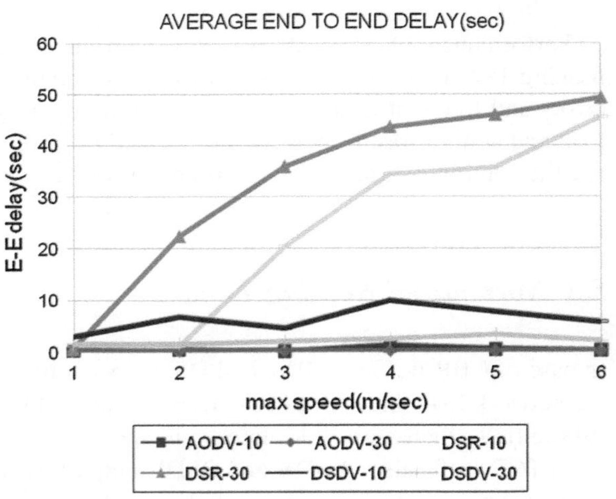

Fig. 8 E-E delay versus speed (CBR Traffic)

In case of TCP traffic, performance of AODV and DSDV is comparable and both outperform DSR in high or low network load conditions and at all speeds as shown in Fig. 9. Over all, AODV is the best choices for real-time delivery of packet in both types of traffic. The average delay in TCP traffic is low as compared to CBR traffic.

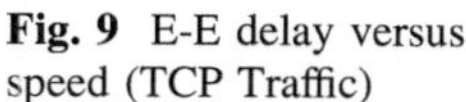

Fig. 9 E-E delay versus speed (TCP Traffic)

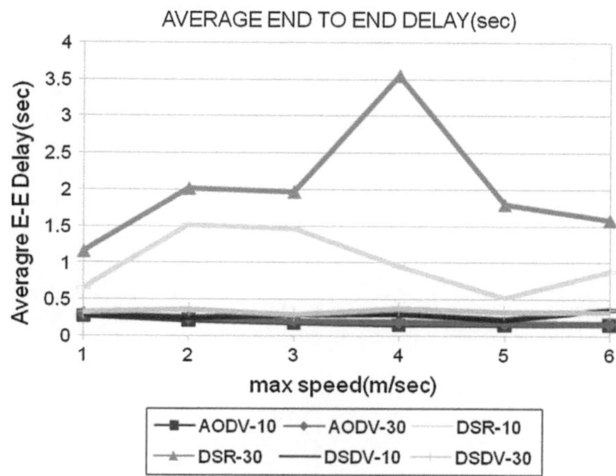

References

Altman E, Jimenez T (2012) NS simulator for beginners. Available at http://citeseer.ist.psu.edu/altman03ns.html

Ariyakhajorn J, Wannawilai P, Sathitwiriyawong C (2006) A comparative study of random waypoint and Gauss-Markov mobility models in the performance evaluation of MANET. In: Proceedings of international symposium on communications and information technologies, (ISCIT '06). IEEE, 18 Oct–20 Sep, pp 894–899

Bindra HS, Maakar SK, Sangal AL (2010) Performance evaluation of two reactive routing protocols of MANET using group mobility model. Int J Comput Sci Issues 7(3, 10):38–43

Corson S, Macker J (1999) Mobile ad hoc networking (MANET): routing protocol performance issues and evaluation considerations. In: Proceedings of IETF RFC 2501, Jan 1999

Feng Q, Cai Z, Yang J, Hu X (2009) A performance comparison of the ad hoc network protocols. Second international workshop on computer science and engineering (IWCSE), vol 2, pp 293–297

Greis M (2010) Marc Gries tutorial. Available at http://www.isi.edu/nsnam/ns/tutorial/

Jayakumar G, Gopinath G (2008a) Performance comparison of MANET protocols based on manhattan grid mobility model. J Mobile Commun 2(1):18–26

Jayakumar G, Gopinath G (2008b) Performance comparison of two on-demand routing protocols for ad-hoc networks based on random way point mobility model. Am J Appl Sci 5(6):659–664

Jeya Kumar MK, Rajesh RS (2009) Performance analysis of MANET routing protocols in different mobility models. IJCSNS Int J Comput Sci Netw Secur 9(2):22–29

Kumar S, Rathy RK, Pandey D (2009) Traffic pattern based performance comparison of two reactive routing protocols for ad hoc networks using NS2. In: Proceedings of 2nd IEEE international conference on computer science and information technology (ICCSIT), pp 369–373

Lin B, Chlamtac I (2000) Wireless and mobile network architectures. Wiley, New Delhi

Sarkar SK, Basawaraju TG, Puttamadappa C (2008) Ad hoc mobile wireless networks: principles, protocols and applications. Auerbach Publications, Boco Raton, p 1

Author Index

S. S. Khangura et al. (eds.), *Proceedings of the International Conference on Research and Innovations in Mechanical Engineering*, Lecture Notes in Mechanical Engineering, DOI: 10.1007/978-81-322-1859-3, © Springer India 2014